Classics in Mathematics

Lars Hörmander

The Analysis of Linear Partial Differential Operators IV

Lars Hörmander

The Analysis of Linear Partial Differential Operators IV

Fourier Integral Operators

Reprint of the 1994 Edition

Lars Hörmander
Department of Mathematics
University of Lund
Box 118
S-221 00 Lund, Sweden
lvh@maths.lth.se

Originally published as Vol. 275 in the series:
Grundlehren der Mathematischen Wissenschaften

ISBN 978-3-642-00117-8 ISBN 978-3-642-00136-9 (eBook)

DOI 10.1007/978-3-642-00136-9

Classics in Mathematics ISSN 1431-0821

Library of Congress Control Number: 2009921797

Mathematics Subject Classification (2000): 35-02; 35S05; 47G30; 58J50; 58J40; 58J32; 47F05; 58J47; 35P25

Cover design: WMXDesign GmbH, Heidelberg, Germany

springer.com

Grundlehren der mathematischen Wissenschaften 275

A Series of Comprehensive Studies in Mathematics

Lars Hörmander

The Analysis of Linear Partial Differential Operators IV

Fourier Integral Operators

Springer-Verlag
Berlin Heidelberg NewYork Tokyo

Lars Hörmander
Department of Mathematics
University of Lund
Box 118
S-221 00 Lund, Sweden

Corrected Second Printing 1994
With 7 Figures

AMS Subject Classification (1980): 35A, G, H, J, L, M, P, S; 47G; 58G

ISBN 978-3-642-00117-8
Springer-Verlag Berlin Heidelberg New York Tokyo

Library of Congress Cataloging in Publication Data
(Revised for volume 4)
Hörmander, Lars.
The analysis of linear partial differential operators.
(Grundlehren der mathematischen Wissenschaften; 256-)
Expanded version of the author's 1 vol. work: Linear partial differential operators.
Includes bibliographies and indexes.
Contents: 1. Distribution theory and Fourier analysis - 2. Differential operators with constant coefficients - 3. Pseudo-differential operators - 4. Fourier integral operators.
1. Differential equations, Partial. 2. Partial differential operators. I. Title. II. Series.
QA377.H578 1983 515.7'242 83-616

Springer-Verlag Berlin Heidelberg New York
a member of BertelsmannSpringer Science+Business Media GmbH

SPIN: 10868044 41/3111 – 5 4 3 2 – Printed on acid-free paper

Preface

to Volumes III and IV

The first two volumes of this monograph can be regarded as an expansion and updating of my book "Linear partial differential operators" published in the Grundlehren series in 1963. However, volumes III and IV are almost entirely new. In fact they are mainly devoted to the theory of linear differential operators as it has developed after 1963. Thus the main topics are pseudo-differential and Fourier integral operators with the underlying symplectic geometry. The contents will be discussed in greater detail in the introduction.

I wish to express here my gratitude to many friends and colleagues who have contributed to this work in various ways. First I wish to mention Richard Melrose. For a while we planned to write these volumes together, and we spent a week in December 1980 discussing what they should contain. Although the plan to write the books jointly was abandoned and the contents have been modified and somewhat contracted, much remains of our discussions then. Shmuel Agmon visited Lund in the fall of 1981 and generously explained to me all the details of his work on long range scattering outlined in the Goulaouic-Schwartz seminars 1978/79. His ideas are crucial in Chapter XXX. When the amount of work involved in writing this book was getting overwhelming Anders Melin lifted my spirits by offering to go through the entire manuscript. His detailed and constructive criticism has been invaluable to me; I as well as the readers of the book owe him a great debt. Bogdan Ziemian's careful proofreading has eliminated numerous typographical flaws. Many others have also helped me in my work, and I thank them all.

Some material intended for this monograph has already been included in various papers of mine. Usually it has been necessary to rewrite these papers completely for the book, but selected passages have been kept from a few of them. I wish to thank the following publishers holding the copyright for granting permission to do so, namely:
Marcel Dekker, Inc. for parts of [41] included in Section 17.2;
Princeton University Press for parts of [38] included in Chapter XXVII;
D. Reidel Publishing Company for parts of [40] included in Section 26.4;
John Wiley & Sons Inc. for parts of [39] included in Chapter XVIII.
(Here [N] refers to Hörmander [N] in the bibliography.)

Finally I wish to thank the Springer-Verlag for all the support I have received during my work on this monograph.

Djursholm in November, 1984 Lars Hörmander

Contents

Introduction

to Volumes III and IV

A great variety of techniques have been developed during the long history of the theory of linear differential equations with variable coefficients. In this book we shall concentrate on those which have dominated during the latest phase. As a reminder that other earlier techniques are sometimes available and that they may occasionally be preferable, we have devoted the introductory Chapter XVII mainly to such methods in the theory of second order differential equations. Apart from that Volumes III and IV are intended to develop systematically, with typical applications, the three basic tools in the recent theory. These are the theory of pseudo-differential operators (Chapter XVIII), Fourier integral operators and Lagrangian distributions (Chapter XXV), and the underlying symplectic geometry (Chapter XXI). In the choice of applications we have been motivated mainly by the historical development. In addition we have devoted considerable space and effort to questions where these tools have proved their worth by giving fairly complete answers.

Pseudo-differential operators developed from the theory of singular integral operators. In spite of a long tradition these played a very modest role in the theory of differential equations until the appearance of Calderón's uniqueness theorem at the end of the 1950's and the Atiyah-Singer-Bott index theorems in the early 1960's. Thus we have devoted Chapter XXVIII and Chapters XIX, XX to these topics. The early work of Petrowsky on hyperbolic operators might be considered as a precursor of pseudo-differential operator theory. In Chapter XXIII we discuss the Cauchy problem using the improvements of the even older energy integral method given by the calculus of pseudo-differential operators.

The connections between geometrical and wave optics, classical mechanics and quantum mechanics, have a long tradition consisting in part of heuristic arguments. These ideas were developed more systematically by a number of people in the 1960's and early 1970's. Chapter XXV is devoted to the theory of Fourier integral operators which emerged from this. One of its first applications was to the study of asymptotic properties of eigenvalues (eigenfunctions) of higher order elliptic operators. It is therefore discussed in Chapter XXIX here together with a number of later developments which give beautiful proofs of the power of the tool. The study by Lax of the propagation of singularities of solutions to the Cauchy problem was one of

L. Hörmander, *Classics in Mathematics*
The Analysis of Linear Partial Differential Operators IV,
DOI: 10.1007/978-3-642-00117-8_Intro, © Springer-Verlag Berlin Heidelberg 2009

the forerunners of the theory. We prove such results using only pseudo-differential operators in Chapter XXIII. In Chapter XXVI the propagation of singularities is discussed at great length for operators of principal type. It is the only known approach to general existence theorems for such operators. The completeness of the results obtained has been the reason for the inclusion of this chapter and the following one on subelliptic operators. In addition to Fourier integral operators one needs a fair amount of symplectic geometry then. This topic, discussed in Chapter XXI, has deep roots in classical mechanics but is now equally indispensible in the theory of linear differential operators. Additional symplectic geometry is provided in the discussion of the mixed problem in Chapter XXIV, which is otherwise based only on pseudo-differential operator theory. The same is true of Chapter XXX which is devoted to long range scattering theory. There too the geometry is a perfect guide to the analytical constructs required.

The most conspicuous omission in these books is perhaps the study of analytic singularities and existence theory for hyperfunction solutions. This would have required another volume – and another author. Very little is also included concerning operators with double characteristics apart from a discussion of hypoellipticity in Chapter XXII. The reason for this is in part shortage of space, in part the fact that few questions concerning such operators have so far obtained complete answers although the total volume of results is large. Finally, we have mainly discussed single operators acting on scalar functions or occasionally determined systems. The extensive work done on for example first order systems of vector fields has not been covered at all.

Chapter XXV. Lagrangian Distributions and Fourier Integral Operators

Summary

In Section 18.2 we introduced the space of conormal distributions associated with a submanifold Y of a manifold X. This is a natural extension of the classical notion of multiple layer on Y. All such distributions have their wave front sets in the normal bundle of Y which is a conic Lagrangian manifold. In Section 25.1 we generalize the notion of conormal distribution by defining the space of Lagrangian distributions associated with an arbitrary conic Lagrangian $\Lambda \subset T^*(X) \smallsetminus 0$. This is the space of distributions u such that there is a fixed bound for the order of $P_1 \dots P_N u$ for any sequence of first order pseudo-differential operators $P_1, \dots, P_N$ with principal symbols vanishing on Λ. This implies that $WF(u) \subset \Lambda$. Symbols can be defined for Lagrangian distributions in much the same way as for conormal distributions. The only essential difference is that the symbols obtained are half densities on the Lagrangian tensored with the Maslov bundle of Section 21.6.

In Section 25.2 we introduce the notion of Fourier integral operator; this is the class of operators having Lagrangian distribution kernels. As in the discussion of wave front sets in Section 8.2 (see also Section 21.2) it is preferable to associate a Fourier integral operator with the canonical relation $\subset (T^*(X) \smallsetminus 0) \times (T^*(Y) \smallsetminus 0)$ obtained by twisting the Lagrangian with reflection in the zero section of $T^*(Y)$. We prove that the adjoint of a Fourier integral operator associated with the canonical relation C is associated with the inverse of C, and that the composition of operators associated with C_1 and C_2 is associated with the composition $C_1 \circ C_2$ when the compositions are defined. Precise results on continuity in the $H_{(s)}$ spaces are proved in Section 25.3 when the canonical relation is the graph of a canonical transformation. We also study in some detail the case where the canonical relation projects into $T^*(X)$ and $T^*(Y)$ with only fold type of singularities.

The real valued C^∞ functions vanishing on a Lagrangian $\subset T^*(X) \smallsetminus 0$ form an ideal with $\dim X$ generators which is closed under Poisson brackets. We define general Lagrangian ideals by taking complex valued functions instead. With suitable local coordinates in X they always have a local system

L. Hörmander, *Classics in Mathematics*
The Analysis of Linear Partial Differential Operators IV,
DOI: 10.1007/978-3-642-00117-8_1, © Springer-Verlag Berlin Heidelberg 2009

of generators of the form

$$x_j - \partial H(\xi)/\partial \xi_j, \quad j=1,\dots,n,$$

just as in the real case. The ideal is called positive if $\operatorname{Im} H \leqq 0$. This condition is crucial in the analysis and turns out to have an invariant meaning. Distributions associated with positive Lagrangian ideals are studied in Section 25.4. The corresponding Fourier integral operators are discussed in Section 25.5. The results are completely parallel to those of Sections 25.1, 25.2 and 25.3 apart from the fact that for the sake of brevity we do not extend the notion of principal symbol.

25.1. Lagrangian Distributions

According to Definition 18.2.6 the space $I^m(X, Y; E)$ of conormal distribution sections of the vector bundle E is the largest subspace of ${}^\infty H^{\mathrm{loc}}_{(-m-n/4)}(X, E)$, $n = \dim X$, which is left invariant by all first order differential operators tangent to the submanifold Y. It follows from Theorem 18.2.12 that it is even invariant under all first order pseudo-differential operators from E to E with principal symbol vanishing on the conormal bundle of Y. The definition is therefore applicable with no change to any Lagrangian manifold:

Definition 25.1.1. Let X be a C^∞ manifold and $\Lambda \subset T^*(X) \smallsetminus 0$ a C^∞ closed conic Lagrangian submanifold, E a C^∞ vector bundle over X. Then the space $I^m(X, \Lambda; E)$ of Lagrangian distribution sections of E, of order m, is defined as the set of all $u \in \mathscr{D}'(X, E)$ such that

$$L_1 \dots L_N u \in {}^\infty H^{\mathrm{loc}}_{(-m-n/4)}(X, E) \tag{25.1.1}$$

for all N and all properly supported $L_j \in \Psi^1(X; E, E)$ with principal symbols L_j^0 vanishing on Λ.

The following lemma allows us to localize the study of $I^m(X, \Lambda; E)$.

Lemma 25.1.2. *If $u \in I^m(X, \Lambda; E)$ then $WF(u) \subset \Lambda$, and $Au \in I^m(X, \Lambda; E)$ if $A \in \Psi^0(X; E, E)$. Conversely, $u \in I^m(X, \Lambda; E)$ if for every $(x_0, \xi_0) \in T^*(X) \smallsetminus 0$ one can find $A \in \Psi^0(X; E, E)$ properly supported and non-characteristic at (x_0, ξ_0) such that $Au \in I^m(X, \Lambda; E)$.*

Proof. If $(x_0, \xi_0) \notin \Lambda$ we can choose $L_1, \dots, L_N$ in (25.1.1) non-characteristic in a conic neighborhood Γ and conclude that $u \in H^{\mathrm{loc}}_{(s)}$ in Γ if $s < N - m - n/4$. Hence $WF(u) \cap \Gamma = \emptyset$. To prove the second statement we observe that

$$L_1 \dots L_N A u = L_1 \dots L_{N-1} A L_N u - L_1 \dots L_{N-1} [A, L_N] u.$$

Here $[A, L_N] \in \Psi^0(X; E, E)$ and $L_N u \in I^m(X, \Lambda; E)$ by Definition 25.1.1. By induction with respect to N we conclude that

$$L_1 \dots L_N A u \in {}^\infty H^{\mathrm{loc}}_{(-m-n/4)}(X, E)$$

for all properly supported $A \in \Psi^0(X; E, E)$ and $L_j \in \Psi^1(X; E, E)$ with principal symbols vanishing on Λ. To prove the converse we choose B according to Lemma 18.1.24 so that $(x_0, \xi_0) \notin WF(BA-I)$. Thus $(x_0, \xi_0) \notin WF(BAu-u)$, and since $BAu \in I^m(X, \Lambda; E)$ it follows that

$$L_1 \dots L_N u \in {}^\infty H^{\mathrm{loc}}_{(-m-n/4)} \quad \text{at } (x_0, \xi_0)$$

if $L_1, \dots, L_N$ satisfy the conditions in Definition 25.1.1. Hence (25.1.1) is fulfilled so $u \in I^m(X, \Lambda; E)$.

Remark. So far we have not used that Λ is Lagrangian. However, if (25.1.1) is fulfilled we have $[L_j, L_k]^N u \in {}^\infty H^{\mathrm{loc}}_{(-m-n/4)}(X, E)$ for any N, so $WF(u)$ is contained in the characteristic set of $[L_j, L_k]$ by the first part of the proof. Hence $WF(u)$ cannot contain an arbitrary point in Λ unless Λ is involutive. The hypothesis that Λ is Lagrangian means that Λ is minimal with this property, or alternatively that we have a maximal set of conditions (25.1.1) which do not imply that u is smooth.

Lemma 25.1.2 reduces the study of distributions $u \in I^m(X, \Lambda; E)$ to the case where $WF(u)$ is contained in a small closed conic neighborhood Γ_0 of some point $(x_0, \xi_0) \in \Lambda$, and supp u is close to x_0. In that case Definition 25.1.1 is applicable even if Λ is just defined in an open conic neighborhood Γ_1 of Γ_0, for only the restriction of the principal symbol of L_j to Γ_1 is relevant. More generally, given a conic Lagrangian submanifold Λ of the open cone $\Gamma_1 \subset T^*(X) \setminus 0$ we shall say that $u \in I^m(X, \Lambda; E)$ at $(x_0, \xi_0) \in \Gamma_1$ if there is an open conic neighborhood $\Gamma_0 \subset \Gamma_1$ of (x_0, ξ_0) such that $Au \in I^m(X, \Lambda; E)$ for all properly supported $A \in \Psi^0$ with $WF(A) \subset \Gamma_0$; it suffices to know this for some such A which is non-characteristic at (x_0, ξ_0).

In view of Theorem 21.2.16 we may thus assume now that $X = \mathbb{R}^n$ and that $\Lambda = \{(H'(\xi), \xi); \xi \in \mathbb{R}^n \setminus 0\}$ where H is a real valued function in $C^\infty(\mathbb{R}^n \setminus 0)$ which is homogeneous of degree 1. We may also assume that E is the trivial bundle, which is then omitted from the notation.

Proposition 25.1.3. *If* $u \in I^m_{\mathrm{comp}}(\mathbb{R}^n, \Lambda)$, $\Lambda = \{(H'(\xi), \xi);\ \xi \in \mathbb{R}^n \setminus 0\}$, *then* $\hat{u}(\xi) = e^{-iH(\xi)} v(\xi)$, $|\xi| > 1$, *where* $v \in S^{m-n/4}(\mathbb{R}^n)$. *Conversely, the inverse Fourier transform of* $e^{-iH} v$ *is in* $I^m(\mathbb{R}^n, \Lambda)$ *if* $v \in S^{m-n/4}(\mathbb{R}^n)$.

Proof. Choose $\chi \in C_0^\infty(\mathbb{R}^n)$ equal to 1 in a neighborhood of 0 and define h by $\hat{h} = \chi \hat{H}_0$ where $H_0 = (1-\chi)H$. Then $\hat{H}_0 - \hat{h} \in \mathscr{S}$ (see the proof of Theorem 7.1.22), so $H_0 - h \in \mathscr{S}$. Thus $h \in S^1$ has the principal symbol H, so it suffices to prove the result with H replaced by h. Set $h_j(\xi) = \partial h(\xi)/\partial \xi_j$. The operator $h_j(D)$ is convolution with the inverse Fourier transform of h_j so it is properly supported. Hence

$$D^\beta \prod (x_j - h_j(D))^{\alpha_j} u \in {}^\infty H_{(-m-n/4)} \quad \text{if } |\beta| = |\alpha| \tag{25.1.2}$$

for $[x_j - h_j(D), D_k] = i\delta_{jk}$ so commuting the factors D^β we obtain a sum of products of operators of the form $(x_j - h_j(D))D_k$ to which (25.1.1) is applicable. Recalling the definition of ${}^\infty H_{(-m-n/4)}$ we obtain

$$\int_{R/2<|\xi|<2R} |\xi^\beta \prod (-D_j - h_j(\xi))^\alpha \hat{u}(\xi)|^2 \, d\xi \leqq C_\alpha R^{2(m+n/4)}; \quad R>1, \ |\beta| = |\alpha|.$$

With the notation $\hat{u}(\xi) = e^{-ih(\xi)} v(\xi)$ this means that

$$\int_{R/2<|\xi|<2R} |\xi|^{2|\alpha|} |D^\alpha v(\xi)|^2 \, d\xi \leqq C_\alpha R^{2(m+n/4)}.$$

If $v_R(\xi) = v(R\xi)/R^{m-n/4}$ then

$$\int_{\frac12<|\xi|<2} |D^\alpha v_R(\xi)|^2 \, d\xi \leqq C_\alpha$$

which by Lemma 7.6.3 gives uniform bounds for $D^\alpha v_R$ when $|\xi|=1$, that is, bounds for $|D^\alpha v(\xi)|\,(1+|\xi|)^{|\alpha|-m+n/4}$. The argument can be reversed to prove the last statement in the proposition, for the passage from the operators $(x_j - h_j(D))\,D_k$ to the general operators in (25.1.1) can be made by the argument preceding Theorem 18.2.7.

A slight modification of the proof gives precise information about the smoothness of elements in I^m. We state the result directly in a global form.

Theorem 25.1.4. *If $U \in I^m(X, \Lambda)$ and $U \in H_{(s_0)}$ at $(x_0, \xi_0) \in \Lambda$, then $U \in I^\mu(X, \Lambda)$ at (x_0, ξ_0) if $\mu + s_0 + n/4 > 0$.*

Proof. Choose $A \in \Psi^0(X)$ properly supported, non-characteristic at (x_0, ξ_0), so that $AU \in H_{(s_0)}$. By Lemma 25.1.2 we have $AU \in I^m$. We can choose A so that $WF(AU)$ is in a small conic neighborhood of (x_0, ξ_0). Writing $u = AU$ we conclude that it is sufficient to prove that $u \in I^\mu$ if $u \in H_{(s_0)}$ and u satisfies the hypotheses in Proposition 25.1.3. With the notation used there we have

$$\int_{\frac12<|\xi|<2} |D^\alpha v_R(\xi)|^2 \, d\xi \leqq C_\alpha, \qquad \int_{\frac12<|\xi|<2} |v_R(\xi)|^2 \, d\xi \leqq C R^{-2(s_0+m+n/4)}.$$

Let $|\xi|=1$ and set $V_{R,\xi}(\eta) = v_R(\xi + \eta/R^\delta) R^{-n\delta/2}$ where $\delta > 0$. Then

$$\int_{|\eta|<1} |D^\alpha V_{R,\xi}(\eta)|^2 \, d\eta \leqq C_\alpha R^{-2|\alpha|\delta}, \qquad \int_{|\eta|<1} |V_{R,\xi}(\eta)|^2 \, d\eta \leqq C R^{-2(s_0+m+n/4)}.$$

Now we use the Sobolev inequality

$$|D^\beta V(0)|^2 \leqq C_\beta \int_{|\eta|<1} \Big(\sum_{|\alpha|=s} |D^{\alpha+\beta} V(\eta)|^2 + |V(\eta)|^2 \Big) \, d\eta$$

where $s > n/2$. This is somewhat more general than (7.6.6) but follows from the same proof. Taking s so large that $s\delta > s_0 + m + n/4$ we obtain

$$|D^\beta V_{R,\xi}(0)| \leqq C' R^{-(s_0+m+n/4)},$$

hence

$$|D^\beta v_R(\xi)| \leq C' R^{\delta(n/2+|\beta|)-(s_0+m+n/4)}, \qquad |\xi|=1,$$
$$|D^\beta v(\xi)| \leq C' |\xi|^{\delta(n/2+|\beta|)-(s_0+n/2+|\beta|)}, \qquad |\xi|>1.$$

For every β we can choose δ so that the exponent is smaller than $\mu-n/4-|\beta|$, and then we obtain $v\in S^{\mu-n/4}$, hence $u\in I^\mu$.

We shall now prove that elements in $I^m(X,\Lambda)$ can also be represented by means of arbitrary phase functions ϕ parametrizing Λ in the sense of Definition 21.2.15. At first we assume that ϕ is non-degenerate.

Proposition 25.1.5. *Let $\phi(x,\theta)$ be a non-degenerate phase function in an open conic neighborhood of $(x_0,\theta_0)\in\mathbb{R}^n\times(\mathbb{R}^N\setminus 0)$ which parametrizes the Lagrangian manifold Λ in a neighborhood of (x_0,ξ_0); $\xi_0=\phi'_x(x_0,\theta_0)$, $\phi'_\theta(x_0,\theta_0)=0$. If $a\in S^{m+(n-2N)/4}(\mathbb{R}^n\times\mathbb{R}^N)$ has support in the interior of a sufficiently small conic neighborhood Γ of (x_0,θ_0), then the oscillatory integral*

$$u(x)=(2\pi)^{-(n+2N)/4}\int e^{i\phi(x,\theta)}a(x,\theta)\,d\theta \tag{25.1.3}$$

defines a distribution $u\in I^m_{\text{comp}}(\mathbb{R}^n,\Lambda)$. If $\Lambda=\{(H'(\xi),\xi)\}$ as in Proposition 25.1.3 then (for $|\xi|>1$)

$$e^{iH(\xi)}\hat{u}(\xi)-(2\pi)^{n/4}a(x,\theta)e^{\pi i/4\,\operatorname{sgn}\Phi}|\det\Phi|^{-\frac{1}{2}}\in S^{m-n/4-1} \tag{25.1.4}$$

where (x,θ) is determined by $\phi'_\theta(x,\theta)=0$, $\phi'_x(x,\theta)=\xi$, and

$$\Phi=\begin{pmatrix}\phi''_{xx} & \phi''_{x\theta}\\ \phi''_{\theta x} & \phi''_{\theta\theta}\end{pmatrix}.$$

Here $a(x,\theta)$ is interpreted as 0 if there is no such point in Γ. $e^{iH(\xi)}\hat{u}(\xi)$ is polyhomogeneous if a is. Conversely, every $u\in I^m(X,\Lambda)$ with $WF(u)$ in a small conic neighborhood of (x_0,ξ_0) can, modulo C^∞, be written in the form (25.1.3).

In the proof we shall need an extension of Lemma 18.1.18.

Lemma 25.1.6. *Let $\Gamma_j\subset\mathbb{R}^{n_j}\times(\mathbb{R}^{N_j}\setminus 0)$, $j=1,2$, be open conic sets and let $\psi:\Gamma_1\to\Gamma_2$ be a C^∞ proper map commuting with multiplication by positive scalars in the second variable. If $a\in S^m(\mathbb{R}^{n_2}\times\mathbb{R}^{N_2})$ has support in the interior of a compactly based cone $\subset\Gamma_2$ then $a\circ\psi\in S^m(\mathbb{R}^{n_1}\times\mathbb{R}^{N_1})$ if the composition is defined as 0 outside Γ_1.*

Proof. The support of $a\circ\psi$ belongs to a compactly based cone $\subset\Gamma_1$ where $\psi(x,\xi)=(y,\eta)$ implies $|\xi|/C<|\eta|<C|\xi|$. The hypothesis on a means that

$$|D^\alpha_{y,\eta}a(y,t\eta)|\leq C_\alpha t^m, \quad 1/C<|\eta|<C.$$

Since $a\circ\psi(x,t\xi)=a(.,t.)\circ\psi(x,\xi)$ by the homogeneity of ψ, we obtain

$$|D^\alpha_{x,\xi}(a\circ\psi)(x,t\xi)|\leq C'_\alpha t^m, \quad |\xi|=1,$$

by using Leibniz' rule. This proves the lemma.

Proof of Proposition 25.1.5. By hypothesis $\phi'_x(x_0,\theta_0)=\xi_0\neq 0$, so the oscillatory integral (25.1.3) is well defined. u has compact support if Γ has a compact base. We shall use the method of stationary phase to evaluate

$$e^{iH(\xi)}\hat{u}(\xi)=(2\pi)^{-(n+2N)/4}\iint e^{i(\phi(x,\theta)+H(\xi)-\langle x,\xi\rangle)}a(x,\theta)\,dx\,d\theta. \tag{25.1.5}$$

The exponent has a critical point if

$$\phi'_x(x,\theta)=\xi,\quad \phi'_\theta=0,$$

which by hypothesis means that $(x,\xi)\in\Lambda$, hence that $x=H'(\xi)$. The critical point is non-degenerate. In fact, the maps

$$C=\{(x,\theta);\ \phi'_\theta=0\}\ni(x,\theta)\mapsto(x,\phi'_x)\in\Lambda \quad\text{and}\quad \Lambda\ni(x,\xi)\mapsto\xi$$

are diffeomorphisms. Hence $C\ni(x,\theta)\mapsto\phi'_x$ is a diffeomorphism, so $d\phi'_x=d\phi'_\theta=0$ implies $dx=d\theta=0$. The matrix Φ is therefore non-singular. If we divide (multiply) the first n (last N) rows (columns) by $|\theta|$ we see that $\det\Phi$ is homogeneous in θ of degree $n-N$. Hence $a(x,\theta)\,|\det\Phi|^{-\frac12}$ is in $S^{m-n/4}$ in a conic neighborhood of C. By Lemma 25.1.6 this remains true for the restriction to C regarded as a function of ξ.

It follows from Theorem 7.7.1 that there is a constant C such that for any N

$$\Big|\int e^{i(\phi(x,\theta)-\langle x,\xi\rangle)}a(x,\theta)\,dx\Big|\leq C_N(|\xi|+|\theta|)^{-N},\qquad \text{if } |\theta|>C|\xi| \ \text{ or } \ |\xi|>C|\theta|. \tag{25.1.6}$$

In fact, $(\phi(x,\theta)-\langle x,\xi\rangle)/(|\xi|+|\theta|)=f(x)$ is homogeneous in (ξ,θ) of degree 0 and bounded in C^∞. If $(x,\theta)\in\operatorname{supp}a$ we have

$$|f'(x)|\geq(|\xi|-C_1|\theta|)/(|\xi|+|\theta|)\geq\tfrac12 \qquad \text{if } |\theta|/|\xi| \text{ is small,}$$
$$|f'(x)|\geq(C_2|\theta|-|\xi|)/(|\xi|+|\theta|)>C_2/2 \qquad \text{if } |\xi|/|\theta| \text{ is small.}$$

We can therefore apply Theorem 7.7.1 with $\omega=|\xi|+|\theta|$.

Choose $\chi\in C_0^\infty(\mathbb{R}^N\setminus 0)$ equal to 1 when $1/C<|\theta|<C$. By (25.1.6) the difference between $e^{iH(\xi)}\hat{u}(\xi)$ and

$$U(\xi)=(2\pi)^{-(n+2N)/4}\iint e^{i(\phi(x,\theta)+H(\xi)-\langle x,\xi\rangle)}\chi(\theta/|\xi|)\,a(x,\theta)\,dx\,d\theta$$

is rapidly decreasing as $\xi\to\infty$. Set $|\xi|=t$, $\xi/t=\eta$ and replace θ by $t\theta$. Then

$$U(\xi)=(2\pi)^{-(n+2N)/4}\iint e^{it(\phi(x,\theta)+H(\eta)-\langle x,\eta\rangle)}\chi(\theta)\,a(x,t\theta)\,t^N\,dx\,d\theta.$$

Here the exponent has only one critical point in the support of the integrand and it is defined by $\phi'_\theta(x,\theta)=0$, $\phi'_x(x,\theta)=\eta$. At that point

$$\phi(x,\theta)=\langle\theta,\phi'_\theta(x,\theta)\rangle=0,\qquad \langle x,\eta\rangle=\langle H'(\eta),\eta\rangle=H(\eta)$$

so the critical value is 0. Using (7.7.13) we obtain an asymptotic expansion of U. Since $\chi=1$ at the critical point, the leading term is

$$(2\pi)^{n/4}\,a(x,t\theta)\,t^{(N-n)/2}\,e^{\pi i/4\,\operatorname{sgn}\Phi}\,|\det\Phi|^{-\frac12},$$

that is, the term displayed in (25.1.4) in view of the homogeneity of $\det \Phi$ already pointed out. The kth term will contain another factor t^{-k} and a linear combination of derivatives of $a(x, t\theta)$ with respect to x, θ, so it is in $S^{m-n/4-k}$. In view of Proposition 18.1.4 it follows that we have an asymptotic series in the sense of Proposition 18.1.3, and this completes the proof of the first part of the proposition.

To prove the converse it is by Proposition 25.1.3 sufficient to consider an element $u \in I^m(X, \Lambda)$ with $v = \hat{u} e^{iH} \in S^{m-n/4}$ having support in a small conic neighborhood of ξ_0. Choose a C^∞ map $(x, \theta) \mapsto \psi(x, \theta) \in \mathbb{R}^n \setminus 0$ in a conic neighborhood of (x_0, θ_0) such that ψ is homogeneous of degree 1 and $\psi(x, \theta) = \partial\phi/\partial x$ when $\partial\phi/\partial\theta = 0$. Let

$$a_0(x, \theta) = (2\pi)^{-n/4} v \circ \psi(x, \theta) e^{-\pi i/4 \operatorname{sgn} \Phi} |\det \Phi|^{\frac{1}{2}} \in S^{m+(n-2N)/4}$$

near C, and define u_0 by (25.1.3) with a replaced by a_0. From the first part of the proposition it follows then that $u - u_0 \in I^{m-1}$. Repeating the argument gives a sequence $a_j \in S^{m+(n-2N)/4-j}$ such that $u - u_0 - \ldots - u_j \in I^{m-j-1}$ if u_j is defined by (25.1.3) with a replaced by a_j. If a is an asymptotic sum of the series $\sum a_j$ it follows that (25.1.3) is valid modulo C^∞. The proof is complete.

We shall now examine what must be changed in the preceding argument if ϕ is just a clean phase function. We still have (25.1.6) so only $U(\xi)$ is important. However, $\phi(x, \theta) + H(\eta) - \langle x, \eta \rangle$ does not satisfy the hypotheses in Theorem 7.7.6. We do know that (locally)

$$C = \{(x, \theta); \partial\phi(x, \theta)/\partial\theta = 0\}$$

is a manifold of dimension $e + n$, where e is the excess, and that the composed map $C \to \Lambda \to \mathbb{R}^n$: $(x, \theta) \mapsto (x, \phi'_x) \mapsto \phi'_x$ has surjective differential, hence a fiber C_η of dimension e over η where $x = H'(\eta)$. The critical points of $\phi(x, \theta) + H(\eta) - \langle x, \eta \rangle$ are defined by $\phi'_\theta = 0$, $\phi'_x = \eta$, that is, $(x, \theta) \in C_\eta$, and $d\phi'_\theta = 0$, $d\phi'_x = 0$ precisely along the tangent space of C_η. Note that we have fixed upper and lower bounds for $|\theta|$ on C_η since $|\phi'_x| = 1$. We can split the θ variables into two groups θ', θ'' so that the number of θ'' variables is e and the projection $C_\eta \ni (x, \theta) \mapsto \theta''$ has bijective differential. Then $d\phi'_\theta = 0$, $d\phi'_x = 0$, $d\theta'' = 0$ implies $dx = d\theta = 0$. Thus the Hessian of $\phi(x, \theta) + H(\eta) - \langle x, \eta \rangle$ with respect to (x, θ') is not 0, so the critical point on C_η when θ'' is fixed is non-degenerate. If we change the definition of Φ to

$$\Phi = \begin{pmatrix} \phi''_{xx} & \phi''_{x\theta'} \\ \phi''_{\theta' x} & \phi''_{\theta'\theta'} \end{pmatrix},$$

an application of Theorem 7.7.6 to the integral $U(\xi)$ with respect to the $n + N - e$ variables x, θ' gives, when we integrate with respect to θ'' afterwards,

$$e^{iH(\xi)} \hat{u}(\xi) - (2\pi)^{n/4 - e/2} \int_{C_\eta} t^{(N+e-n)/2} a(x, t\theta) e^{\pi i/4 \operatorname{sgn} \Phi} |\det \Phi|^{-\frac{1}{2}} d\theta'' \in S^{m+e/2-n/4-1}.$$

Note that the order has increased by $e/2$ since the stationary phase evaluation is applied to e variables less. For the same reason a factor $(2\pi)^{e/2}$ is lost. If we introduce $t\theta$ as a new variable, noting that $\det \Phi$ is homogeneous of degree $n-N+e$ now, we obtain

Proposition 25.1.5'. *Let $\phi(x,\theta)$ be a clean phase function with excess e in an open conic neighborhood of $(x_0,\theta_0)\in\mathbb{R}^n\times(\mathbb{R}^N\setminus 0)$ which parametrizes the Lagrangian manifold Λ in a neighborhood of (x_0,ξ_0); $\xi_0=\phi'_x(x_0,\theta_0)$, $\phi'_\theta(x_0,\theta_0)=0$. If $a\in S^{m+(n-2N-2e)/4}(\mathbb{R}^n\times\mathbb{R}^N)$ has support in the interior of a sufficiently small conic neighborhood Γ of (x_0,θ_0) then the oscillatory integral*

$$u(x)=(2\pi)^{-(n+2N-2e)/4}\int e^{i\phi(x,\theta)}a(x,\theta)\,d\theta \tag{25.1.3$'$}$$

defines a distribution $u\in I^m_{\mathrm{comp}}(\mathbb{R}^n,\Lambda)$. If $\Lambda=\{(H'(\xi),\xi)\}$ as in Proposition 25.1.3 then

$$e^{iH(\xi)}\hat{u}(\xi)-(2\pi)^{n/4}\int_{C_\xi} a(x,\theta)\,e^{\pi i/4\,\mathrm{sgn}\,\Phi}\,|\det\Phi|^{-\frac{1}{2}}\,d\theta''\in S^{m-n/4-1}. \tag{25.1.4$'$}$$

Here $C_\xi=\{(x,\theta);\ \phi'_\theta(x,\theta)=0,\ \phi'_x(x,\theta)=\xi\}$; $\theta=(\theta',\theta'')$ is a splitting of the θ variables in two groups such that $C_\xi\ni(x,\theta)\mapsto\theta''$ has bijective differential; and

$$\Phi=\begin{pmatrix}\phi''_{xx} & \phi''_{x\theta'}\\ \phi''_{\theta' x} & \phi''_{\theta'\theta'}\end{pmatrix}.$$

Conversely, modulo C^∞ every $u\in I^m(X,\Lambda)$ with $WF(u)$ in a small conic neighborhood of (x_0,ξ_0) can be written in the form (25.1.3)$'$.

Remark. If $f\in C^\infty(Y)$ has a critical point at $y_0\in Y$ then $|\det f''(y_0)|^{\frac{1}{2}}$ transforms as a density at y_0. This is why in the standard stationary phase formula the density in the integrand is transformed to a scalar in the asymptotic expansion. If on the other hand f is critical on a submanifold $Z\subset Y$ and is non-degenerate in transversal directions, then the square root of the determinant of the Hessian in transversal planes defines a density in the normal bundle. Dividing a density in Y by it gives a density on Z. This confirms the invariant meaning of the integrand in (25.1.4)$'$.

There is no difficulty in performing a change of local coordinates x in the representation (25.1.3) of an element in $I^m(X,\Lambda)$, so Proposition 25.1.3 contains all that is needed to define a principal symbol isomorphism for I^m extending Theorem 18.2.11. However, it is instructive to establish first a theorem on limits of elements in I^m which connects the definitions in this section with those given in the linear case in Section 21.6.

Proposition 25.1.7. *Let $u\in I^m_{\mathrm{comp}}(\mathbb{R}^n,\Lambda)$, $\Lambda=\{(H'(\xi),\xi),\ \xi\in\mathbb{R}^n\setminus 0\}$, and set $e^{iH}\hat{u}=(2\pi)^{n/4}v$, $v\in S^{m-n/4}$. If $\psi\in C^\infty(\mathbb{R}^n)$ is real valued, $\psi(x_0)=0$, $\psi'(x_0)=\xi_0\neq 0$, $(x_0,\xi_0)\in\Lambda$, then as $t\to+\infty$*

$$t^{-2m-n/2}(ue^{-it^2\psi})(x_0+x/t)-v(t^2\xi_0)t^{-2m+n/2}u^{\psi}_{x_0,\xi_0}(x)\to 0\quad\text{in }\mathscr{D}', \tag{25.1.7}$$

where

$$u^{\psi}_{x_0,\xi_0}(x)=(2\pi)^{-3n/4}\int \exp iQ^{\psi}_{x_0,\xi_0}(x,\xi)\,d\xi, \tag{25.1.8}$$
$$Q^{\psi}_{x_0,\xi_0}(x,\xi)=\langle x,\xi\rangle-\langle\psi''(x_0)x,x\rangle/2-\langle H''(\xi_0)\xi,\xi\rangle/2.$$

Note that the factor $t^{-n/2}$ in the left-hand side of (25.1.7) means that $ue^{-it^2\psi}$ is pulled back as a half density by the map $x\mapsto x_0+x/t$. The other factor t^{-2m} reflects that u is of order m and is examined near the frequency $t^2\xi_0$.

Proof of Proposition 25.1.7. By Fourier's inversion formula

$$u(x)=(2\pi)^{-3n/4}\int e^{i(\langle x,\xi\rangle-H(\xi))}v(\xi)\,d\xi.$$

Replacing ξ by $t^2\xi_0+t\xi$ we obtain if $\chi\in C_0^\infty$

$$t^{-2m-n/2}\langle(ue^{-it^2\psi})(x_0+./t),\chi\rangle =(2\pi)^{-3n/4}\iint e^{iE_t(x,\xi)}v(t^2\xi_0+t\xi)\,t^{-2m+n/2}\chi(x)\,dx\,d\xi, \tag{25.1.9}$$

where

$$E_t(x,\xi)=\langle x_0+x/t,\,t^2\xi_0+t\xi\rangle-t^2\psi(x_0+x/t)-t^2H(\xi_0+\xi/t).$$

Now $H(\xi_0)=\langle H'(\xi_0),\xi_0\rangle=\langle x_0,\xi_0\rangle$, $H'(\xi_0)=x_0$, $\psi(x_0)=0$, $\psi'(x_0)=\xi_0$, so

$$E_t(x,\xi)=Q^{\psi}_{x_0,\xi_0}(x,\xi)+O(1/t)$$

uniformly on compact sets. Hence

$$\int e^{iE_t(x,\xi)}\chi(x)\,dx\to\int\exp(iQ^{\psi}_{x_0,\xi_0}(x,\xi))\,\chi(x)\,dx \tag{25.1.10}$$

uniformly for ξ in a compact set. If $x\in\operatorname{supp}\chi$ then

$$|\partial E_t(x,\xi)/\partial x|=|t\xi_0+\xi-t\psi'(x_0+x/t)|\geq|\xi|-C\geq(|\xi|+1)/2$$

if $|\xi|>2C+1$, so Theorem 7.7.1 shows that the left-hand side of (25.1.10) has a bound $C_N(1+|\xi|)^{-N}$ independent of t, for every N. Thus

$$(2\pi)^{-3n/4}\iint e^{iE_t(x,\xi)}\chi(x)\,dx\,d\xi\to\langle u^{\psi}_{x_0,\xi_0},\chi\rangle,$$

and (25.1.7) follows if we show that for large N

$$\int|v(t^2\xi_0+t\xi)-v(t^2\xi_0)|\,t^{-2m+n/2}(1+|\xi|)^{-N}\,d\xi\to 0,\qquad t\to\infty.$$

The integrand can be estimated by

$$|t\xi|(t^2)^{m-n/4-1}t^{-2m+n/2}(1+|\xi|)^{-N}\leq t^{-1}|\xi|(1+|\xi|)^{-N},$$

if $|\xi|<t|\xi_0|/2$, so this part of the integral is $O(1/t)$. When $|\xi|>t|\xi_0|/2$ the bound $(1+|\xi|)^{4|m|+n/2-N}$ for the integrand is obvious, which completes the proof.

Remark. If $u\in\mathscr{D}'(\mathbb{R}^n)$, $(x_0,\xi_0)\notin WF(u)$ and $0\neq\xi_0=\psi'(x_0)$, then

$$t^N(ue^{-it^2\psi})(x_0+x/t)\to 0\quad\text{in }\mathscr{D}'$$

for every N. In fact, replacing u by χu where $\chi \in C_0^\infty$ is 1 in a neighborhood of x_0 and supported in another small neighborhood we can assume that $u \in \mathscr{E}'$ and that $\hat{u}$ is rapidly decreasing in a conic neighborhood of ξ_0. We may also assume that $\psi(x) = \langle x, \xi_0 \rangle$ for if ρ vanishes of second order at x_0 then $t^2 \rho(x_0 + x/t) \to \langle \rho''(x_0)x, x \rangle/2$, $t \to \infty$. The Fourier transform of $t^{N+1}(ue^{-it^2\psi})(x_0 + x/t)$ is then $t^{N+n+1}\hat{u}(t^2\xi_0 + t\xi)e^{it\langle x_0, \xi\rangle}$ which is bounded when $|\xi|/t$ is small, and uniformly bounded by a power of $(1+|\xi|)$ elsewhere.

If $v \in S_{\text{phg}}^{m-n/4}$ it follows from (25.1.7) that

$$t^{-2m-n/2}(ue^{-it^2\psi})(x_0 + x/t) \to v_0(\xi_0)\, u^\psi_{x_0,\xi_0}(x) \quad \text{in } \mathscr{D}', \tag{25.1.7$'$}$$

where v_0 is the principal symbol of v. At first sight it might seem that the limit is strongly tied to the specific local coordinates x, but in fact it is not:

Lemma 25.1.8. *Let u_t be distributions in a neighborhood of 0 in $\mathbb{R}^n$ such that $M_t^* u_t \to U$ in $\mathscr{D}'$ as $t \to 0$, where $M_t(x) = tx$. If θ is a local diffeomorphism at 0 with $\theta(0)=0$, it follows then that*

$$M_t^* \theta^* u_t \to \theta_0^* U, \quad t \to 0,$$

where $\theta_0(x) = \theta'(0)x$ is the differential of θ at 0.

Proof. We can write $M_t^* \theta^* u_t = M_t^* \theta^* M_{1/t}^* M_t^* u_t$. Since

$$M_{1/t} \circ \theta \circ M_t(x) = t^{-1}\theta(tx) \to \theta_0(x)$$

in C^∞ as $t \to 0$, it follows that $M_t^* \theta^* u_t \to \theta_0^* U$.

The existence of the limit U is thus coordinate independent. If we regard u as a distribution on a manifold, the limit is a distribution on the tangent space at 0. If u_t is transformed as a half density distribution under a change of variables, we obtain of course a factor $|\det \theta'(0)|^{\frac{1}{2}}$, so the limit is a half density on the tangent space.

Let us now return to (25.1.7)$'$ where $v \in S_{\text{phg}}^{m-n/4}$ and v_0 is the principal symbol. If u is thought of as a half density $u(x)|dx|^{\frac{1}{2}}$ in a manifold X, expressed in the local coordinates x, we conclude that the limit $v_0(\xi_0)\, u^\psi_{x_0,\xi_0}$ is a half density on the tangent space $T_{x_0}(X)$. In the tangent space $S = T_{x_0,\xi_0}(T^*(X))$ the tangent planes λ_1 and λ of the graphs of ψ' and of Λ are given by $\xi = \psi''(x_0)x$ and $x = H''(\xi_0)\xi$ in our local coordinates. In S the tangent space of the fiber defined by $x=0$ is a distinguished Lagrangian plane λ_0. If we compare (25.1.8) with (21.6.5) and (21.6.6) it follows that $v(\xi_0)u^\psi_{x_0,\xi_0} \in I(\lambda, \lambda_1)$ defines an element in $I(\lambda)$ independent of the choice of ψ, hence an element in the tensor product $M_\lambda \otimes \Omega_\lambda^{\frac{1}{2}}$ where M_λ is the fiber over λ of the Maslov bundle defined on $T^*(X)$ by the tangents of the fibers, and $\Omega_\lambda^{\frac{1}{2}}$ is the fiber of the half density bundle on λ. With the trivialization of the Maslov bundle given by the Lagrangian planes $\xi=0$ in the local coordinates used in Propositions 25.1.5 and 25.1.7, the half density in the tangent space at $(H'(\xi_0), \xi_0)\in\Lambda$ is $v(\xi_0)|d\xi|^{\frac12}$ when ξ parametrizes Λ by $x=H'(\xi)$. Thus we obtain an invariant definition of a section of $M_\Lambda\otimes\Omega_\Lambda^{\frac12}$ which is homogeneous of degree $m+n/4$. (For the definition of homogeneity see the discussion preceding Definition 18.2.10.) It will be called the *principal symbol* of u.

The preceding discussion motivates our definition of the principal symbol for general $u\in I^m(X,\Lambda)$, but it will actually only depend on Proposition 25.1.5. As a preliminary to the definition we must extend Definition 18.2.10 to symbols on conic manifolds. On any conic manifold V there is defined a multiplication M_t by real numbers $t>0$, satisfying the conditions in Definition 21.1.8. We define $S^m(V)$ as the set of all $a\in C^\infty(V)$ such that the functions $t^{-m}M_t^*a$ are uniformly bounded in $C^\infty(V)$ when $t\geqq 1$. If for some compact set $K\subset V$ the support of a is contained in $\bigcup_{t\geqq 1} M_tK$, and V is an open subset of $\mathbb{R}^n\times(\mathbb{R}^N\setminus 0)$ with M_t defined as multiplication by t in the second variable, then the proof of Lemma 25.1.6 shows that this definition agrees with our earlier ones. An advantage of the present definition is that it is applicable also if say a is a half density on V. Let a_0 be a fixed positive half density on V which is homogeneous of degree μ, that is, $M_t^*a_0=t^\mu a_0$. For example, if $V=\mathbb{R}^n\times(\mathbb{R}^N\setminus 0)$ with variables x, θ, then $|dx|^{\frac12}|d\theta|^{\frac12}$ is a half density which is homogeneous of degree $\mu=N/2$. We can now write every $a\in S^m(V,\Omega_V^{\frac12})$ in the form $a=a_0b$ where $b\in S^{m-\mu}(V)$ is a scalar symbol, and conversely all such products are in $S^m(V,\Omega_V^{\frac12})$.

We return now to the definition of the principal symbol of a general $u\in I^m(X,\Lambda)$ where X is a C^∞ manifold and $\Lambda\subset T^*(X)\setminus 0$ is a C^∞ conic Lagrangian manifold. For any $(x_0,\xi_0)\in\Lambda$ we can choose local coordinates x at x_0 such that a conic neighborhood Γ of (x_0,ξ_0) in Λ is defined in the local coordinates by $x=H'(\xi)$ as in Proposition 25.1.3. If $\Gamma_1\subset\Gamma$ is a compactly generated cone we can use Lemma 25.1.2 to split u into a sum u_1+u_2 where $u_j\in I^m(X,\Lambda)$ and $WF(u_1)\subset\Gamma$, $WF(u_2)\cap\Gamma_1=\emptyset$. We can take u_1 with compact support in the coordinate patch. For the Fourier transform in the local coordinates we have by Proposition 25.1.3

$$e^{iH(\xi)}\hat{u}_1(\xi)=(2\pi)^{n/4}\,v(\xi)\in S^{m-n/4}. \tag{25.1.11}$$

If $u=\tilde{u}_1+\tilde{u}_2$ is another decomposition with the same properties, we have $WF(\tilde{u}_1-u_1)\cap\Gamma_1=\emptyset$. Since $WF(\tilde{u}_1-u_1)\subset\Gamma$ and $\Gamma_1=\{(H'(\xi),\xi),\ \xi\in\gamma_1\}$ for some closed cone $\gamma_1\subset\mathbb{R}^n\setminus 0$ we conclude that the Fourier transform of $\tilde{u}_1-u_1$ is rapidly decreasing in a conic neighborhood of γ_1. Hence the class of v in $S^{m-n/4}(\gamma_1)/S^{-\infty}(\gamma_1)$ does not depend on the decomposition of u, and we can consider $v|d\xi|^{\frac12}$ as an element in $S^{m+n/4}(\Gamma_1,\Omega_{\Gamma_1}^{\frac12})/S^{-\infty}(\Gamma_1,\Omega_{\Gamma_1}^{\frac12})$ in view of the isomorphism $\gamma_1\ni\xi\mapsto(H'(\xi),\xi)\in\Gamma_1$. We shall now study to what extent the residue class mod $S^{m+n/4-1}$ depends on the choice of local coordinates. It is convenient to do so by examining the symbol definition just made when u is defined by (25.1.3) in terms of a non-degenerate phase function but the local coordinates are fixed. Note that (25.1.11) gives

$$u_1(x)=(2\pi)^{-3n/4}\int e^{i(\langle x,\xi\rangle - H(\xi))}v(\xi)\,d\xi \tag{25.1.11'}$$

which is a special case of (25.1.3) with $\phi(x,\xi)=\langle x,\xi\rangle - H(\xi)$ and $N=n$.

From (25.1.4) it follows that if $u\in I^m_{\text{comp}}(X,\Lambda)$ and $e^{iH}\hat{u}=(2\pi)^{n/4}v$, then

$$v(\xi)|d\xi|^{\frac12}-a(x,\theta)\,e^{\pi i/4\,\mathrm{sgn}\,\Phi}|\det\Phi|^{-\frac12}|d\xi|^{\frac12}\in S^{m+n/4-1}(\Lambda,\Omega^{\frac12}_\Lambda) \tag{25.1.12}$$

where $\phi'_\theta(x,\theta)=0$, $\phi'_x(x,\theta)=\xi$ defines (x,θ) as a function of ξ. Apart from the Maslov factor $\exp(\pi i/4\,\mathrm{sgn}\,\Phi)$ we can interpret (25.1.12) as follows. (Compare (21.6.17)'.) Set as before

$$C=\{(x,\theta);\phi'_\theta(x,\theta)=0\}.$$

The pullback $d_C=\delta(\phi'_\theta)$ of the δ function in $\mathbb{R}^N$ by the map $(x,\theta)\mapsto\phi'_\theta\in\mathbb{R}^N$ is a density on C given by

$$d_C=|d\lambda|\,|D(\lambda,\phi'_\theta)/D(x,\theta)|^{-1}$$

if $\lambda=(\lambda_1,\dots,\lambda_n)$ denote arbitrary local coordinates on C extended to C^∞ functions in a neighborhood and $|d\lambda|$ is the Lebesgue density. This follows from (6.1.1). In particular we can take $\lambda=\phi'_x$ when Λ is parametrized by ξ. Then we obtain $d_C=|d\xi|\,|\det\Phi|^{-1}$, hence

$$v(\xi)|d\xi|^{\frac12}\equiv a(x,\theta)\,d_C^{\frac12}e^{\pi i/4\,\mathrm{sgn}\,\Phi} \mod S^{m+n/4-1}(\Lambda,\Omega^{\frac12}) \tag{25.1.13}$$

where C is identified with Λ by the map $(x,\theta)\mapsto(x,\phi'_x)$.

If we now introduce new coordinates $\tilde{x}$ and transform u as a half density, that is, $\tilde{u}(\tilde{x})=|Dx/D\tilde{x}|^{\frac12}u(x)$, then (25.1.3) gives

$$\tilde{u}(\tilde{x})=(2\pi)^{-(n+2N)/4}\int e^{i\tilde{\phi}(\tilde{x},\theta)}\tilde{a}(\tilde{x},\theta)\,d\theta,$$
$$\tilde{\phi}(\tilde{x},\theta)=\phi(x,\theta),\quad \tilde{a}(\tilde{x},\theta)=|Dx/D\tilde{x}|^{\frac12}a(x,\theta).$$

With the obvious identification of the manifolds C and $\tilde{C}$ defined by ϕ and by $\tilde{\phi}$, we have $d_C=|Dx/D\tilde{x}|\,d_{\tilde{C}}$ so

$$\tilde{a}d^{\frac12}_{\tilde{C}}=ad^{\frac12}_C. \tag{25.1.14}$$

The half density $v(\xi)|d\xi|^{\frac12}$ is thus invariant under a change of local coordinates apart from a Maslov factor of absolute value 1. Every non-singular $(n+N)\times(n+N)$ matrix Φ has signature congruent to $n+N \mod 2$ so if $\tilde{\Phi}$ is the matrix replacing Φ in the new coordinate system then the Maslov factor $e^{\pi i/4(\mathrm{sgn}\,\Phi-\mathrm{sgn}\,\tilde{\Phi})}$ which occurs is a power of the imaginary unit i. This means that (25.1.13) gives a principal symbol $\in S^{m+n/4}(\Lambda,\Omega^{\frac12}_\Lambda)$ for the element $u\in I^m$ defined by (25.1.3), which is uniquely determined modulo $S^{m+n/4-1}(\Lambda,\Omega^{\frac12}_\Lambda)$ and is multiplied by a power of i when the local coordinates are changed. For every $u\in I^m(X,\Lambda)$ we therefore get a principal symbol in

$$S^{m+n/4}(\Lambda,M_\Lambda\otimes\Omega^{\frac12}_\Lambda)/S^{m+n/4-1}(\Lambda,M_\Lambda\otimes\Omega^{\frac12}_\Lambda),$$

where M_Λ is a locally constant line bundle. It is defined by a covering $\Lambda=\bigcup\Gamma_j$ of Λ with open cones Γ_j and transition functions which are just powers of i. The discussion of the polyhomogeneous case above or just comparison of (25.1.13) and (21.6.17)′ identifies M_Λ with the Maslov bundle defined more geometrically in Section 21.6. In particular, it follows from (21.6.18) that if we have another local representation

$$u(x)=(2\pi)^{-(n+2\tilde{N})/4}\int e^{i\tilde{\phi}(x,\tilde{\theta})}\,\tilde{a}(x,\tilde{\theta})\,d\tilde{\theta}$$

in addition to (25.1.3) then

$$\tilde{a}d_{\tilde{C}}^{\frac{1}{2}}-e^{\pi is/4}ad_C^{\frac{1}{2}}\in S^{m+n/4-1}(\Lambda,\Omega_\Lambda^{\frac{1}{2}})$$

in the common domain of definition on Λ if

$$s=(\operatorname{sgn}\phi''_{\theta\theta}(x,\theta)-\operatorname{sgn}\tilde{\phi}''_{\tilde{\theta}\tilde{\theta}}(x,\tilde{\theta}))$$

where $\phi'(x,\theta)=\tilde{\phi}'(x,\tilde{\theta})=0$ and $\phi'_x(x,\theta)=\tilde{\phi}'_x(x,\tilde{\theta})=\xi$. Here the integer s is locally constant, and the x coordinates are now arbitrary. This connects with the definition of the Maslov bundle indicated after (21.6.17).

Summing up, we have now proved the following extension of Theorem 18.2.11 where we allow again the presence of a general vector bundle:

Theorem 25.1.9. *Let X be a C^∞ manifold, $\Lambda\subset T^*(X)\setminus 0$ a C^∞ conic Lagrangian submanifold, and E a C^∞ complex vector bundle over X. Then we have an isomorphism*

$$I^m(X,\Lambda;\Omega_X^{\frac{1}{2}}\otimes E)/I^{m-1}(X,\Lambda;\Omega_X^{\frac{1}{2}}\otimes E)$$
$$\to S^{m+n/4}(\Lambda,M_\Lambda\otimes\Omega_\Lambda^{\frac{1}{2}}\otimes\hat{E})/S^{m+n/4-1}(\Lambda,M_\Lambda\otimes\Omega_\Lambda^{\frac{1}{2}}\otimes\hat{E}).$$

Here $\hat{E}$ is the lifting of the bundle E to Λ. The image under the map is called the principal symbol.

Proof. By Lemma 25.1.2 this only has to be verified locally. For suitable fixed local coordinates the statement follows from Proposition 25.1.3. The Maslov bundle has been defined so that it is independent of the local coordinates chosen. - We shall often write E instead of $\hat{E}$ when no confusion seems possible.

Under the hypotheses in Proposition 25.1.5′ the principal symbol of u expressed in terms of the local coordinates there is equal to

$$|d\xi|^{\frac{1}{2}}\int_{C_\xi}a(x,\theta)e^{\pi i/4\operatorname{sgn}\Phi}|\det\Phi|^{-\frac{1}{2}}\,d\theta''.$$

This follows from (25.1.4)′, and C_ξ,Φ have been defined in Proposition 25.1.5′. We want to compare this with the definitions in Section 21.6. To every $(x_1,\theta_1)\in C_\xi$ the Hessian Q of $\phi/2$ at (x_1,θ_1) defines

$$U=a(x_1,\theta_1)(2\pi)^{-(n+2N-2e)/4}\int e^{iQ(x,\theta)}\,d\theta\in I(\lambda,\lambda_1)\otimes\Omega(R)$$

where R is the radical of Q_{x_1,θ_1} in the θ direction, λ is the tangent plane of Λ at $(H'(\xi),\xi)$ and λ_1 is the horizontal Lagrangian plane defined by $d\xi=0$ there. By hypothesis $R\ni\theta\mapsto\theta''$ is bijective, and the symbol of U as defined in Section 21.6 with the local coordinates x, ξ is

$$|d\xi|^{\frac{1}{2}} a(x_1, \theta_1) e^{\pi i/4 \operatorname{sgn} \Phi} |\det \Phi|^{-\frac{1}{2}} |d\theta''|.$$

Since $C_\xi \ni (x, \theta) \mapsto \theta''$ is bijective, $|d\theta''|$ is a positive density on C_ξ. Thus the symbol of u is the integral over C_ξ of the density on C_ξ with values in $(M_\Lambda \otimes \Omega_\Lambda^{\frac{1}{2}})_{(H'(\xi), \xi)}$ defined according to Section 21.6.

The phase function $-\phi$ defines the Lagrangian $\check{\Lambda} = i\Lambda$ where i: $T^*(X) \to T^*(X)$ is defined by $i(x, \xi) = (x, -\xi)$. From (25.1.4) it follows that the principal symbol of $\bar{u}$ is defined on $\check{\Lambda}$ by

$$\overline{a(x, \theta)}\, e^{-\pi i/4 \operatorname{sgn} \Phi} |\det \Phi|^{-\frac{1}{2}}, \tag{25.1.15}$$

that is, we obtain the pullback by i of the complex conjugate of the principal symbol of u. Now the complex conjugate of a section of $M_\Lambda \otimes \Omega_\Lambda^{\frac{1}{2}}$ is a section of $M_\Lambda^{-1} \otimes \Omega_\Lambda^{\frac{1}{2}}$, and $i^* M_\Lambda^{-1} = M_{\check{\Lambda}}$ by (21.6.5) since $i^* \sigma = -\sigma$. (This is just another way of expressing the complex conjugation of the Maslov factor in (25.1.15).) The pullback of a section of $M_\Lambda^{-1} \otimes \Omega_\Lambda^{\frac{1}{2}}$ can thus be identified with a section of $M_{\check{\Lambda}} \otimes \Omega_{\check{\Lambda}}^{\frac{1}{2}}$. Summing up, we have

Theorem 25.1.10. *Let the hypotheses of Theorem 25.1.9 be fulfilled and let j: $E \to F$ be an antilinear bundle map. Then $u \in I^m(X, \Lambda; \Omega_X^{\frac{1}{2}} \otimes E)$ implies $ju \in I^m(X, \check{\Lambda}; \Omega_X^{\frac{1}{2}} \otimes F)$ if $\check{\Lambda} = i\Lambda$, $i(x, \xi) = (x, -\xi)$; and $i^* ja \in S^{m+n/4}(X, \check{\Lambda};\ M_{\check{\Lambda}} \otimes \Omega_{\check{\Lambda}}^{\frac{1}{2}} \otimes \hat{F})$ is a principal symbol of ju if $a \in S^{m+n/4}(X, \Lambda; M_\Lambda \otimes \Omega_\Lambda^{\frac{1}{2}} \otimes \hat{E})$ is one for u.*

As in Section 18.1 we could have used considerably more general symbols in the preceding discussion. Lemma 25.1.6 remains valid for the symbol spaces $S_\rho^m = S_{\rho, 1-\rho}^m$. More generally, we can define $S_\rho^m(V)$ if V is a conic manifold as the set cf $a \in C^\infty(V)$ such that when $t \geqq 1$

$$t^{-m-(1-\rho)k} M_t^* a$$

is uniformly bounded in $C^k(V)$ for every $k \geqq 0$. We abandon now the intrinsic Definition 25.1.1 and define I_ρ^m when $\rho > \frac{1}{2}$ as the set of distributions which are microlocally of the form (25.1.11)$'$ with $v \in S_\rho^{m-n/4}$. An analogue of Proposition 25.1.5 follows with no essential change of the proof, and it leads to a principal symbol isomorphism

$$I_\rho^m(X, \Lambda; \Omega_X^{\frac{1}{2}} \otimes E)/I_\rho^{m+1-2\rho}(X, \Lambda; \Omega_X^{\frac{1}{2}} \otimes E) \to$$
$$S_\rho^{m+n/2}(\Lambda, M_\Lambda \otimes \Omega_\Lambda^{\frac{1}{2}} \otimes \hat{E})/S_\rho^{m+n/2+1-2\rho}(\Lambda, M_\Lambda \otimes \Omega_\Lambda^{\frac{1}{2}} \otimes \hat{E}).$$

It would also have been possible to define I_ρ^m by the condition (25.1.1) in Definition 25.1.1 for all $L_j \in \Psi_\rho^{2\rho-1}(X; E, E)$ with principal symbol vanishing on Λ. However, the proof of the analogue of Proposition 25.1.3 becomes somewhat longer since, we must consider on one hand operators with symbols of the form $|\xi|^\rho \chi((x - h'(\xi))|\xi|^{1-\rho})(x_j - h_j(\xi))$ and on the other hand operators with symbols of the form $|\xi|^{2\rho-1}(1 - \chi((x - h'(\xi))|\xi|^{1-\rho}))$. Here $\chi \in C_0^\infty$ is equal to 1 in a neighborhood of the origin. We leave the details for the interested and energetic reader.

25.2. The Calculus of Fourier Integral Operators

Let X and Y be two C^∞ manifolds and E, F two complex vector bundles on X, Y. Then every $A\in\mathscr{D}'(X\times Y,\Omega^{\frac12}_{X\times Y}\otimes\operatorname{Hom}(F,E))$ defines a continuous map

$$\mathscr{A}\colon\ C_0^\infty(Y,\Omega^{\frac12}_Y\otimes F)\to\mathscr{D}'(X,\Omega^{\frac12}_X\otimes E)$$

and conversely. (See Section 5.2 and, for the role of the half densities, also Section 18.1.) Here the fiber of the vector bundle $\operatorname{Hom}(F,E)$ at (x,y) consists of the linear maps $F_y\to E_x$. In particular, if Λ is a closed conic Lagrangian submanifold of $T^*(X\times Y)\smallsetminus 0$ we can identify $I^m(X\times Y,\Lambda;\Omega^{\frac12}_{X\times Y}\otimes\operatorname{Hom}(F,E))$ with a space of such maps. If we have

$$\Lambda\subset(T^*(X)\smallsetminus 0)\times(T^*(Y)\smallsetminus 0) \tag{25.2.1}$$

then it follows from Theorem 8.2.13 and Lemma 25.1.2 that $\mathscr{A}$ is even a continuous map from $C_0^\infty(Y)$ to $C^\infty(X)$ which can be extended to a continuous map from $\mathscr{E}'(Y)$ to $\mathscr{D}'(X)$ with

$$WF(\mathscr{A}u)\subset C(WF(u)),\quad u\in\mathscr{E}'(Y,\Omega^{\frac12}_Y\otimes F), \tag{25.2.2}$$

where

$$C=\Lambda'=\{(x,\xi,y,-\eta)\in(T^*(X)\smallsetminus 0)\times(T^*(Y)\smallsetminus 0);(x,\xi,y,\eta)\in\Lambda\}$$

is a canonical relation from $T^*(Y)\smallsetminus 0$ to $T^*(X)\smallsetminus 0$. (See Definition 21.2.12.) As in Section 21.2 we call $\Lambda=C'$ the twisted canonical relation. The Maslov bundle M_Λ can be regarded as a bundle M_C on C defined by C and the product symplectic form $\sigma_X-\sigma_Y$.

Definition 25.2.1. Let C be a homogeneous canonical relation from $T^*(Y)\smallsetminus 0$ to $T^*(X)\smallsetminus 0$ which is closed in $T^*(X\times Y)\smallsetminus 0$, and let E, F be vector bundles on X, Y. Then the operators with kernel belonging to $I^m(X\times Y,C';\Omega^{\frac12}_{X\times Y}\otimes\operatorname{Hom}(F,E))$ are called Fourier integral operators of order m from sections of F to sections of E, associated with the canonical relation C.

Let E^* be the vector bundle with fiber E^*_x at $x\in X$ antidual to the fiber E_x of E. Then we have a pairing

$$u,v\mapsto\int(u,v)(x);\quad u\in C_0^\infty(X,\Omega^{\frac12}_X\otimes E),\quad v\in\mathscr{D}'(X,\Omega^{\frac12}_X\otimes E^*),$$

and a similar one for Y and F. If $A\in\mathscr{D}'(X\times Y,\Omega^{\frac12}_{X\times Y}\otimes\operatorname{Hom}(F,E))$ then the adjoint of the map $C_0^\infty(Y,\Omega^{\frac12}_Y\otimes F)\to\mathscr{D}'(X,\Omega^{\frac12}_X\otimes E)$ defined by A is defined by $A^*\in\mathscr{D}'(Y\times X,\Omega^{\frac12}_{Y\times X}\otimes\operatorname{Hom}(E^*,F^*))$. If s is the map $Y\times X\to X\times Y$ interchanging the two factors then A^* is obtained by composing s^*A with the antilinear bundle map $\operatorname{Hom}(F,E)\to\operatorname{Hom}(E^*,F^*)$ given by taking adjoints. If

$$A\in I^m(X\times Y,C';\Omega^{\frac12}_{X\times Y}\otimes\operatorname{Hom}(F,E))$$

then

$$s^*A\in I^m(Y\times X,s^*C';\Omega^{\frac12}_{Y\times X}\otimes\operatorname{Hom}(F,E));$$

if a is the principal symbol of A then s^*a is the principal symbol of s^*A. This is obvious by the invariance of our constructions. With the same notation i for the reflection in the cotangent bundle as in Theorem 25.1.10 we have $i^*s^*C'=(C^{-1})'$ where C^{-1} is the inverse canonical relation obtained by interchanging $T^*(X)$ and $T^*(Y)$. Thus we obtain in view of Theorem 25.1.10:

Theorem 25.2.2. *Let C be a homogeneous canonical relation from $T^*(Y)\setminus 0$ to $T^*(X)\setminus 0$ which is closed in $T^*(X\times Y)\setminus 0$, and let E, F be vector bundles on X, Y. If $A\in I^m(X\times Y, C'; \Omega^{\frac12}_{X\times Y}\otimes \operatorname{Hom}(F,E))$, identified with the corresponding linear operator, then $A^*\in I^m(Y\times X, (C^{-1})'; \Omega^{\frac12}_{Y\times X}\otimes\operatorname{Hom}(E^*,F^*))$. If $a\in S^{m+n/4}(C; M_C\otimes\Omega^{\frac12}_C\otimes\operatorname{Hom}(F,E))$ is a principal symbol for A, where $n=\dim(X\times Y)$, then $s^*a^*\in S^{m+n/4}(C^{-1}, M_{C^{-1}}\otimes\Omega^{\frac12}_{C^{-1}}\otimes\operatorname{Hom}(E^*,F^*))$ is a principal symbol for A^*. Here s is the interchanging map $Y\times X\to X\times Y$.*

Note that we have here chosen to regard the principal symbol as defined on C rather than on C'. This is usually more convenient in connection with Fourier integral operators and should cause no confusion.

We shall now discuss products, so let C_1 be a homogeneous canonical relation from $T^*(Y)\setminus 0$ to $T^*(X)\setminus 0$ and C_2 another from $T^*(Z)\setminus 0$ to $T^*(Y)\setminus 0$ where X, Y, Z are three manifolds, with vector bundles, E, F, G. Let

$$A_1\in I^{m_1}(X\times Y, C_1'; \Omega^{\frac12}_{X\times Y}\otimes\operatorname{Hom}(F,E)),$$
$$A_2\in I^{m_2}(Y\times Z, C_2'; \Omega^{\frac12}_{Y\times Z}\otimes\operatorname{Hom}(G,F))$$

and assume that both are properly supported so that the composition A_1A_2 of the corresponding operators is defined. We want to show that it is associated with the composition C of the canonical relations C_1 and C_2 provided that the composition is clean, proper and connected in a sense which we shall now define. Already after the statement of Theorem 21.2.14 we defined the composition to be *clean* if $C_1\times C_2$ intersects $T^*(X)\times\Delta(T^*(Y))\times T^*(Z)$ cleanly, that is, in a manifold $\hat C$ with tangent plane everywhere equal to the intersection of the tangent planes of the intersecting manifolds. We shall say that the composition is *proper* if the map

$$\hat C\to T^*(X\times Z)\setminus 0$$

is proper. (When Y is compact this is automatically true since C_1 and C_2 are closed in $T^*(X\times Y)\setminus 0$ and $T^*(Y\times Z)\setminus 0$ respectively but contained in $(T^*(X)\setminus 0)\times(T^*(Y)\setminus 0)$ and $(T^*(Y)\setminus 0)\times(T^*(Z)\setminus 0)$.) Then the range C is a closed subset of $T^*(X\times Z)\setminus 0$ contained in $(T^*(X)\setminus 0)\times(T^*(Z)\setminus 0)$. The inverse image C_γ in $\hat C$ of $\gamma\in C$ is a compact manifold of dimension equal to the excess e of the clean intersection. To avoid self-intersections of C we assume that the composition is *connected* in the sense that C_γ is connected for every $\gamma\in C$. Then it follows from Theorem 21.2.14 that C is also a canonical relation. We shall prove that

(25.2.3) $$A_1A_2\in I^{m_1+m_2+e/2}(X\times Z, C; \Omega^{\frac12}_{X\times Z}\otimes \operatorname{Hom}(G,E))$$

and compute the principal symbol. Note that when the composition is *transversal*, that is, the excess $e=0$, then $A_1A_2\in I^{m_1+m_2}$. The normalizations introduced in Section 25.1 were to a large extent motivated by our wish to maintain this natural property of the order of differential and pseudo-differential operators.

By a partition of unity we can reduce the proof of (25.2.3) to the local case where $X\subset\mathbb{R}^{n_X}$, $Y\subset\mathbb{R}^{n_Y}$, $Z\subset\mathbb{R}^{n_Z}$, the bundles E, F, G are trivial and

$$A_1(x,y)=(2\pi)^{-(n_X+n_Y+2N_1)/4}\int e^{i\phi(x,y,\theta)}a_1(x,y,\theta)\,d\theta,$$
$$A_2(y,z)=(2\pi)^{-(n_Y+n_Z+2N_2)/4}\int e^{i\psi(y,z,\tau)}a_2(y,z,\tau)\,d\tau.$$

Here ϕ is a non-degenerate phase function in a conic neighborhood of $(x_0,y_0,\theta_0)\in X\times Y\times(\mathbb{R}^{N_1}\setminus 0)$ parametrizing C_1 in a conic neighborhood of (x_0,ξ_0,y_0,η_0), thus

$$\phi'_\theta=0,\quad \phi'_x=\xi_0,\quad \phi'_y=-\eta_0\quad \text{at } (x_0,y_0,\theta_0).$$

Similarly ψ is a non-degenerate phase function in a conic neighborhood of $(y_0,z_0,\tau_0)\in Y\times Z\times(\mathbb{R}^{N_2}\setminus 0)$ parametrizing C_2 in a conic neighborhood of (y_0,η_0,z_0,ζ_0), thus

$$\psi'_\tau=0,\quad \psi'_y=\eta_0,\quad \psi'_z=-\zeta_0\quad \text{at } (y_0,z_0,\tau_0).$$

The amplitudes a_1, a_2 have supports in small conic neighborhoods of (x_0,y_0,θ_0) and (y_0,z_0,τ_0) respectively, and

(25.2.4) $$a_1\in S^{m_1+(n_X+n_Y-2N_1)/4},\quad a_2\in S^{m_2+(n_Y+n_Z-2N_2)/4}.$$

If $a_j\in S^{-\infty}$ then $A=A_1A_2$ is given by

(25.2.5) $$\begin{aligned}A(x,z)&=\int A_1(x,y)A_2(y,z)\,dy\\ &=(2\pi)^{-(n_X+n_Z+2(n_Y+N_1+N_2))/4}\iiint e^{i\Phi(x,z,y,\theta,\tau)}a(x,z,y,\theta,\tau)\,dy\,d\theta\,d\tau\end{aligned}$$

where

$$\Phi(x,z,y,\theta,\tau)=\phi(x,y,\theta)+\psi(y,z,\tau);$$
$$a(x,z,y,\theta,\tau)=a_1(x,y,\theta)a_2(y,z,\tau).$$

From Proposition 21.2.19 we know that Φ is a clean phase function defining C, in a conic neighborhood of $(x_0,z_0,y_0,\theta_0,\tau_0)$. This will lead to a proof of (25.2.3) when we have proved that the integration can be restricted to a set where $|\theta|$ and $|\tau|$ have the same order of magnitude so that a is a well behaved symbol. This is not the case for the function a as it stands since, for example, differentiation with respect to θ only improves the magnitude by a factor $1/(1+|\theta|)$ and not by $1/(1+|\theta|+|\tau|)$.

We assume that a_1 and a_2 have supports in compactly generated cones Γ_1 and Γ_2 where $\partial\phi(x,y,\theta)/\partial y$ and $\partial\psi(y,z,\tau)/\partial y$ never vanish. Then we can find C_1 and C_2 such that

$$C_1|\tau|<|\theta|<C_2|\tau| \quad \text{if } (x,y,\theta)\in\Gamma_1,\ (y,z,\tau)\in\Gamma_2$$

and

$$\partial\phi(x,y,\theta)/\partial y+\partial\psi(y,z,\tau)/\partial y=0.$$

In fact, if $|\theta|=1$ for example we have an upper bound for

$$|\partial\phi(x,y,\theta)/\partial y|=|\partial\psi(y,z,\tau)/\partial y|,$$

hence an upper bound for $|\tau|$. We choose a homogeneous function $\chi(\theta,\tau)$ of degree 0 which is equal to 1 when $C_1|\tau|/2<|\theta|<2C_2|\tau|$ and has support in the cone where $C_1|\tau|/3<|\theta|<3C_2|\tau|$. With the notation

$$b(x,z,y,\theta,\tau)=\chi(\theta,\tau)\,a(x,z,y,\theta,\tau),$$
$$r(x,z,y,\theta,\tau)=(1-\chi(\theta,\tau))\,a(x,z,y,\theta,\tau)$$

we have

$$|\theta|+|\tau|<C|\partial\Phi(x,z,y,\theta,\tau)/\partial y| \quad \text{in supp}\, r, \tag{25.2.6}$$

$$C_1|\tau|/3<|\theta|<3C_2|\tau| \quad \text{in supp}\, b. \tag{25.2.7}$$

From (25.2.4) and (25.2.6) it follows that the repeated integral

$$R(x,z)=(2\pi)^{-n}\iint d\theta\, d\tau \int e^{i\Phi(x,z,y,\theta,\tau)}\, r(x,z,y,\theta,\tau)\, dy,$$

$n=(n_X+n_Z+2(n_Y+N_1+N_2))/4$, exists and is a C^∞ function in $X\times Z$ depending continuously on a_1 and a_2. In fact, by Theorem 7.7.1 and (25.2.6) the inner integral can be estimated by any power of $(1+|\theta|+|\tau|)^{-1}$, and this remains true after any number of differentiations with respect to x or z.

From (25.2.4) and (25.2.7) it follows that

$$b\in S^\mu((X\times Y\times Z)\times\mathbb{R}^{N_1+N_2}), \tag{25.2.8}$$
$$\mu=m_1+m_2+(n_X+n_Z+2(n_Y-N_1-N_2))/4.$$

However, we wish to consider y as one of the parameters so we take say

$$\omega=((|\theta|^2+|\tau|^2)^{\frac12}y,\theta,\tau)\in\mathbb{R}^{n_Y+N_1+N_2}\setminus 0$$

as a new variable. Then

$$D\omega/D(y,\theta,\tau)=(|\theta|^2+|\tau|^2)^{n_Y/2}$$

so we have

$$b(x,z,y,\theta,\tau)\,D(y,\theta,\tau)/D\omega\in S^{\mu-n_Y}((X\times Z)\times\mathbb{R}^{n_Y+N_1+N_2}) \tag{25.2.8$'$}$$

where y, θ, τ are regarded as functions of ω. Hence Proposition 25.1.5′ shows that

$$B(x,z)=(2\pi)^{-n}\int e^{i\Phi(x,z,y,\theta,\tau)}\, b(x,z,y,\theta,\tau)\, dy\, d\theta\, d\tau\in I^{m_1+m_2+e/2}(X\times Z,C') \tag{25.2.9}$$

where e is the excess of the intersection, and B depends continuously on a_1 and a_2 when they vary in the symbol spaces (25.2.4). Since

$$A_1 A_2 = B + R$$

when a_j are of order $-\infty$ we conclude that this equality is always valid, which implies (25.2.3).

To compute the principal symbol of $A_1 A_2$ we can use (25.1.4)'. This formula means that when we split the variables θ into two groups θ', θ'' in such a way that Φ is a non-degenerate phase function for fixed θ'' and there are e such variables, then the principal symbol is the integral with respect to θ'' of that defined by the integral for fixed θ''. We should really apply this to the ω variables, but the invariance of (25.1.4)' under coordinate changes shows that this is irrelevant. Now we know from Proposition 21.2.19 that the function Φ in (25.2.9) is a clean phase function parametrizing C. Let Φ be non-degenerate in (y', θ', τ') while the e variables (y'', θ'', τ'') parametrize the sets C_γ, $\gamma = (x, \xi, z, \zeta) \in C$. Denote by $B_{y'', \theta'', \tau''}$ the kernel obtained when we only integrate with respect to y', θ', τ' in (25.2.9).

Next observe that, for example, the principal symbol of A_1 at the point $(x, \phi'_x, y, -\phi'_y) \in C_1$ corresponding to a point with $\phi'_\theta = 0$, and the tangent plane of C_1 there, are determined by the Hessian of ϕ as in the discussion of the linear case in Section 21.6. The same goes for A_2 and $B_{y'', \theta'', \tau''}$. The product of the principal symbols of A_1 and A_2 defined in Theorem 21.6.7 is therefore equal to the density $dy'' d\theta'' d\tau''$ times the principal symbol of $B_{y'', \theta'', \tau''}$. Integrating with respect to y'', θ'', τ'' we now obtain the second main theorem in the calculus of Fourier integral operators (see the discussion following Theorem 25.1.5'):

Theorem 25.2.3. *Let C_1 be a C^∞ homogeneous canonical relation from $T^*(Y) \setminus 0$ to $T^*(X) \setminus 0$ and C_2 another from $T^*(Z) \setminus 0$ to $T^*(Y) \setminus 0$ where X, Y, Z are three C^∞ manifolds, with C^∞ vector bundles E, F, G. Let*

$$A_1 \in I^{m_1}(X \times Y, C_1'; \Omega^{\frac{1}{2}}_{X \times Y} \otimes \operatorname{Hom}(F, E)),$$
$$A_2 \in I^{m_2}(Y \times Z, C_2'; \Omega^{\frac{1}{2}}_{Y \times Z} \otimes \operatorname{Hom}(G, F))$$

and assume that both are properly supported. Assume that the composition $C = C_1 \circ C_2$ is clean, with excess e, proper and connected. For $\gamma \in C$ denote by C_γ the compact e dimensional fiber over γ of the intersection of $C_1 \times C_2$ and $T^*(X) \times \Delta(T^*(Y)) \times T^*(Z)$. Then

$$A_1 A_2 \in I^{m_1 + m_2 + e/2}(X \times Z, C'; \Omega^{\frac{1}{2}}_{X \times Z} \otimes \operatorname{Hom}(G, E)),$$

and for the principal symbols a_1, a_2, a of $A_1, A_2, A_1 A_2$ we have

$$a = \int_{C_\gamma} a_1 \times a_2. \tag{25.2.10}$$

Here $a_1 \times a_2$ is the density on C_γ with values in the fiber of $M_C \otimes \Omega^{\frac{1}{2}}_{X \times Z} \otimes \operatorname{Hom}(G, E)$ defined in Theorem 21.6.7, extended by tensor product with $\operatorname{Hom}(G, F)$ and $\operatorname{Hom}(F, E)$.

We recall from Theorem 21.6.7 that the half density associated with a is $(2\pi)^{-e/2}$ times the half density associated purely geometrically with a_1 and a_2 through (21.6.21), (21.6.22). Some authors choose not to include this factor in the definition of the product but put it in the right-hand side of (25.2.10). There is of course no unique best procedure but one should be careful to have the factor at precisely one place.

Theorem 25.2.3 covers a range of apparently different situations. For example, if $X=Y=Z$, $C_1=C_2=$identity, we have recovered Theorem 18.1.23. With $X=Y$, Z reduced to a point, $C_1=$identity and C_2 equal to the conormal bundle of a submanifold of X we have Theorem 18.2.12. For more general Lagrangians C_2 we have a result first encountered in Lemma 25.1.2. Some other cases will be discussed in Section 25.3. However, Theorem 25.2.3 is unsatisfactory for some C_j because the principal symbol is always 0. Then $A_1A_2\in I^{m_1+m_2-1+e/2}$, and one should compute the principal symbol of A_1A_2 as operator of this lower order. In principle this can always be done by means of (25.2.9). To avoid lengthy geometrical arguments we shall just consider a simple case which will be important later on. Even so we need a preliminary discussion of how vector fields act on densities.

Let M be a manifold and v a real C^∞ vector field on M. Then v generates a local one parameter group of C^∞ maps ϕ^t in M, defined by

$$d\phi^t(x)/dt=v(\phi^t(x)),\quad \phi^0(x)=x,\quad x\in M.$$

If $a\in\Omega^\kappa(M)$ then we define the *Lie derivative* $\mathscr{L}_v a$ along v by

$$\mathscr{L}_v a=\frac{d}{dt}(\phi^t)^*a|_{t=0}.$$

Let $x_1,\dots,x_N$ be local coordinates in M and write $a=u|dx|^\kappa$. Then $(\phi^t)^*a=u_t|dx|^\kappa$,

$$u_t(x)=u(\phi^t(x))(D\phi^t(x)/Dx)^\kappa.$$

The derivative of the Jacobian is $\operatorname{Tr}(\partial v_j/\partial x_k)=\operatorname{div} v$ when $t=0$, hence

$$\mathscr{L}_v(u|dx|^\kappa)=(\sum v_j\partial u/\partial x_j+\kappa(\operatorname{div} v)u)|dx|^\kappa. \tag{25.2.11}$$

We take this as definition of the Lie derivative if v is a complex vector field; it is clear that the definition is independent of the choice of local coordinates since this is true when v is real.

We can now state a theorem on the product in a case where Theorem 25.2.3 can be improved.

Theorem 25.2.4. *Let $P\in\Psi^m_{\mathrm{phg}}(X)$ be properly supported, with principal symbol p and subprincipal symbol c. Assume that C is a homogeneous canonical relation from $T^*(Y)\setminus 0$ to $T^*(X)\setminus 0$ such that p vanishes on the projection of C in $T^*(X)\setminus 0$. If $A\in I^{m'}(X\times Y, C'; \Omega(X\times Y)^{\frac12})$ and $a\in S^{m'+(n_X+n_Y)/4}(C', M_C\otimes\Omega^{\frac12}_C)$ is a principal symbol of A, it follows that $PA\in I^{m+m'-1}(X\times Y, C'; \Omega(X\times Y)^{\frac12})$ has*

$$i^{-1}\mathscr{L}_{H_p}a+ca \tag{25.2.12}$$

as principal symbol. Here H_p is the Hamilton field of p, lifted to a function on $(T^*(X)\setminus 0)\times(T^*(Y)\setminus 0)$, so H_p is tangential to C.

Note that since M is locally constant the presence of the Maslov factors can be ignored in the computation of the Lie derivative.

In the proof of Theorem 25.2.4 we shall use local coordinates in X and in Y such that C' can locally be represented in the form $\{(\partial H/\partial\xi, \xi, \partial H/\partial\eta, \eta)\}$ where $H(\xi,\eta)$ is a homogeneous function of degree 1 in (ξ,η). (This could also have been done in the proof of Theorem 25.2.3 but the simplification would have been marginal so we preferred to use general phase functions.) To show that such coordinates exist although we do not allow general coordinate changes in $X\times Y$ as in Theorem 21.2.16, we need the following lemma.

Lemma 25.2.5. *If $\lambda\subset T^*(\mathbf{R}^N)$ is a Lagrangian plane, then λ is transversal to the Lagrangian plane $\lambda_a=\{(x,\xi);\ \xi_j=a_j x_j\}$ for almost all $a\in\mathbf{R}^N$.*

Proof. Choose a_1 so that $(1,0,\dots,0,a_1,0,\dots,0)\notin\lambda$. Let V be the line generated by this vector and V^σ the σ orthogonal space defined by $\xi_1=a_1x_1$. Then V^σ/V is isomorphic to $T^*\mathbf{R}^{N-1}$ and $\lambda'=(\lambda\cap V^\sigma)$ is Lagrangian there by Proposition 21.2.13. If the lemma is proved for lower dimensions we can choose $a_2,\dots,a_N$ so that λ' is transversal to the plane defined by $\xi_j=a_jx_j$, $j>1$. Since

$$\lambda\cap\{(x,\xi);\ \xi_j=a_jx_j,\ j=1,\dots,N\}\subset\lambda'$$

we obtain $x_2=\dots=\xi_n=0$ in the intersection, hence $x_1=\xi_1=0$ too.

Proof of Theorem 25.2.4. As in the proof of Theorem 25.2.3 we can argue locally. By Theorem 21.2.16 and Lemma 25.2.5 we can choose local coordinates in X, Y so that

$$A(x,y)=(2\pi)^{-3(n_X+n_Y)/4}\iint e^{i(\langle x,\xi\rangle+\langle y,\eta\rangle-H(\xi,\eta))}a(\xi,\eta)\,d\xi\,d\eta$$

where $a\in S^{m'-(n_X+n_Y)/4}(\mathbf{R}^{n_X+n_Y})$ has support in a conic neighborhood of a point (ξ_0,η_0) where H is in C^∞. If $v\in C_0^\infty$ it follows that the Fourier transform of Av is

$$(2\pi)^{n_X}(2\pi)^{-3(n_X+n_Y)/4}\int e^{-iH(\xi,\eta)}a(\xi,\eta)\hat{v}(-\eta)\,d\eta,$$

so

$$PAv(x)=(2\pi)^{-3(n_X+n_Y)/4}\iint e^{i(\langle x,\xi\rangle-H(\xi,\eta))}P(x,\xi)a(\xi,\eta)\hat{v}(-\eta)\,d\xi\,d\eta$$

where $P(x,\xi)$ is the full symbol of P. Hence the kernel of PA is

$$(2\pi)^{-3(n_X+n_Y)/4}\iint e^{i(\langle x,\xi\rangle+\langle y,\eta\rangle-H(\xi,\eta))}P(x,\xi)a(\xi,\eta)\,d\xi\,d\eta.$$

Since $p(x,\xi)=0$ on C we can write

$$p(x,\xi)=\sum p_j(x,\xi,\eta)(x_j-\partial H/\partial\xi_j)$$

where p_j is homogeneous of degree m with respect to (ξ, η). We may assume that a vanishes in a neighborhood of 0 and write $P=p+r$. Then an integration by parts gives

$$(PA)(x,y)=(2\pi)^{-3(n_X+n_Y)/4}\iint e^{i(\langle x,\xi\rangle+\langle y,\eta\rangle-H(\xi,\eta))}(ra-\sum D_{\xi_j}(p_j a))\,d\xi\,d\eta.$$

The principal symbol of A is $a(\xi,\eta)|d\xi|^{\frac12}|d\eta|^{\frac12}$ and that of PA is

$$(ra-\sum D_{\xi_j}(p_j a))_{x=\partial H/\partial\xi}|d\xi|^{\frac12}|d\eta|^{\frac12}$$

when (ξ,η) are taken as parameters on C. Now

$$H_p=\sum \partial p/\partial\xi_j\,\partial/\partial x_j-\sum \partial p/\partial x_j\,\partial/\partial\xi_j$$

is tangential to C. Functions on C are restrictions of functions of the form $F(\xi,\eta)$, so in terms of the parameters (ξ,η) on C this vector field has the form

$$-\sum \partial p/\partial x_j|_{x=\partial H/\partial\xi}\,\partial/\partial\xi_j.$$

Hence the principal symbol of PA is $(i^{-1}\mathscr{L}_{H_p}+\gamma)(a|d\xi|^{\frac12}|d\eta|^{\frac12})$ where

$$\gamma=r-\sum D_{\xi_j}p_j+\tfrac12\sum D_{\xi_j}p_j(\partial H/\partial\xi,\xi)$$

evaluated for $x=\partial H/\partial\xi$. Here the last sum is caused by the divergence term in (25.2.11). Thus γ is the value for $x=\partial H/\partial\xi$ of

$$r(x,\xi)-\sum D_{\xi_j}p_j(x,\xi)+\tfrac12\sum D_{\xi_j}p_j(x,\xi)+\tfrac12\sum D_{x_k}p_j(x,\xi)\,\partial^2H/\partial\xi_k\partial\xi_j.$$

Since

$$\sum \partial^2 p/\partial x_k\partial\xi_k=\sum \partial p_k/\partial\xi_k-\sum \partial p_j/\partial x_k\,\partial^2H/\partial\xi_j\partial\xi_k$$

when $x=\partial H/\partial\xi$, this proves that (25.2.12) is the principal symbol of PA.

In the applications of the calculus developed in this section we shall often have Fourier integral operators defined only microlocally. It is obvious how these microlocal versions of Theorems 25.2.2, 25.2.3 and 25.2.4 should be stated; they are really the statements which occurred in the proofs.

The results proved in this section remain valid for I^m_ρ instead of I^m, $\rho>\frac12$, but the remainder terms in the calculus are of course only $2\rho-1$ units lower in order.

25.3. Special Cases of the Calculus, and L^2 Continuity

Next to pseudo-differential operators, which have kernels in $I^m(X\times X, C')$ where C is the diagonal Λ^* in $(T^*(X)\setminus 0)\times(T^*(X)\setminus 0)$, the simplest Fourier integral operators are those defined by

$$A\in I^m(X\times Y, C'; \Omega^{\frac12}_{X\times Y}\otimes \operatorname{Hom}(F,E))$$

where C is the graph of a canonical transformation from $T^*(Y)\setminus 0$ to $T^*(X)\setminus 0$. In that case C is a symplectic manifold with the symplectic form

$$\sigma_C=\pi_X^*\sigma_X=\pi_Y^*\sigma_Y$$

where σ_X, σ_Y are the symplectic forms in $T^*(X)$, $T^*(Y)$ and π_X, π_Y are the projections from C to $T^*(X)$, $T^*(Y)$. The manifolds X and Y must have the same dimension n then. We can factor out the natural symplectic half density $\Omega_C^{\frac{1}{2}}$ from the symbol of A. This is a half density of order $n/2$, so the order $m+(n+n)/4$ of the half density valued principal symbol is reduced to m. Thus the principal symbol a is now regarded as an element in $S^m(C, M_C\otimes \operatorname{Hom}(F,E))$. (That the order is reduced to m here is of course no coincidence; we have chosen our normalizations to make this true for pseudo-differential operators.) From the remarks following Theorem 21.6.7 it follows that if $B\in I^{m'}(Y\times Z,\ C_2';\ \Omega_{Y\times Z}^{\frac{1}{2}}\otimes \operatorname{Hom}(G,F))$ and A, B are properly supported then $AB\in I^{m+m'}(X\times Z,\ (C\circ C_2)';\ \Omega_{X\times Z}^{\frac{1}{2}}\otimes \operatorname{Hom}(G,E))$ and the principal symbol is obtained by multiplying that of B by a, with C identified with $T^*(Y)$, and then identifying C_2 with $C\circ C_2$.

From the calculus it is easy to deduce L^2 continuity. We content ourselves with stating it in the scalar case but nothing needs to be changed in the case of bundle maps.

Theorem 25.3.1. *Assume that C is locally the graph of a canonical transformation from $T^*(Y)\setminus 0$ to $T^*(X)\setminus 0$ and let $A\in I^0(X\times Y,\ C';\ \Omega_{X\times Y}^{\frac{1}{2}})$. Then A defines a continuous map from $L^2_{\mathrm{comp}}(Y,\Omega_Y^{\frac{1}{2}})$ to $L^2_{\mathrm{loc}}(X,\Omega_X^{\frac{1}{2}})$. It is compact if the principal symbol of A tends to 0 at ∞ over every compact subset of $X\times Y$.*

Proof. If the symbol of A has support in a sufficiently small cone then A has compact support in $X\times Y$ and $A^*A\in I^0(Y\times Y,\Delta^*;\ \Omega_{Y\times Y}^{\frac{1}{2}})$, by Theorems 25.2.2 and 25.2.3. Thus A^*A is a pseudo-differential operator of order 0, hence L^2 continuous by Theorem 18.1.29, so

$$(Au,Au)=(A^*Au,u)\leq C(u,u),\quad u\in L^2(Y,\Omega_Y^{\frac{1}{2}}).$$

If a is a principal symbol of A, then $|a|^2$ is a principal symbol for A^*A when considered as a function in $T^*(Y)$. If $a\to 0$ at ∞ it follows that A^*A is compact. This implies that A is compact, for if u_j is bounded in L^2 and A^*Au_j is convergent then Au_j is convergent since

$$\|A(u_j-u_k)\|^2=(A^*A(u_j-u_k),\ u_j-u_k)\to 0.$$

Corollary 25.3.2. *Assume that C is locally the graph of a canonical transformation from $T^*(Y)\setminus 0$ to $T^*(X)\setminus 0$ and let $A\in I^m(X\times Y, C';\ \Omega_{X\times Y}^{\frac{1}{2}})$. Then A defines a continuous map from $H^{\mathrm{comp}}_{(s)}(Y,\Omega_Y^{\frac{1}{2}})$ to $H^{\mathrm{loc}}_{(s-m)}(X,\Omega_X^{\frac{1}{2}})$ for every real s.*

Proof. Let B be a properly supported pseudo-differential operator of order $s-m$ in X, and let B_1, B_2 be such operators of order s and $-s$ in Y with

B_2B_1-I of order $-\infty$. Then

$$BA=(BAB_2)B_1+BA(I-B_2B_1).$$

The last term is continuous from $\mathscr{E}'$ to C^∞, B_1 is continuous from $H_{(s)}$ to L^2 and BAB_2 is continuous from L^2 to L^2 so BA is continuous from $H_{(s)}$ to L^2 as claimed.

The calculus always allows one to pass from L^2 continuity to $H_{(s)}$ continuity as in the Corollary, and we shall therefore usually only discuss L^2 continuity in what follows.

For Fourier integral operators associated with a canonical transformation it is sometimes convenient to use a representation which is close to that of pseudo-differential operators

$$Au(x)=(2\pi)^{-n}\int e^{i\phi(x,\eta)}a(x,\eta)\hat{u}(\eta)\,d\eta, \quad u\in C_0^\infty(\mathbb{R}^n). \tag{25.3.1}$$

This means that the kernel is

$$A(x,y)=(2\pi)^{-n}\int e^{i(\phi(x,\eta)-\langle y,\eta\rangle)}a(x,\eta)\,d\eta. \tag{25.3.2}$$

This formula does indeed define $A\in I^m(\mathbb{R}^{2n}, C')$ if $a\in S^m$ and

$$C=\{(x,\phi'_x,\phi'_\eta,\eta)\}. \tag{25.3.3}$$

C is (locally) the graph of a canonical transformation if and only if $\det(\partial^2\phi/\partial x\,\partial\eta)\neq 0$, for this is the condition for the maps $(x,\eta)\mapsto(x,\phi'_x)$ and $(x,\eta)\mapsto(\phi'_\eta,\eta)$ to be local diffeomorphisms. The graph of any homogeneous canonical transformation is locally of this form:

Proposition 25.3.3. *Let C be the graph of a local homogeneous canonical transformation from a neighborhood of $(y_0,\eta_0)\in T^*(Y)\setminus 0$ to a neighborhood of $(x_0,\xi_0)\in T^*(X)\setminus 0$. Then one can choose local coordinates y at y_0 such that C is of the form (25.3.3) in a neighborhood of (x_0,ξ_0,y_0,η_0) and $\det\partial^2\phi/\partial x\,\partial\eta\neq 0$ at (x_0,η_0). One calls ϕ a generating function of C.*

Proof. We can choose the y coordinates so that

$$C\ni(x,\xi,y,\eta)\mapsto(x,\eta)$$

is a local diffeomorphism at (x_0,ξ_0,y_0,η_0). In fact, C^{-1} maps the Lagrangian fiber $T^*_{x_0}$ in $T^*(X)\setminus 0$ to a conic Lagrangian through (y_0,η_0). By Theorem 21.2.16 we can choose the local coordinates so that it is transversal to the plane $\eta=\eta_0$, and then the map $C\ni(x,\xi,y,\eta)\mapsto(x,\eta)$ has injective differential. By Theorem 21.2.18 we can now find a homogeneous function $S(x,\eta)$ in a neighborhood of $(x_0,-\eta_0)$ such that

$$C'=\{(x,-\partial S(x,\eta)/\partial x,\ \partial S(x,\eta)/\partial\eta,\eta)\}.$$

With $\phi(x,\eta)=-S(x,-\eta)$ the proposition is proved.

If $A\in I^m(X\times Y, C')$ and $WF'(A)$ is sufficiently close to (x_0,ξ_0,y_0,η_0), it follows from Proposition 25.1.5 that modulo C^∞ one can write A in the form

$$A=(2\pi)^{-n}\int e^{i(\phi(x,\eta)-\langle y,\eta\rangle)}a(x,y,\eta)\,d\eta \tag{25.3.2$'$}$$

where $a\in S^m$. On $\{(x,y,\eta);\ \phi'_\eta(x,\eta)-y=0\}$ we can take (x,η) as parameters and the density $\delta(y-\phi'_\eta)$ is then $|dx|\,|d\eta|$. Hence $a(x,\phi'_\eta(x,\eta),\eta)\,|dx|^{\frac12}|d\eta|^{\frac12}$ is a principal symbol of A with the parameters (x,η) on C and the Maslov bundle trivialized by the phase function $\phi(x,\eta)-\langle y,\eta\rangle$. We have the same principal symbol if $a(x,y,\eta)$ is replaced by $a_0(x,\eta)=a(x,\partial\phi/\partial\eta,\eta)$. Thus

$$A_1=(2\pi)^{-n}\int e^{i(\phi(x,\eta)-\langle y,\eta\rangle)}(a(x,y,\eta)-a_0(x,\eta))\,d\eta\in I^{m-1}$$

and we can apply the same argument to A_1. Iterating the argument we obtain $a_j\in S^{m-j}$ such that

$$(2\pi)^{-n}\int e^{i(\phi(x,\eta)-\langle y,\eta\rangle)}(a(x,y,\eta)-\sum_{j<N}a_j(x,\eta))\,d\eta\in I^{m-N}.$$

If $b\sim\sum_0^\infty a_j\in S^m$ then

$$A-(2\pi)^{-n}\int e^{i(\phi(x,\eta)-\langle y,\eta\rangle)}b(y,\eta)\,d\eta\in C^\infty$$

which proves that modulo C^∞ every $A\in I^m$ with $WF'(A)$ close to (x_0,ξ_0,y_0,η_0) is of the form (25.3.2), (25.3.1).

To compute the principal symbol of A with the half density removed we must use the symplectic coordinates $(x,\xi)=(x,\phi'_x(x,\eta))$. Since

$$|dx|\,|d\xi|=|\det\phi''_{x\eta}(x,\eta)|\,|dx|\,|d\eta|$$

division by $|dx|^{\frac12}|d\xi|^{\frac12}$ leaves as principal symbol of (25.3.2)$'$

$$a(x,\phi'_\eta(x,\eta),\eta)\,|\det\phi''_{x\eta}(x,\eta)|^{-\frac12},$$

with the Maslov bundle trivialized by the phase function in (25.3.2)$'$.

One can work almost as easily with Fourier integral operators belonging to a canonical transformation as with pseudo-differential operators. The notion of elliptic operator is well defined:

Definition 25.3.4. If C is (the graph of) a homogeneous canonical transformation from $T^*(Y)\setminus 0$ to $T^*(X)\setminus 0$ and $A\in I^m(X\times Y, C';\ \Omega^{\frac12}_{X\times Y}\otimes\mathrm{Hom}(F,E))$, then A is called non-characteristic at $(x_0,\xi_0,y_0,\eta_0)\in C$ if the principal symbol has an inverse $\in S^{-m}(C, M_C^{-1}\otimes\mathrm{Hom}(E,F))$ in a conic neighborhood. A is called elliptic if it is non-characteristic at every point in C.

The proof of Theorem 18.1.24 can be used again to show that if C^{-1} is also a graph and A is everywhere elliptic and properly supported then A has a parametrix $B\in I^{-m}(Y\times X, (C^{-1})';\ \Omega^{\frac12}_{Y\times X}\otimes\mathrm{Hom}(E,F))$, that is, $BA-I$ and $AB-I$ have C^∞ kernels. We have such operators in mind in the following theorem of Egorov:

Theorem 25.3.5. *Let C be the graph of a homogeneous canonical bijection from $T^*(Y)\setminus 0$ to $T^*(X)\setminus 0$ and let $A\in I^m(X\times Y, C'; \Omega^{\frac{1}{2}}_{X\times Y})$, and $B\in I^{-m}(Y\times X, (C^{-1})'; \Omega^{\frac{1}{2}}_{Y\times X})$ be properly supported. If P is a properly supported pseudo-differential operator in X it follows that $Q=BPA$ is a properly supported pseudo-differential operator in Y. If c is a principal symbol of BA and p is a principal symbol of P, then $c\,(p\circ\chi)$ is a principal symbol of Q if C is the graph of the canonical transformation χ.*

Proof. If $P\in\Psi^\mu(X)$ then $PA\in I^{m+\mu}(X\times Y, C')$ hence $BPA\in\Psi^\mu(Y)$. A principal symbol of PA is obtained when that of A is multiplied by p lifted to C by the projection to $T^*(X)\setminus 0$. This is equivalent to multiplication by the function $p\circ\chi$ on $T^*(Y)\setminus 0$ lifted by the projection to $T^*(Y)\setminus 0$, thus $PA-AR\in I^{m+\mu-1}$ if $R\in\Psi^\mu(Y)$ has principal symbol $p\circ\chi$. Hence $BPA-BAR\in\Psi^{\mu-1}$ which proves the statement.

There are of course also local forms of the preceding results. When using Theorem 25.3.5 for example we choose B and A with $(y_0,\eta_0)\notin WF(BA-I)$ which only requires χ to be defined near (y_0,η_0). The result then says that a pseudo-differential operator with principal symbol $p\circ\chi$ is microlocally conjugate to one with principal symbol p. Together with the results of Section 21.3 and arguments to remove the terms of lower order, this will be important in the study of pseudo-differential operators in Chapter XXVI. In what follows we shall also use the related observation that the microlocal study of operators $A\in I^m(X\times Y, C')$ with a general canonical relation C can be simplified by composition right and left with elliptic Fourier integral operators belonging to canonical transformations. Thus we may replace C by $C_1\circ C\circ C_2$ where C_1 and C_2 are graphs of canonical transformations. This means that we can simplify C by making arbitrary homogeneous canonical changes of coordinates in $T^*(X)\setminus 0$ and $T^*(Y)\setminus 0$. To discuss the simplifications which this can yield we start as usual with the linear case.

Lemma 25.3.6. *Let S_1, S_2 be symplectic vector spaces with symplectic forms σ_1, σ_2, and let $G\subset S_1\oplus S_2$ be Lagrangian for $\sigma_1-\sigma_2$. Then there are symplectically orthogonal decompositions of S_1 and S_2,*

$$S_1=S_{11}\oplus S_{12},\quad S_2=S_{21}\oplus S_{22}$$

such that $G=\lambda_1\oplus\tilde{G}\oplus\lambda_2$ where λ_j is Lagrangian in S_{jj} and $\tilde{G}$ is the graph of a linear symplectic transformation $S_{21}\to S_{12}$.

Proof. Let $\lambda_1=\{\gamma\in S_1;\ (\gamma,0)\in G\}$, $\lambda_2=\{\gamma\in S_2;\ (0,\gamma)\in G\}$. These are isotropic planes and $G\subset\lambda_1^\sigma\oplus\lambda_2^\sigma$ since G is Lagrangian. If we write

$$\lambda_1^\sigma=\lambda_1\oplus S_{12}$$

then $S_{12}\cong\lambda_1^\sigma/\lambda_1$ is symplectic and $S_1=S_{11}\oplus S_{12}$ where the symplectically orthogonal complement S_{11} of S_{12} contains λ_1. We define S_{21} and S_{22} in the same way and obtain $G = \lambda_1 \oplus \tilde{G} \oplus \lambda_2$ where $\tilde{G} \subset S_{12} \oplus S_{21}$ has bijective projection on S_{12} and S_{21}. Since $\dim S_{jj}=2\dim\lambda_j$ it follows that λ_j is Lagrangian in S_{jj}.

Let σ_G be the form σ_1 lifted to G or equivalently the form σ_2 lifted to G. It is equal to $\sigma_{\tilde{G}}$ in $\tilde{G}$ and vanishes in λ_1 and λ_2 which are also σ_G orthogonal to G. Hence the rank of σ_G is $\dim S_{12}$,

$$\operatorname{corank}\sigma_G=\dim\lambda_1+\dim\lambda_2. \tag{25.3.4}$$

This number will play a role in the L^2 estimates below. However, we shall first use Lemma 25.3.6 to put quite general canonical relations in a convenient form:

Proposition 25.3.7. *Let S_1 and S_2 be conic symplectic manifolds and C a homogeneous canonical relation in a neighborhood of $(\gamma_1,\gamma_2)\in S_1\times S_2$. Assume that the radial direction of S_j at γ_j is not a tangent of C, $j=1,2$. Then one can choose homogeneous symplectic coordinates (x,ξ) in S_1 at γ_1 and (y,η) in S_2 at γ_2, such that with a splitting $x=(x',x''),\dots,(\eta',\eta'')$ of the x,ξ,y,η variables with n variables in the first group we have $x=y=0$, $\xi=(1,\dots,0)$, $\eta=(1,0,\dots,0)$ at (γ_1,γ_2), and the tangent plane of C there is defined by*

$$dx'=dy',\quad d\xi'=d\eta',\quad d\xi''=0,\quad d\eta''=0.$$

Here $2n$ is the rank of σ_C at (γ_1,γ_2).

Proof. We apply Lemma 25.3.6 to the tangent space $G=T_{\gamma_1,\gamma_2}(C)$ contained in $T_{\gamma_1}(S_1)\oplus T_{\gamma_2}(S_2)$. By hypothesis the radial vector $\rho_j\in T_{\gamma_j}(S_j)$ is not in λ_j. However, $(\rho_1,\rho_2)\in G$ since C is conic, so

$$\sigma_j(\rho_j,t_j)=0 \quad \text{if } t_j\in\lambda_j,$$

and in general

$$\sigma_1(\rho_1,t_1)=\sigma_2(\rho_2,t_2) \quad \text{if } (t_1,t_2)\in G.$$

Set $\dim S_j=2n_j$ and $\dim S_{12}=\dim S_{21}=2n$ with the notation in Lemma 25.3.6. Then λ_j is of dimension n_j-n so we can choose a basis $e_{n+1},\dots,e_{n_1}$ for λ_1 and a basis $\tilde{e}_{n+1},\dots,\tilde{e}_{n_2}$ for λ_2. Since $\rho_1,e_{n+1},\dots,e_{n_1}$ are linearly independent we can use the beginning of the proof of Theorem 21.1.9 to choose $e_1,\dots,e_n$ and $\varepsilon_2,\dots,\varepsilon_{n_1}$ satisfying (i)′, (iv), (ii), (iii) there with $b_k=\delta_{1k}$. With $\varepsilon_1=\rho_1$ we get a symplectic basis. Since $e_1,\dots,e_n$, $\varepsilon_1,\dots,\varepsilon_n$ are σ orthogonal to λ_1 we can find $\tilde{e}_1,\dots,\tilde{e}_n$, $\tilde{\varepsilon}_1,\dots,\tilde{\varepsilon}_n$ such that $(e_1,\tilde{e}_1),\dots,(\varepsilon_n,\tilde{\varepsilon}_n)$ are in G. These new vectors then satisfy the same commutation relations as $e_1,\dots,e_n$, $\varepsilon_1,\dots,\varepsilon_n$, including the symplectic product with ρ_2 resp. ρ_1. All of them are orthogonal to $e_{n+1},\dots,e_{n_2}$. We can take $\tilde{\varepsilon}_1=\rho_2$. Omitting $\tilde{\varepsilon}_1$ at first we can use the extension argument in the beginning of the proof of Theorem 21.1.9 again to complete the symplectic basis $\tilde{e}_1,\dots,\tilde{\varepsilon}_{n_2}$. The last vector $\tilde{\varepsilon}_1$ chosen is necessarily ρ_2 which restores the vectors chosen originally. Thus G is spanned by $(e_j,\tilde{e}_j)$, $(\varepsilon_j,\tilde{\varepsilon}_j)$, $j=1,\dots,n$, and by $(e_j,\tilde{e}_j)$ for $j>n$. Now the end of the proof of Theorem 21.1.9 shows that we can find homogeneous symplectic coordinates (x,ξ) in $T^*(X)\setminus 0$, equal to $(0,(1,0))$ at γ_1, such that $H_{x_j}=-\varepsilon_j$ and $H_{\xi_j}=e_j$ there. Then C has the stated form if we choose (y,η) in a similar way.

The location of the tangent plane of C in Proposition 25.3.7 shows in particular that x', x'', η', y'' can be used as local coordinates on C. As in the proof of Proposition 25.3.3 an application of Theorem 21.2.18 shows that we can find a homogeneous function $S(x',x'',y'',\eta')$ in a conic neighborhood of $\eta'=(-1,0,\dots,0)$, $x'=0$, $x''=0$, $y''=0$, such that

$$C'=\{(x',x'',-\partial S/\partial x', -\partial S/\partial x''; \partial S/\partial\eta', y'', \eta', -\partial S/\partial y'')\}.$$

If we put $\phi(x',x'',y'',\eta')=-S(x',x'',y'',-\eta')$ then

$$C=\{(x',x'',\partial\phi/\partial x', \partial\phi/\partial x''; \partial\phi/\partial\eta', y'', \eta', -\partial\phi/\partial y'')\}$$

which shows that C is locally parametrized by the phase function

$$\phi(x',x'',y'',\eta')-\langle y',\eta'\rangle.$$

If $A\in I^m(\mathbb{R}^{n_1}\times\mathbb{R}^{n_2}, C')$ and $WF'(A)$ belongs to a small conic neighborhood of $(0,\varepsilon_1,0,\tilde\varepsilon_1)$ where $\varepsilon_1=(1,0,\dots,0)\in\mathbb{R}^{n_1}$ and $\tilde\varepsilon_1=(1,0,\dots,0)\in\mathbb{R}^{n_2}$ it follows that we can write A modulo C^∞ in the form

$$A=(2\pi)^{-(n_1+n_2+2n)/4}\int e^{i(\phi(x',x'',y'',\eta')-\langle y',\eta'\rangle)}a(x',x'',y',y'',\eta')\,d\eta'.$$

Here the dependence of a on y' can be eliminated as in the discussion of (25.3.2)′ above, and then we have

(25.3.5) $$Au=(2\pi)^{-(n_1+n_2+2n)/4}\iint e^{i\phi(x',x'',y'',\eta')}a(x',x'',y'',\eta')\hat u(\eta',y'')\,d\eta'\,dy''$$

where $\hat u(\eta',y'')$ is the Fourier transform of $u\in C_0^\infty(\mathbb{R}^{n_2})$ with respect to the n variables y', and $a\in S^{m+(n_1+n_2-2n)/4}$ has support in a small conic neighborhood of $(0,0,\tilde\varepsilon_1)$. At that point itself the difference $\phi(x',x'',y'',\eta')-\langle x',\eta'\rangle$ vanishes of third order so A is fairly close to a pseudo-differential operator in x' depending on the parameters x'', y''. This will be used later on. At the moment we just observe that for small x'', y'' we do have a Fourier integral operator associated with the canonical transformation with generating function ϕ, where x'' and y'' have been fixed. The order is $m+(n_1+n_2-2n)/4$ $=m+(\operatorname{corank}\sigma_C)/4$. By Theorem 25.3.1 we therefore obtain L^2 continuity when $m\leqq-(\operatorname{corank}\sigma_C)/4$.

Theorem 25.3.8. *Let $A\in I^m(X\times Y, C'; \Omega^{\frac12}_{X\times Y})$ where C is a homogeneous canonical relation to which the radial vectors of $T^*(X)\setminus 0$ and $T^*(Y)\setminus 0$ are never tangential. Then A defines a continuous operator from $L^2_{\mathrm{comp}}(Y,\Omega^{\frac12}_Y)$ to $L^2_{\mathrm{loc}}(X,\Omega^{\frac12}_X)$ if $m\leqq-(\operatorname{corank}\sigma_C)/4$. Here σ_C is the two form on C obtained by lifting the symplectic form in $T^*(X)$ or $T^*(Y)$ by the projection from C.*

Proof. The distribution A can be microlocalized so we may assume that $WF'(A)$ is in a small conic neighborhood of $(\gamma_1,\gamma_2)\in C$. If $n_1=\dim X$, and

$n_2=\dim Y$, we can choose a homogeneous canonical transformation from a conic neighborhood of $(0,\tilde{\varepsilon}_1)\in T^*(\mathbb{R}^{n_2})$ to a conic neighborhood of γ_2, with graph C_2, and another homogeneous canonical transformation, with graph C_1, from a conic neighborhood of γ_1 to a conic neighborhood of $(0,\varepsilon_1)\in T^*(\mathbb{R}^{n_1})$ such that $C_1\circ C\circ C_2$ has the form in Proposition 25.3.7. Let F_1 and F_2 be Fourier integral operators of order 0 associated with C_1 and C_2 which are elliptic at $(0,\varepsilon_1;\gamma_1)$ and $(\gamma_2;0,\varepsilon_1)$ respectively. Then $F_1AF_2\in I^m(\mathbb{R}^{n_1}\times\mathbb{R}^{n_2};(C_1\circ C\circ C_2)')$ is of the form discussed above, hence L^2 continuous. We can choose G_1 and G_2 of order 0 associated with C_1^{-1} and C_2^{-1} so that

$$(\gamma_1,\gamma_1)\notin WF'(G_1F_1-I),\quad (\gamma_2,\gamma_2)\notin WF'(F_2G_2-I).$$

Then the operator $G_1F_1AF_2G_2$ is L^2 continuous and $G_1F_1AF_2G_2-A\in C^\infty$ if $WF'(A)$ is sufficiently close to (γ_1,γ_2). The proof is complete.

The conclusion in Theorem 25.3.8 is optimal if σ_C has a constant rank $2n$. In fact, in that case the projection $C\to T^*(X)\setminus 0$ has constant rank equal to $n+\dim X$, by Lemma 25.3.6. The range is therefore locally a submanifold Σ_1 of that dimension (Proposition C.3.3), and Lemma 25.3.6 also shows that Σ_1 is involutive and that the tangent plane is not orthogonal to the radial vector. By Theorem 21.2.4 it follows that we can choose homogeneous symplectic coordinates such that Σ_1 is defined by $\xi''=0$ and the projection of C in $T^*(Y)\setminus 0$ is defined by $\eta''=0$. The generating function ϕ constructed above is then independent of x'' and y'' so we have operators A which are independent of x'' and y'' if the amplitude a is so chosen. For them it is obvious that Theorem 25.3.8 cannot be improved. However, in general the rank of σ_C may only drop from its maximum value on a small set, so the following converse of Theorem 25.3.8 leaves a gap in general.

Theorem 25.3.9. *Assume that every $A\in I^m(X\times Y, C'; \Omega^{\frac12}_{X\times Y})$ defines a continuous operator from $L^2_{\text{comp}}(Y,\Omega^{\frac12}_Y)$ to $L^2_{\text{loc}}(X,\Omega^{\frac12}_X)$ and that the radial vectors of $T^*(X)\setminus 0$ and $T^*(Y)\setminus 0$ are never tangential to the homogeneous canonical relation C. Then* corank $\sigma_C\leqq -12m$.

Proof. In view of Proposition 25.3.7 and the argument in the proof of Theorem 25.3.8 it is sufficient to prove that if n''_X and n''_Y denote the number of x'' and y'' variables then $n''_X+n''_Y\leqq -12m$ if (25.3.5) is L^2 continuous when

$$a(x',x'',y'',\eta')=(1+|\eta'|^2)^{\mu/2},\quad \mu=m+(n''_X+n''_Y)/4,$$

in a conic neighborhood of $x'=x''=y''=0$, $\eta'=\theta=(1,0,\dots,0)$ where

$$(25.3.6)\quad \begin{aligned}&|\phi(x',x'',y'',\eta')-\langle x',\eta'\rangle|\\ &\leqq C((|x'|^3+|x''|^3+|y''|^3)|\eta'|+(|\eta_2|^3+\dots+|\eta_n|^3)/|\eta'|^2).\end{aligned}$$

We apply A to

$$u_t(y)=u(ty',t^\kappa y'')e^{it^2\langle y',\theta\rangle},\quad \kappa=\tfrac23,$$

where $u \in C_0^\infty(\mathbb{R}^{n_2})$. For large t the support of u_t is close to 0 and the spectrum is concentrated at $t^2\theta$. The L^2 norm of u_t is $t^{-(n+\kappa n_Y'')/2}\|u\|_{L^2}$, and the Fourier transform with respect to y' is

$$\eta' \to t^{-n}\hat{u}(\eta'/t - t\theta, t^\kappa y'').$$

Hence $Au_t(x) = f_t(tx', t^\kappa x'')$ where

$$f_t(x) = t^{-\kappa n_Y''}(2\pi)^{-(n_1+n_2+2n)/4} \iint e^{i\phi(x'/t, x''/t^\kappa, y''/t^\kappa, t^2\theta + t\eta')}$$
$$\cdot a(x'/t, x''/t^\kappa, y''/t^\kappa, t^2\theta + t\eta')\hat{u}(\eta', y'')\,d\eta'\,dy''.$$

Since $x = \frac{2}{3}$ we obtain using (25.3.6)

$$|t^2\phi(x'/t, x''/t^\kappa, y''/t^\kappa, \theta + \eta'/t) - t^2\langle x'/t, \theta + \eta'/t^2\rangle|$$
$$\leqq C(|x'|^3/t + |x''|^3 + |y''|^3 + |\eta'|^3/t) \quad \text{if } |\eta'| < t/2.$$

The integrand can be estimated by

$$C|\hat{u}(\eta', y'')|(1 + |t^2\theta + t\eta'|)^\mu \leqq C' t^{2\mu}|\hat{u}(\eta', y'')|(1 + |\eta'|)^{2|\mu|}$$

if we consider separately the cases where $|\eta'| < t/2$ and $|\eta'| > t/2$. Hence

$$f_t(x)t^{-2\mu+\kappa n_Y''}e^{-it\langle x',\theta\rangle} \to (2\pi)^{-(n_1+n_2+2n)/4}\iint e^{i\langle x',\eta'\rangle + iQ(x'',y'')}\hat{u}(\eta', y'')\,d\eta'\,dy''$$
$$= (2\pi)^{-(n_X''+n_Y'')/4}\int e^{iQ(x'',y'')}u(x', y'')\,dy''$$

where Q is the third order polynomial in (x'', y'') in the Taylor expansion of ϕ at $(0,0,0,\theta)$. For suitable u this is not 0 so

$$t^{2\mu - \kappa n_Y'' - (n+\kappa n_X'')/2} \leqq C\|Au_t\|, \qquad \|u_t\| = \|u\|t^{-(n+\kappa n_Y'')/2}.$$

Hence the continuity of A implies that

$$2\mu \leqq \kappa(n_X'' + n_Y'')/2 = (n_X'' + n_Y'')/3,$$

so $m \leqq (n_X'' + n_Y'')(\frac{1}{6} - \frac{1}{4}) = -(n_X'' + n_Y'')/12$, which proves the theorem.

Remark. Using non-homogeneous canonical transformations one can eliminate the condition on the radial vectors in Theorems 25.3.8 and 25.3.9.

We shall now discuss the canonical relations with folds in Theorem 21.4.11 where equality will actually be attained in the inequality proved in Theorem 25.3.9. Substitution of Theorem 21.4.11 for Proposition 25.3.7 in the proof of Theorem 25.3.8 shows that it is sufficient to study the canonical relation $C \subset T^*(\mathbb{R}^n \setminus 0) \times T^*(\mathbb{R}^n \setminus 0)$ defined by the phase function

$$\phi(x, y, s, \xi) = \langle x - y, \xi\rangle + s\xi_1 - s^3\xi_n/3.$$

We recall that $\phi_\xi' = \phi_s' = 0$ means that

$$x_1 - y_1 + s = 0; \quad x_n - y_n - s^3/3 = 0; \quad x_j = y_j, \quad 1 < j < n; \quad \xi_1 = s^2\xi_n,$$

so ϕ parametrizes the canonical relation

$$C = \{(x, s^2\xi_n, \xi', x_1 + s, x_2, \ldots, x_{n-1}, x_n - s^3/3, s^2\xi_n, \xi'), \xi_n > 0\}.$$

Here x, $\xi'=(\xi_2,\ldots,\xi_n)$ and $s\in\mathbb{R}$ are parameters on C, $\xi_n>0$. We have $\delta(\phi'_\xi,\phi'_s)=|dx|\,|d\xi'|\,|ds|$ with these parameters since

$$D(x,\xi',s,\phi'_\xi,\phi'_s)/D(x,y,\xi,s)=\pm 1.$$

To obtain a homogeneous system of parameters we must replace s by $\xi_n s$ say, and this will be important to keep in mind since it influences the order of the corresponding Fourier integral operators.

If $A\in I^m(\mathbb{R}^n\times\mathbb{R}^n,C')$ and $WF'(A)$ is in a compactly generated cone $\subset C$, it follows from Proposition 25.1.5 that modulo C^∞

$$A(x,y)=(2\pi)^{-(2n+2(n+1))/4}\iint e^{i\phi(x,y,s,\xi)}a(x,y,s,\xi)\,ds\,d\xi \tag{25.3.7}$$

where $a\in S^{m+1+(2n-2(n+1))/4}=S^{m+\frac12}$. Here we have taken into account that replacing s by a homogeneous parameter reduces the degree by one unit as in the proof of Theorem 25.2.3. The principal symbol of A is

$$a_0(x,s,\xi')\,|dx|^{\frac12}\,|d\xi'|^{\frac12}\,|ds|^{\frac12}, \tag{25.3.8}$$
$$a_0(x,s,\xi')=a(x,x_1+s,x_2,\ldots,x_{n-1},x_n-s^3/3,s,s^2\xi_n,\xi')\in S^{m+\frac12}$$

when x, s, ξ' are parameters on C and the Maslov bundle is trivialized by ϕ. We obtain the same principal symbol when a is replaced by a_0. As in the discussion of (25.3.2)' above we can therefore conclude that modulo C^∞ it is possible to represent A with an amplitude $a(x,s,\xi')\in S^{m+\frac12}$ independent of y and ξ_1 in (25.3.7), thus

$$Au(x)=(2\pi)^{-n-\frac12}\iint e^{i(\langle x,\xi\rangle+s\xi_1-s^3\xi_n/3)}a(x,s,\xi')\,\hat u(\xi)\,d\xi\,ds. \tag{25.3.7$'$}$$

Using Theorem 7.7.18 we could show that integration with respect to s gives modulo C^∞ kernels

$$Au(x)=(2\pi)^{-n+\frac12}\int e^{i\langle x,\xi\rangle}\tilde a(x,\xi)\,\hat u(\xi)\,d\xi, \tag{25.3.9}$$

$$\tilde a(x,\xi)=Ai(-\xi_1\xi_n^{-\frac13})\,b_0(x,\xi)+Ai'(-\xi_1\xi_n^{-\frac13})\,b_1(x,\xi). \tag{25.3.10}$$

Here Ai is the Airy function (7.6.16) and $b_0\in S^{m+\frac16}(\mathbb{R}^n\times\mathbb{R}^n)$, $b_1\in S^{m-\frac16}(\mathbb{R}^n\times\mathbb{R}^n)$ and b_j have support where $|\xi|<C_1\xi_n$ for some C_1. However, we shall not rely on this method of proof, which would actually require a slight extension of Theorem 7.7.18. Instead we shall prove that (25.3.9), (25.3.10) with such restrictions on the supports implies that $A\in I^m(\mathbb{R}^n\times\mathbb{R}^n,C')$ and show that elements of this form have principal symbols restricted only by the support conditions. As in the discussion of (25.3.2)' above this allows us to show by successive reduction of the degree that any element in I^m with sufficiently small wave front set has the representation (25.3.9), (25.3.10).

Inserting the definition of the Airy function in (25.3.10) we obtain the oscillatory integral

$$\tilde a(x,\xi)=(2\pi)^{-1}\int_{-\infty}^{\infty} e^{i(s^3/3-s\xi_1\xi_n^{-\frac13})}(b_0(x,\xi)+isb_1(x,\xi))\,ds \tag{25.3.11}$$
$$=(2\pi)^{-1}\int_{-\infty}^{\infty} e^{i(s\xi_1-s^3\xi_n/3)}(b_0(x,\xi)\xi_n^{\frac13}-is\xi_n^{\frac23}b_1(x,\xi))\,ds.$$

By hypothesis we have $|\xi_1|<C_1\xi_n$ in $\operatorname{supp} b_j$. Hence

$$\left|\frac{d}{ds}(s\xi_1-s^3\xi_n/3)\right|=|\xi_1-s^2\xi_n|>s^2\xi_n/2 \quad \text{if } s^2>2C_1.$$

Choose $\chi\in C_0^\infty(\mathbb{R})$ equal to 1 in $(-\sqrt{2C_1}, \sqrt{2C_1})$. Then it follows from Theorem 7.7.1 that the integral obtained by inserting a factor $1-\chi(s)$ in the last member of (25.3.11) defines a function in $S^{-\infty}$. (It does not really matter that the integral is taken over a non-compact set for already a first integration by parts with respect to s will give an integrable factor $<C/(s^2\xi_n)$.) Hence, modulo C^∞, the kernel of (25.3.9) is

$$A(x,y)=(2\pi)^{-n-\frac{1}{2}}\iint e^{i(\langle x-y,\xi\rangle+s\xi_1-s^3\xi_n/3)}\chi(s)(b_0(x,\xi)\xi_n^{\frac{1}{4}}-isb_1(x,\xi)\xi_n^{\frac{3}{4}})\,ds\,d\xi,$$

and the principal symbol is by (25.3.8)

$$(25.3.12)\qquad \begin{gathered} a(x,s,\xi')|dx|^{\frac{1}{2}}|d\xi'|^{\frac{1}{2}}|ds|^{\frac{1}{2}},\\ a(x,s,\xi')=b_0(x,s^2\xi_n,\xi')\xi_n^{\frac{1}{4}}-isb_1(x,s^2\xi_n,\xi')\xi_n^{\frac{3}{4}}. \end{gathered}$$

Here $\chi(s)$ has been omitted since $\chi(s)=1$ in the support of the other factor. The right-hand side is in $S^{m+\frac{1}{4}}$ and every function $a\in S^{m+\frac{1}{4}}$ can be written in this way when the support is appropriate. In fact, splitting a in a sum of an even and an odd function with respect to s we can write

$$a(x,s,\xi')=a_0(x,s^2,\xi')+sa_1(x,s^2,\xi')$$

where a_0 and a_1 are symbols of the same order. (See Theorem C.4.4 in the appendix.) We can then take

$$b_0(x,\xi)=\xi_n^{-\frac{1}{4}}a_0(x,\xi_1/\xi_n,\xi'),\qquad b_1(x,\xi)=i\xi_n^{-\frac{3}{4}}a_1(x,\xi_1/\xi_n,\xi')$$

which have the required properties than.

$\tilde{a}$ is not very well behaved as a symbol. In fact, from (7.6.20) it follows that $|Ai^{(k)}(t)|\leq C_k(1+|t|)^{k/2-\frac{1}{4}}$ and this is the best possible estimate of its kind. Differentiation with respect to ξ will therefore not improve the behavior at infinity. However, if $m+\frac{1}{6}\leq 0$ we have for every α

$$|D_x^\alpha\tilde{a}(x,\xi)|\leq C_\alpha.$$

By Theorem 18.1.11′ this implies L^2 continuity if a vanishes for x outside some compact set. Summing up, we have now proved:

Theorem 25.3.10. *If $b_0\in S^{m+\frac{1}{4}}(\mathbb{R}^n\times\mathbb{R}^n)$ and $b_1\in S^{m-\frac{1}{4}}(\mathbb{R}^n\times\mathbb{R}^n)$ have supports where $|\xi|<C_1\xi_n$, then* (25.3.9), (25.3.10) *define an element $A\in I^m(\mathbb{R}^n\times\mathbb{R}^n, C')$ where C is the canonical relation* (21.4.20), (21.4.20)′. *With the Maslov bundle trivialized by the phase function* (21.4.19) *the principal symbol is given by* (25.3.12). *Every $A\in I^m(\mathbb{R}^n\times\mathbb{R}^n, C')$ with $|\xi|<C_2\xi_n$ in the wave front set has such a representation, and A is L^2 continuous if $m\leq-\frac{1}{6}$.*

From Theorem 25.3.10 and Theorem 21.4.11 we obtain using also the arguments in the proof of Corollary 25.3.2:

Theorem 25.3.11. *Let $C\subset(T^*(X)\setminus 0)\times(T^*(Y)\setminus 0)$ be a homogeneous canonical relation which is closed in $T^*(X\times Y)\setminus 0$, such that the projections $C\to T^*(X)\setminus 0$ and $C\to T^*(Y)\setminus 0$ have at most fold singularities. Assume that there is no point in C where the canonical one forms in $T^*(X)\setminus 0$ and $T^*(Y)\setminus 0$ both vanish in the tangent space of C. Then it follows that every $A\in I^m(X\times Y, C'; \Omega^{\frac12}_{X\times Y})$ defines a continuous map*

$$H^{\text{comp}}_{(s)}(Y,\Omega^{\frac12}_Y)\to H^{\text{loc}}_{(s-m-\frac16)}(X,\Omega^{\frac12}_X).$$

Note in particular that Theorem 25.3.11 shows that Theorem 25.3.9 cannot be improved.

25.4. Distributions Associated with Positive Lagrangian Ideals

Let S be a C^∞ symplectic manifold with $\dim S=2n$, and let $\Lambda\subset S$ be a C^∞ Lagrangian submanifold of S. Then the set

$$J=\{u\in C^\infty(S,\mathbb{R}),\ u=0 \text{ on } \Lambda\}$$

is an ideal in $C^\infty(S,\mathbb{R})$ with the following three properties:

(i) J is closed under Poisson brackets, that is,

$$\{u,v\}\in J \quad \text{if } u,v\in J. \tag{25.4.1}$$

(ii) For every point in S there is a neighborhood V and functions $u_1,\dots,u_n\in J$ such that $du_1,\dots,du_n$ are linearly independent at every point in V and $u_1,\dots,u_n$ generate J in V, that is, every $u\in J$ with $\operatorname{supp} u\subset V$ can be written

$$u=\sum_1^n a_j u_j, \quad a_j\in C_0^\infty(V).$$

(iii) If $fu\in J$ for every $f\in C_0^\infty$ then $u\in J$.

Conversely, assume that J is an ideal in $C^\infty(S,\mathbb{R})$ with the properties (i), (ii) and (iii), and let Λ be the set of common zeros of the functions in J. From (ii) it follows that Λ is a manifold of dimension n, and (25.4.1) shows that Λ is involutive, hence Lagrangian. By a partition of unity we find using (iii) that J consists of all functions in $C^\infty(S,\mathbb{R})$ vanishing on Λ. Identifying Lagrangian manifolds with the corresponding ideals we can now define complex Lagrangians by simply removing the restriction to real valued functions.

Definition 25.4.1. Let S be a C^∞ symplectic manifold. An ideal $J\subset C^\infty(S,\mathbb{C})$ satisfying the conditions (i), (ii), (iii) above, with $du_1,\dots,du_n$ now linearly independent over $\mathbb{C}$, is called a complex Lagrangian ideal. If S is conic we say that J is conic provided that in condition (ii) one can choose V conic and the generators u_j homogeneous.

The set of zeros of J,

$$J_{\mathbb{R}}=\{\gamma\in S;\ u(\gamma)=0 \text{ for all } u\in J\}$$

is not necessarily a manifold but we can define a tangent plane $T_\gamma(J)$ in the complexification of $T_\gamma(S)$ for every $\gamma\in J_{\mathbb{R}}$ as the intersection of the planes $du_j=0$ where $u_1,\dots,u_n$ are local generators of J. It is clear that this definition does not depend on the choice of generators. $T_\gamma(J)$ is Lagrangian since $\{u_j,u_k\}=0$ at γ, $j,k=1,\dots,n$, so the real points in $T_\gamma(J)$ form an isotropic plane. The following is an analogue of Theorem 21.2.16.

Proposition 25.4.2. *Let X be a C^∞ manifold and let J be a complex conic Lagrangian ideal in $C^\infty(T^*(X)\setminus 0)$. For every $\gamma_0\in J_{\mathbb{R}}$ the local coordinates $x_1,\dots,x_n$ at the projection $x_0\in X$ of γ_0 can be chosen so that, with the corresponding coordinates (x,ξ) in $T^*(X)$, the Lagrangian plane defined in the complexification of $T_{\gamma_0}(T^*(X))$ by the equations $d\xi_j=0$ is transversal to $T_{\gamma_0}(J)$. If $\gamma_0=(x_0,\xi_0)$ one can choose $H\in C^\infty$, homogeneous of degree 0 in a conic neighborhood of ξ_0 such that J is generated by $x_j-\partial H(\xi)/\partial\xi_j$ in a conic neighborhood of γ_0. Conversely such functions always generate a complex conic Lagrangian ideal. If $\tilde H$ is another function with the same property then we have for every N in a conic neighborhood of ξ_0*

$$|H'(\xi)-\tilde H'(\xi)|\leqq C_N|\operatorname{Im} H'(\xi)|^N,\quad |H(\xi)-\tilde H(\xi)|\leqq C_N|\xi|\,|\operatorname{Im} H'(\xi)|^N.$$

Proof. If λ is a Lagrangian plane in $T^*(\mathbb{C}^n)$ it follows from the complex version of Corollary 21.2.11 that one can find a complex Lagrangian plane transversal to λ and the plane $x=0$, hence of the form

$$\xi=Bx$$

where B is a complex symmetric matrix. That this plane is transversal to λ means that a certain determinant involving B is not 0. Since a polynomial does not vanish identically for real arguments if it is not 0, one can always choose B real. But then the proof of Theorem 21.2.16 shows that the local coordinates can be chosen so that the desired transversality is obtained.

With such coordinates for $\lambda=T_{\gamma_0}(J)$ let $u_j(x,\xi)$, $j=1,\dots,n$, be local generators of J at γ_0 which are homogeneous of degree 0. The transversality means that the equations $du_j=0$, $d\xi_j=0$, $j=1,\dots,n$, imply $dx=0$, that is, $\det(\partial u_j/\partial x_k)\neq 0$. By Lemma 7.5.9 we can then in a neighborhood of γ_0 find generators for J of the form $x_j-h_j(\xi)$, $j=1,\dots,n$. Restricting h_j to the unit sphere and extending by homogeneity from there we find that h_j can be taken homogeneous of degree 0. By (25.4.1)

$$\{x_j-h_j(\xi),x_k-h_k(\xi)\}=\partial h_k/\partial\xi_j-\partial h_j/\partial\xi_k\in J.$$

Set $H(\xi)=\sum \xi_j h_j(\xi)$. Since h_k is homogeneous of degree 0 we have

$$\partial H/\partial \xi_k - h_k = \sum \xi_j(\partial h_j/\partial \xi_k - \partial h_k/\partial \xi_j) \in J,$$

so $x_k - \partial H/\partial \xi_k \in J$. Hence $x_k - \partial H/\partial \xi_k$, $k=1,\dots,n$ are generators of J by Lemma 7.5.8. To prove the last statement we use that the homogeneity gives

$$H(\xi)-\tilde{H}(\xi)=\sum \xi_j(\partial H/\partial \xi_j - \partial \tilde{H}/\partial \xi_j).$$

Since $\partial H/\partial \xi_j - \partial \tilde{H}/\partial \xi_j \in J$ we have by Lemma 7.5.10

$$|\partial H/\partial \xi - \partial \tilde{H}/\partial \xi| \leqq C_N |\partial \operatorname{Im} H/\partial \xi|^N$$

for every N, which completes the proof.

Guided by Proposition 25.1.3 we shall define $I^m(X,J)$ microlocally as the set of inverse Fourier transforms of functions of the form $e^{-iH}v$ where $v \in S^{m-n/4}(\mathbb{R}^n)$. However, this will not be a temperate distribution unless $\operatorname{Im} H \leqq 0$. If this condition is fulfilled, then $J_{\mathbb{R}}$ is defined by

$$\{(H'(\xi), \xi);\ \operatorname{Im} H(\xi)=0\}$$

in a neighborhood of γ_0, for $\operatorname{Im} H=0$ implies $\operatorname{Im} H'=0$ since $\operatorname{Im} H \leqq 0$, and $\operatorname{Im} H'=0$ implies $\operatorname{Im} H=0$ since H is homogeneous of degree 1. By Lemma 7.7.2 we have in a conic neighborhood of ξ_0

$$|\operatorname{Im} H'(\xi)|^2 \leqq -C \operatorname{Im} H(\xi)/|\xi|$$

for both sides are homogeneous of degree 0. If $\tilde{H}$ is another function defining J we obtain from Proposition 25.4.2

$$|\operatorname{Im} H(\xi) - \operatorname{Im} \tilde{H}(\xi)| \leqq C|\operatorname{Im} H(\xi)|^2/|\xi| < |\operatorname{Im} H(\xi)|/2$$

in a conic neighborhood of ξ_0, since $\operatorname{Im} H(\xi_0)=0$. Thus $\operatorname{Im} \tilde{H} \leqq 0$ also. The sign is also preserved when the local coordinates are changed. To prove this we shall more generally consider how J can be defined by means of phase functions. The following definition replaces Definition 21.2.15. For the sake of simplicity we restrict ourselves to the non-degenerate case.

Definition 25.4.3. Let X be a C^∞ manifold, $x_0 \in X$, and let $\phi(x,\theta)$ be a C^∞ function in an open conic neighborhood $\Gamma \subset X \times (\mathbb{R}^N \setminus 0)$ of (x_0, θ_0), which is homogeneous of degree 1 in θ. Then ϕ is called a non-degenerate phase function of positive type at (x_0, θ_0) if $\phi'_\theta(x_0, \theta_0)=0$ and

(i) $\phi'_x(x_0, \theta_0) \neq 0$;

(ii) the differentials $d(\partial\phi/\partial\theta_1), \dots, d(\partial\phi/\partial\theta_N)$ are linearly independent over the complex numbers at (x_0, θ_0);

(iii) $\operatorname{Im} \phi(x,\theta) \geqq 0$ in Γ.

A phase function of positive type always parametrizes a Lagrangian:

Proposition 25.4.4. *If ϕ is a non-degenerate phase function of positive type at $(x_0, \theta_0) \in X \times (\mathbb{R}^N \setminus 0)$ then $\phi'_x(x_0, \theta_0) = \xi_0 \in T^*_{x_0} \setminus 0$. If $x_1, \dots, x_n$ are local* *coordinates in X and $\xi_1,\dots,\xi_n$ corresponding coordinates in $T^*(X)$ then the functions independent of θ in the ideal $\hat{J}$ generated by*

$$\partial\phi(x,\theta)/\partial\theta_j,\quad j=1,\dots,N;\qquad \partial\phi/\partial x_k-\xi_k,\quad k=1,\dots,n, \tag{25.4.2}$$

form a conic Lagrangian ideal J in a conic neighborhood of (x_0,ξ_0). If this is regarded as a conic set in $T^*(X)\setminus 0$ then J does not depend on the choice of local coordinates, and $T_\gamma(J)$ is a positive Lagrangian plane at every point $\gamma\in J$. When the coordinates are chosen as in Proposition 25.4.2 the ideal is generated by $x_j-\partial H/\partial\xi_j$, $j=1,\dots,n$, where $H\in C^\infty$ is positively homogeneous and $\operatorname{Im} H\leqq 0$ in a conic neighborhood of ξ_0. We have

$$\phi(x,\theta)-\langle x,\xi\rangle+H(\xi)\in\hat{J}^2.$$

Proof. Since $\phi(x_0,\theta_0)=\langle\phi'_\theta(x_0,\theta_0),\theta_0\rangle=0$ and $\operatorname{Im}\phi\geqq 0$, we have $\operatorname{Im}\phi'_x(x_0,\theta_0)=0$ so $\xi_0=\phi'_x(x_0,\theta_0)\in T^*_{x_0}\setminus 0$ by (i). The differentials of the generators (25.4.2) are linearly independent at (x_0,θ_0,ξ_0), for if

$$\sum a_j d\,\partial\phi/\partial\theta_j+\sum b_k d(\partial\phi/\partial x_k-\xi_k)=0$$

then $b_k=0$ since ξ_k is an independent variable, hence $a_j=0$ since ϕ is non-degenerate. Let $\hat{J}$ be the ideal generated by the functions (25.4.2) in a neighborhood of (x_0,θ_0,ξ_0), and let J be the set of functions in $\hat{J}$ which are independent of θ. It is clear that J is invariantly defined on $T^*(X)$ in a neighborhood of (x_0,ξ_0). To examine J it is convenient to use local coordinates such that $\det\Phi\neq 0$ at (x_0,θ_0) if

$$\Phi=\begin{pmatrix}\phi''_{xx} & \phi''_{x\theta}\\ \phi''_{\theta x} & \phi''_{\theta\theta}\end{pmatrix}. \tag{25.4.3}$$

The proof of Proposition 25.1.5 shows with no essential change that this is true if and only if the plane $d\xi=0$ in the complexification of $T_{x_0,\xi_0}(T^*(X))$ is transversal to the complex Lagrangian plane λ which is the image under $d(x,\phi'_x(x,\theta))$ of the subspace of the complexified tangent plane of $X\times(\mathbb{R}^N\setminus 0)$ at (x_0,θ_0) where $d\phi'_\theta=0$. In the proof of Proposition 25.4.2 we saw that this transversality can be obtained by a suitable choice of local coordinates at x_0 in X. Now the fact that $\det\Phi\neq 0$ means by Theorem 7.5.9 that generators for $\hat{J}$ can be chosen of the form

$$x_j-X_j(\xi),\qquad \theta_j-\Theta_j(\xi). \tag{25.4.2$'$}$$

We can choose X_j and Θ_j homogeneous of degree 0 and 1 respectively, hence defined in a conic neighborhood of ξ_0.

Since the generators (25.4.2) are the derivatives of

$$f(x,\theta,\xi)=\phi(x,\theta)-\langle x,\xi\rangle$$

with respect to x and θ, it follows from Lemma 7.7.8 that there is a function $f^0(\xi)$ with $f(x,\theta,\xi)-f^0(\xi)\in\hat{J}^2$ and

$$\operatorname{Im} f^0(\xi)\geqq C(|\operatorname{Im} X(\xi)|^2+|\operatorname{Im}\Theta(\xi)|^2) \tag{25.4.4}$$

in a neighborhood of ξ_0. Restricting to $|\xi| = |\xi_0|$ and extending by homogeneity we can make f^0 homogeneous of degree 1. Since

$$\hat{J} \ni \partial(f(x,\theta,\xi) - f^0(\xi))/\partial\xi_j = -x_j - \partial f^0(\xi)/\partial\xi_j$$

it follows that we can replace the generators $x_j - X_j(\xi)$ in (25.4.2)′ by $x_j + \partial f^0(\xi)/\partial\xi_j$. Every C^∞ function $g(x,\xi)$ in a neighborhood of (x_0,ξ_0) can be written in the form

$$g(x,\xi) = \sum q_j(x,\xi)(x_j + \partial f^0(\xi)/\partial\xi_j) + r(\xi).$$

If $g \in J$ then $r \in J$ so in a neighborhood of ξ_0 we have for every N

$$|r(\xi)| \leqq C_N(|\operatorname{Im} X(\xi)| + |\operatorname{Im} \Theta(\xi)|)^N$$

by Lemma 7.5.10. Since

$$|\operatorname{Im} f^0(\xi)| = |\operatorname{Im} \langle \partial f^0(\xi)/\partial\xi, \xi\rangle| \leqq C|\xi|\,|\operatorname{Im} \partial f^0(\xi)/\partial\xi|$$

it follows in view of (25.4.4) that for every N

$$|r(\xi)| \leqq C'_N |\operatorname{Im} \partial f^0(\xi)/\partial\xi|^N.$$

Hence Theorem 7.5.12 shows that r is in the ideal generated by $x_j + \partial f^0(\xi)/\partial\xi_j$, $j=1,\dots,n$. If we set $H(\xi) = -f^0(\xi)$ it follows that $\operatorname{Im} H \leqq 0$ and that J is generated by $x_j - \partial H(\xi)/\partial\xi_j$, $j=1,\dots,n$.

The complex Lagrangian plane λ at the beginning of the proof is the image of $(dx, d\xi)$ when $d(\xi_k - \partial\phi/\partial x_k) = 0$, $k=1,\dots,n$, and $d(\partial\phi/\partial\theta_j) = 0$, $j=1,\dots,N$. These conditions are equivalent to $d(x_k - \partial H/\partial\xi_k) = 0$ and $d(\theta_j - \Theta_j(\xi)) = 0$, $k=1,\dots,n$; $j=1,\dots,N$. Hence $\lambda = T_{x_0,\xi_0}(J)$ so the transversality condition in the proof is identical to that in Proposition 25.4.2. That $T_{x_0\xi_0}(J)$ is positive follows from the fact that $\operatorname{Im} H''(\xi_0) \leqq 0$ (cf. the proof of Proposition 21.5.9). The proof is complete.

It is sometimes useful to decrease the number of θ-variables:

Corollary 25.4.5. *Let the hypotheses in Proposition 25.4.4 be fulfilled, and assume that with a splitting of the θ variables in two groups θ', θ'' we have $\det \phi''_{\theta''\theta''} \neq 0$ at (x_0,θ_0). Then $\theta'_0 \neq 0$, and there is a homogeneous function $\phi'(x,\theta')$ in a neighborhood of (x_0,θ'_0) such that $\phi'(x,\theta') - \phi(x,\theta)$ is in the square of the ideal generated by the components of $\partial\phi(x,\theta)/\partial\theta''$. The function ϕ' is a phase function of positive type at (x_0,θ_0) defining the same ideal J as ϕ.*

Proof. The homogeneity of ϕ gives $\phi''_{\theta''\theta}\theta = 0$, so $\theta'_0 = 0$ would give $\theta''_0 = 0$, hence a contradiction. The existence of ϕ' follows from the discussion preceding Lemma 7.7.8. The homogeneity is obtained by restricting to the sphere $|\theta'| = |\theta'_0|$ and extending by homogeneity. That $\operatorname{Im} \phi' \geqq 0$ is a consequence of Lemma 7.7.8. Since $\phi' - \phi \in \hat{J}^2$, with the notation in Proposition 25.4.4, we have

$$\partial\phi'/\partial x_j - \xi_j \in \hat{J}, \quad \partial\phi'/\partial\theta'_j \in \hat{J}.$$

Hence the Lagrangian ideal J' defined by ϕ' is contained in J. Since the number of generators of J and of J' are both equal to $\dim X$, it follows from Lemma 7.5.8 that the ideals J and J' must be identical. This completes the proof.

Remark. From Corollary 25.4.5 it is easy to obtain an extension of Theorem 21.2.18. In fact, let $x=(x',x'')$ be a splitting of the variables and set $\Phi(x,\theta,y'',\eta'')=\phi(x',y'',\theta)+\langle x''-y'',\eta''\rangle$. Under the hypotheses in Proposition 25.4.4 it is immediately seen that Φ is a non-degenerate phase function of positive type at $(x_0,\theta_0,x''_0,\xi''_0)$ which also defines J. (We can homogenize the parameters (y'',θ,η'') as in the proof of Theorem 25.2.2.) Φ has a non-degenerate critical point at $(x_0,\theta_0,x''_0,\xi''_0)$ as a function of (y'',θ) if $dx=0$, $d\phi'_\theta(x',y'',\theta)=0$, $d\phi'_{y''}(x',y'',\theta)=0$, $d\eta''=0$ implies $dy''=d\theta=0$, that is, if *the map* $T_{x_0,\xi_0}(J)\ni(dx,d\xi)\mapsto(dx',d\xi'')$ *is injective*. Then Corollary 25.4.5 shows that we can eliminate the (y'',θ) variables and obtain a new non-degenerate phase function of positive type which has the form $\phi_1(x',\eta'')+\langle x'',\eta''\rangle$. Proposition 25.4.4 is the special case where no ξ' variables occur.

Definition 25.4.6. A conic Lagrangian ideal J is said to be positive at $(x_0,\xi_0)\in J_{\mathbb{R}}$ if it can be parametrized in a neighborhood by a non-degenerate homogeneous phase function of positive type.

By Proposition 25.4.4 an equivalent definition is that for some local coordinates the ideal J is locally generated by

$$x_j-\partial H(\xi)/\partial\xi_j,\qquad j=1,\dots,n,$$

where H is homogeneous and $\operatorname{Im} H\leqq 0$. This is then true for all local coordinates such that the plane $d\xi=0$ at (x_0,ξ_0) is transversal to $T_{x_0,\xi_0}(J)$.

We shall now prove an analogue of Proposition 25.1.5 which will allow us to define $I^m(X,J)$ when J is a positive conic Lagrangian ideal.

Theorem 25.4.7. *Let ϕ, J and H be as in Proposition 25.4.4 and let $a\in S^{m+(n-2N)/4}(\mathbb{R}^n\times\mathbb{R}^N)$ have support in a small conic neighborhood Γ of (x_0,θ_0). Then*

$$u(x)=(2\pi)^{-(n+2N)/4}\int e^{i\phi(x,\theta)}a(x,\theta)\,d\theta\in\mathscr{E}'(\mathbb{R}^n)\tag{25.4.5}$$

is defined as an oscillatory integral and

$$WF(u)\subset\{(x,\phi'_x(x,\theta));\ (x,\theta)\in\Gamma,\ \phi'_\theta(x,\theta)=0\}\subset J_{\mathbb{R}}.\tag{25.4.6}$$

One can find v, $v_0\in S^{m-n/4}(\mathbb{R}^n)$ with support in a small conic neighborhood of ξ_0 and $v-v_0\in S^{m-n/4-1}(\mathbb{R}^n)$ such that $(x_0,\xi_0)\notin WF(u-u_1)$ if

$$u_1(x)=(2\pi)^{-n}\int e^{i(\langle x,\xi\rangle-H(\xi))}v(\xi)\,d\xi,\tag{25.4.7}$$

and

$$v_0(\xi)-(2\pi)^{n/4}a(x,\theta)(\det\Phi/i)^{-\frac12}\in\hat{J}\tag{25.4.8}$$

in a neighborhood of (x_0,θ_0,ξ_0). Here $\hat{J}$ is the ideal generated by (25.4.2) and Φ is defined by (25.4.3). Conversely, if u_1 is defined by (25.4.7) with

$v \in S^{m-n/4}(\mathbb{R}^n)$ *then one can find* $a \in S^{m+(n-2N)/4}$ *such that* $(x_0, \xi_0) \notin WF(u-u_1)$ *if* u *is defined by* (25.4.5).

As a preparation for the proof we must discuss the condition (25.4.8).

Lemma 25.4.8. *For every* $a \in S^\mu(\mathbb{R}^n \times \mathbb{R}^N)$ *one can find* $v \in S^\mu(\mathbb{R}^n)$ *such that*

$$v(\xi) - a(x, \theta) \in \hat{J} \tag{25.4.9}$$

at infinity in a conic neighborhood of (x_0, θ_0, ξ_0) *independent of* a. *Conversely, given* $v \in S^\mu$ *we can find* $a \in S^\mu$ *such that* (25.4.9) *is valid in such a set.*

Proof. That $a \in S^\mu$ means that the functions

$$(x, \theta) \mapsto a(x, t\theta) t^{-\mu}$$

are bounded in C^∞ when $t > 1$, say. A moment's reflection on the proof of Lemma 7.5.7 then shows that in a fixed neighborhood V of (x_0, θ_0, ξ_0) we can write

$$a(x, t\theta) t^{-\mu} = \sum b_j(x, \theta, \xi, t)(x_j - X_j(\xi)) + \sum c_j(x, \theta, \xi, t)(\theta_j - \Theta_j(\xi)) + v(\xi, t)$$

where $X_j(\xi)$ and $\Theta_j(\xi)$ are defined as in the proof of Proposition 25.4.4 and b_j, c_j, v are bounded in $C^\infty(V)$. If $\chi(\xi) \in C^\infty$ has support in a small neighborhood of ξ_0, the product of the two sides by χ will be equal in a conic neighborhood of (x_0, θ_0, ξ_0); hence we have there for $t > 1$

$$\begin{aligned} \chi(\xi/t) a(x, \theta) = t^\mu \chi(\xi/t) (&\textstyle\sum b_j(x, \theta/t, \xi/t, t)(x_j - X_j(\xi)) \\ &+ \textstyle\sum c_j(x, \theta/t, \xi/t)(\theta_j - \Theta_j(\xi))/t + v(\xi/t, t)). \end{aligned}$$

We can choose χ so that

$$\int_0^\infty \chi(\xi/t)\, dt/t = 1$$

in a conic neighborhood of ξ_0, for this integral is a homogeneous function of ξ of degree 0 (cf. Section 3.2). Now an integration with respect to dt/t from 1 to ∞ gives if $|\xi|$ is large enough and (x, θ, ξ) is in a sufficiently small conic neighborhood of (x_0, θ_0, ξ_0)

$$a(x, \theta) = \sum B_j(x, \theta, \xi)(x_j - X_j(\xi)) + \sum C_j(x, \theta, \xi)(\theta_j - \Theta_j(\xi)) + v(\xi)$$

where $B_j \in S^\mu$, $C_j \in S^{\mu-1}$ and $v \in S^\mu$. This proves the first statement. To prove the second one we just divide $v(t\xi) t^{-\mu}$ in the same way by $\partial\phi/\partial\theta - \xi$ with a function of (x, θ) as remainder, multiply by a function of (x, θ), change variables and integrate. There is no need to repeat the details.

Remark. If $v \in S \cap J$ it follows from Lemma 7.5.10 that $|D^\alpha v(\xi)|$ can be estimated by $|\xi|^{\mu-|\alpha|-N} |\operatorname{Im} H(\xi)|^N$ when $|\xi| > 1$, for all N and α. Since $|\operatorname{Im} H|^N e^{\operatorname{Im} H}$ is bounded it follows that the corresponding distribution (25.4.7) is in C^∞. Thus the choice of v_0 in (25.4.8) is irrelevant mod C^∞.

Proof of Theorem 25.4.7. The distribution (25.4.5) was defined in Theorem 7.8.2, and (25.4.6) follows from Theorem 8.1.9. In the rest of the proof we can follow that of Proposition 25.1.5, starting from the representation (25.1.5) of $\hat{u}(\xi)$, which we divide by e^{iH} since this is an exponentially increasing function. As before it follows that in a conic neighborhood of ξ_0 we have $\hat{u}(\xi)=U_0(\xi)+U_1(\xi)$ where U_0 is rapidly decreasing and with $|\xi|=t$, $\xi/t=\eta$, $\chi\in C_0^\infty(\mathbf{R}^N\setminus 0)$

$$U_1(\xi)=(2\pi)^{-(n+2N)/4}\iint e^{it(\phi(x,\theta)-\langle x,\eta\rangle)}\chi(\theta)\,a(x,t\theta)t^N\,dx\,d\theta. \tag{25.4.10}$$

We assume that $|\xi_0|=1$. When $\eta=\xi_0$ the exponent has only the critical point (x_0,θ_0) in the support of the integrand so for η close to ξ_0 we can by Theorem 7.7.1 choose χ with support as close to θ_0 as we please. In view of (25.4.6) we know that U_1 is rapidly decreasing outside any given conic neighborhood of ξ_0 if the support of a is in a sufficiently small conic neighborhood of (x_0,θ_0). Thus we may require that η is close to ξ_0. We can then apply Theorem 7.7.12, with $\omega=t$, noting that the error term there is uniform with respect to u when u and the various residue classes are bounded in C^∞. The exponent $\phi(x,\theta)-\langle x,\eta\rangle$ is congruent to $-H(\eta)$ mod $\hat{J}^2$. The coefficients of the operator $L_{f,j}$ in (7.7.23) are homogeneous in θ of degree $-j$, hence congruent mod $\hat{J}$ to homogeneous functions of η of degree $-j$. Replacing η by ξ/t and multiplying by the factor t^{-j} in (7.7.23) we obtain homogeneous functions of ξ of degree $-j$. By Lemma 25.4.8 $a^{(\beta)}_{(\alpha)}(x,\theta)$ is congruent mod $\hat{J}$ to a function $a_{\alpha\beta}(\xi)\in S^{\mu-|\beta|}$ in a conic neighborhood of (x_0,θ_0,ξ_0), if $\mu=m+(n-2N)/4$. Hence $a^{(\beta)}_{(\alpha)}(x,t\theta)$ is congruent to $a_{\alpha\beta}(t\eta)$ modulo the ideal generated by the x, θ derivatives of $\phi(x,\theta)-\langle x,\eta\rangle$, so

$$\partial_x^\alpha\partial_\theta^\beta a(x,t\theta)\equiv t^{|\beta|}a_{\alpha\beta}(t\eta)$$

modulo this ideal. Since $|\xi|^{|\beta|}a_{\alpha\beta}(\xi)\in S^\mu$ the jth term in the sum in (7.7.23) is therefore in $S^{\mu-j}$; if we take ν terms the error term can be estimated by

$$C_\nu t^{-\nu-(n+N)/2}t^{\mu+N}=C_\nu t^{m-n/4-\nu}.$$

If $\Delta(\xi)$ is homogeneous of degree $n-N$ and congruent to $\det(\Phi/i)$ modulo $\hat{J}$, then the determinant factor in (7.7.23) becomes

$$(t/2\pi)^{-(N+n)/2}(\Delta(\eta))^{-\frac12}=(2\pi)^{(N+n)/2}\Delta(\xi)^{-\frac12}t^{-N}$$

where t^{-N} cancels the factor t^N in (25.4.10). Thus we obtain from (7.7.23) a sequence $v_j\in S^{\mu+(N-n)/2-j}=S^{m-n/4-j}$ such that

$$|U_1(\xi)-\sum_{j<\nu}v_j(\xi)e^{-iH(\xi)}|\le C_\nu|\xi|^{m-n/4-\nu}.$$

If $v\sim\sum_0^\infty v_j\in S^{m-n/4}$ it follows that

$$U_2(\xi)=U_1(\xi)-v(\xi)e^{-iH(\xi)}$$

is rapidly decreasing in a conic neighborhood of ξ_0, so $(x_0,\xi_0)\notin WF(u_2)$ if $\hat{u}_2=U_2$. This completes the proof of the first statement in the theorem.

To prove the converse we observe that we can choose $a_1 \in S^{m+(n-2N)/4}$ so that

(25.4.11) $$v(\xi)-(2\pi)^{n/4} a_1(x,\theta)(\det(\Phi(x,\theta)/i))^{-\frac{1}{2}} \in \hat{J}.$$

If u is defined by (25.4.5) with $a=a_1$ then $u_2=u_1-u$ is mod C^∞ of the form (25.4.7) with v equal to some $R_1 \in S^{m-n/4-1}$ in a conic neighborhood of ξ_0. We can iterate this procedure and obtain successively $a_j \in S^{m+(n-2N)/4+1-j}$ such that if u is defined by (25.4.5) with $a=a_1+\ldots+a_j$ then u_1-u is of the form (25.4.7) with v equal to a function $R_j \in S^{m-n/4-j}$ in a fixed conic neighborhood of ξ_0. When $a \sim \Sigma a_j$ the assertion in the theorem is proved.

Definition 25.4.9. Let J be a positive conic Lagrangian ideal in $T^*(X)\setminus 0$ where X is a C^∞ manifold. If E is a vector bundle on X then $I^m(X,J;E)$ is the space of distribution sections U of E on X such that for every $(x_0,\xi_0) \in J_{\mathbb{R}}$ there is a homogeneous phase function $\phi(x,\theta)$ of positive type parametrizing J near (x_0,ξ_0), a local trivialization of E and a distribution u of the form (25.4.5) in a neighborhood of x_0, with $a \in S^{m+(n-2N)/4}$, such that $(x_0,\xi_0) \notin WF(U-u)$. Here $n=\dim X$ and N is the number of θ variables.

By Theorem 25.4.7 it would be equivalent to require this condition for *every* phase function ϕ of positive type. Starting from (25.4.8) we could also introduce a notion of principal symbol. However, this requires not only an extension of Section 21.6 but also a study of the effect of the damping factor e^{-iH} in (25.4.8) so we content ourselves with referring the interested reader to the literature listed in the notes at the end of the chapter.

The condition on U in Definition 25.4.9 is a microlocal one. If we just have J defined in an open cone $\subset T^*(X)\setminus 0$ it is therefore clear what it means to say that $U \in I^m(X,J;E)$ there.

25.5. Fourier Integral Operators with Complex Phase

In this section we shall make the discussion quite brief since it is essentially a repetition of arguments from Section 25.2 with reference to Section 25.4 instead of Section 25.1. Let X and Y be two C^∞ manifolds. If J is a positive conic Lagrangian ideal in $C^\infty(T^*(X\times Y)\setminus 0)$ then

$$J'=\{f; i^*f \in J\}, \quad i(x,\xi,y,\eta)=(x,\xi,y,-\eta)$$

is called a positive conic canonical (or twisted Lagrangian) ideal. We have of course $J=(J')'$.

Definition 25.5.1. Let $J \subset C^\infty(T^*(X\times Y)\setminus 0)$ be a positive conic canonical ideal with $J_{\mathbb{R}} \subset (T^*(X)\setminus 0)\times(T^*(Y)\setminus 0)$, and let E, F be vector bundles on X, Y. Then the operators with kernel in $I^m(X\times Y, J'; \Omega^{\frac{1}{2}}_{X\times Y}\otimes \mathrm{Hom}(F,E))$ are called Fourier integral operators of order m from sections of $\Omega_Y^{\frac{1}{2}}\otimes F$ to sections of $\Omega_X^{\frac{1}{2}}\otimes E$, associated with the canonical ideal J.

In the following theorem J^{-1} denotes the ideal on $T^*(Y\times X)\setminus 0$ obtained by swapping the factors $T^*(Y)$ and $T^*(X)$, and $\bar{J}^{-1}$ is the complex conjugate.

Theorem 25.5.2. *If* $A\in I^m(X\times Y, J'; \Omega^{\frac{1}{2}}_{X\times Y}\otimes \operatorname{Hom}(F,E))$ *then*

$$A^*\in I^m(Y\times X, (\bar{J}^{-1})'; \Omega^{\frac{1}{2}}_{Y\times X}\otimes \operatorname{Hom}(E^*,F^*)).$$

Proof. It suffices to consider the local case where E, F are trivial bundles and X, Y subsets of Euclidean spaces. We can write

$$A(x,y)=\int a(x,y,\theta)\, e^{i\phi(x,y,\theta)}\, d\theta$$

where $a\in S^{m+(n-2N)/4}$, n is the dimension of $X\times Y$ and N is the number of θ variables. Then

$$A^*(y,x)=\int a(x,y,\theta)^*\, e^{-i\overline{\phi(x,y,\theta)}}\, d\theta.$$

J consists of the functions independent of θ in the ideal generated by

$$\partial\phi(x,y,\theta)/\partial x_j-\xi_j, \quad \partial\phi(x,y,\theta)/\partial y_k+\eta_k, \quad \partial\phi(x,y,\theta)/\partial\theta_l$$

while A^* is associated with the ideal generated by

$$-\overline{\partial\phi(x,y,\theta)}/\partial y_k-\eta_k, \quad -\overline{\partial\phi(x,y,\theta)}/\partial x_j+\xi_j, \quad -\overline{\partial\phi(x,y,\theta)}/\partial\theta_l.$$

This is the ideal $\bar{J}$, or considered as an ideal of functions on $T^*(Y\times X)$ the ideal $\bar{J}^{-1}$. This proves the theorem.

We shall only discuss composition in the transversal case. Thus let X, Y, Z be C^∞ manifolds and let

$$(x_0,\xi_0,y_0,\eta_0)\in(T^*(X)\setminus 0)\times(T^*(Y)\setminus 0),$$
$$(y_0,\eta_0,z_0,\zeta_0)\in(T^*(Y)\setminus 0)\times(T^*(Z)\setminus 0) \tag{25.5.1}$$

be points in $J_{1\mathbb{R}}$ resp. $J_{2\mathbb{R}}$, where J_1 and J_2 are positive conic canonical ideals defined in some conic neighborhoods. If T_{x_0,ξ_0} is the complexified tangent space of $T^*(X)$ at (x_0,ξ_0) and T_{y_0,η_0}, T_{z_0,ζ_0} are similarly defined, then the tangent planes $T(J_1)$ and $T(J_2)$ are linear subspaces of $T_{x_0,\xi_0}\times T_{y_0,\eta_0}$ and $T_{y_0,\eta_0}\times T_{z_0,\zeta_0}$. When $T(J_1)\times T(J_2)$ intersects $T_{x_0,\xi_0}\times\Delta(T_{y_0,\eta_0})\times T_{z_0,\zeta_0}$ transversally, $\Delta(T_{y_0,\xi_0})$ denoting the diagonal in $T_{y_0,\eta_0}\times T_{y_0,\eta_0}$, the composition is said to be *transversal* at the point in question.

Proposition 25.5.3. *Let J_1 and J_2 be positive conic canonical ideals living on $T^*(X\times Y)\setminus 0$ and $T^*(Y\times Z)\setminus 0$ respectively, and assume that the composition is transversal at the point (25.5.1). Then the functions in a neighborhood of (x_0,ξ_0,z_0,ζ_0) in $T^*(X)\times T^*(Z)$ which lifted to*

$$\tilde{\Delta}=T^*(X)\times\Delta(T^*(Y))\times T^*(Z)$$

are restrictions to $\tilde{\Delta}$ of elements in the ideal generated by $J_1 \otimes 1$ and $1 \otimes J_2$ form a local positive conic canonical ideal $J_1 \circ J_2$.

Proof. Lemma 25.2.5 remains valid if λ is complex. In view of Proposition 25.4.2 we can therefore choose local coordinates in X, Y and Z at x_0, y_0 and z_0 such that J_1 is defined by the phase function

$$\phi_1(x,y,\xi,\eta) = \langle x,\xi\rangle - \langle y,\eta\rangle - H_1(\xi,\eta)$$

and J_2 is defined by

$$\phi_2(y,z,\eta,\zeta) = \langle y,\eta\rangle - \langle z,\zeta\rangle - H_2(\eta,\zeta)$$

which are both of positive type. This means that

$$x_j - \partial H_1(\xi,\eta)/\partial\xi_j, \quad y_k + \partial H_1(\xi,\eta)/\partial\eta_k$$

are local generators for J_1 and that

$$y_k - \partial H_2(\eta,\zeta)/\partial\eta_k, \quad z_l + \partial H_2(\eta,\zeta)/\partial\zeta_l$$

are local generators for J_2. The tangent planes $T(J_1)$ and $T(J_2)$ are defined by the vanishing of the differentials of these functions. Since $T(J_1)\times T(J_2)$ is Lagrangian and $T(\tilde{\Delta})$ is involutive, the transversality means that at (25.5.1) $T(J_1)\times T(J_2)$ does not contain any element $\neq 0$ with 0 components in the tangent spaces of $T^*(X)$ and $T^*(Z)$ and equal components in the tangent space of $T^*(Y)$. Thus the equations

$$dx_j = d\xi_j = 0, \quad dz_l = d\zeta_l = 0, \quad d(x_j - \partial H_1(\xi,\eta)/\partial\xi_j) = 0,$$
$$d(z_l + \partial H_2(\eta,\zeta)/\partial\zeta_l) = 0, \quad d(\partial H_1(\xi,\eta)/\partial\eta_k + \partial H_2(\eta,\zeta)/\partial\eta_k) = 0$$

must imply $d\eta = 0$. This means precisely that

$$\partial^2 H_1/\partial\xi\,\partial\eta\,\eta' = 0, \quad \partial^2 H_2/\partial\zeta\,\partial\eta\,\eta' = 0, \quad \partial^2(H_1+H_2)/\partial\eta^2\,\eta' = 0$$

must imply $\eta' = 0$, that is, that

(25.5.2) the differentials $d(\partial(H_1(\xi,\eta)+H_2(\eta,\zeta))/\partial\eta_k)$ are linearly independent.

Now the ideal generated by $J_1 \otimes 1$ and $1 \otimes J_2$ is generated by the functions

$$x_j - \partial H_1(\xi,\eta)/\partial\xi_j, \quad y_k + \partial H_1(\xi,\eta)/\partial\eta_k, \quad y'_k - \partial H_2(\eta',\zeta)/\partial\eta'_k,$$
$$z_l + \partial H_2(\eta',\zeta)/\partial\zeta_l$$

if (y',η') are the coordinates in the second copy of $T^*(Y)$. We do not change the restriction to the diagonal if we add the generators $y_k - y'_k$ and $\eta_k - \eta'_k$. The ideal then obtained is generated by

$$x_j - \partial H_1(\xi,\eta)/\partial\xi_j, \quad \partial H_1(\xi,\eta)/\partial\eta_k + \partial H_2(\eta,\zeta)/\partial\eta_k, \quad y_k + \partial H_1(\xi,\eta)/\partial\eta_k,$$
$$z_l + \partial H_2(\eta,\zeta)/\partial\zeta_l \quad \text{and} \quad y_k - y'_k,\ \eta_k - \eta'_k.$$

The restriction to $\tilde{\Delta}$ with coordinates $x, \xi, y, \eta, z, \zeta$ is therefore precisely the ideal generated by these functions with $y_k - y'_k$ and $\eta_k - \eta'_k$ omitted. Thus $J_1 \circ J_2$ consists of the functions independent of y, η in this ideal.

Set with $\theta = (y, \xi, \eta, \eta', \zeta)$

$$\begin{aligned}(25.5.3)\qquad \Phi(x,z,\theta) &= \phi_1(x,y,\xi,\eta) + \phi_2(y,z,\eta',\zeta)\\ &= \langle x,\xi\rangle - \langle y,\eta\rangle - H_1(\xi,\eta) + \langle y,\eta'\rangle - \langle z,\zeta\rangle - H_2(\eta',\zeta).\end{aligned}$$

As in the proof of Theorem 25.2.3 we should think of $y, \xi, \eta, \eta', \zeta$ as homogeneous functions of degree $0, 1, \dots, 1$ of some other variable ω; which makes Φ a homogeneous function. However, it is more convenient never to display this variable explicitly. It is clear that $\operatorname{Im}\Phi \geqq 0$, and we shall now prove that Φ is non-degenerate. Since

$$\begin{gathered}\partial\Phi/\partial y_k = \eta'_k - \eta_k, \qquad \partial\Phi/\partial\xi_j = x_j - \partial H_1(\xi,\eta)/\partial\xi_j,\\ \partial\Phi/\partial\eta_k = -y_k - \partial H_1(\xi,\eta)/\partial\eta_k,\\ \partial\Phi/\partial\eta'_k = y_k - \partial H_2(\eta',\zeta)/\partial\eta'_k, \qquad \partial\Phi/\partial\zeta_l = -z_l - \partial H_2(\eta',\zeta)/\partial\zeta_l\end{gathered}$$

and x_j, y_k, z_l are independent variables, the linear independence of all these differentials follows if we show that at (25.5.1)

$$\sum a_k(d\eta'_k - d\eta_k) + \sum b_k d(\partial H_1(\xi,\eta)/\partial\eta_k + \partial H_2(\eta',\zeta)/\partial\eta'_k) = 0$$

implies $a_k = b_k = 0$. Writing $d\eta' = d\eta + d\eta' - d\eta$ in the second sum we obtain $b_k = 0$ by (25.5.2), and then it follows at once that $a_k = 0$ too.

The positive conic canonical ideal J defined by Φ according to Proposition 25.4.4 consists of the functions of x, ξ, z, ζ in the ideal generated by $\partial\Phi/\partial\xi_j$, $\partial\Phi/\partial y_k$, $\partial\Phi/\partial\eta_k$, $\partial\Phi/\partial\eta'_k$, $\partial\Phi/\partial\zeta_l$. Since $\partial\Phi/\partial y_k = \eta'_k - \eta_k$, a function is in this ideal if and only if this is true for the restriction to the set where $\eta' = \eta$. But this means that J coincides with $J_1 \circ J_2$ which completes the proof.

In the proof of Proposiiton 25.5.3 we have used very special defining phase functions for J_1 and J_2. However, a defining phase function for $J_1 \circ J_2$ can be obtained in a natural way from arbitrary defining phase functions for J_1 and J_2.

Proposition 25.5.4. *Let the hypotheses in Proposition 25.5.3 be fulfilled and let $\phi(x,y,\theta)$, $\psi(y,z,\tau)$ be homogeneous phase functions of positive type at (x_0,y_0,θ_0) and (y_0,z_0,τ_0) respectively which define J_1 and J_2 near (25.5.1). Then*

$$(25.5.3)' \qquad \Phi(x,z,y,\theta,\tau) = \phi(x,y,\theta) + \psi(y,z,\tau)$$

is a phase function of positive type defining $J_1 \circ J_2$.

Here Φ is a homogeneous function of a suitable parameter, say

$$\omega = ((|\theta|^2 + |\tau|^2)^{\frac{1}{4}} y, \theta, \tau),$$

but it is more convenient to use the parameters (y, θ, τ) in the proof.

Proof. First note that since the number of generators of a canonical ideal in $T^*(X\times Z)\setminus 0$ is always $\dim(X\times Z)$, it follows from Lemma 7.5.8 that if one such ideal is contained in another one then they are identical. Hence it suffices to show that $J_1\circ J_2$ is contained in the canonical ideal defined by Φ. Now J_1 consists of functions independent of θ in the ideal generated by

$$\partial\phi(x,y,\theta)/\partial x_j-\xi_j,\quad \partial\phi(x,y,\theta)/\partial y_k+\eta_k,\quad \partial\phi(x,y,\theta)/\partial\theta_\nu$$

while J_2 consists of functions independent of τ in the ideal generated by

$$\partial\psi(y,z,\tau)/\partial y_k-\eta_k,\quad \partial\psi(y,z,\tau)/\partial z_l+\zeta_l,\quad \partial\psi(y,z,\tau)/\partial\tau_\mu.$$

Thus $J_1\otimes 1$ and $1\otimes J_2$ restricted to the diagonal are contained in the ideal generated by these functions. If $f(x,\xi,z,\zeta)\in J_1\circ J_2$ we can then write locally

$$f=\sum a_j(\partial\phi/\partial x_j-\xi_j)+\sum b_k(\partial\phi/\partial y_k+\eta_k)+\sum c_\nu\,\partial\phi/\partial\theta_\nu$$
$$+\sum d_k\,\partial(\phi+\psi)/\partial y_k+\sum e_l(\partial\psi/\partial z_l+\zeta_l)+\sum f_\mu\,\partial\psi/\partial\tau_\mu,$$

where $a_j,\dots,f_\mu$ are functions of $x,\ \xi,\ z,\ \zeta,\ y,\ \theta,\ \tau,\ \eta$. We can make all coefficients except b_k independent of η by reducing them modulo the ideal generated by $\partial\phi/\partial y_k+\eta_k$, $k=1,\dots,\dim Y$, and changing b_k to take care of the quotients. By Lemma 7.5.10 it follows then, since $R=\sum b_k(\partial\phi/\partial y_k+\eta_k)$ is independent of η, that for every N

$$|R(x,\xi,z,\zeta,y,\theta,\tau)|\le C_N|\operatorname{Im}\partial\phi/\partial y|^N. \tag{25.5.4}$$

Lemma 7.5.11 gives the same estimate for the derivatives of R. We can estimate $|\partial\operatorname{Im}\phi/\partial y|^2$ by $\operatorname{Im}\phi$ locally, and since $\operatorname{Im}\phi=\langle\partial\operatorname{Im}\phi/\partial\theta,\theta\rangle$ it follows that we can replace $\partial\phi/\partial y$ by $\partial\phi/\partial\theta$ in (25.5.4). Hence Theorem 7.5.12 shows that R is in the ideal generated by $\partial\phi/\partial\theta_\nu$. Thus we may drop the sum $\sum b_k(\partial\phi/\partial y_k+\eta_k)$ altogether, which means that f is in the canonical ideal defined by Φ.

It is now easy to study composition of Fourier integral operators. In fact the proof of Theorem 25.2.3 does not require any change at all since Theorem 7.7.1 is equally valid for phase functions of positive type and Proposition 25.5.4 shows that the phase function Φ occurring in the proof defines the composition ideal. Hence we obtain

Theorem 25.5.5. *Let J_1 and J_2 be positive conic canonical ideals in, respectively, $T^*(X\times Y)\setminus 0$ and $T^*(Y\times Z)\setminus 0$ such that*

(i) $J_{1\mathbb{R}}\subset(T^*(X)\setminus 0)\times(T^*(Y)\setminus 0)$, $J_{2\mathbb{R}}\subset(T^*(Y)\setminus 0)\times(T^*(Z)\setminus 0)$;

(ii) *the composition is transversal at every point in the intersection of*

$$J_{1\mathbb{R}}\times J_{2\mathbb{R}}\quad\textit{and}\quad T^*(X)\times\Delta(T^*(Y))\times T^*(Z);$$

(iii) *the map*

$$(J_{1\mathbb{R}}\times J_{2\mathbb{R}})\cap(T^*(X)\times\Delta(T^*(Y))\times T^*(Z))\to T^*(X\times Z)\setminus 0$$

is injective and proper.

Then $J_1 \circ J_2$, defined locally by Proposition 25.5.3, is a positive conic canonical ideal in $T^*(X\times Z)\smallsetminus 0$. Let

$$A_1\in I^{m_1}(X\times Y, J_1'; \Omega^{\frac12}_{X\times Y}\otimes \mathrm{Hom}(F,E)),$$
$$A_2\in I^{m_2}(Y\times Z, J_2'; \Omega^{\frac12}_{Y\times Z}\otimes \mathrm{Hom}(G,F))$$

be properly supported, E, F, G being vector bundles on X, Y, Z. Then

$$A_1A_2\in I^{m_1+m_2}(X\times Z, (J_1\circ J_2)'; \Omega^{\frac12}_{X\times Z}\otimes \mathrm{Hom}(G,E)).$$

Finally we shall prove an analogue of Theorem 25.3.1.

Theorem 25.5.6. *Let $J\subset C^\infty(T^*(X\times Y)\smallsetminus 0)$ be a positive conic canonical ideal with $J_{\mathbb{R}}\subset (T^*(X)\smallsetminus 0)\times(T^*(Y)\smallsetminus 0)$. Then every $A\in I^0(X\times Y, J'; \Omega^{\frac12}_{X\times Y})$ defines a continuous map from $L^2_{\mathrm{comp}}(Y,\Omega^{\frac12})$ to $L^2_{\mathrm{loc}}(X,\Omega^{\frac12})$ if and only if there is no $\gamma\in J_{\mathbb{R}}$ such that $T_\gamma(J)$ has a real element with only one of the components in $T(T^*(X))$ and $T(T^*(Y))$ equal to 0.*

Proof of the Necessity. Let $\gamma=(x_0,\xi_0,y_0,\eta_0)\in J_{\mathbb{R}}$ and choose local coordinates at x_0 in X and at y_0 in Y such that J is defined by a phase function of the form

$$\langle x,\xi\rangle-\langle y,\eta\rangle-H(\xi,\eta)$$

in a conic neighborhood of γ. Thus $\operatorname{Im} H\leqq 0$ there, and if the coordinates are chosen so that $x_0=0$, $y_0=0$ then $H'_\xi(\xi_0,\eta_0)=0$, $H'_\eta(\xi_0,\eta_0)=0$. Set $n=n_X+n_Y$ where $n_X=\dim X$, $n_Y=\dim Y$. If $a\in S^{(n-2n)/4}(\mathbb{R}^n)$ has support in a small conic neighborhood of (ξ_0,η_0) then the oscillatory integral

$$A(x,y)=\iint e^{i(\langle x,\xi\rangle-\langle y,\eta\rangle-H(\xi,\eta))}a(\xi,\eta)\,d\xi\,d\eta$$

is in $I^0(X\times Y, J')$ in a neighborhood of (x_0,y_0) and in C^∞ elsewhere, so A is supposed to be the kernel of a continuous operator from $L^2_{\mathrm{comp}}(\mathbb{R}^{n_Y})$ to $L^2_{\mathrm{loc}}(\mathbb{R}^{n_X})$. We choose $a(\xi,\eta)=(1+(|\xi|^2+|\eta|^2)/(|\xi_0|^2+|\eta_0|^2))^{-n/8}$ at infinity in a conic neighborhood of (ξ_0,η_0). With $u\in C_0^\infty(\mathbb{R}^{n_Y})$ and $v\in C_0^\infty(\mathbb{R}^{n_X})$ we set

$$u_t(y)=e^{it^2\langle y,\eta_0\rangle}u(ty)t^{n_Y/2},\quad v_t(x)=e^{it^2\langle x,\xi_0\rangle}v(tx)t^{n_X/2}.$$

Since $\|u_t\|=\|u\|$, $\|v_t\|=\|v\|$, the norms being L^2 norms, and since the supports of u_t and v_t tend to $\{0\}$, the L^2 continuity gives for some M

$$\varlimsup_{t\to\infty}|(Au_t,v_t)|\leqq M\|u\|\,\|v\|.$$

A direct computation gives

$$\hat u_t(\eta)=t^{-n_Y/2}\hat u((\eta-t^2\eta_0)/t),\quad \hat v_t(\xi)=t^{-n_X/2}\hat v((\xi-t^2\xi_0)/t),$$
$$(Au_t,v_t)=t^{n/2}\iint e^{-iH(t^2\xi_0+t\xi,t^2\eta_0+t\eta)}\hat u(\eta)\overline{\hat v(\xi)}\,a(t^2\xi_0+t\xi,t^2\eta_0+t\eta)\,d\xi\,d\eta.$$

The exponential is bounded by 1 in absolute value, and

$$t^2H(\xi_0+\xi/t,\eta_0+\eta/t)\to Q(\xi,\eta),\quad t\to\infty,$$

where Q is the quadratic part in the Taylor expansion of H at (ξ_0,η_0). (Recall that $dH=0$ at (ξ_0,η_0) which implies $H=0$ by the homogeneity.) We have

$$|t^{n/2}a(t^2\xi_0+t\xi,t^2\eta_0+t\eta)-1|\leqq C(1+|\xi|+|\eta|)^{(n+2)/2}/t$$

by the mean value theorem if $(|\xi|+|\eta|)/t$ is small and by the boundedness of a otherwise. Letting $t\to\infty$ we therefore obtain

$$|\iint e^{-iQ(\xi,\eta)}\hat{u}(\eta)\overline{\hat{v}(\xi)}\,d\xi\,d\eta|\leqq M\|u\|\,\|v\|,$$

that is, $\|A_0\|\leqq M$ if A_0 is the operator with kernel

$$A_0(x,y)=\iint e^{i(\langle x,\xi\rangle-\langle y,\eta\rangle-Q(\xi,\eta))}\,d\xi\,d\eta.$$

(Note that the argument so far is parallel to the proof of Proposition 25.1.7.) The tangent plane $T(J)$ is the linear canonical relation

$$\{(\partial Q/\partial\xi,\xi,-\partial Q/\partial\eta,\eta)\}\subset T^*(\mathbb{C}^{n_X})\times T^*(\mathbb{C}^{n_Y}). \tag{25.5.5}$$

Assume now that $(\tilde{x},\tilde{\xi};0,0)$ is real and belongs to the canonical relation (25.5.5). Then it is symplectically orthogonal, that is,

$$\langle\partial Q/\partial\xi,\tilde{\xi}\rangle-\langle\tilde{x},\xi\rangle=0\quad\text{for all }(\xi,\eta).$$

With the notation $\phi(x,\xi,y,\eta)=\langle x,\xi\rangle-\langle y,\eta\rangle-Q(\xi,\eta)$ this implies that

$$\begin{aligned}(\langle\tilde{x},D_x\rangle-\langle x,\tilde{\xi}\rangle)A_0(x,y)&=\iint(\langle\tilde{x},\xi\rangle-\langle x,\tilde{\xi}\rangle)e^{i\phi}\,d\xi\,d\eta\\&=\iint\langle-D_\xi,\tilde{\xi}\rangle e^{i\phi}\,d\xi\,d\eta=0.\end{aligned}$$

Choose $u\in C_0^\infty(\mathbb{R}^{n_Y})$ with $A_0u\neq0$. Since $A_0u\in L^2$ and

$$(\langle\tilde{x},D_x\rangle-\langle x,\tilde{\xi}\rangle)(A_0u)=0$$

we obtain $\tilde{\xi}=0$ if $\tilde{x}=0$, for A_0u cannot be supported by a hyperplane. If $\tilde{x}\neq0$ and ψ is a real valued solution of the equation $\langle\tilde{x},\partial\psi/\partial x\rangle=\langle x,\tilde{\xi}\rangle$ then $e^{-i\psi}A_0u$ is constant in the direction x, which contradicts that $0\neq A_0u\in L^2$. Hence $\tilde{x}=\tilde{\xi}=0$. Taking the adjoint of A_0 we conclude that (25.5.5) cannot have a real element of the form $(0,0;y,\eta)$ either, which completes the proof of the necessity.

Proof of the Sufficiency. We can localize by multiplying A to the left and right by partitions of unity in X or Y. More generally, we can microlocalize by using pseudo-differential partitions of unity, so A may be assumed to have support near $(x_0,y_0)\in\mathbb{R}^{n_X+n_Y}$ while $WF'(A)$ is contained in a small conic neighborhood of the ray through $\gamma=(x_0,\xi_0,y_0,\eta_0)$. We shall prove that A^*A is a pseudo-differential operator of type $\frac12,\frac12$ which will make the proof of Theorem 25.3.1 applicable.

If $\phi(x,y,\theta)$ is a non-degenerate phase function of positive type at (x_0,y_0,θ_0) defining J at γ, we can write A modulo C^∞ in the form

$$A(x,y)=\int e^{i\phi(x,y,\theta)}a(x,y,\theta)\,d\theta$$

where $a\in S^{(n-2N)/4}(\mathbb{R}^n\times\mathbb{R}^N)$, $n=n_X+n_Y$, and a has support in a small conic neighborhood of (x_0,y_0,θ_0),

$$\phi'_\theta=0,\quad \phi'_x=\xi_0,\quad \phi'_y=-\eta_0\quad \text{at } (x_0,y_0,\theta_0).$$

By Theorem 25.5.2 the kernel of the adjoint operator

$$A^*(y,x)=\overline{A(x,y)}=\int e^{-i\overline{\phi(x,y,\theta)}}\,\overline{a(x,y,\theta)}\,d\theta$$

is in $I^0(Y\times X,\bar{J}^{-1})$ where $\bar{J}^{-1}$ is defined by the phase function $-\overline{\phi(x,y,\theta)}$. The composition $\bar{J}^{-1}\circ J$ is transversal, for $t\in T_\gamma(J)\cap\overline{T_\gamma(J)}$ implies $t=0$ if the component t_Y of t in $T_{y_0,\eta_0}(T^*(Y))_{\mathbb{C}}$ is 0. Hence $A^*A\in I^0(Y\times Y,(\bar{J}^{-1}\circ J)')$, the support of A^*A is close to (y_0,y_0) and $WF'(A^*A)$ is in a small conic neighborhood of (y_0,η_0,y_0,η_0). The composition $\bar{J}^{-1}\circ J$ is defined there by the non-degenerate phase function of positive type

$$\phi(x,y,\theta)-\overline{\phi(x,z,\tau)} \tag{25.5.6}$$

where z denotes the variable in the copy of Y to the left.

There is also a defining phase function of the form $\tilde{\phi}(z,\eta)-\langle y,\eta\rangle$ with $\operatorname{Im}\tilde{\phi}\geqq 0$. By the remark following Corollary 25.4.5 this will be proved if we show that the map

$$T_{y_0,\eta_0,y_0,\eta_0}(\bar{J}^{-1}\circ J)\ni(dz,d\zeta,dy,d\eta)\mapsto(dz,d\eta)$$

is injective. If (t'_Y,t''_Y) is in the kernel, then $\sigma(t'_Y,\bar{t}'_Y)=\sigma(t''_Y,\bar{t}''_Y)=0$ since the z and η components are equal to 0. We can find $t_X\in T_{x_0,\xi_0}(T^*(X))_{\mathbb{C}}$ such that

$$(t_X,t''_Y)\in T_\gamma(J),\qquad(\bar{t}_X,\bar{t}'_Y)\in T_\gamma(J),$$

hence by the positivity of $T_{x_0,\xi_0,y_0,\eta_0}(J)$

$$i(\sigma(\bar{t}_X,t_X)-\sigma(\bar{t}''_Y,t''_Y))\geqq 0,\quad i(\sigma(\bar{t}_X,t_X)-\sigma(\bar{t}'_Y,t'_Y))\leqq 0.$$

It follows that $\sigma(\bar{t}_X,t_X)=0$ too, hence

$$(\bar{t}_X,\bar{t}''_Y)\in T_\gamma(J),\qquad(t_X,t'_Y)\in T_\gamma(J)$$

by virtue of Proposition 21.5.10. It follows that $(0,\operatorname{Re}(t'_Y-t''_Y))\in T_\gamma(J)$ and that $(0,\operatorname{Im}(t'_Y-t''_Y))\in T_\gamma(J)$, so $t'_Y=t''_Y$. Since the z components of t'_Y and the η components of t''_Y are 0, it follows that $t'_Y=t''_Y=0$, which proves the assertion.

Set

$$\Phi(z,\eta,x,y,\theta,\tau)=\phi(x,y,\theta)-\bar{\phi}(x,z,\tau)+\langle y,\eta\rangle \tag{25.5.7}$$

and let $\hat{J}$ be the ideal generated by the derivatives of Φ with respect to x, y, θ, τ. Then $\tilde{\phi}-\Phi\in\hat{J}^2$ by Corollary 25.4.5 and the remark following it. The ideal $\hat{J}$ has also generators of the form

$$x_j-X_j(z,\eta),\quad y_j-Y_j(z,\eta),\quad \theta_j-\Theta_j(z,\eta),\quad \tau_j-T_j(z,\eta)$$

where X_j, Y_j, Θ_j, T_j are homogeneous of degree 0, 0, 1, 1 respectively, and $X=x_0$, $Y=y_0$, $\Theta=T=\theta_0$ at (y_0,η_0). By Lemma 7.7.8 we have near (y_0,η_0)

(25.5.8) $$|\operatorname{Im}(X, Y, \Theta, T)(z,\eta)|^2 \leqq C \operatorname{Im}\tilde{\phi}(z,\eta).$$

Since $\tilde{\phi}-\Phi\in\hat{J}^2$ it follows if $X^r=\operatorname{Re}X, \ldots, T^r=\operatorname{Re}T$ that

$$\operatorname{Im}\Phi(z,\eta,X^r(z,\eta),\ldots,T^r(z,\eta))\leqq C_1 \operatorname{Im}\tilde{\phi}(z,\eta),$$

for the generators of $\hat{J}$ are $\leqq C|\operatorname{Im}(X,Y,\Theta,T)|$ at $(z,\eta,X^r,\ldots,T^r)$. If $\phi=\phi_1+i\phi_2$ then the left-hand side is equal to

$$\phi_2(X^r, Y^r, \Theta^r)+\phi_2(X^r, z, T^r)$$

and we conclude in view of Lemma 7.7.2 that near (y_0,η_0)

(25.5.9) $$|\phi_2'(X^r, Y^r, \Theta^r)|+|\phi_2'(X^r, z, T^r)|\leqq C_2(\operatorname{Im}\tilde{\phi}(z,\eta))^{\frac12}.$$

With the notation

$$f(x,y,\theta)=(\partial\phi_1(x,y,\theta)/\partial(x,\theta),\ \partial\phi_2(x,y,\theta)/\partial(x,y,\theta))$$

we have in a neighborhood of (y_0,η_0)

(25.5.10) $$|f(X^r(z,\eta), Y^r(z,\eta), \Theta^r(z,\eta))-f(X^r(z,\eta), z, T^r(z,\eta))|\leqq C_3(\operatorname{Im}\tilde{\phi}(z,\eta))^{\frac12}.$$

For the components of f involving ϕ_2 this follows from (25.5.9). By the definition of $\hat{J}$ we have $\partial\Phi/\partial x\in\hat{J}$, $\partial\Phi/\partial\theta-\partial\Phi/\partial\tau\in\hat{J}$, hence

$$|\phi_x'(X^r, Y^r, \Theta^r)-\bar{\phi}_x'(X^r, z, T^r)|+|\phi_\theta'(X^r, Y^r, \Theta^r)-\phi_\tau'(X^r, z, T^r)|\leqq C_4(\operatorname{Im}\tilde{\phi}(z,\eta))^{\frac12}.$$

This gives the estimate (25.5.10) of the components involving ϕ_1 since we have already estimated those involving ϕ_2.

The Jacobian matrix $\partial f(x,y,\theta)/\partial(y,\theta)$ is injective at (x_0,y_0,θ_0). Indeed, if $dx=0$ and $df=0$ then $d\phi_\theta'=0$ so $(dx,d\xi,dy,d\eta)\in T_\gamma(J)$ if $\xi=\phi_x'$ and $\eta=-\phi_y'$. If dy and $d\theta$ are real then $d\eta$ is also real since f contains $\partial\phi_2/\partial y$. Thus $dx=0$, $d\xi=0$, dy and $d\eta$ are real so $dy=d\eta=0$ by our hypothesis on J. This proves the injectivity since f is real. Hence Taylor's formula gives in a neighborhood of (x_0,y_0,θ_0) that

$$|y-z|+|\theta-\tau|\leqq C|f(x,y,\theta)-f(x,z,\tau)|.$$

If we combine this estimate with (25.5.10) we have proved that

(25.5.11) $$|Y^r(z,\eta)-z|+|\Theta^r(z,\eta)-T^r(z,\eta)|\leqq C_5(\operatorname{Im}\tilde{\phi}(z,\eta))^{\frac12}.$$

Set $\psi(z,\eta)=\tilde{\phi}(z,\eta)-\langle z,\eta\rangle$. To estimate the derivatives of ψ we shall use again that

$$\psi(z,\eta)+\langle z,\eta\rangle-\Phi(z,\eta,x,y,\theta,\tau)\in\hat{J}^2.$$

The first order derivatives are in $\hat{J}$, hence bounded by $C(\operatorname{Im}\tilde{\phi}(z,\eta))^{\frac12}$ at $(X^r, Y^r, \Theta^r, T^r)$. We have

$$\frac{\partial}{\partial z}(\langle z,\eta\rangle-\Phi(z,\eta,x,y,\theta,\tau))=\eta+\phi_y'(x,y,\theta)+(\bar{\phi}_z'(x,z,\tau)-\phi_y'(x,y,\theta)),$$

$$\frac{\partial}{\partial\eta}(\langle z,\eta\rangle-\Phi(z,\eta,x,y,\theta,\tau))=z-y.$$

Here $\eta+\phi'_y(x,y,\theta)\in\hat{J}$. Taking $x=X^r,\dots,\tau=T^r$ we now obtain using (25.5.9) and (25.5.11)

$$|\partial\psi(z,\eta)/\partial z|+|\partial\psi(z,\eta)/\partial\eta|\leqq C(\operatorname{Im}\psi(z,\eta))^{\frac12}$$

in a neighborhood of (y_0,η_0). In view of the homogeneity it follows that

$$|\partial\psi(z,\eta)/\partial z|^2/|\eta|+|\partial\psi(z,\eta)/\partial\eta|^2\,|\eta|\leqq C\operatorname{Im}\psi(z,\eta) \tag{25.5.12}$$

in a conic neighborhood.

(25.5.12) has the important consequence that in a conic neighborhood of (y_0,η_0)

$$|D_y^\alpha D_\eta^\beta e^{i\psi(y,\eta)}|\leqq C_\alpha|\eta|^{(|\alpha|-|\beta|)/2}e^{-\operatorname{Im}\psi(y,\eta)/2},\quad |\eta|>1. \tag{25.5.13}$$

In particular, $e^{i\psi}\in S^0_{\frac12,\frac12}$. To prove (25.5.13) we observe that $D_y^\alpha D_\eta^\beta e^{i\psi}$ is a linear combination of terms of the form

$$(D_y^{\alpha_1}D_\eta^{\beta_1}\psi)\dots(D_y^{\alpha_k}D_\eta^{\beta_k}\psi)e^{i\psi}.$$

When $|\alpha_j+\beta_j|\geqq 2$ we just use the estimate

$$|D_y^{\alpha_j}D_\eta^{\beta_j}\psi|\leqq C_{\alpha\beta}|\eta|^{1-|\beta_j|}\leqq C_\alpha|\eta|^{(|\alpha_j+\beta_j|-2|\beta_j|)/2},\quad |\eta|>1,$$

and when $|\alpha_j+\beta_j|=1$ we use that by (25.5.12)

$$|D_y^{\alpha_j}D_\eta^{\beta_j}\psi|\leqq C|\eta|^{(|\alpha_j|-|\beta_j|)/2}(\operatorname{Im}\psi)^{\frac12}.$$

Since $(\operatorname{Im}\psi)^N e^{-(\operatorname{Im}\psi)/2}$ is bounded for every N, the estimate (25.5.13) follows. (Conversely, it is easy to see that (25.5.13) implies (25.5.12).)

Now we have proved that $A^*A\in\operatorname{Op}(S^0_{\frac12,\frac12})$. Hence A^*A is L^2 continuous by Theorem 18.6.3, and this proves that A is L^2 continuous (see also the proof of Theorem 25.3.1).

Notes

Operators of the form now called Fourier integral operators were introduced by Lax [3] to study the singularities of solutions of hyperbolic differential equations. The study was purely local but some global considerations were added by Ludwig [1]. The constructions of Lax were taken up again in Hörmander [22] in order to prove equally precise results on the spectral function of higher order elliptic operators as those based on the Hadamard construction in Chapter XVII. A systematic discussion of a global theory was given in Hörmander [26] after an announcement in [26a]. As emphasized by Maslov the theory has much in common with his canonical operators. The interested reader might consult Maslov [1] to explore this relationship.

The definition in Hörmander [26] of Lagrangian distributions - called Fourier integral distributions there - was based on representations with non-degenerate phase functions. Following a suggestion by Melrose (cf. Melrose [1]) we have chosen a different definition which is obviously global and invariant. It is of course equivalent to the one using phase functions. Following Duistermaat-Guillemin [1] we have also allowed clean phase functions here. Section 25.2 follows Hörmander [26] and so does Section 25.3 essentially apart from Theorem 25.3.10 which comes from Taylor [3].

Positive Lagrangians and corresponding distributions were introduced by Melin-Sjöstrand [1]. We have followed their paper in Section 25.4 but changed the definitions by introducing Lagrangian ideals instead of almost analytic continuations. Section 25.5 contains the discussion of L^2 continuity in Melin-Sjöstrand [2] with some improvements from Hörmander [43] where precise estimates of norms can also be found.

Chapter XXVI. Pseudo-Differential Operators of Principal Type

Summary

In Section 10.4 we saw that the strength of a differential operator with constant coefficients in $\mathbb{R}^n$ is determined by the principal part p if and only if $p=0$ implies $dp \neq 0$ in $\mathbb{R}^n \setminus 0$. Such operators were said to be of principal type. The purpose of this chapter is to study general operators $P \in \Psi^m_{\text{phg}}(X)$ on a manifold X assuming that the condition $dp \neq 0$ when $p=0$ is valid in a suitably strengthened form which makes the properties of P independent of lower order terms.

At first we assume that the principal symbol p is real valued. In the constant coefficient case we know then from Section 8.3 that singularities of solutions of the equation $Pu=f$ travel along bicharacteristic curves, that is, integral curves of the Hamilton field H_p of p with $p=0$, unless they are disturbed by singularities of f. In Theorem 23.2.9 and remarks at the beginning of Section 24.2 the result was extended to second order differential operators, and in Section 26.1 it is proved for pseudo-differential operators. After P is reduced to first order by multiplication with an elliptic operator of order $1-m$ we use the homogeneous Darboux theorem in Chapter XXI to reduce p locally to a coordinate ξ_1 by a homogeneous canonical transformation χ. The calculus of Fourier integral operators in Chapter XXV then shows that conjugation of P by a suitable Fourier integral operator associated with χ reduces p microlocally to the operator D_1 for which the propagation of singularities is quite obvious. Thus we obtain the desired extension of the theorem on propagation of singularities; it is non-trivial provided that H_p is non-radial at the characteristic points which is also required for the application of the homogeneous Darboux theorem. Existence theorems for the adjoint operator on a compact subset K of X follow when K is non-trapping for bicharacteristics of p, that is, no bicharacteristic remains forever over K. When this is true for every compact subset K of X we say that P is of principal type in X; locally this just means that dp is not proportional to the canonical one form $\langle \xi, dx \rangle$ at the characteristic points. Under appropriate conditions on convexity of X with respect to the bicharacteristic flow, related to those in Section 10.8, we can also construct global two sided parametrices of P.

L. Hörmander, *Classics in Mathematics* 54
The Analysis of Linear Partial Differential Operators IV,
DOI: 10.1007/978-3-642-00117-8_2,

The situation is much more complicated when p is complex valued. This complexity is already seen in the geometry of the characteristic set $p^{-1}(0)$ which first of all may not be a manifold, and secondly may be complicated from the symplectic point of view since the rank of the symplectic form restricted to the characteristic set is variable. Two simple extreme cases are studied first. In Section 26.2 we assume that $p^{-1}(0)$ is an involutive manifold of codimension 2, thus $\{\operatorname{Re} p, \operatorname{Im} p\}=0$ when $p=0$. As in the real case we can then reduce P microlocally to the Cauchy-Riemann operator D_1+iD_2. This commutes with operators with symbol analytic in x_1+ix_2 which leads to a proof that if $Pu\in C^\infty$ then the regularity function

$$s_u^*(x,\xi)=\sup\{s; u\in H_{(s)} \text{ at } (x,\xi)\}$$

is superharmonic in the leaves of the foliation of the involutive manifold $p^{-1}(0)$. (These have a natural analytic structure defined by the complex tangent vector field H_p; the solutions of $H_p w=0$ are the analytic functions.)

In Section 26.3 we study the opposite extreme case where $\{\operatorname{Re} p, \operatorname{Im} p\} \neq 0$ which implies that $p^{-1}(0)$ is a symplectic manifold of codimension 2. A famous example is the Lewy operator

$$P=D_1+iD_2+i(x_1+ix_2)D_3$$

in $\mathbb{R}^3$. It appears as the tangential Cauchy-Riemann operator on the boundary of the strictly pseudo-convex domain

$$\Omega=\{(z_1,z_2)\in\mathbb{C}^2; |z_1|^2+2\operatorname{Im} z_2<0\}.$$

In fact, $\partial/\partial\bar z_1+a\partial/\partial\bar z_2$ is tangential to $\partial\Omega$ if and only if on $\partial\Omega$

$$0=(\partial/\partial\bar z_1+a\partial/\partial\bar z_2)(z_1\bar z_1-iz_2+i\bar z_2)=z_1+ai.$$

Writing $z_1=x_1+ix_2$, $z_2=x_3+ix_4$ and taking x_1,x_2,x_3 as parameters in $\partial\Omega$ we obtain the Lewy operator multiplied by $\frac{1}{2}i$. The fact that Ω is strictly pseudo-convex implies that for any point in $\partial\Omega$ one can find U analytic in $\bar\Omega$ except at the given point. Indeed, if $a\in\mathbb{C}$ then

$$\operatorname{Re}(z_1\bar a+z_2/i-|a|^2/2)\leq\operatorname{Re}(z_1\bar a-|z_1|^2/2-|a|^2/2)\leq 0, \quad z\in\bar\Omega,$$

with strict inequality except when $z_1=a$ and $\operatorname{Im} z_2=-|a|^2/2$. Hence, if $b\in\mathbb{R}$

$$U(z)=1/(z_1\bar a+z_2/i-|a|^2/2+ib)$$

is analytic in $\bar\Omega$ except at $z_1=a$, $z_2=b-i|a|^2/2$. The boundary value u satisfies the equation $Pu=0$ and has a singularity which does not propagate. If $(x,\xi)\in WF(u)$ it is clear that $p(x,\xi)=\xi_1+i\xi_2+i(x_1+ix_2)\xi_3=0$, that is, $\xi_1=x_2\xi_3$ and $\xi_2=-x_1\xi_3$, and since u is a boundary value of a function analytic in z_2 in a lower half plane we must have $\xi_3<0$. Noting that

$$\{\operatorname{Re} p, \operatorname{Im} p\}=\{\xi_1-x_2\xi_3, \xi_2+x_1\xi_3\}=2\xi_3<0$$

we are led to the result proved in Section 26.3 that for every pseudodifferential operator P and characteristic point (x,ξ) with $\{\operatorname{Re} p, \operatorname{Im} p\}(x,\xi)<0$ one can find u with $Pu \in C^\infty$ and $WF(u)$ equal to the ray through (x, ξ). An essentially dual fact, first observed by Hans Lewy for the Lewy operator, is that the equation $Pu=f$ cannot be solved for most f if there is a characteristic point with $\{\operatorname{Re} p, \operatorname{Im} p\}>0$; in fact, it is then usually impossible to solve the equation microlocally at (x, ξ). In the proofs of these facts we shall use Fourier integral operators to reduce to the model operator $D_1+ix_1D_2$, sometimes called the Mizohata operator, which is somewhat simpler than the Lewy operator. The existence and regularity of solutions of the equation $(D_1+ix_1D_2)u=f$ can be studied quite explicitly. At the same time we discuss the equation $(D_1+ix_1^kD_2)u=f$ for every positive integer k. When k is even the properties are quite close to those of the Cauchy-Riemann operator $(k=0)$ and for all odd k we have properties similar to those of the Lewy operator.

The results of Section 26.3 suggest that solvability of the inhomogeneous equation $Pu=f$ requires that $\operatorname{Im} p$ has no sign change from $-$ to $+$ along bicharacteristics of $\operatorname{Re} p$. This condition was originally conjectured by Nirenberg and Treves and called condition (Ψ) by them. Section 26.4 is devoted to the proof of this conjecture by means of an idea of R. Moyers, after the functional analytic aspects of various notions of solvability have been discussed at some length and the condition (Ψ) has been given an appropriate global form invariant under multiplication by non-vanishing functions.

It is still unknown if condition (Ψ) is sufficient for solvability. From Section 26.5 on we therefore assume the stronger condition (P) which rules out all sign changes of $\operatorname{Im} p$ on bicharacteristics of $\operatorname{Re} p$. (For differential operators this is equivalent to (Ψ).) Condition (P) leads to considerably simplified properties of the characteristic set discussed in Section 26.5. The main point is that the flowout along $H_{\operatorname{Re} p}$ and $H_{\operatorname{Im} p}$ of the set of characteristic points with $d\operatorname{Re} p$ and $d\operatorname{Im} p$ linearly independent is an involutive manifold N_2^e of codimension 2. Thus N_2^e is foliated by two dimensional leaves where a degenerate Cauchy-Riemann equation is defined by the Hamilton field H_p. The propagation of singularities along bicharacteristics of $\operatorname{Re} p$ which leave the characteristic set at some time is discussed in Section 26.6 by means of energy integral estimates. Similar estimates are the basis of the study in Section 26.7 of degenerate Cauchy-Riemann equations

$$(D_1+ia(x)D_2)u=f$$

with $a \geqq 0$, which implies condition (P). The results show in particular that there is an analytic structure in the leaves B of N_2^e, or rather in the sets $\tilde{B}$ obtained by collapsing to a point every embedded one dimensional bicharacteristic curve, that is, any curve where H_p is proportional to the tangent. In Section 26.9 we show that with this structure the superharmonicity of the regularity function s_u^* proved in Section 26.2 for the non-degenerate case remains valid in N_2^e. An essential ingredient in the proof is another version of the energy integral estimates, due to Nirenberg and Treves, which is given in Section 26.8. This estimate together with the advanced calculus of pseudo-differential operators in Section 18.5 leads also to the proof in Section 26.10 that when $Pu\in C^\infty$ then s_u^* is quasi-concave on any one dimensional bicharacteristic, that is, the minimum in any interval is taken at an end point.

All the results on singularities established in Sections 26.6–26.10 are combined in Section 26.11 to an existence theorem for a pseudo-differential operator P satisfying condition (P). It states that if no complete one or two dimensional bicharacteristic is trapped over the compact set K then the equation $Pu=f$ can be solved in a neighborhood of K for any f which is orthogonal to the finite dimensional space of solutions $v\in C_0^\infty(K)$ of the equation $P^*v=0$. When no bicharacteristic is trapped over a compact subset of X, we say that P is of principal type in X and have semi-global existence theorems for arbitrary lower order terms.

26.1. Operators with Real Principal Symbols

It was proved in Section 8.3 that the singularities of solutions of differential equations with constant coefficients and real principal part propagate along the bicharacteristics. We shall now show how the symplectic geometry and operator theory developed in Chapters XXI and XXV allow one to extend the result to variable coefficients. In doing so we shall start from scratch and do not rely on the results of Section 8.3.

Theorem 26.1.1. *Let X be a C^∞ manifold and let $P\in\Psi^m(X)$ be properly supported and have a principal symbol p which is real and homogeneous of degree m. If $u\in\mathscr{D}'(X)$ and $Pu=f$, it follows that $WF(u)\setminus WF(f)$ is contained in $\operatorname{Char}(P)=p^{-1}(0)$ and is invariant under the flow defined there by the Hamilton vector field H_p.*

By Theorem 18.1.28 we have

$$WF(u)\subset WF(f)\cup\operatorname{Char}(P)$$

so only the invariance under the Hamilton flow has to be proved. At a point where $H_p=0$ or H_p has the radial direction this invariance is also obvious, so in the proof we may assume that H_p and the radial direction are linearly independent. We shall prove the theorem by reducing it to the special case $P=D_1$ in $\mathbb{R}^n$ where it follows by explicit solution of the equation $Pu=f$. The study of this special case as well as the reduction will at the same time prepare for the construction of a parametrix later on in this section, so we shall also include some material which will be required then.

By E_1^+ and E_1^- we denote the forward and backward fundamental solutions of the operator D_1, the kernels of which are defined by

$$E_1^+=iH(x_1-y_1)\otimes\delta(x'-y'),\quad E_1^-=-iH(y_1-x_1)\otimes\delta(x'-y').$$

Here H is the Heaviside function, $H(t)=1$ for $t>0$ and $H(t)=0$ for $t<0$, and we have used the notation $x=(x_1, x')$ and $y=(y_1, y')$ for points in $\mathbb{R}^n$. Note that $E_1^+ - E_1^- = i\delta(x'-y')$ or, in Fourier integral form,

$$(E_1^+ - E_1^-)(x, y) = (2\pi)^{-(n-1)} \int e^{i\langle x'-y', \theta\rangle} i\, d\theta. \tag{26.1.1}$$

This is a conormal distribution with respect to $\{(x, y); x'=y'\}$, and the order is $-\frac{1}{2}$ since there are $n-1$ phase variables and $(2n-2(n-1))/4=\frac{1}{2}$. Thus we have $E_1^+ - E_1^- \in I^{-\frac{1}{2}}(\mathbb{R}^n \times \mathbb{R}^n, C_1')$ where

$$C_1 = \{(x, \xi, y, \eta);\ x'=y', \xi'=\eta' \neq 0, \xi_1=\eta_1=0\} \tag{26.1.2}$$

is the corresponding canonical relation. It follows that $\chi E_1^\pm$ belongs to $I^{-\frac{1}{2}}(\mathbb{R}^n \times \mathbb{R}^n, C_1')$ if $\chi \in C^\infty(\mathbb{R}^n \times \mathbb{R}^n)$ vanishes in a neighborhood of the diagonal, for if $x \neq y$ then either E_1^+ or E_1^- vanishes in a neighborhood of (x, y). In particular, we conclude that $WF'(E_1^\pm)$ is contained in C_1 except over the diagonal in $\mathbb{R}^n \times \mathbb{R}^n$. Since $(D_{x_j}+D_{y_j})E_1^\pm = 0$ for $j=1, \ldots, n$ we have $\xi=\eta$ in $WF'(E_1^\pm)$ (see also (8.2.15)), and

$$WF'(E_1^\pm) \supset WF'(D_{x_1} E_1^\pm) = WF'(\delta(x-y)).$$

The right hand side is the diagonal in $(T^*(\mathbb{R}^n)\setminus 0) \times (T^*(\mathbb{R}^n)\setminus 0)$ (Theorem 8.1.5) so we have proved

Proposition 26.1.2. *Let E_1^+ and E_1^- be the forward and the backward fundamental solutions of $D_1 = -i\partial/\partial x_1$ in $\mathbb{R}^n$. Then we have*

(i) *$WF'(E_1^\pm)$ is the union of the diagonal in $(T^*(\mathbb{R}^n)\setminus 0) \times (T^*(\mathbb{R}^n)\setminus 0)$ and the part of the canonical relation C_1 defined by (26.1.2) where $x_1 \gtrless y_1$.*

(ii) *$E_1^+ - E_1^- \in I^{-\frac{1}{2}}(\mathbb{R}^n \times \mathbb{R}^n, C_1')$, and $\chi E_1^\pm \in I^{-\frac{1}{2}}(\mathbb{R}^n \times \mathbb{R}^n, C_1')$ if χ is in $C^\infty(\mathbb{R}^n \times \mathbb{R}^n)$ and vanishes near the diagonal.*

The statement (i) is a more elementary analogue of Theorem 8.3.7 for the operator D_1. It is all that is needed to prove Theorem 26.1.1 for $P=D_1$ by repeating the proof of Theorem 8.3.3; this is left for the reader to do.

In the general proof of Theorem 26.1.1 we may assume that P is a first order operator, for if Q is an elliptic pseudo-differential operator with positive principal part, homogeneous of degree $1-m$, then $Pu=f$ implies $(QP)u=Qf$ where QP has the same characteristics and bicharacteristics as P, and $WF(Qf)=WF(f)$. As already pointed out it is also sufficient to consider characteristics (x_0, ξ_0) of P where H_p does not have the radial direction. This makes Theorem 21.3.1 applicable so we can find a homogeneous canonical transformation χ from an open conic neighborhood of $(0, \varepsilon_n) \in T^*(\mathbb{R}^n)\setminus 0$ to an open conic neighborhood of (x_0, ξ_0) such that $\chi^* p = \xi_1$. This geometrical construction can be lifted to the operator level:

Proposition 26.1.3. *Let $P \in \Psi^1(X)$ have real and homogeneous principal part p, let $p(x_0, \xi_0)=0$ and assume that the Hamilton field H_p at (x_0, ξ_0) and the radial direction are linearly independent. Let χ be any homogeneous canonical* *transformation from an open conic neighborhood of* $(0, \varepsilon_n) \in T^*(\mathbb{R}^n) \setminus 0$ *to a conic neighborhood of* (x_0, ξ_0) *such that* $\chi^* p = \xi_1$. *For any* $\mu \in \mathbb{R}$ *one can then find properly supported Fourier integral operators* $A \in I^\mu(X \times \mathbb{R}^n, \Gamma')$ *and* $B \in I^{-\mu}(\mathbb{R}^n \times X, (\Gamma^{-1})')$, *where* Γ *is the graph of* χ, *such that*

(i) $WF'(A)$ *and* $WF'(B)$ *are in small conic neighborhoods of* $(x_0, \xi_0, 0, \varepsilon_n)$ *and* $(0, \varepsilon_n, x_0, \xi_0)$ *respectively.*

(ii) $(x_0, \xi_0, x_0, \xi_0) \notin WF'(AB - I)$; $(0, \varepsilon_n, 0, \varepsilon_n) \notin WF'(BA - I)$.

(iii) $(x_0, \xi_0, x_0, \xi_0) \notin WF'(AD_1 B - P)$; $(0, \varepsilon_n, 0, \varepsilon_n) \notin WF'(BPA - D_1)$.

Thus D_1 *and* P *are microlocally conjugate to each other.*

Proof. Choose any $A_1 \in I^\mu(X \times \mathbb{R}^n, \Gamma')$ such that $WF'(A_1)$ is close to $(x_0, \xi_0, 0, \varepsilon_n)$ and A_1 is non-characteristic there. As observed after Definition 25.3.4 we can then choose $B_1 \in I^{-\mu}(\mathbb{R}^n \times X, (\Gamma^{-1})')$ so that (ii) is fulfilled. Then it follows from Theorem 25.3.5 that

$$(0, \varepsilon_n) \notin WF(B_1 P A_1 - D_1 - Q)$$

for some $Q \in \Psi^0(\mathbb{R}^n)$. We shall prove in a moment that there exist elliptic pseudo-differential operators $A_2, B_2 \in \Psi^0(\mathbb{R}^n)$ such that

$$B_2 A_2 - I \in \Psi^{-\infty}, \quad B_2(D_1 + Q) A_2 - D_1 \in \Psi^{-\infty}. \tag{26.1.3}$$

Admitting this for a moment we set $A = A_1 A_2$ and $B = B_2 B_1$. Then

$$(0, \varepsilon_n) \notin WF(B_2(B_1 A_1 - I) A_2) = WF(BA - I),$$
$$(0, \varepsilon_n) \notin WF(B_2(B_1 P A_1 - D_1 - Q) A_2) = WF(BPA - D_1)$$

which proves the second half of (ii) and (iii). The first half follows at once if we multiply left and right by A and by B.

To solve (26.1.3) we observe that by Theorem 18.1.24 one can for every elliptic A_2 of order 0 find $B_2 \in \Psi^0$ with $B_2 A_2 - I \in \Psi^{-\infty}$ and $A_2 B_2 - I \in \Psi^{-\infty}$, so (26.1.3) is equivalent to the condition $(D_1 + Q) A_2 - A_2 D_1 \in \Psi^{-\infty}$ for some elliptic A_2, that is,

$$[D_1, A_2] + Q A_2 \in \Psi^{-\infty}. \tag{26.1.3$'$}$$

If q^0 is a principal symbol of Q and a^0 is a principal symbol of A_2, then the principal symbol of (26.1.3)$'$ vanishes if

$$i^{-1}\{\xi_1, a^0\} + q^0 a^0 = 0,$$

that is, $\partial a^0 / \partial x_1 = -i q^0 a^0$. This equation is solved by

$$a^0(x, \xi) = \exp\left(-i \int_0^{x_1} q^0(t, x', \xi)\, dt\right)$$

which is an element of S^0 by Lemma 18.1.10. Choosing A^0 with principal symbol a^0 we can now successively choose $A^j \in \Psi^{-j}(\mathbb{R}^n)$ so that for every j

$$[D_1, A^0 + \ldots + A^j] + Q(A^0 + \ldots + A^j) = R_j \in \Psi^{-j-1}.$$

In fact, this only requires that a principal symbol a^j of A^j satisfies the equation

$$i^{-1}\partial a^j/\partial x_1 + q^0 a^j = -r^0_{j-1}$$

where r^0_{j-1} is a principal symbol of R_{j-1}. The solution

$$a^j(x,\xi) = a^0(x,\xi)\int_0^{x_1} -i r^0_{j-1}(t,x',\xi)/a^0(x',t,\xi)\,dt$$

is in S^{-j} since r^0_{j-1}/a^0 is. If the symbol of A_2 is chosen as the asymptotic sum of the symbols of $A^0, A^1, \dots$ we have satisfied (26.1.3)′.

Proof of Theorem 26.1.1. First recall that we have reduced the proof to the case $m=1$ and that the theorem has been proved for the operator D_1. So suppose that $m=1$ and let $(x_0,\xi_0)\in WF(u)\setminus WF(f)$, hence $p(x_0,\xi_0)=0$. As already pointed out we may assume that $H_p(x_0,\xi_0)$ and the radial direction are linearly independent. We then choose A, B according to Proposition 26.1.3 and set $v=Bu\in\mathscr{D}'(\mathbb{R}^n)$. Since

$$D_1 v = (D_1 - BPA)Bu + BP(AB-I)u + Bf$$

it follows from (ii) and (iii) in Proposition 26.1.3 that $(0,\varepsilon_n)\notin WF(D_1 v)$. On the other hand, $(0,\varepsilon_n)\in WF(v)$ since $(x_0,\xi_0)\in WF(u)$ and

$$u=(I-AB)u+Av, \qquad (x_0,\xi_0)\notin WF((I-AB)u).$$

Thus $(x_1,0,\varepsilon_n)\in WF(v)$ for small $|x_1|$, and since $WF(v)\subset\chi^{-1}WF(u)$ it follows that $WF(u)$ contains the image of this curve under χ. Now the definition of the Hamilton field is symplectically invariant so this means that $WF(u)$ contains a neighborhood of (x_0,ξ_0) on the bicharacteristic curve through (x_0,ξ_0) which completes the proof.

Theorem 26.1.1 can be given a more precise form if we take into account the $H_{(s)}$ classes of u and f. First recall that $f\in H^{loc}_{(s)}$ at (x_0,ξ_0) means that $Af\in L^2_{loc}$ for some $A\in\Psi^s$ which is non-characteristic at (x_0,ξ_0). If $f=Pu$ this means that $u\in H^{loc}_{(s+m)}$ at (x_0,ξ_0) if $(x_0,\xi_0)\notin\operatorname{Char}(P)$. The $H_{(s)}$ regularity in the characteristic set propagates along the bicharacteristics:

Theorem 26.1.4. *Let P satisfy the hypotheses in Theorem 26.1.1, let I be an interval on a bicharacteristic curve where $f=Pu$ is in $H^{loc}_{(s)}$. If $u\in H^{loc}_{(s+m-1)}$ at some point on I it follows that this is true on all of I.*

Proof. The $H_{(s)}$ continuity properties of pseudo-differential and Fourier integral operators allow us to reduce the proof to the case $m=1$ and then, using Proposition 26.1.3 as before with $\mu=-s$, to the case $P=D_1$, $s=0$ and $(x_0,\xi_0)=(0,\varepsilon_n)$. Since $E_1^{\pm}$ maps L^2_{comp} to L^2_{loc} the proof works as before in this situation if the wave front set of a distribution is replaced throughout by the set of points in $T^*(\mathbb{R}^n)\setminus 0$ where it is not in L^2.

We shall now prove that bicharacteristics do carry the singularities of some solutions provided that they do not close on the cosphere bundle.

Theorem 26.1.5. *Assume that $P\in\Psi^m(X)$ is properly supported and has a real principal part p which is homogeneous of degree m. Let I be a compact interval on a bicharacteristic of p which has an injective projection to the cosphere bundle of X, let Γ be the cone generated by I in $T^*(X)\setminus 0$ and let Γ' be the cone generated by the end points of I. For any $s\in\mathbb{R}$ one can then find $u\in\mathscr{D}'(X)$ so that $u\in H^{\mathrm{loc}}_{(t)}(X)$ for every $t<s$ and*

$$WF(Pu)=\Gamma',\quad WF(u)=\Gamma,\quad u\notin H^{\mathrm{loc}}_{(s)} \text{ at } (x,\xi) \text{ if } (x,\xi)\in I.$$

If $X=\mathbb{R}^n$, $P=D_1$, $I=\{(x_1,0,\varepsilon_n), x_1\in\mathbb{R}\}$ then we can simply take

$$u(x)=(x_2^2+\ldots+x_{n-1}^2+0-ix_n)^{-n/4}.$$

Since the measure of $\{x'\in\mathbb{R}^{n-1}, |x_2^2+\ldots+x_{n-1}^2-ix_n|\leq t\}$ is $Ct^{(n-2)/2+1}=Ct^{n/2}$ for reasons of homogeneity, and $\int_0^1 t^{-a}d(t^{n/2})<\infty$ if and only if $a<n/2$, we have $u\in L^p_{\mathrm{loc}}$ if and only if $p<2$. Hence it follows from Theorem 7.1.13 that $\widehat{(\phi u)}\in L^q$ for every $q>2$ if $\phi\in C_0^\infty$, so $u\in H^{\mathrm{loc}}_{(t)}$ if $t<0$. It is clear that $D_1u=0$, and $WF(u)\subset\{(x,t\varepsilon_n), x'=0, t>0\}$ by Theorem 8.1.6. Since the projection sing supp u of $WF(u)$ in $\mathbb{R}^n$ is equal to the x_1 axis this inclusion is an equality and u is not in L^2 at any point on I.

If as in Theorem 26.1.5 we have a finite interval $I=\{(x_1,0,\varepsilon_n); a\leq x_1\leq b\}$ we shall cut off the function u at a and b with some care so that the wave front set does not grow. To do so we choose functions $\psi_j\in C_0^\infty((a,b)\times\mathbb{R}^{n-1})$ with $\sum_{-\infty}^{\infty}\psi_j=1$ in a neighborhood of $(a,b)\times\{0\}$ and supp $\psi_j\to\{a\}$ resp. $\{b\}$ as $j\to-\infty$ resp. $+\infty$. We can choose a regularization v_j of $u_j=\psi_j u$ such that if $U_j=u_j-v_j$ then supp $U_j\to\{a\}$ or $\{b\}$ as $j\to+\infty$ or $-\infty$, $\|U_j\|_{(-1/|j|)}\leq 2^{-|j|}$ and

$$|\hat{U}_j(\xi)|\leq(1+|\xi|)^{-|j|}\quad\text{when } |\xi_1|+\ldots+|\xi_{n-1}|\geq\xi_n/|j|.$$

In fact, $\hat{U}_j(\xi)=\hat{u}_j(\xi)(1-\hat{\chi}(\delta_j\xi))$ where $\chi\in C_0^\infty$, $\hat{\chi}(0)=1$. We have

$$\int|\hat{u}_j(\xi)|^2(1+|\xi|)^t\,d\xi<\infty,\quad t<0,$$

and $|\hat{u}_j(\xi)|(1+|\xi|)^{|j|}\to 0$ at ∞ outside any conic neighborhood of ε_n, so we just have to take δ_j small enough. Now we obtain $U=\sum U_j\in H_{(t)}$ for every $t<0$, $WF(U)\subset I$ and $U-u\in C^\infty$ at $(x_1,0)$ if $a<x_1<b$, so U is not in $H_{(0)}$ at any point on I. Since D_1U is only singular at $(a,0)$ and $(b,0)$ and since $WF(D_1U)\subset WF(U)=I$, it follows that U has the properties required in Theorem 26.1.5.

To prove Theorem 26.1.5 in general we need a global version of Theorem 21.3.1 and of Proposition 26.1.3 allowing us to conjugate P to D_1.

Proposition 26.1.6. *Let X be a C^∞ manifold and p a real valued C^∞ function in $T^*(X)\smallsetminus 0$ which is homogeneous of degree 1. Let I be a compact interval on $\mathbb{R}$ and $\gamma: I\to T^*(X)\smallsetminus 0$ a bicharacteristic, thus*

$$p\circ\gamma=0, \quad \gamma'=H_p\circ\gamma.$$

We assume that the composition of γ and the projection $\pi: T^*(X)\smallsetminus 0\to S^*(X)$ on the cosphere bundle is injective. Then one can find a conic neighborhood V of $\{(x_1, 0, \varepsilon_n);\ x_1\in I\}$ and a C^∞ homogeneous canonical transformation χ from V to an open conic neighborhood $\chi(V)\subset T^*(X)\smallsetminus 0$ of $\gamma(I)$ such that $\chi(x_1, 0, \varepsilon_n)=\gamma(x_1)$ and $\chi^* p=\xi_1$.

Proof. Assume to simplify notation that $0\in I$. We can use Theorem 21.3.1 to find a homogeneous canonical transformation χ from a convex conic neighborhood V_0 of $(0,\varepsilon_n)$ to a conic neighborhood of $\gamma(0)$ such that $\chi^* p=\xi_1$ and $\chi(0,\varepsilon_n)=\gamma(0)$. Then χ_* maps the Hamilton field $\partial/\partial x_1$ of ξ_1 to H_p, so

$$\partial\chi(x,\xi)/\partial x_1=H_p(\chi(x,\xi)). \tag{26.1.4}$$

When $x'=0$ and $\xi=\varepsilon_n$ we also have the solution $\gamma(x_1)$, $x_1\in I$, with the same initial value when $x_1=0$. Hence we can uniquely extend χ to a conic neighborhood V of $I\times\{(0,\varepsilon_n)\}$, which is convex in the x_1 direction, so that (26.1.4) remains valid. The projected curves $x_1\mapsto\pi\chi(x,\xi)$ are the integral curves of the vector field induced by H_p on $S^*(X)$. (Functions on $S^*(X)$ can be identified with homogeneous functions f of degree 0 on $T^*(X)\smallsetminus 0$, and $H_p f=\{p,f\}$ is then also homogeneous of degree 0.) Since χ is also homogeneous it follows from the hypothesis on $\pi\circ\gamma$ that χ is a diffeomorphism if V is small enough. If we write

$$\chi^{-1}=(X_1,\dots,X_n,\Xi_1,\dots,\Xi_n),$$

then the fact that χ_*^{-1} maps H_p to $\partial/\partial x_1$ means that

$$H_pX_1=1, \quad H_pX_j=0 \ \text{ if } j>1, \quad H_p\Xi_k=0 \ \text{ for all } k.$$

Hence the Poisson brackets $\{X_i,X_j\}$, $\{X_i,\Xi_k\}$, $\{\Xi_k,\Xi_l\}$ are constant along the orbits of the Hamilton field H_p, by the Jacobi identity. They vanish at some point since we started from a canonical transformation, so they vanish identically, which proves that also the extended map χ is canonical.

The following extension of Proposition 26.1.3 follows with the same proof:

Proposition 26.1.3′. *Let $P\in\Psi^1(X)$ have real and homogeneous principal part p, and let $\gamma: I\to T^*(X)\smallsetminus 0$ be a bicharacteristic with the properties assumed in Proposition 26.1.6. If Γ is the graph of a canonical transformation χ from a* *conic neighborhood of $J=I\times(0,\varepsilon_n)$ to a conic neighborhood of $\gamma(I)$, satisfying the conclusion in Proposition 26.1.6, then one can for any $\mu\in\mathbb{R}$ find properly supported Fourier integral operators $A\in I^{\mu}(X\times\mathbb{R}^n,\Gamma')$ and $B\in I^{-\mu}(\mathbb{R}^n\times X,(\Gamma^{-1})')$ such that*

(i) *$WF'(A)$ and $WF'(B)$ lie in small conic neighborhoods of the graph of χ restricted to J and its inverse respectively,*

(ii) $\gamma(I)\cap WF(AB-I)=\emptyset$, $J\cap WF(BA-I)=\emptyset$.

(iii) $\gamma(I)\cap WF(AD_1B-P)=\emptyset$, $J\cap WF(BPA-D_1)=\emptyset$.

Proof of Theorem 26.1.5. We may again assume in the proof that $m=1$. Changing notation in Theorem 26.1.5 so that I is replaced by $\gamma(I)$, $I\subset\mathbb{R}$, we have precisely the situation in Proposition 26.1.3′. Choose A and B according to Proposition 26.1.3′ with $\mu=-s$. We have already constructed a distribution U in $\mathbb{R}^n$ with $WF(U)$ generated by J, $WF(D_1U)$ generated by the end points of J, $U\in H^{\mathrm{loc}}_{(t)}$ for every $t<0$ and $U\notin H^{\mathrm{loc}}_{(0)}$ at (x,ξ) for every $(x,\xi)\in J$. If we set $u=AU$ then $U\equiv Bu \mod C^\infty$, $Pu=PAU\equiv ABPAU \equiv AD_1U \mod C^\infty$, so u has the required properties.

We shall now discuss existence theorems for the equation $Pu=f$ which follow from Theorems 26.1.4 and 26.1.5 applied to the adjoint P^* combined with abstract functional analysis. At first we shall only consider solvability on compact sets. All operators will tacitly be assumed to act on half densities so that the adjoints are well defined and of the same kind.

Theorem 26.1.7. *Assume that $P\in\Psi^m(X)$ is properly supported and has a real principal part p which is homogeneous of degree m. Let K be a compact subset of X such that no complete bicharacteristic curve is contained in K. Then it follows that*

$$N(K)=\{v\in\mathscr{E}'(K),\ P^*v=0\}$$

is a finite dimensional subspace of $C_0^\infty(K)$ orthogonal to $P\mathscr{D}'(X)$. If $f\in H^{\mathrm{loc}}_{(s)}(X)$ for some $s\in\mathbb{R}$ (resp. $f\in C^\infty(X)$) and if f is orthogonal to $N(K)$, then one can find $u\in H^{\mathrm{loc}}_{(s+m-1)}(X)$ (resp. $u\in C^\infty(X)$) so that $Pu=f$ in a neighborhood of K.

Proof. The principal part of P^* is also p. Hence $N(K)\subset C^\infty$ by Theorem 26.1.1, for if $v\in N(K)$ and $(x,\xi)\in WF(v)$ then the bicharacteristic starting at (x,ξ) would have to remain over K. By the closed graph theorem the L^2 topology in $N(K)$ is equivalent to the C^∞ topology, which shows that the unit ball in the L^2 topology is compact. Thus $\dim N(K)<\infty$.

The hypotheses of the theorem are also fulfilled if K is replaced by a sufficiently small compact neighborhood K'. To prove this we may assume that $m=1$ and can then consider the bicharacteristics as curves in the cosphere bundle. Since this is compact over K', we would obtain a bicharacteristic staying over K for all values of the parameter if there is one over K'

for every compact neighborhood K' of K. This proves the assertion. Since $\dim N(K')$ decreases with K' and is finite, it is also clear that $N(K')=N(K)$ if K' is sufficiently close to K.

Let $\|\ \|_{(t)}$ denote a norm which defines the $H_{(t)}$ topology for distributions with support in an arbitrary fixed compact subset of X. Since $v\in\mathscr{E}'(K)$, $P^*v\in H_{(t)}$ implies $v\in H_{(t+m-1)}$ by Theorem 26.1.4, it follows from the closed graph theorem that

$$\|v\|_{(t+m-1)}\leqq C(\|P^*v\|_{(t)}+\|v\|_{(t+m-2)}),\quad v\in C_0^\infty(K). \tag{26.1.5}$$

Let V be a supplementary space of $N(K)$ in $H_{(t+m-1)}\cap\mathscr{E}'(K)$. Then there is another constant C_1 such that

$$\|v\|_{(t+m-1)}\leqq C_1\|P^*v\|_{(t)},\quad v\in V\cap C_0^\infty(K). \tag{26.1.6}$$

In fact, if this were false we could select a sequence $v_j\in V$ with

$$\|v_j\|_{(t+m-1)}=1,\quad \|P^*v_j\|_{(t)}\to 0.$$

A weakly convergent subsequence must converge strongly in $H_{(t+m-2)}$ to a limit $v\in V$ with $P^*v=0$ and $1\leqq C\|v\|_{(t+m-2)}$, by (26.1.5). Hence v is a non-zero element of $N(K)$ belonging to V, which is a contradiction.

If $f\in H_{(s)}^{\mathrm{loc}}(X)$ is orthogonal to $N(K)$ we set $t=1-m-s$ and have by (26.1.6) for some C

$$|(f,v)|\leqq C\|P^*v\|_{(t)},\quad v\in C_0^\infty(K), \tag{26.1.7}$$

for this is true if $v\in V\cap C_0^\infty(K)$ and neither side changes if an element of $N(K)$ is added to v. By the Hahn-Banach theorem it follows that the anti-linear form $P^*v\mapsto(f,v)$, $v\in C_0^\infty(K)$, can be extended to an anti-linear form on $H_{(t)}^{\mathrm{comp}}$ which is continuous for $\|\ \|_{(t)}$. Thus there is a distribution $u\in H_{(-t)}^{\mathrm{loc}}=H_{(s+m-1)}^{\mathrm{loc}}$ such that

$$(f,v)=(u,P^*v),\quad v\in C_0^\infty(K),$$

which implies that $Pu=f$ in the interior of K. If we apply this conclusion to a suitable neighborhood K' of K, we obtain $Pu=f$ in a neighborhood of K.

To prove the C^∞ case of the theorem we denote by $C^\infty(K)$ the quotient of $C^\infty(X)$ by the subspace of functions vanishing of infinite order on K. The dual space of this Fréchet space is $\mathscr{E}'(K)$ (Theorem 2.3.3). To show that the range of the map $C^\infty(X)\to C^\infty(K)$ defined by P is the orthogonal space of $N(K)$ we have to show that $P^*\mathscr{E}'(K)$ is weakly closed in $\mathscr{E}'(X)$, or equivalently that the intersection of $P^*\mathscr{E}'(K)$ and the unit ball in $H_{(t)}\cap\mathscr{E}'(K_1)$ is weakly closed for every real t and compact $K_1\subset X$. (See Lemmas 16.5.8 and 16.5.9.) Now $v\in\mathscr{E}'(K)$, $P^*v\in H_{(t)}$ implies $v\in H_{(t+m-1)}$ by Theorem 26.1.4, and by (26.1.6) we have $v=v_1+v_2$ where $v_1\in N(K)$ and $\|v_2\|_{(t+m-1)}\leqq C$. Since the set of such $v_2\in\mathscr{E}'(K)$ is weakly compact and $P^*v=P^*v_2$, the assertion is proved.

Remark 1. When K consists of a point x_0 we conclude that for every $f \in C^\infty$ one can choose $u \in C^\infty$ so that $Pu = f$ in a neighborhood of x_0, provided that H_p does not have the radial direction at any characteristic point (x_0, ξ).

Remark 2. The condition on the bicharacteristics made in Theorem 26.1.7 is merely sufficient and in no way necessary for the conclusion to be valid. For example, if P is a differential operator with constant coefficients our assumption means that P is of principal type (Definition 10.4.11) but the conclusion is always valid in the C^∞ case and holds in the $H_{(s)}$ spaces also for example if P is the heat operator, which has multiple characteristics. Even when the characteristics are simple the condition is not necessary in the variable coefficient case. For example, the conclusions of Theorem 26.1.7 are valid for

$$P = x_2 \partial/\partial x_1 - x_1 \partial/\partial x_2 + c$$

in $X = \{(x_1, x_2);\ 1 < x_1^2 + x_2^2 < 2\}$ if c is a real constant $\neq 0$, although (the normals of) the circles $x_1^2 + x_2^2 = r^2$ are bicharacteristics. Thus the lower order terms may in general be essential. However, they are irrelevant when the hypotheses of Theorem 26.1.7 are fulfilled, and just as in Definition 10.4.11 we introduce a terminology which refers to this fact:

Definition 26.1.8. Let $P \in \Psi^m(X)$ be a properly supported pseudo-differential operator. We shall say that P is of real principal type in X if P has a real homogeneous principal part p of order m and no complete bicharacteristic strip of P stays over a compact set in X.

We shall now discuss global solvability of the equation $Pu = f$ modulo C^∞. The results should be compared with Sections 10.6 and 10.7 in the constant coefficient case.

Theorem 26.1.9. *Let P be of real principal type in the manifold X. Then the following conditions are equivalent:*

(a) *P defines a surjective map from $\mathscr{D}'(X)$ to $\mathscr{D}'(X)/C^\infty(X)$.*

(b) *For every compact set $K \subset X$ there is another compact set $K' \subset X$ such that*

$$u \in \mathscr{E}'(X), \quad \operatorname{sing\,supp} P^* u \subset K \Rightarrow \operatorname{sing\,supp} u \subset K'.$$

(c) *For every compact set $K \subset X$ there is another compact set $K' \subset X$ such that every bicharacteristic interval with respect to P having endpoints over K must lie entirely over K'.*

Proof. (b) $\Rightarrow$ (c) with the same K' by Theorem 26.1.5. Using Theorem 26.1.1 we shall also prove that (c) $\Rightarrow$ (b). In doing so we may assume that P is of order 1 since we can multiply P by an elliptic operator of order $1-m$ without affecting these conditions. When the degree is 1 the bicharacteristic strips can be considered as integral curves of a vector field on the cosphere bundle which is an advantage since the fibers are then compact.

Assuming that (c) is valid, that $u\in\mathscr{E}'(X)$, $\operatorname{sing\,supp} P^*u\subset K$, $(x,\xi)\in WF(u)$, we shall show that there is a contradiction if $x\notin K'$. By Theorem 26.1.1 the bicharacteristic through (x,ξ) stays in $WF(u)$ until it reaches a point lying over K. In view of (c) and the assumption that $x\notin K'$ at least one half ray γ of the bicharacteristic starting at (x,ξ) contains no point above K, so $\gamma\subset WF(u)$. Choose (x_0,ξ_0) so that its projection in the cosphere bundle is a limit point of γ at infinity, which is possible since γ lies over the compact set $\operatorname{supp} u$. Then the entire bicharacteristic strip with initial data (x_0,ξ_0) must stay over $\operatorname{supp} u$, which contradicts the hypothesis that P is of principal type.

Since P is of principal type we know that $u\in C^\infty$ if $u\in\mathscr{E}'$ and $Pu\in C^\infty$. Combined with the purely functional analytic arguments in the proof of Theorem 10.7.8 this gives that (b) $\Rightarrow$ (a).

It remains to show that (a) $\Rightarrow$ (c). Assume that (c) is not valid. For some compact set $K\subset X$ we can then find a sequence of compact intervals $I_1, I_2, \dots$ on bicharacteristic strips with end points lying over K and points $(x_j,\xi_j)\in I_j$ with $x_j\to\infty$ in X, that is, only finitely many contained in any compact subset. We may assume that the intervals I_j are disjoint even when considered in the cosphere bundle. Let (y_j,η_j) be one end point of I_j and let Γ_j be the cone $\subset T^*(X)\setminus 0$ generated by the bicharacteristic between (y_j,η_j) and (x_j,ξ_j) while Γ_j' consists of the rays through these points. Now use Theorem 26.1.5 to determine $u_j\in\mathscr{E}'(X)$ such that

$$WF(u_j)=\Gamma_j,\quad WF(Pu_j)=\Gamma_j',\quad u_j\notin H_{(-j)} \text{ at any point in } \Gamma_j.$$

We can write $Pu_j=f_j+g_j$ where $WF(f_j)$ and $WF(g_j)$ are the rays through (x_j,ξ_j) and (y_j,η_j) respectively. In doing so we can take the support of f_j so close to x_j that the supports of the distributions f_j are locally finite. We can then form

$$f=\sum f_j.$$

Now we shall prove that $Pu-f$ is not in C^∞ for any $u\in\mathscr{D}'(X)$, which means that (a) is not valid. To do so we choose s so large negative that $u\in H_{(s)}^{\mathrm{loc}}$ in a neighborhood of K. When $-j\leqq s$ it follows that $u-u_j$ is not in $H_{(s)}^{\mathrm{loc}}$ at any point on I_j close to (y_j,η_j) whereas $u-u_j\in H_{(s)}^{\mathrm{loc}}$ at the other end point of I_j. By Theorem 26.1.4 this shows since $m=1$ that $P(u-u_j)$ is not in $H_{(s)}$ at every point in the interior of I_j. However,

$$P(u-u_j)=Pu-f+\sum_{k\neq j} f_k-g_j$$

and the interior of I_j does not meet the wave front set of the sum nor that of g_j. Hence $Pu-f$ is not in $H_{(s)}$ at every point on I_j, which completes the proof.

When convexity conditions similar to those of Section 10.6 are fulfilled one can improve Theorem 26.1.9 to existence of genuine solutions. However, this does not differ very much from the discussion in Section 10.6 so we leave for the reader to contemplate such results or consult the references at the end of the chapter. Instead we shall study global parametrices for operators satisfying the condition in Theorem 26.1.9, for which it is convenient to introduce a name:

Definition 26.1.10. If P is of real principal type in X we shall say that X is pseudo-convex with respect to P when condition (c) in Theorem 26.1.9 is fulfilled.

To clarify the geometric properties of the Hamilton field on the characteristic set we need two lemmas on vector fields satisfying conditions like (c) in Theorem 26.1.9.

Lemma 26.1.11. *Let M be a C^∞ manifold and v a C^∞ vector field on M. Then the following conditions are equivalent:*

(a) *No complete integral curve of v is relatively compact, and for every compact set K in M there is another K' containing every compact interval on an integral curve of v with end points in K.*

(b) *v has no periodic integral curves, and the relation R consisting of all $(y_1, y_2)\in M\times M$ with y_1 and y_2 on the same integral curve of v is a closed C^∞ submanifold of $M\times M$.*

(c) *There exists a manifold M_0, an open neighborhood M_1 of $M_0\times 0$ in $M_0\times\mathbb{R}$ which is convex in the $\mathbb{R}$ direction, and a diffeomorphism $M\to M_1$ which carries v into the vector field $\partial/\partial t$ if points in $M_0\times\mathbb{R}$ are denoted by (y_0, t).*

Proof. Let us first show that the first part of (a) implies

(a′) No integral curve of v defined for all positive or all negative values of the parameter is relatively compact.

In fact, suppose that $\mathbb{R}_+\ni t\mapsto y(t)$ is an integral curve of v with compact closure K. Then we can find a sequence $t_j\to+\infty$ such that $x=\lim y(t_j)$ exists. Since $t\mapsto y(t_j+t)$ is an integral curve for $t\in(-t_j,\infty)$ it follows that K contains a complete relatively compact integral curve starting at x, which contradicts the first part of (a). (This argument was already used to prove that (c) $\Rightarrow$ (b) in Theorem 26.1.9.)

Next we prove that (a) $\Rightarrow$ (b). Denote the v flow by ϕ so that $t\mapsto\phi(y,t)$ is the solution of the equation $dx/dt=v(x)$ with $x(0)=y$, defined on a maximal open interval $\subset\mathbb{R}$. If D_ϕ is the domain of ϕ, then

$$R=\{(\phi(y,t),y);\ (y,t)\in D_\phi\}.$$

The map $(y,t)\mapsto(\phi(y,t),y)$ is injective since there are no closed integral curves, and it is clear that the differential is injective. To prove that R is a closed C^∞ submanifold it suffices therefore to show that the map is proper. Let $(y_j,t_j)\in D_\phi$ and assume that $y_j\to y$, $\phi(y_j,t_j)\to x$ as $j\to\infty$. We have to show that (y_j,t_j) has a limit point in D_ϕ. In doing so we may assume that $t_j\to T\in[-\infty,\infty]$. By the second part of condition (a) there is a compact set

K' such that $\phi(y_j, t) \in K'$ when $t \in [0, t_j]$. If $T = \pm\infty$ it follows that $\phi(y, s) \in K'$ for $s \geqq 0$ or for $s \leqq 0$. But this contradicts (a′) so T is finite and $(y_j, t_j) \to (y, T) \in D_\phi$.

(b) ⇒ (c). It follows from (b) that the quotient space $M_0 = M/R$ is a Hausdorff space, and identifying a neighborhood of the equivalence class of y with a manifold transversal to v at y we obtain a structure of C^∞ manifold in M_0. The map $M \to M_0$ has a C^∞ cross section $M_0 \to M$. This is obvious locally and using a partition of unity in M_0 we can piece local sections together to a global one, for only an affine structure is required to form averages. We can now take

$$M_1 = \{(y, t);\ y \in M_0,\ (y, t) \in D_\phi\}$$

and the map $M_1 \to M$ given by ϕ. Since the implication (c) ⇒ (a) is trivial, this completes the proof.

In our applications of Lemma 26.1.11 we shall have a conic manifold M and a vector field v commuting with multiplication by positive scalars as is the case for the Hamilton field of a function which is homogeneous of degree 1. Thus vu is homogeneous of degree m if u is. In particular, if M_s is the quotient of M by multiplication with $\mathbb{R}_+$, then v induces a vector field v_s on M_s, as already observed in the proof of Proposition 26.1.6.

Lemma 26.1.12. *Let M be a conic manifold and v a C^∞ vector field on M commuting with multiplication by positive scalars, such that the vector field v_s induced on M_s has the properties in Lemma 26.1.11. Then there exists a C^∞ manifold M'_0, an open neighborhood M' of $M'_0 \times 0$ in $M'_0 \times \mathbb{R}$ which is convex in the direction of $\mathbb{R}$, and a diffeomorphism $M \to M' \times \mathbb{R}_+$, commuting with multiplication by positive scalars (defined as identity in M' and standard multiplication in $\mathbb{R}_+$) such that v is mapped to the vector field $\partial/\partial t$ if (y_0, t, r) denotes the variables in $M'_0 \times \mathbb{R} \times \mathbb{R}_+$.*

Proof. First note that by a partition of unity we can construct a positive C^∞ function $r(y)$ on M which is homogeneous of degree 1. If π is the projection of M on M_s we obtain a diffeomorphism

$$M \ni y \mapsto (\pi(y),\ r(y)) \in M_s \times \mathbb{R}_+$$

commuting with multiplication by $\mathbb{R}_+$. From condition (c) in Lemma 26.1.11 applied to v_s we now obtain a diffeomorphism $M \to M' \times \mathbb{R}_+$ with M' as in that lemma, which transforms v to a vector field of the form

$$v_1 = \partial/\partial t + a(y_0, t)\, r\, \partial/\partial r$$

since it is equal to $\partial/\partial t$ for functions independent of r. Now solve the equation

$$\partial b(y_0, t)/\partial t + a(y_0, t) = 0$$

with initial condition $b=0$ when $t=0$ for example. Then $b\in C^\infty(M')$, and if $R=r\exp b(y_0,t)$ we have $v_1R=0$. If we take R as a new radial variable instead of r, there will be no term $\partial/\partial R$ in the new expression of v so the lemma is proved.

Remark 1. Under the hypotheses in Lemma 26.1.12 the vector field $(v,0)$ on $M\times M$ defines a vector field $\tilde{v}$ on the relation manifold R (see Lemma 26.1.11 (b)) which satisfies the same conditions. This is obvious when the vector field is put in the form given by Lemma 26.1.12.

Remark 2. Let v satisfy the conditions in Lemma 26.1.12 and let $c\in C^\infty(M)$ be homogeneous of degree 0. Then the equation $(v+c)u=f$ has a solution $u\in S^m(M)$ for every $f\in S^m(M)$. In fact, if $c=0$ we just have to integrate f with respect to t from $t=0$ with the coordinates given by Lemma 26.1.12. For a general c we first obtain in this way a homogeneous function C with $vC=c$, and multiplication by e^C reduces to the case $c=0$.

Let us now return to an operator $P\in\Psi^m(X)$ of real principal type, with principal symbol p, assuming that X is pseudo-convex with respect to P. Denote by N the set of zeros of p in $T^*(X)\smallsetminus 0$. This is a conic manifold, and the Hamilton field H_p is tangential to N. The integral curves are the bicharacteristics of P, and we define the bicharacteristic relation C of P by

$$C=\{((x,\xi),(y,\eta))\in N\times N;\ (x,\xi)\text{ and }(y,\eta)\text{ lie on the same bicharacteristic}\}. \tag{26.1.8}$$

The construction is invariant under the action of canonical transformations on p since the definition of the Hamilton field is. Multiplication of p by a non-vanishing function will change the parameter on the bicharacteristics but not affect C. Note that the set C_1 defined by (26.1.2) is the bicharacteristic relation of D_1.

By the preceding remarks we may assume that P is of degree 1 when studying C. By hypothesis the vector field induced by H_p on N_s satisfies condition (a) in Lemma 26.1.11 so Lemma 26.1.12 is applicable. It follows at once that C is a closed conic submanifold of $N\times N$, and since the positive homogeneous function r is constant along the bicharacteristics it is clear that C is also closed in $T^*(X\times X)\smallsetminus 0$. Since C_1 is a canonical relation, that is, the product symplectic form vanishes in C_1, it follows in view of Proposition 26.1.6 that C is a canonical relation. In fact, if $((x,\xi),(y,\eta))\in C$ we can by a canonical transformation reduce p to ξ_1 in a neighborhood of the bicharacteristic between (x,ξ) and (y,η). Thus we have proved:

Proposition 26.1.13. *Assume that P is of real principal type in X and that X is pseudo-convex with respect to P. Then the bicharacteristic relation C of P is a homogeneous canonical relation from $T^*(X)\smallsetminus 0$ to $T^*(X)\smallsetminus 0$ which is closed in $T^*(X\times X)\smallsetminus 0$.*

If Δ_N is the diagonal in N, then $C \setminus \Delta_N$ is the disjoint union $C^+ \cup C^-$ of the forward (backward) bicharacteristic relations C^+ and C^- defined as the set of all $((x, \xi), (y, \eta)) \in N \times N$ such that (x, ξ) lies after (resp. before) (y, η) on a bicharacteristic. These are open subsets of C and inverse relations. The definition is invariant under multiplication of p by positive functions but C^+ and C^- are interchanged if we multiply by a negative function. The importance of these sets is suggested by Proposition 26.1.2 which we shall now extend as follows:

Theorem 26.1.14. *Let $P \in \Psi^m(X)$ be of real principal type in X and assume that X is pseudo-convex with respect to P. Then there exist parametrices E^+ and E^- of P with*

$$WF'(E^+) = \Delta^* \cup C^+, \quad WF'(E^-) = \Delta^* \cup C^- \tag{26.1.9}$$

where Δ^* is the diagonal in $(T^*(X) \setminus 0) \times (T^*(X) \setminus 0)$. Any left or right parametrix E with $WF'(E)$ contained in $\Delta^* \cup C^+$ resp. $\Delta^* \cup C^-$ must be equal to E^+ resp. E^- modulo C^∞. For every $s \in \mathbb{R}$ the parametrices E^+ and E^- define continuous maps from $H_{(s)}^{\text{comp}}(X)$ to $H_{(s+m-1)}^{\text{loc}}(X)$. Finally

$$E^+ - E^- \in I^{\frac{1}{2}-m}(X \times X, C'), \tag{26.1.10}$$

and $E^+ - E^-$ is non-characteristic at every point of C'.

Before the proof we recall that a continuous operator $E: C_0^\infty(X) \to \mathscr{D}'(X)$ is called a right parametrix if

$$PE = I + R$$

where I is the identity and R has a C^∞ kernel. If instead $EP = I + R'$ with $R' \in C^\infty$ then E is called a left parametrix. We shall say that E is a parametrix if E is both a right and a left parametrix. Note that the theorem is an extension of Theorem 8.3.7 also.

Proof of Theorem 26.1.14. We begin with a proof of the uniqueness. Assume for example that E_1 is a right and E_2 a left parametrix with $WF'(E_j) \subset \Delta^* \cup C^+$, which implies that they map C_0^∞ to C^∞. To prove that $E_1 - E_2 \in C^\infty$ we would like to argue that $E_2 P E_1$ is congruent both to E_1 and to E_2 mod C^∞ (cf. the proof of Theorem 18.1.9), but this is in no way obvious since E_1 and E_2 are not properly supported. However, we do know that $E_2 B E_1$ is defined if B is a pseudo-differential operator with kernel of compact support in $X \times X$, for B maps $\mathscr{D}'(X)$ to $\mathscr{E}'(X)$ then. If $(x, \xi, y, \eta) \in WF'(E_2 B E_1)$ but (x, ξ) and (y, η) are both in the complement of $WF(B)$ it follows that $(x, \xi, z, \zeta) \in C^+$ and that $(z, \zeta, y, \eta) \in C^+$ for some $(z, \zeta) \in WF(B)$. This implies that (x, ξ), (y, η), (z, ζ) are on the same bicharacteristic strip, with (z, ζ) between the other points. Let K and K' be as in condition (c) in Theorem 26.1.9. If $WF(B)$ has no point over K' it follows that $WF'(E_2 B E_1)$ has no point in $K \times K$. Now choose $\phi \in C_0^\infty(X)$ equal to 1 near K' and form

$$E_2\phi PE_1-E_2P\phi E_1=E_2(\phi P-P\phi)E_1.$$

The wave front set of the right-hand side contains no point over $K\times K$, so the same is true of $E_2\phi-\phi E_1$. Since K is arbitrary it follows that $E_2-E_1\in C^\infty$.

Since $PE=I+R$ is equivalent to $E^*P^*=I+R^*$ and P^* has the same principal symbol as P, the existence of left parametrices with the properties listed in the theorem follows from the existence of right parametrices for P^*. To prove the theorem it is therefore sufficient to construct a right parametrix with the required regularity properties. In doing so we may assume that the order of P is 1, for P can otherwise be replaced by the product with an elliptic operator Q of degree $1-m$ with positive homogeneous principal symbol; Q has a pseudo-differential parametrix by Theorem 18.1.24.

The first step in the construction is local in the cotangent bundle near the diagonal.

Lemma 26.1.15. *Let $P\in\Psi^1(X)$ satisfy the hypotheses of Theorem 26.1.14 and let $(x_0,\xi_0)\in T^*(X)\smallsetminus 0$, $p(x_0,\xi_0)=0$. Choose A and B according to Proposition 26.1.3 with $\mu=0$ and set $F_1^\pm=\psi E_1^\pm$ where $\psi\in C^\infty(\mathbb{R}^{2n})$ is equal to 1 in a neighborhood of the diagonal. If ψ vanishes outside a sufficiently small neighborhood of the diagonal, $T\in\Psi^0(X)$ has its wave front set in a sufficiently small conic neighborhood of (x_0,ξ_0), and $F^\pm=AF_1^\pm BT$, then*

(i) $WF'(F^\pm)\subset\Delta^*\cup C^\pm$,

(ii) $PF^\pm=T+R^\pm$ *where* $R^\pm\in I^{-\frac12}(X\times X,C')$ *and* $WF'(R^\pm)\subset C^\pm$,

(iii) $F^+-F^-\in I^{-\frac12}(X\times X,C')$.

Proof. Conditions (i) and (iii) follow immediately from the corresponding conditions in Proposition 26.1.2. To prove (ii) we form

$$PF^\pm=PAF_1^\pm BT=(PA-AD_1)F_1^\pm BT+AD_1F_1^\pm BT. \tag{26.1.11}$$

By (iii) in Proposition 26.1.3 we have

$$(x_0,\xi_0,0,\varepsilon_n)\notin WF'(PA-AD_1)\subset\Gamma.$$

It follows that there is a conical neighborhood V of $(0,\varepsilon_n)$ such that $(PA-AD_1)v\in C^\infty$ if $WF(v)\subset V$. Since $WF'(F_1^\pm)$ can be made arbitrarily close to the diagonal in $(T^*(\mathbb{R}^n)\smallsetminus 0)\times(T^*(\mathbb{R}^n)\smallsetminus 0)$ by choosing the support of ψ close to the diagonal in $\mathbb{R}^n\times\mathbb{R}^n$, we can choose ψ and a conic neighborhood W of $(0,\varepsilon_n)$ such that $WF(F_1^\pm v)\subset V$ if $WF(v)\subset W$. If $WF(T)\subset\chi(W)$ it follows that the first term in the right-hand side of (26.1.11) is in C^∞. To study the last term in (26.1.11) we note that $D_1F_1^\pm=I+R_1^\pm$ where

$$R_1^\pm=(D_{x_1}\psi(x,y))E_1^\pm\in I_1^{-\frac12}(\mathbb{R}^n\times\mathbb{R}^n,C_1'),\quad WF'(R_1^\pm)\subset C_1^\pm.$$

Since $ABT-T=(AB-I)T\in C^\infty$ if $WF(T)$ is sufficiently close to (x_0,ξ_0), it follows that $PF^\pm=T+R^\pm$ where $R^\pm-AR_1^\pm BT\in C^\infty$, which proves (ii).

End of Proof of Theorem 26.1.14. If $(x_0, \xi_0) \in T^*(X) \setminus 0$ and $p(x_0, \xi_0) \neq 0$ then Theorem 18.1.24′ gives a stronger result than Lemma 26.1.15: we can find a pseudo-differential operator F such that $PF = T + R$ where $R \in C^\infty$ and $WF(F) = WF(T)$. Now choose a locally finite covering $\{V_i\}$ of $T^*(X) \setminus 0$ by open cones V_i such that either Lemma 26.1.15 or the preceding observation is applicable when $WF(T) \subset V_i$. We can choose V_i so that the projections W_i in X are also locally finite and can then write $I = \sum T_i$ where $WF(T_i) \subset V_i$ and the support of the kernel of T_i belongs to $W_i \times W_i$. For every i we choose $F_i^\pm$ according to Lemma 26.1.15 or as indicated above, with $\operatorname{supp} F_i^\pm \subset W_i \times W_i$. Then the sum

$$F^\pm = \sum F_i^\pm$$

is defined, (26.1.9) and (26.1.10) are satisfied by these operators, and $F^\pm$ maps $H_{(s)}^{\text{comp}}(X)$ continuously into $H_{(s)}^{\text{loc}}(X)$ for every s. In fact, this is true for $F_1^\pm = \psi E_1^\pm$ if ψ is taken as a function of $x - y$, for the operator $F_1^\pm$ is then convolution by a measure of compact support. All other factors are $H_{(s)}$ continuous by Corollary 25.3.2.

So far we just have

$$PF^\pm = I + R^\pm \quad \text{where } R^\pm \in I_1^{-\frac{1}{2}}(X \times X, C'), \quad WF'(R^\pm) \subset C^\pm.$$

However, by Lemma 26.1.16 below we can choose $G^\pm \in I^{-\frac{1}{2}}(X \times X, C')$ so that

$$PG^\pm - R^\pm \in C^\infty(X \times X), \quad WF'(G^\pm) \subset C^\pm \circ WF'(R^\pm).$$

Since corank $\sigma_C = 2$ it follows from Theorem 25.3.8 that $G^\pm$ is continuous from $H_{(s)}^{\text{comp}}(X)$ to $H_{(s)}^{\text{loc}}(X)$ for every s, so $E^\pm = F^\pm - G^\pm$ is a right parametrix which has this continuity property. The construction shows that $F^+ - F^-$ and therefore $E^+ - E^-$ is non-characteristic at the diagonal of N (cf. (26.1.1)). Since $P(E^+ - E^-) \in C^\infty$ it follows from Theorem 25.2.4 that the principal symbol satisfies a first order homogeneous differential equation along the bicharacteristics starting there. Hence $E^+ - E^-$ is non-characteristic everywhere. (Using Proposition 26.1.3′ instead of Proposition 26.1.3 we could in fact have computed the principal symbol directly at any point in C.) This implies that $WF'(E^+ - E^-) = C$, and since $WF'(E^\pm) \subset \Delta^* \cup C^\pm$ we conclude that $WF'(E^\pm) \supset C^\pm$. Since

$$\Delta^* = WF'(I) = WF'(PE^\pm) \subset WF'(E^\pm)$$

the proof of Theorem 26.1.14 will be completed by the following

Lemma 26.1.16. *If $F \in I^s(X \times X, C')$ and $WF'(F) \subset C^\pm$, then one can find $A \in I^s(X \times X, C')$ with*

$$PA - F \in C^\infty, \quad WF'(A) \subset C^\pm \circ WF'(F) \subset C^\pm.$$

Proof. If a_0 and f are the principal symbols of A_0 and of F, then it follows from Theorem 25.2.4 that $PA_0 - F \in I^{s-1}$ if

$$i^{-1} \mathscr{L}_{H_p} a_0 + c a_0 = f,$$

where $c\in S^0$. Let ω be a non-vanishing section of $M_C\otimes\Omega_C^{\frac{1}{2}}$ which is homogeneous of degree $n/2$. (As a complex vector bundle $M_C\otimes\Omega_C^{\frac{1}{2}}$ is trivial.) If we set $a_0=\omega u$ and $f=\omega g$, the equation is of the form

$$i^{-1}H_p u+c'u=g$$

where c' is homogeneous of degree 0 and u, g are scalar symbols of degree s. It follows from the remarks after Lemma 26.1.12 that this equation has a unique solution $u\in S^s$ vanishing on the diagonal in N, and the support is contained in $C^{\pm}\circ WF'(F)$. The same argument can be applied to PA_0-F. Hence we obtain a sequence $A_j\in I^{s-j}(X\times X, C')$ with

$$WF'(A_j)\subset C^{\pm}\circ WF'(F)$$

and

$$P(A_0+\ldots+A_j)-F\in I^{s-j-1}(X\times X, C').$$

If we choose A so that $A-A_0-\ldots-A_j\in I^{s-j-1}$ for every j, the lemma is proved.

Theorem 26.1.14 can be generalized when the characteristic set N is not connected. In fact, if $N=N_+\cup N_-$ with N_+ and N_- disjoint and open, then we can find E^+ and E^- as in Theorem 26.1.14 with (26.1.9) replaced by

$$WF'(E^{\pm})=\Delta^*\cup(C^{\pm}\cap(N_+\times N_+))\cup(C^{\mp}\cap(N_-\times N_-)). \tag{26.1.9$'$}$$

The very slight modification of the proof is left as an exercise for the reader. Important examples of this situation are the advanced and retarded fundamental solutions of the wave operator.

The most noteworthy feature of Theorem 26.1.14 is that a two sided parametrix is obtained. In the following sections we shall prove far reaching extensions of Theorem 26.1.4 concerning the propagation of singularities, and this will lead to existence theorems similar to Theorem 26.1.7. However, we do not have any general methods for constructing two-sided parametrices.

26.2. The Complex Involutive Case

The study of pseudo-differential operators $P\in\Psi^m(X)$ with homogeneous principal symbol p is far more intricate when p is complex valued than in the real case discussed in Section 26.1. Already the geometry of the characteristic set $N=p^{-1}(0)$ may then be very complicated even if $dp\neq 0$. At first we shall therefore only consider the subset

$$N_2=\{(x,\xi)\in T^*(X)\setminus 0;\ p(x,\xi)=0,\ d\operatorname{Re}p(x,\xi) \text{ and } d\operatorname{Im}p(x,\xi) \text{ are linearly independent}\} \tag{26.2.1}$$

which is a conic manifold of codimension 2. Section 26.3 will be devoted to the open subset

(26.2.2) $N_{2s}=\{(x,\xi)\in T^*(X)\setminus 0;\ p(x,\xi)=0,\ \{\operatorname{Re} p, \operatorname{Im} p\}(x,\xi)\neq 0\}$

which is a symplectic manifold. The purpose of the present section is to study the interior N_{2i} of $N_2\setminus N_{2s}$ which is an involutive manifold. We recall from Section 21.2 that as involutive manifold N_{2i} is foliated by 2 dimensional leaves Γ. In analogy with the real case we shall call them bicharacteristics of P. The Hamilton vector field

$$H_p=H_{\operatorname{Re} p}+iH_{\operatorname{Im} p}$$

is tangential to any leaf Γ and has linearly independent real and imaginary parts so it defines an analytic structure in Γ where the analytic functions are the solutions of the equation $H_p u=0$. By Theorem 21.2.7 a leaf Γ is either conic or else the radial direction is never tangential to Γ. We shall postpone the discussion of the first case until Section 26.7 and only discuss here the open subset N_{2i}^0 of N_{2i} where

(26.2.3) $H_{\operatorname{Re} p}, H_{\operatorname{Im} p}$ and the radial direction are linearly independent.

Whereas Theorem 26.1.4 reflects the fact that the equation $H_p u=0$ in the real case has only constant solutions on a bicharacteristic, we shall now have to take into account that this equation has a large solution space in the two dimensional bicharacteristic Γ. To state an analogous result we recall from Section 18.1 that if $u\in\mathscr{D}'(X)$ then the regularity of u at (x,ξ) can be measured by the function

$$s_u^*(x,\xi)=\sup\{t;\ u\in H_{(t)} \text{ at } (x,\xi)\},\quad (x,\xi)\in T^*(X)\setminus 0,$$

which is lower semi-continuous and positively homogeneous of degree 0. We have by (18.1.38)

(26.2.4) $$s_{Au}^*\geq s_u^*-\mu$$

if A is a pseudo-differential operator of order μ, and by (18.1.39) there is equality in (26.2.4) where A is non-characteristic. If more generally A is a Fourier integral operator of order μ belonging to a canonical transformation χ then (26.2.4) is just modified to

(26.2.4)′ $$\chi^* s_{Au}^*\geq s_u^*-\mu,$$

with equality at the non-characteristic points.

The following is an analogue Theorem 26.1.4:

Theorem 26.2.1. *Let $u\in\mathscr{D}'(X)$, $Pu=f$, and let $\Gamma\subset T^*(X)\setminus 0$ be an open subset of a leaf in the foliation of N_{2i}^0. If s is a superharmonic function in Γ such that $s_f^*\geq s$ then*

$$\min(s_u^*, s+m-1)$$

is superharmonic in Γ.

When $\Gamma \cap WF(f)=\emptyset$ we can take $s=+\infty$ and conclude that s_u^* is superharmonic. Since a superharmonic function in an open connected set is identically $+\infty$ if it is $+\infty$ in an open subset, we obtain by applying Theorem 26.2.1 to all leaves close to a given one:

Corollary 26.2.2. *If $u \in \mathscr{D}'(X)$ and $Pu=f$, then*

$$(N_{2i}^0 \cap WF(u)) \setminus WF(f)$$

is invariant under the bicharacteristic foliation in $N_{2i}^0 \setminus WF(f)$.

Proof of Theorem 26.2.1. Choose a homogeneous function a of degree $1-m$ with $a(x_0, \xi_0) \neq 0$ at a given point in N_{2i}^0 and a homogeneous canonical transformation χ as in Theorem 21.3.2 such that

$$\chi^*(ap)=\xi_1+i\xi_2$$

in a conic neighborhood of $(0, \varepsilon_n)$. If $Q \in \Psi^{1-m}$ has principal symbol a, we can now repeat the proof of Proposition 26.1.3 to construct Fourier integral operators A and B of order 0 satisfying the conditions (i), (ii) there as well as (iii) with D_1 replaced by D_1+iD_2 and P replaced by QP. The only change is that to construct A_2 we must solve a Cauchy-Riemann equation in each step, and this can be done by Cauchy's integral formula. If $v=Bu$ and $(D_1+iD_2)v=g$ we obtain using (26.2.4)' as in the proof of Theorem 26.1.1 or (26.1.4) that

$$s_v^*=\chi^* s_u^*, \qquad s_g^*=\chi^* s_f^*+m-1$$

in a neighborhood of $(0, \varepsilon_n)$. This reduces the proof to the special case $P=D_1+iD_2$ and the leaf through $(0, \varepsilon_n)$. It will then be made in three steps.

a) If $u \in \mathscr{E}'(\mathbb{R}^n)$ and $(D_1+iD_2)u=f \in L^2$ then $u \in L^2$. This is a very special case of Theorem 10.3.2. A direct proof follows from the fact that $u=E*f$ where the fundamental solution $E=(2\pi)^{-1}(x_1+ix_2)^{-1}\,\delta(x_3, \dots, x_n)$ is a measure.

b) (Localization) Let $u \in \mathscr{E}'(\mathbb{R}^n)$, $(D_1+iD_2)u=f$, and assume that for some compact set $K \subset \mathbb{R}^2$ we have, 0 denoting the origin in $\mathbb{R}^{n-2}$,

$$u \in L^2 \quad \text{at} \quad \partial K \times \{0\} \times \varepsilon_n, \qquad f \in L^2 \quad \text{at} \quad K \times \{0\} \times \varepsilon_n.$$

Then $u \in L^2$ at $K \times \{0\} \times \varepsilon_n$. For the proof we set

$$v=\chi_1(x_1, x_2)\,\chi_2(x_3, \dots, x_n)\,\chi_3(D)u$$

where $\chi_1 \in C_0^\infty(\mathbb{R}^2)$ is equal to 1 in K, $\chi_2 \in C_0^\infty(\mathbb{R}^{n-2})$ and $\chi_2(0)=1$, $\chi_3 \in S^0(\mathbb{R}^n)$ and $\chi_3(t\varepsilon_n)=1$ when $t>1$. Then $(D_1+iD_2)v=g$ where

$$g=\chi_1\chi_2\chi_3(D)f+(D_1\chi_1+iD_2\chi_1)\chi_2\chi_3(D)u.$$

If $\operatorname{supp}\chi_1$ is sufficiently close to K, $\operatorname{supp}\chi_2$ is sufficiently close to 0 and $\operatorname{supp}\chi_3$ is in a sufficiently small conic neighborhood of ε_n then $g \in L^2$ so $v \in L^2$, by a), which proves the assertion.

c) Let $u\in\mathscr{D}'(\mathbb{R}^n)$, $(D_1+iD_2)u=f$ and assume that for a compact set $K\subset\mathbb{R}^2$ and an entire function ϕ in $z=x_1+ix_2$ we have

$$\min(s_u^*, s)>\operatorname{Re}\phi \quad\text{at}\quad \partial K\times\{0\}\times\varepsilon_n \quad\text{and}\quad s_f^*\geqq s \quad\text{at}\quad K\times\{0\}\times\varepsilon_n,$$

where $s(x_1, x_2)$ is superharmonic in a neighborhood of K. Hence $s>\operatorname{Re}\phi$ in K, and by Proposition 16.1.4 the superharmonicity of $\min(s_u^*, s)$ will follow if we show that $s_u^*\geqq\operatorname{Re}\phi$ at $K\times\{0\}\times\varepsilon_n$. Choose $\chi\in C_0^\infty(\mathbb{R}^n)$ equal to 1 in a neighborhood of $K\times\{0\}$, and set $U=a(x, D)(\chi u)$ where

$$a(x,\xi)=\chi(x)(1+|\xi|^2)^{\phi(z)/2}.$$

If $\operatorname{Re}\phi<\mu$ at x, then $a\in S^\mu$ in a neighborhood since differentiation with respect to z can only give factors $\log(1+|\xi|^2)$. We have $(D_1+iD_2)U=F$,

$$F=a(x, D)\chi f+[D_1+iD_2, a(x, D)\chi]\,u.$$

The commutator is of order $-\infty$ in a neighborhood of $K\times\{0\}$ since $\chi=1$ there and ϕ is analytic. Hence

$$U\in L^2 \quad\text{at}\quad \partial K\times\{0\}\times\varepsilon_n \quad\text{and}\quad F\in L^2 \quad\text{at}\quad K\times\{0\}\times\varepsilon_n$$

so b) gives that $U\in L^2$ at $K\times\{0\}\times\varepsilon_n$. Set

$$b(x,\xi)=\chi(x)(1+|\xi|^2)^{-\phi(z)/2}.$$

Then

$$V=b(x, D)\,U=\chi^3 u+c(x, D)u$$

where $c\in S^{\varepsilon-1}$ for any $\varepsilon>0$ in the neighborhood of $K\times\{0\}$ where $\chi=1$. Since $\chi^3+c(x, D)$ is non-characteristic there we obtain $s_u^*=s_V^*$, hence

$$s_u^*\geqq\operatorname{Re}\phi-\delta \quad\text{in}\quad K\times\{0\}\times\varepsilon_n$$

for any $\delta>0$. This completes the proof.

For the operator D_1+iD_2 in $\mathbb{R}^n$, $n\geqq 3$, we shall now prove an analogue of Theorem 26.1.5 which proves that the superharmonicity in Theorem 26.2.1 is exactly the right condition. The result can immediately be carried over locally to the leaves of N_{2i}^0 for a general P, by the argument used to prove Theorem 26.1.5 with Proposition 26.1.3′ replaced by the modification of Proposition 26.1.3 at the beginning of the proof of Theorem 26.2.1. A global form of the result can be proved by working more directly with the operator P, but for this we refer to the literature indicated at the end of the chapter.

Theorem 26.2.3. *Let Ω be an open connected subset of $\mathbb{R}^2$ with boundary $\partial\Omega$, and set $\Gamma=\overline{\Omega}\times\{0\}\times\mathbb{R}_+\varepsilon_n$, $\Gamma'=\partial\Omega\times\{0\}\times\mathbb{R}_+\varepsilon_n$ where 0 is the origin in $\mathbb{R}^{n-2}$. Let s be a lower semi-continuous function in $\mathbb{R}^2$ with values in $(-\infty,+\infty]$ which is $+\infty$ in $\complement\overline{\Omega}$, superharmonic and not identically $+\infty$ in*

Ω. Then one can find $u\in\mathscr{D}'(\mathbb{R}^n)$ with $WF(u)=\Gamma$, $WF((D_1+iD_2)u)\subset\Gamma'$ and

$$s_u^*=\pi^*s \quad \text{in } \Gamma\setminus\Gamma', \qquad s_u^*\geqq\pi^*s \quad \text{in } \Gamma'. \tag{26.2.5}$$

Here π is the projection $\Gamma\to\bar\Omega$.

The proof is similar to that of Theorem 8.3.8, although more technical, so the reader may wish to recall that proof first. Using a functional analytic argument we shall show that u can be found so that $s_u^*\geqq\pi^*s$ with equality in a countable subset E of Γ. This will give (26.2.5) if E is suitably chosen:

Lemma 26.2.4. *For every lower semi-continuous function s in an open set $\Omega\subset\mathbb{R}^N$ there is a countable subset E of Ω such that for every lower semi-continuous function s' in Ω with $s'\leqq s$ in E we have $s'\leqq s$ in Ω.*

Proof. Let V_j be an enumeration of the closed balls with rational center and radius which are contained in Ω. Choose $x_j\in V_j$ such that $s(x_j)=\min_{V_j}s$ which is possible since s is lower semi-continuous, and let $E=\{x_j\}$. If now s' is lower semi-continuous and $s'(x)>s(x)$ for some $x\in\Omega$, we can find V_j with $x\in V_j$ such that $s'(y)>s(x)$ for every $y\in V_j$. Hence $s'(x_j)>s(x)\geqq s(x_j)$. This proves the lemma.

Remark. The choice of E here can be quite unique. For example, if $N=1$ and $s(x)=0$ for irrational x, $s(p/q)=-1/|q|$ when p/q is a reduced fraction, then $s/2\leqq s$ at all irrational points but not at the rational ones. It is easily seen that E must in fact contain all rational points in this case.

We shall also need an analogue of Theorem 15.1.1 for open subsets of $\mathbb{C}$. (In Section 15.1.1 we only considered the whole of $\mathbb{C}^n$ to avoid technical difficulties which occur otherwise when $n>1$.)

Lemma 26.2.5. *Let ω be an open set in $\mathbb{C}$ and $\phi\in C^2(\omega)$ a strictly subharmonic function, that is, $\Delta\phi>0$. If $f\in L^2(\omega, e^{-\phi}(\Delta\phi)^{-1}d\lambda)$, where $d\lambda$ is the Lebesgue measure, then one can find $u\in L^2(\omega, e^{-\phi}d\lambda)$ with $\partial u/\partial\bar z=f$ and*

$$\int|u|^2e^{-\phi}d\lambda\leqq 4\int|f|^2e^{-\phi}(\Delta\phi)^{-1}d\lambda. \tag{26.2.6}$$

Proof. As in the proof of Theorem 15.1.1 we set

$$(u,v)_\phi=\int_\omega u\bar v e^{-\phi}d\lambda; \qquad u,v\in L^2_\phi=L^2(\omega, e^{-\phi}d\lambda).$$

The equation $\partial u/\partial\bar z=f$ means that

$$(f,w)_\phi=-(u,\delta w)_\phi, \qquad w\in C_0^\infty(\omega),$$
$$\delta w=e^\phi\partial(e^{-\phi}w)/\partial z=\partial w/\partial z-w\partial\phi/\partial z.$$

Now

$$\begin{aligned}\|\delta w\|_\phi^2&=-(\partial/\partial\bar z\,\delta w, w)_\phi=(\partial^2\phi/\partial z\partial\bar z\, w, w)_\phi+\|\partial w/\partial\bar z\|_\phi^2\\ &\geqq 4^{-1}((\Delta\phi)w,w)_\phi, \qquad w\in C_0^\infty(\omega).\end{aligned}$$

Hence

$$|(f, w)_\phi| \leqq M \|\delta w\|_\phi, \quad w \in C_0^\infty; \quad M^2 = 4\int |f|^2 e^{-\phi} (\Delta\phi)^{-1} d\lambda,$$

so the lemma follows from the Hahn-Banach theorem if we extend the map $\delta w \mapsto (f, w)_\phi$ to an antilinear map on L^2_ϕ without increasing the norm.

Just as in Section 15.1 we can use Lemma 26.2.5 to construct analytic functions with appropriate bounds:

Lemma 26.2.6. *Let ϕ, ω be as in Lemma 26.2.5, and let $z_0 \in \omega_0 \Subset \omega$. If t is a large positive number we can then find an analytic function f_t in ω such that*

$$f_t(z_0) = t^{\phi(z_0)}, \quad |f_t(z)| \leqq 2t^{\phi(z)}, \quad z \in \omega_0. \tag{26.2.7}$$

There are constants C_α such that for all non-negative integers α

$$|D_z^\alpha f_t(z)| \leqq C_\alpha (\log t)^\alpha \, t^{\phi(z)}, \quad z \in \omega_0. \tag{26.2.8}$$

Proof. Taylor's formula shows that

$$\phi(z) = \operatorname{Re} g(z) + \partial^2 \phi(z_0)/\partial z \partial \bar{z} |z - z_0|^2 + o(|z - z_0|^2)$$

where g is the analytic polynomial

$$g(z) = \phi(z_0) + 2(z - z_0)\partial\phi(z_0)/\partial z + (z - z_0)^2 \partial^2 \phi(z_0)/\partial z^2.$$

If b and δ are sufficiently small positive numbers it follows that

$$\phi(z) \geqq \operatorname{Re} g(z) + b|z - z_0|^2, \quad |z - z_0| < \delta.$$

Now choose $\chi \in C_0^\infty(\{z; |z - z_0| < \delta\})$ with $\chi(z) = 1$ when $|z - z_0| < \delta/2$, and set

$$f_t(z) = \chi(z) t^{g(z)} - (z - z_0) u(z).$$

f_t is analytic if

$$\partial u/\partial \bar{z} = t^{g(z)} (z - z_0)^{-1} \partial\chi/\partial\bar{z} = h_t. \tag{26.2.9}$$

With $\varepsilon = b\delta^2/4 > 0$ we have

$$\int |h_t|^2 t^{-2\phi} d\lambda \leqq C t^{-2\varepsilon}.$$

Shrinking ω if necessary we may assume that $\Delta\phi$ is bounded from below in ω and conclude, using Lemma 26.2.5 with ϕ replaced by $2\phi \log t$, that (26.2.9) has a solution with

$$\int |u|^2 t^{-2\phi} d\lambda \leqq t^{-2\varepsilon}.$$

An application of Lemma 15.1.8 with $r = 1/\log t$ now gives

$$|u(z)| \leqq C' \log t \, t^{-\varepsilon + \phi(z)}, \quad z \in \omega',$$

where $\omega_0 \Subset \omega' \Subset \omega$. This implies (26.2.7) for large t and $z \in \omega'$. Cauchy's inequality in discs with radius $1/\log t$ and center in ω_0 then proves (26.2.8).

Proof of Theorem 26.2.3. Let F be the Fréchet space of all $u\in\mathscr{D}'(\mathbb{R}^n)$ with $WF(u)\subset\Gamma$ and $s_u^*\geqq\pi^*s$ in Γ; the topology is the weakest one making the maps

$$F\ni u\mapsto Bu\in L^2_{\mathrm{loc}}$$

continuous for every properly supported $B\in\Psi^\mu$ with $\mu<\pi^*s$ in $WF(B)\cap\Gamma$. (It suffices to use countably many operators B, so the topology is metrizable, and it is a routine exercise to verify the completeness.) By Lemma 26.2.4 we can choose a countable subset E of $\Gamma\setminus\Gamma'$ such that $u\in F$ and $s_u^*\leqq\pi^*s$ in E implies $s_u^*=\pi^*s$ in $\Gamma\setminus\Gamma'$.

The subset F_0 of F where $WF((D_1+iD_2)u)\subset\Gamma'$ is also a Fréchet space with the weakest topology making the inclusion $F_0\to F$ and the map

$$F_0\ni u\mapsto(D_1+iD_2)u\in C^\infty(\Omega\times\mathbb{R}^{n-2})$$

continuous. We shall prove that if $\gamma\in E$ and $T\in\Psi^{s(\pi\gamma)}$ is properly supported, with homogeneous principal symbol which does not vanish at γ, then

$$\{u\in F_0,\ Tu\in L^2_{\mathrm{loc}}\} \tag{26.2.10}$$

is of the first category. If we use this fact for a countable number of operators T with $WF(T)$ shrinking to γ it follows that

$$\{u\in F_0,\ s_u^*(\gamma)>s(\pi\gamma)\} \tag{26.2.11}$$

is of the first category. Hence $s_u^*\leqq\pi^*s$ in E for all $u\in F_0$ except a set of the first category, and this will prove Theorem 26.2.3.

Suppose now that (26.2.10) is not of the first category. Then it follows from the closed graph theorem that the map

$$F_0\ni u\mapsto Tu\in L^2_{\mathrm{loc}}$$

is continuous. Let $x_0=(\pi\gamma,0)$ be the projection of γ in $\mathbb{R}^n$ and let K be a compact neighborhood of x_0. Then we have

$$\|Tu\|_{L^2(K)}\leqq\|\chi(D_1+iD_2)u\|_{(M)}+\sum\|B_ju\|_{L^2(K_j)} \tag{26.2.12}$$

where $\chi\in C_0^\infty(\Omega\times\mathbb{R}^{n-2})$, M is a large integer, $B_j\in\Psi^{\mu_j}$ is properly supported, $\mu_j<\pi^*s$ in $\Gamma_j=WF(B_j)\cap\Gamma$, and the sum is finite. Let K_Ω be the union of $\pi\gamma$ and the projection of $\operatorname{supp}\chi$ in Ω. We shall choose u carefully near K_Ω so that the first term on the right-hand side drops out.

Choose open sets ω and ω_0 in $\mathbb{R}^2$ with

$$K_\Omega\subset\omega_0\Subset\omega\Subset\Omega$$

and then choose $\phi\in C^2(\omega)$ with $\Delta\phi>0$ and

$$\phi>-s \text{ in } \omega,\quad \phi<-\mu_j \text{ in } \omega\cap\pi\Gamma_j. \tag{26.2.13}$$

As in the proof of Theorem 15.1.6, for example, we can achieve this by regularizing $-s$ and adding a small multiple of $x_1^2+x_2^2$, for $-s<-\mu_j$ in $\pi\Gamma_j$ and $-s$ is semi-continuous from above. Choose $\chi_0\in C_0^\infty(\omega_0)$ equal to 1 in a neighborhood of K_Ω, $\Phi \in C_0^\infty(\mathbb{R}^{n-2})$ with $\Phi(0)=1$, and set

$$a_t(x_1,x_2,\xi'')=\chi_0(x_1,x_2)\,f_t(z)\,t^{(2-n)/2}\,\Phi((\xi''/t-\varepsilon_n'')\log t). \tag{26.2.14}$$

Here $\xi''=(\xi_3,\dots,\xi_n)$, $z=x_1+ix_2$ and f_t is given by Lemma 26.2.6 with $z_0=\pi\gamma$. Thus $|\xi''/t-\varepsilon_n''|<C/\log t$ in $\operatorname{supp} a_t$, and since differentiation with respect to ξ'' will give a factor $\log t/t$, we obtain in view of (26.2.8)

$$|D_\xi^\alpha D_x^\beta a_t(x_1,x_2,\xi'')|\leq C_{\alpha\beta}(\log(2+|\xi|))^{|\alpha+\beta|}(1+|\xi''|)^{\phi(z)+(2-n)/2-|\alpha|}. \tag{26.2.15}$$

If $t=2^\nu$ where ν is an integer $>N_0$, say, the supports are disjoint so

$$A_N=\sum_{N_0}^{N} a_{2^\nu}$$

also has the bound (26.2.15). Note that any conic neighborhood of $\operatorname{supp}\chi_0 \times \varepsilon_n''$ contains the supports of all terms except a finite number. Thus A_N is uniformly of order $-\mu$ outside such a neighborhood, for any μ.

We shall prove that (26.2.12) is not valid for the corresponding conormal distributions (with respect to the x_1x_2 plane)

$$u_N(x)=\int A_N(x_1,x_2,\xi'')\,e^{i\langle x'',\xi''\rangle}\,d\xi''$$

when $N\to\infty$. First of all we have $\chi(D_1+iD_2)u_N=0$ since $\chi_0=1$ in $\operatorname{supp}\chi$. This means that the first term in the right hand side of (26.2.12) vanishes when $u=u_N$. By (26.2.13) we have $\phi+\mu_j<-\varepsilon_j<0$ in $\omega\cap\pi\Gamma_j$ so (26.2.15) implies that A_N is bounded in $S^{-\varepsilon_j-\mu_j+(2-n)/2}$, in a neighborhood of $\operatorname{supp}\chi_0\cap\pi\Gamma_j$. Using (18.2.16) we now conclude that

$$B_ju_N=\int B_{jN}(x_1,x_2,\xi'')\,e^{i\langle x'',\xi''\rangle}\,d\xi''$$

where B_{jN} is bounded in $S^{-\varepsilon_j+(2-n)/2}$ as $N\to\infty$. If Λ is the conormal bundle of the x_1x_2 plane this means that B_ju_N is bounded in $I^{-\varepsilon_j-n/4}(\mathbb{R}^n,\Lambda)$ (Proposition 25.1.5). Hence $B_j u_N$ in bounded in ${}^\infty H^{\mathrm{loc}}_{(\varepsilon_j)}$ which proves that $\|B_ju_N\|_{L^2(K_j)}$ is bounded. From (26.2.12) it follows now that Tu_N is bounded in $L^2(K)$, so $Tu_\infty\in L^2$ in a neighborhood of x_0. Now Tu_∞ can also be calculated by (18.2.16). If $t(x,\xi)$ is the principal symbol of T, which is homogeneous of degree $s(\pi\gamma)$, then the symbol of Tu_∞ is

$$t(x_1,x_2,0,\xi'')\,A_\infty(x_1,x_2,\xi'')$$

plus lower order terms. At $(\pi\gamma, 2^\nu\varepsilon_n'')$ the symbol is thus asymptotically equal to

$$t(\gamma)(2^\nu)^{s(\pi\gamma)+\phi(\pi\gamma)+(2-n)/2}.$$

However, since $Tu_\infty\in{}^\infty H_{(0)}$ in a neighborhood of x_0 it follows from Theorem 25.1.4 that the symbol must be in $S^{\varepsilon+(2-n)/2}$ for every $\varepsilon>0$ in a neighborhood of $\pi\gamma$. This contradicts that $\phi+s>0$ by (26.2.13). The proof is now complete.

26.3. The Symplectic Case

In this section we shall study the pseudo-differential operator P with principal symbol p in the symplectic characteristic manifold N_{2s} defined by (26.2.2). By Theorem 21.3.3 we can locally in N_{2s} reduce p to $\xi_1+ix_1\xi_n$ by multiplication with a non-vanishing factor and composition with a canonical transformation. In Theorem 21.3.5 we have also given an invariant description of a more general situation where we can reduce p to the form $\xi_1+ix_1^k\xi_n$ where k is a positive integer. As in Proposition 26.1.3 we can lift these transformations to the operator level; in doing so we only consider the polyhomogeneous case for the sake of simplicity.

Proposition 26.3.1. *Let $P\in\Psi^m_{\text{phg}}(X)$ have principal symbol p with $p(x^0,\xi^0)=0$, and assume that there is a homogeneous function a of degree $1-m$ in a conic neighborhood of (x^0,ξ^0) with $a(x^0,\xi^0)\neq 0$, and a homogeneous canonical transformation χ from a conic neighborhood of $(0,\pm\varepsilon_n)\in T^*(\mathbb{R}^n)\setminus 0$ to a conic neighborhood of $(x^0,\xi^0)\in T^*(X)\setminus 0$ such that $\chi^*(ap)=\xi_1+ix_1^k\xi_n$. Then we can find properly supported Fourier integral operators $A\in I^{1-m}_{\text{phg}}(X\times\mathbb{R}^n,\Gamma')$ and $B\in I^0_{\text{phg}}(\mathbb{R}^n\times X,(\Gamma^{-1})')$, where Γ is the graph of χ, such that*

(i) *$WF'(A)$ and $WF'(B)$ are in small conic neighborhoods of $(x^0,\xi^0,0,\pm\varepsilon_n)$ and $(0,\pm\varepsilon_n,x^0,\xi^0)$ respectively.*

(ii) *$BA\in\Psi^{1-m}(\mathbb{R}^n)$ is non-characteristic at $(0,\pm\varepsilon_n)$*

(iii) *$(0,\pm\varepsilon_n)\notin WF(BPA-D_1-ix_1^kD_n)$.*

Proof. Choose any $A_1\in I^{1-m}_{\text{phg}}(X\times\mathbb{R}^n,\Gamma')$ and $B_1\in I^0_{\text{phg}}(\mathbb{R}^n\times X,(\Gamma^{-1})')$ such that the principal symbol of A_1B_1 is equal to a in a neighborhood of (x^0,ξ^0). Then the principal symbol of B_1PA_1 is equal to $\xi_1+ix_1^k\xi_n$ in a neighborhood of $(0,\pm\varepsilon_n)$. Replacing P by B_1PA_1 it is then as in the proof of Proposition 26.1.3 sufficient to prove the theorem when $X=\mathbb{R}^n$, $m=1$ and the principal symbol of P is equal to $\xi_1+ix_1^k\xi_n$. The full symbol is then $\xi_1+ix_1^k\xi_n+p_0(x,\xi)+p_{-1}(x,\xi)+\dots$. We want to find pseudo-differential operators A and C of order 0, non-characteristic at $(0,\pm\varepsilon_n)$ such that the symbol of

$$PA-C(D_1+ix_1^kD_n) \tag{26.3.1}$$

is of order $-\infty$ in a conic neighborhood of $(0,\pm\varepsilon_n)$. If B is defined so that the symbol of $BC-I$ is of order $-\infty$ in another such neighborhood, we shall then have all statements in the proposition.

Let the symbols of A and C be $a_0+a_{-1}+\dots$ and $c_0+c_{-1}+\dots$. The leading symbol of (26.3.1) vanishes if $a_0=c_0$. The next term vanishes if

$$-i\{\xi_1+ix_1^k\xi_n,a_0\}+p_0a_0+(\xi_1+ix_1^k\xi_n)(a_{-1}-c_{-1})=0, \tag{26.3.2}$$

that is,

$$\begin{aligned}-i(\partial/\partial x_1+ix_1^k\partial/\partial x_n-ikx_1^{k-1}\xi_n\partial/\partial\xi_1)a_0+p_0a_0\\ +(\xi_1+ix_1^k\xi_n)(a_{-1}-c_{-1})=0.\end{aligned} \tag{26.3.2$'$}$$

It suffices to solve this equation when $\xi_n=1$ and extend the solution by homogeneity after cutting it off outside a neighborhood of the origin. To do so we choose a_0 so that

$$-i(\partial/\partial x_1+ix_1^k\partial/\partial x_n-ikx_1^{k-1}\xi_n\partial/\partial\xi_1)a_0+p_0a_0 \tag{26.3.3}$$

vanishes of infinite order when $x_1=0$, and $a_0=1$ there. This means that (26.3.3) and all the x_1 derivatives shall vanish when $x_1=0$, which successively determines $\partial^j a_0/\partial x_1^j$ when $x_1=0$ for every j. By Theorem 1.2.6 we can choose a_0 with these derivatives. The quotient r of (26.3.3) by $\xi_1+ix_1^k\xi_n$ is then a C^∞ function r, homogeneous of degree -1, and (26.3.2) is valid if $a_{-1}-c_{-1}=-r$. Using this equation to eliminate c_{-1} from the next equation, it becomes an inhomogeneous equation of the form (26.3.2)′ which can be solved in the same way. Repeating the argument we obtain a solution of (26.3.1), and this completes the proof.

From Proposition 26.3.1 it follows as in the proof of Theorems 26.1.4 and 26.2.1 that any microlocal statement on the singularities of the equation

$$(D_1+ix_1^kD_n)u=f \tag{26.3.4}$$

at $(0,\pm\varepsilon_n)$ can be carried over to the equation $Pu=f$ at (x^0,ξ^0). We shall therefore study the equation (26.3.4) carefully. For odd values of k it will turn out that its properties differ significantly from those of the constant coefficient operators which served as models in Sections 26.1 and 26.2. Fourier transform of (26.3.4) with respect to x_n leads to an ordinary differential equation which we shall examine first.

Lemma 26.3.2. *If $u\in C_0^\infty(\mathbb{R})$ and k is an integer $\geqq 0$ then*

$$\int(|u'|^2+(1+x^{2k})|u|^2)\,dx\leqq C_k\int|u'-x^ku|^2\,dx. \tag{26.3.5}$$

Proof. We may assume that u is real valued. With $f=u'-x^ku$ we have

$$f^2=u'^2+x^{2k}u^2-x^k(u^2)',$$

hence

$$\int f^2\,dx=\int(u'^2+(x^{2k}+kx^{k-1})u^2)\,dx.$$

When k is *odd* the terms in the right hand side are all positive. If we just integrate for $|x|>1$ we also obtain then

$$u(-1)^2+u(1)^2\leqq\int_{|x|>1}f^2\,dx.$$

An integration by parts gives

$$\begin{aligned}\int_{-1}^{1}u^2\,dx&=u(1)^2+u(-1)^2-2\int x\,uu'\,dx\\&\leqq u(1)^2+u(-1)^2+\int_{-1}^{1}u^2\,dx/2+\int_{-1}^{1}2u'^2\,dx,\end{aligned}$$

hence

$$\int_{-1}^{1} u^2\,dx \leqq 2(u(1)^2+u(-1)^2)+4\int u'^2\,dx \leqq 6\int f^2\,dx,$$

so (26.3.5) is valid with $C_k=7$. When k is *even* we first observe that

$$\int_1^\infty f^2\,dx \geqq \int_1^\infty (u'^2+x^{2k}u^2)\,dx+u(1)^2.$$

Set $ue^{-x^{k+1}/(k+1)}=v$, $fe^{-x^{k+1}/(k+1)}=g$. Then $v'=g$, so

$$\int_{-2}^{1} v^2\,dx=\int_{-2}^{1} v^2\,d(x+3)=4v(1)^2-v(-2)^2-2\int_{-2}^{1} vv'(x+3)\,dx$$

$$\leqq 4v(1)^2-v(-2)^2+\int_{-2}^{1} v^2\,dx/2+32\int_{-2}^{1} v'^2\,dx,$$

which gives

$$2v(-2)^2+\int_{-2}^{1} v^2\,dx \leqq 8v(1)^2+64\int_{-2} g^2\,dx.$$

Since $|x|^{k+1}/(k+1)$ is bounded in $(-2,1)$ we have now proved that

$$\int_{-2}^{\infty} u^2\,dx+u(-2)^2 \leqq C_k'\int_{-2}^{\infty} f^2\,dx.$$

If we note that

$$\int_{-\infty}^{-2} f^2\,dx=\int_{-\infty}^{-2} (u'^2+(x^{2k}+kx^{k-1})u^2)\,dx-2^k u(-2)^2$$

and that $x^{2k}+kx^{k-1} \geqq x^{2k}(1-k/2^{k+1}) \geqq 3x^{2k}/4$ if $x<-2$, the estimate (26.3.5) follows.

If we replace x by θx in (26.3.5) we obtain

(26.3.5)′ $\quad \int (|u'|^2+|\theta^{k+1}x^k u|^2+|\theta u|^2)\,dx \leqq C_k\int |u'-\theta^{k+1}x^k u|^2\,dx, \quad u\in C_0^\infty(\mathbb{R}).$

Here θ^{k+1} can be an arbitrary real number if k is even but must be positive when k is odd. This is significant for there is no estimate of the form (26.3.5) with $u'-x^k u$ replaced by $u'+x^k u$ when k is odd. In fact, the equation $u'+x^k u=0$ then has the solution $u(x)=e^{-x^{k+1}}/(k+1)$ in $\mathscr{S}$, and cutting u off far away we find that no such estimate exists. This distinction between even and odd values of k will be crucial in what follows and was in fact already observed in a geometric context in Theorem 21.3.5. From (26.3.5)′ we obtain the following estimate

Proposition 26.3.3. *Let $a(\xi)\in C^\infty(\mathbb{R}^n)$ be homogeneous of degree $1/(k+1)$ and assume that $|\xi|<K\xi_n$ in* supp a *for some constant K. Then we have with L^2 norms*

(26.3.6) $\qquad \|a(D)u\| \leqq C\|(D_1+ix_1^k D_n)u\|, \quad u\in C_0^\infty(\mathbb{R}^n).$

Proof. If $(D_1+ix_1^k D_n)u=f$ then

$$(D_1+ix_1^k\xi_n)U=F$$

where U and F are the Fourier transforms of u and f with respect to x_n. When $\xi_n>0$ it follows from (26.3.5)′ with $\theta=\xi_n^{1/(k+1)}$ that

$$\int|\xi_n^{1/(k+1)}U|^2\,dx_1\leqq C_k\int|F|^2\,dx_1.$$

If $\hat{u}, \hat{f}$ are the Fourier transforms in all the variables it follows that

$$\int_{\xi_n>0}|\xi_n^{1/(k+1)}\hat{u}(\xi)|^2\,d\xi\leqq C_k\int|\hat{f}(\xi)|^2\,d\xi.$$

Since $a(\xi)/\xi_n^{1/(k+1)}$ is bounded in supp a the estimate (26.3.6) is proved.

The estimate (26.3.6) leads directly to a result on hypoellipticity:

Proposition 26.3.4. *The operator* $D_1+ix_1^k D_n$ *is microhypoelliptic where* $\xi_n>0$ *(and also where* $\xi_n<0$ *if* k *is even). More precisely, if* (26.3.4) *is valid and* $f\in H_{(s)}$ *at* (x^0,ξ^0), *then* $u\in H_{(s+1/(k+1))}$ *at* (x^0,ξ^0) *if* $\xi_n^0>0$ *(or* $\xi_n^0<0$ *and* k *is even).*

Proof. Assume first that $u\in L^2_{\text{comp}}$, $f\in L^2_{\text{comp}}$. Choose $\chi\in C_0^\infty$ with $\chi\geqq0$ and $\int\chi\,dx=1$, and form the regularizations

$$u_\varepsilon=u*\chi_\varepsilon=\hat{\chi}(\varepsilon D)u.$$

Then $\|u_\varepsilon\|\leqq\|u\|$ and

$$(D_1+ix_1^k D_n)u_\varepsilon=f*\chi_\varepsilon+i[x_1^k D_n,\hat{\chi}(\varepsilon D)]\,u$$

is also bounded in L^2 as $\varepsilon\to0$ since the symbol of the commutator

$$-\sum_{0<j\leqq k}\varepsilon^{j-1}(D_1^j\hat{\chi})(\varepsilon\xi)\,\varepsilon\xi_n x_1^{k-j}\binom{k}{j}$$

is bounded in S^0_{loc} (Proposition 18.1.2). Hence (26.3.6) shows that $a(D)u_\varepsilon$ is bounded in L^2 as $\varepsilon\to0$, so $a(D)u\in L^2$ if a satisfies the condition in Proposition 26.3.3. This proves that $u\in H_{(1/(k+1))}$ when $\xi_n>0$; replacing x by $-x$ we obtain the same result when $\xi_n<0$ if k is even.

To prove the general statement assume that we already know that $u\in H_{(t)}$ at (x^0,ξ^0) for a certain $t\leqq s$. If $q(x,D)$ is of order t, q has compact support in x, and $WF(q)$ is in a sufficiently small conic neighborhood of (x^0,ξ^0), then $v=q(x,D)u\in L^2_{\text{comp}}$ and

$$(D_1+ix_1^k D_n)v=q(x,D)f+[D_1+ix_1^k D_n,q(x,D)]\,u\in L^2$$

since the commutator is also of order t. Hence $v\in H_{(1/(k+1))}$ by the first part of the proof. If q is chosen non-characteristic at (x^0,ξ^0) it follows that $u\in H_{(t+1/(k+1))}$ at (x^0,ξ^0). By iterating the argument a finite number of times we obtain $u\in H_{(s+1/(k+1))}$ at (x^0,ξ^0), which completes the proof.

In view of Theorems 21.3.3 and 21.3.5 we obtain from Propositions 26.3.1 and 26.3.4

Theorem 26.3.5. *Let $P\in\Psi^m_{phg}(X)$ have principal part p, and let (x^0,ξ^0) be a point where*

$$p(x^0,\xi^0)=0, \qquad \{\operatorname{Re} p, \operatorname{Im} p\}(x^0,\xi^0)>0.$$

If $Pu=f\in H_{(s)}$ at (x^0,ξ^0) it follows then that $u\in H_{(s+m-\frac12)}$ at (x^0,ξ^0). More generally, $Pu=f\in H_{(s)}$ at (x^0,ξ^0) implies $u\in H_{(s+m-k/(k+1))}$ at (x^0,ξ^0) if $p(x^0,\xi^0)=0$, $H_{\operatorname{Re}p}(x^0,\xi^0)\neq 0$, and $\operatorname{Im} p$ has just a zero of order exactly k near (x^0,ξ^0) on each bicharacteristic of $\operatorname{Re} p$ starting near (x^0,ξ^0), with a change of sign from $-$ to $+$ or no sign change at all. In particular, P is then microhypoelliptic at (x^0,ξ^0).

At a non-characteristic point we have of course the "elliptic" result that $f\in H_{(s)}$ implies $u\in H_{(s+m)}$. Thus Theorem 26.3.5 gives a loss of $k/(k+1)$ derivatives compared to the elliptic case. One calls P subelliptic with a loss of $k/(k+1)<1$ derivatives. A complete discussion of subellipticity will be given in Chapter XXVII. In particular we shall then see that the constant $k/(k+1)$ in Theorem 26.3.5 cannot be decreased, which is also easy to prove by tracing the proof of Proposition 26.3.4 backwards.

When the sign change from $+$ to $-$ ruled out in Theorem 26.3.5 occurs, there is no microhypoellipticity at (x^0,ξ^0). Moreover, non-propagating singularities may appear there.

Theorem 26.3.6. *Let $P\in\Psi^m_{phg}(X)$ have principal part p, let*

$$p(x^0,\xi^0)=0, \qquad H_{\operatorname{Re}p}(x^0,\xi^0)\neq 0,$$

and assume that $\operatorname{Im} p$ on every bicharacteristic of $\operatorname{Re} p$ starting near (x^0,ξ^0) has a zero near (x^0,ξ^0) of order exactly k where the sign of $\operatorname{Im} p$ changes from $+$ to $-$. For any $s\in\mathbb{R}$ one can then find $u\in\mathscr{D}'(X)$ with $Pu\in C^\infty(X)$, $WF(u)$ generated by (x^0,ξ^0), and $u\in H^{loc}_{(t)}$ if and only if $t<s$.

Proof. By Proposition 26.3.1 it suffices to prove the theorem when $P=D_1+ix_1^kD_n$, $(x^0,\xi^0)=(0,-\varepsilon_n)$, and k is odd. Choose $\psi\in C^\infty(\mathbb{R})$ equal to 1 on $(2,\infty)$ and 0 on $(-\infty,1)$, and set for real a

$$u_a(x)=\int e^{-\theta(ix_n+x_1^{k+1}/(k+1)+|x''|^2)}\,\theta^a\psi(\theta)\,d\theta$$

where $x''=(x_2,\dots,x_{n-1})$. By Theorem 8.1.9 we have

$$WF(u_a)\subset\{(0,-\theta\varepsilon_n),\theta>0\}.$$

Partial integration shows that u_a and all its derivatives are rapidly decreasing when $x\to\infty$, so $u_a\in H_{(t)}$ at $(0,-\varepsilon_n)$ if and only if $u_a\in H_{(t)}(\mathbb{R}^n)$. Moreover, $\hat u_a$ is rapidly decreasing outside any conic neighborhood of $(0,-\varepsilon_n)$. Denote the Fourier transform of $\exp(-x_1^{k+1}/(k+1))$ by Φ, thus $\Phi\in\mathscr{S}$. If $\varepsilon=1/(k+1)$

then the Fourier transform of u_a with respect to x_1, x'' becomes

$$\int e^{-i\theta x_n} \Phi(\xi_1/\theta^\varepsilon)\theta^{-\varepsilon} e^{-|\xi''|^2/4\theta}(\pi/\theta)^{(n-2)/2}\,\theta^a\,\psi(\theta)\,d\theta$$

which means that

$$\hat{u}(\xi_1,\ldots,\xi_{n-1},-\xi_n)=2\pi^{n/2}\,\Phi(\xi_1/\xi_n^\varepsilon)\,e^{-|\xi''|^2/4\xi_n}\,\xi_n^{a-\varepsilon-(n-2)/2}\,\psi(\xi_n), \qquad \xi_n>0.$$

The product by $(1+|\xi|^2)^{t/2}$ is square integrable when $|\xi_1|<\xi_n$, $|\xi''|<\xi_n$ if and only if

$$2(a-\varepsilon-(n-2)/2)+\varepsilon+(n-2)/2+2t<-1.$$

If we choose a so that

$$2a-\varepsilon-(n-2)/2+2s=-1,$$

the theorem is proved.

In Section 26.4 we shall prove a general form of Theorem 26.3.6 where hypotheses are only made on a single bicharacteristic of $\operatorname{Re} p$. At the same time it will be proved that there is an intimate connection between the existence of non-propagating singularities as in Theorem 26.3.6 and non-existence theorems for the adjoint operator.

As in Section 26.1 we shall finally give parametrix constructions, particularly for the model equation (26.3.4). First we assume that k is *even*. It is then easy to construct a twosided fundamental solution for (26.3.4) reduces to the Cauchy-Riemann equation if $x_1^{k+1}/(k+1)$ is introduced as a new variable instead of x_1. To simplify notation we first assume that $n=2$ and set for $x, y \in \mathbb{R}^2$

$$E(x,y)=\frac{i}{2\pi}(x_1^{k+1}/(k+1)+ix_2-y_1^{k+1}/(k+1)-iy_2)^{-1}. \tag{26.3.7}$$

This is a continuous function of x (or y) with values in L^1_{loc}, and a slight modification of the proof of (3.1.12) gives

$$(D_{x_1}+ix_1^k D_{x_2})E(x,y)=(-D_{y_1}-iy_1^k D_{y_2})E(x,y)=\delta(x-y). \tag{26.3.8}$$

In fact, if $u\in C_0^\infty(\mathbb{R}^n)$ then

$$\int_{|x-y|>\varepsilon} E(x,y)(-D_1u(y)-iy_1^k D_2u(y))\,dy=-\int_{|x-y|=\varepsilon} E(x,y)\,u(y)(y_1^k\,dy_1+i\,dy_2)$$

with the contour integral taken in the positive sense. The argument variation of $y_1^{k+1}/(k+1)+iy_2-x_1^{k+1}/(k+1)-ix_2$ around the circle is 2π, which gives the second part of (26.3.8). The first part follows since $E(x,y)=-E(y,x)$. For the operator E with kernel $E(x,y)$ we obtain

$$(D_1+ix_1^k D_2)Eu=E(D_1+ix_1^k D_2)u=u, \qquad u\in C_0^\infty(\mathbb{R}^2). \tag{26.3.9}$$

It is obvious from (26.3.7) that sing supp E is in the diagonal of $\mathbb{R}^2\times\mathbb{R}^2$. If $(x,\xi,y,\eta)\in WF'(E)$ and $(x,\xi)\neq(y,\eta)$ then it follows from (26.3.9) that $\xi_1=\eta_1=0$, hence $\xi_2=\eta_2\neq 0$ since $(D_{x_2}+D_{y_2})E(x,y)=0$, and therefore $\xi_2=\eta_2$.

Thus $WF'(E)$ is equal to the diagonal in $(T^*(\mathbb{R}^2)\setminus 0)\times(T^*(\mathbb{R}^2)\setminus 0)$ after all. – In case $n>2$ we have the fundamental solution

$$(26.3.7)'\quad E(x,y)=\frac{i}{2\pi}(x_1^{k+1}/(k+1)+ix_n-y_1^{k+1}/(k+1)-iy_n)^{-1}\otimes\delta(x''-y'')$$

where $x''=(x_2,\dots,x_{n-1})$ and $y''=(y_2,\dots,y_{n-1})$. It is clear that

$$(26.3.9)'\qquad (D_1+ix_1^kD_n)Eu=E(D_1+ix_1^kD_n)u=u,\qquad u\in C_0^\infty(\mathbb{R}^n).$$

By Theorem 8.2.9 we have

$$(26.3.10)\qquad WF'(E)=\{(x,\xi;y,\eta)\in(T^*(\mathbb{R}^n)\setminus 0)\times(T^*(\mathbb{R}^n)\setminus 0);\ (x,\xi)=(y,\eta)\text{ or } x''=y'',\ \xi=\eta,\ \xi_1=\xi_n=0\}.$$

From Proposition 26.3.4 it follows that E maps $H_{(s)}^{\mathrm{comp}}$ into $H_{(s+1/(k+1))}^{\mathrm{loc}}$ microlocally where $\xi_n\neq 0$.

The preceding results are essentially familiar from the Cauchy-Riemann equation. However, we shall now see that the situation changes drastically when k is *odd*. At first we assume again that $n=2$. The kernel $E(x,y)$ defined by (26.3.7) now has a singularity both for $x=(y_1,y_2)$ and $x=(-y_1,y_2)$. Instead of (26.3.8) we obtain, say,

$$(-D_{y_1}-iD_{y_2})E(x,y)=\delta(x-(|y_1|,y_2))-\delta(x-(-|y_1|,y_2)).$$

The definition must therefore be changed.

Let us first try to solve the equation

$$(D_1+ix_1^kD_2)u=f\in C_0^\infty(\mathbb{R}^2)$$

by introducing the Fourier transforms U and F of u and f with respect to x_2. This gives the equation $(D_1+ix_1^k\xi_2)U(x_1,\xi_2)=F(x_1,\xi_2)$ or

$$\partial_1(U(x_1,\xi_2)\exp(-x_1^{k+1}\xi_2/(k+1)))=iF(x_1,\xi_2)\exp(-x_1^{k+1}\xi_2/(k+1)).$$

Since the exponential tends to 0 when $x_1\to\infty$ if $\xi_2>0$, the equation cannot have a solution in $\mathscr{S}$ unless the integral of the right-hand side vanishes. For a general F we take the L^2 orthogonal projection on this subspace, so we form

$$(26.3.11)\qquad F(x_1,\xi_2)-c(\xi_2)\exp(-x_1^{k+1}\xi_2/(k+1))$$

where $c(\xi_2)$ is determined by

$$c(\xi_2)I(\xi_2)=\int_{-\infty}^{\infty}F(y_1,\xi_2)\exp(-y_1^{k+1}\xi_2/(k+1))\,dy_1,\qquad \xi_2>0.$$

Here

$$I(\xi_2)=\int_{-\infty}^{\infty}\exp(-2y_1^{k+1}\xi_2/(k+1))\,dy_1=\xi_2^{-1/(k+1)}I(1).$$

($I(1)$ can of course be expressed in terms of $\Gamma(1/(k+1))$.) Let Q_+f be the inverse Fourier transform of the term removed from F in (26.3.11),

(26.3.12) $Q_+ f(x)=(2\pi)^{-1} \iint_{\xi_2>0} e^{i\xi_2\phi(x,y)} f(y)\,dy\,d\xi_2/I(\xi_2), \quad f\in C_0^\infty(\mathbb{R}^2)$

where $\phi(x,y)=x_2-y_2+i(x_1^{k+1}+y_1^{k+1})/(k+1)$. It is the orthogonal projection in L^2 on solutions of the homogeneous equation $D_1 u - i x_1^k D_2 u=0$. An elementary computation gives that the kernel is

(26.3.12)′ $Q_+(x,y)=(2\pi)^{-1} 2^{-k/(k+1)}(k+1)^{-1/(k+1)}(\phi(x,y)/i)^{-(k+2)/(k+1)}.$

The inverse Fourier transform of the solution of the differential equation with F replaced by (26.3.11) for $\xi_2\geqq 0$, and 0 for $\xi_2<0$, is

(26.3.13) $E_+ f(x)=\dfrac{i}{2\pi}\int_0^\infty d\xi_2 \int e^{i\xi_2\psi(x,y)}(H(x_1-y_1)-G(x_1,\xi_2))\, f(y)\,dy$

where H is the Heaviside function, $\psi(x,y)=x_2-y_2+i(y_1^{k+1}-x_1^{k+1})/(k+1)$, and

(26.3.14) $G(x_1,\xi_2)=\int_{-\infty}^{x_1} e^{-2t^{k+1}\xi_2/(k+1)}\,dt/I(\xi_2)=G(x_1\xi_2^{1/(k+1)},1).$

In view of the elementary estimates valid for $\xi_2>0$,

(26.3.15) $$|G(x_1,\xi_2)|<Ce^{-2x_1^{k+1}\xi_2/(k+1)}, \quad x_1<0,$$
$$|1-G(x_1,\xi_2)|<Ce^{-2x_1^{k+1}\xi_2/(k+1)}, \quad x_1>0$$

it follows by partial integration with respect to ξ_2 that the inner integral in (26.3.13) is rapidly decreasing when $\xi_2\to\infty$. In fact, $\operatorname{Im}\psi\geqq 0$ unless $|x_1|>|y_1|$ and then we have $H(x_1-y_1)=0$ if $x_1<0$ and $H(x_1-y_1)=1$ if $x_1>0$. Thus (26.3.13) defines a continuous map from C_0^∞ to C. From the definitions above we obtain

(26.3.16) $$E_+(D_1+ix_1^kD_2)f=H(D_2)f, \qquad f\in C_0^\infty,$$

(26.3.17) $$(D_1+ix_1^kD_2)E_+f=H(D_2)f-Q_+f, \quad f\in C_0^\infty.$$

Passing to adjoints in (26.3.16), (26.3.17) we obtain

$$(D_1-ix_1^kD_2)E_+^*f=H(D_2)f, \quad E_+^*(D_1-ix_1^kD_2)f=H(D_2)f-Q_+^*f, \quad f\in C_0^\infty.$$

We change the sign for x_2 which changes the adjoints to

(26.3.18) $$Q_-(x,y)=(2\pi)^{-1}\int_{\xi_2<0} e^{i\xi_2\overline{\phi(x,y)}}\,d\xi_2/I(-\xi_2)$$

(26.3.19) $$E_-(x,y)=-\frac{i}{2\pi}\int_{\xi_2<0} e^{i\xi_2\psi(x,y)}(H(y_1-x_1)-G(y_1,-\xi_2))\,d\xi_2.$$

If we set $E=E_++E_-$ and note that $H(D_2)+H(-D_2)$ is the identity, we have

(26.3.20) $$(D_1+ix_1^kD_2)Ef=f-Q_+f, \quad f\in C_0^\infty,$$

(26.3.21) $$E(D_1+ix_1^kD_2)f=f-Q_-f, \quad f\in C_0^\infty.$$

The wave front sets of the kernels of these operators are easily determined. First of all we have

(26.3.22) $WF'(Q_\pm)=\{(x,\xi,y,\eta);\, x_1=y_1=\xi_1=\eta_1=0, \quad x_2=y_2,\, \xi_2=\eta_2 \gtrless 0\}$,

for $WF'(Q_\pm)$ is contained in the right hand side by (26.3.12), (26.3.18) and Theorem 8.1.9, and $Q_\pm$ is singular at (x,y) if $x_1=y_1=0$, $x_2=y_2$ by (26.3.12)′ and its analogue for Q_-. For the kernels (26.3.20), (26.3.21) mean that

$$(D_{x_1}+ix_1^k D_{x_2})\,E(x,y)=\delta(x-y)-Q_+(x,y),$$
$$(-D_{y_1}-iy_1^k D_{y_2})E(x,y)=\delta(x-y)-Q_-(x,y),$$

and the translation invariance in x_2 gives in addition

$$(D_{x_2}+D_{y_2})\,E(x,y)=0.$$

The common characteristics of these operators are defined by $\xi_1=\eta_1=0$, $\xi_2=-\eta_2 \neq 0$, $x_1=y_1=0$, and at these points one of the operators is microhypoelliptic by Proposition 26.3.4. It follows that $WF'(E)$ is contained in the diagonal, and since $WF'(Q_+)\cup WF'(Q_-)$ is nowhere dense there we must have equality. We are now ready to prove

Proposition 26.3.7. *Let E, Q_+, Q_- be defined as above. Then* (26.3.20), (26.3.21) *are valid for $f\in\mathscr{E}'$, $WF'(E)$ is the diagonal in $(T^*(\mathbb{R}^2)\setminus 0)\times(T^*(\mathbb{R}^2)\setminus 0)$, and $WF'(Q_\pm)$ is the subset defined by* (26.3.22). *If $f\in H_{(s)}$ at (x^0,ξ^0) then $Q_\pm f\in H_{(s)}$ and $Ef\in H_{(s+1/(k+1))}$ at (x^0,ξ^0), thus $H_{(s)}^{\mathrm{comp}}$ is mapped into $H_{(s)}^{\mathrm{loc}}$ and $H_{(s+1/(k+1))}^{\mathrm{loc}}$ by these operators.*

Proof. Only the continuity statements remain to be proved. We know already that $Q_\pm$ as orthogonal projections in L^2 are bounded there. Let J be the positive canonical ideal defined by the phase function $\xi_2\phi(x,y)$, $\xi_2>0$. It is generated by the functions $\phi(x,y)$, $\xi_1-i\xi_2x_1^k$, $\eta_1-i\eta_2y_1^k$ and $\xi_2-\eta_2$. Then $Q_+\in I^{1/(k+1)-\frac12}(\mathbb{R}^4,J')$ is non-characteristic in the real set $J'_{\mathbb{R}}$. It follows that every $Q\in I^{1/(k+1)-\frac12}(\mathbb{R}^4,J')$ defines an operator which is continuous from L^2_{comp} to L^2_{loc}, for the corresponding operator can be written in the form $QA \bmod C^\infty$, where A is a pseudo-differential operator of order 0. (Note that this follows from Theorem 25.5.6 if $k=1$ but not for larger values of k.) The $H_{(s)}$ continuity of Q_+ now follows immediately (see for example the proof of Corollary 25.3.2). Hence Q_+^* and therefore Q_- is $H_{(s)}$ continuous. If $f\in H_{(s)}^{\mathrm{comp}}$ then

$$(D_1+ix_1^kD_2)Ef=f-Q_+f\in H_{(s)}^{\mathrm{loc}}$$

so $Ef\in H_{(s+1/(k+1))}$ at (x^0,ξ^0) by Proposition 26.3.4 unless this is a characteristic point with $\xi_2^0<0$. For E^* we have the same result except at the characteristic points with $\xi_2^0>0$, and this completes the proof.

The operators Q_+ and Q_- are hermitian symmetric and

(26.3.23) $$(D_1-ix_1^kD_2)Q_+=0, \quad (D_1+ix_1^kD_2)Q_-=0.$$

They are the projection operators on the cokernel and on the kernel of $D_1+ix_1^kD_2$. We shall draw some important conclusions from this after introducing the extra parameters which occur in the n dimensional case. From now on we therefore redefine E, Q_+, Q_- by substituting x_n, y_n for x_2, y_2 and taking the tensor product with $\delta(x''-y'')$, $x''=(x_2,\dots,x_{n-1})$.

Proposition 26.3.7′. *For the distributions E, Q_+, $Q_-\in\mathscr{D}'(\mathbb{R}^{2n})$ and the corresponding operators we have*

$$(D_1+ix_1^kD_n)Ef=f-Q_+f,\quad f\in\mathscr{E}', \tag{26.3.20′}$$

$$E(D_1+ix_1^kD_n)f=f-Q_-f,\quad f\in\mathscr{E}', \tag{26.3.21′}$$

$$(D_1-ix_1^kD_n)Q_+f=0,\quad (D_1+ix_1^kD_n)Q_-f=0,\quad f\in\mathscr{E}', \tag{26.3.23′}$$

$$\begin{aligned}WF'(Q_\pm)=\{(x,\xi,y,\eta)\in(T^*(\mathbb{R}^n)\setminus 0)\times(T^*(\mathbb{R}^n)\setminus 0);\\ (x,\xi)=(y,\eta),\ x_1=\xi_1=0,\ \xi_n\gtrless 0\\ \text{or } x''=y'',\ \xi=\eta,\ \xi_1=\xi_n=0\},\end{aligned} \tag{26.3.22′}$$

$$\begin{aligned}WF'(E)=\{(x,\xi,y,\eta)\in(T^*(\mathbb{R}^n)\setminus 0)\times(T^*(\mathbb{R}^n)\setminus 0);\\ (x,\xi)=(y,\eta) \text{ or } x''=y'',\ \xi=\eta,\ \xi_1=\xi_n=0\}.\end{aligned} \tag{26.3.24}$$

If $f\in\mathscr{E}'(\mathbb{R}^n)$ and $f\in H_{(s)}$ at (x^0,ξ^0) and $\xi_n^0\neq 0$, then $Q_\pm f\in H_{(s)}$ and $Ef\in H_{(s+1/(k+1))}$ at (x^0,ξ^0).

Proof. (26.3.22)′ and (26.3.24) are immediate consequence of Proposition 26.3.7 and Theorem 8.2.9. They show that E and $Q_\pm$ are continuous from $\mathscr{E}'$ to $\mathscr{D}'$, so (26.3.20)′, (26.3.21)′ follow since they hold in C_0^∞ by (26.3.20), (26.3.21). It also follows that $(x^0,\xi^0)\notin WF(Ef)\cup WF(Q_\pm f)$ if $(x^0,\xi^0)\notin WF(f)$ and $\xi_n^0\neq 0$. When proving the last statement we may therefore assume that $f\in H_{(s)}$ and that $WF(f)$ is in a small conic neighborhood of (x^0,ξ^0), thus

$$\hat a(D_n)f\in L^2$$

if $a\in\mathscr{E}'$ and $\hat a\in S^s(\mathbb{R})$. But $\hat a(D_n)$ commutes with E and $Q_\pm$ so it follows from Proposition 26.3.7 that $\hat a(D_n)Q_\pm f\in L^2_{\text{loc}}$, $\hat a(D_n)Ef\in H^{\text{loc}}_{(1/(k+1))}$. The statement follows if we multiply by $b(D)\chi(x)$ where $\chi\in C_0^\infty$ and $b\in S^0$ is chosen so that $|\xi|/|\xi_n|$ is bounded in $\operatorname{supp} b$, for $b(D)\chi(x)\hat a(D_n)$ is then a pseudo-differential operator which can be chosen non-characteristic at (x^0,ξ^0).

Proposition 26.3.7′ immediately gives back Proposition 26.3.4. Indeed, if $u\in\mathscr{E}'$ and $(D_1+ix_1^kD_2)u=f$, then

$$u=Ef+Q_-u.$$

If $f\in H_{(s)}$ at (x^0,ξ^0) and $\xi_n^0>0$ it follows that $u\in H_{(s+1/(k+1))}$ at (x^0,ξ^0) since $(x^0,\xi^0)\notin WF(Q_-u)$. We can also obtain Theorem 26.3.6 for the model operator if we observe that $(D_1+ix_1^kD_2)u=0$ when $u=Q_-f$. We choose $f\in\mathscr{E}'$ with $WF(f)$ equal to a ray in Σ_- where

$$\Sigma_\pm=\{(x,\xi)\in T^*(\mathbb{R}^n);\ x_1=\xi_1=0,\ \xi_n\gtrless 0\}.$$

Then $WF(u)$ is in the same ray by (26.3.22)'. We can choose f so that u is not smooth and then give u the desired regularity by applying a suitable convolution operator in D_n. We can also determine completely when the equation (26.3.4) can be solved microlocally at (x^0,ξ^0), provided that $\xi_n^0 \neq 0$. Let $f\in\mathscr{E}'$ and assume that

$$(x^0,\xi^0)\notin WF((D_1+ix_1^k D_n)u-f) \tag{26.3.25}$$

for some $u\in\mathscr{D}'$. We can of course take $u\in\mathscr{E}'$ then. Since $Q_+(D_1+ix_1^k D_2)=0$ by (26.3.23)', because Q_+ is hermitian symmetric, it follows that

$$(x^0,\xi^0)\notin WF(Q_+f). \tag{26.3.26}$$

Conversely, if (26.3.26) is valid then (26.3.25) is satisfied by $u=Ef$ in view of (26.3.20)', so we obtain

Proposition 26.3.8. *If $f\in\mathscr{E}'$, and $\xi_n^0\neq 0$, then one can find $u\in\mathscr{D}'$ satisfying (26.3.25) if and only if (26.3.26) is fulfilled.*

The condition (26.3.26) is of course automatically fulfilled if $(x^0,\xi^0)\notin\Sigma_+$. However, if $(x^0,\xi^0)\in\Sigma_+$ we can as indicated above for Q_- find f so that $WF(Q_+f)$ is generated by (x^0,ξ^0) and Q_+f has a prescribed regularity. Using Theorem 26.3.1 we can immediately carry this result over to operators satisfying the condition there. When $k=1$ we obtain in particular

Theorem 26.3.9. *For every $(x^0,\xi^0)\in T^*(X)\smallsetminus 0$ where $p=0$ and $\{\operatorname{Re}p, \operatorname{Im}p\}>0$ and for any given s one can find $f\in H_{(s)}^{\mathrm{loc}}(X)$ with $WF(f)$ generated by (x^0,ξ^0) and $(x^0,\xi^0)\in WF(Pu-f)$ for every $u\in\mathscr{D}'(X)$.*

It is also easy to extend Proposition 26.3.7' microlocally to operators satisfying the conditions in Proposition 26.3.1. In case $N_{2\mathrm{s}}$, defined by (26.2.2), is the full characteristic variety one can also give a global version of Proposition 26.3.7'. To do so one just combines the local constructions with a pseudo-differential partition of unity placed to the right (left) except near $\Sigma_-(\Sigma_+)$,

$$\Sigma_\pm=\{(x,\xi)\in T^*(X)\smallsetminus 0;\ p(x,\xi)=0,\ \{\operatorname{Re}p, \operatorname{Im}p\}(x,\xi)\gtrless 0\}. \tag{26.3.27}$$

These constructions fit together in the complement of $\Sigma_-\cup\Sigma_+$ since E is uniquely determined there mod C^∞. The details are left for the reader who might also consult the references at the end of the chapter where it is shown that $Q_\pm$ and E become unique mod C^∞ if $Q_\pm$ are required to be hermitian symmetric.

26.4. Solvability and Condition (Ψ)

Let P be a properly supported pseudo-differential operator in a C^∞ manifold X of dimension n, and let K be a compact subset of X. In this section we shall prove a necessary condition for the equation $Pu=f$ to be solvable at K in a very weak sense suggested by Theorem 26.1.7.

Definition 26.4.1. We shall say that P is solvable at K if for every f in a subspace of $C^\infty(X)$ of finite codimension there is a distribution u in X such that

$$Pu=f \tag{26.4.1}$$

in a neighborhood of K.

In the definition we have not assumed that the neighborhood where (26.4.1) is valid or the order of the distribution u can be chosen independently of f. However, using Baire's theorem we shall now show that this is always possible. At the same time we shall show that solvability is equivalent to a solvability condition mod C^∞.

Theorem 26.4.2. *The following conditions on the properly supported pseudo-differential operator P in X and the compact set $K\subset X$ are equivalent:*

(i) *P is solvable at K.*

(ii) *There is an integer N and an open neighborhood $Y\subset X$ of K such that for every $f\in H^{\mathrm{loc}}_{(N)}(X)$ there is a distribution $u\in H^{\mathrm{loc}}_{(-N)}(X)$ such that $Pu-f\in C^\infty(Y)$.*

(iii) *There is an integer N such that for every $f\in H^{\mathrm{loc}}_{(N)}(X)$ there is a distribution u in X such that $Pu-f\in C^\infty$ in a neighborhood of K.*

(iv) *There is an integer N such that for every $f\in H^{\mathrm{loc}}_{(N)}(X)$ we can find $u\in\mathscr{D}'(X)$ with $Pu-f\in H^{\mathrm{loc}}_{(N+1)}$ in some neighborhood of K.*

(v) *There is an integer N and an open neighborhood $Y\subset X$ of K such that for every f in a subspace $W\subset H^{\mathrm{loc}}_{(N)}(X)$ of finite codimension the equation (26.4.1) is valid in Y for some $u\in H^{\mathrm{loc}}_{(-N)}(X)$.*

Proof. The implications (ii) $\Rightarrow$ (iii) $\Rightarrow$ (iv) and (v) $\Rightarrow$ (i) are obvious. We shall now prove that (i) $\Rightarrow$ (ii). Let $\|\ \|_{(s)}$ denote a norm in $H^{\mathrm{comp}}_{(s)}(X)$ which defines the topology in $H^c_{(s)}(M)=H^{\mathrm{loc}}_{(s)}\cap\mathscr{E}'(M)$ for every compact set $M\subset X$. Choose a fundamental decreasing system of open neighborhoods of K,

$$K\Subset\ldots\Subset Y_2\Subset Y_1\Subset X.$$

Since P is properly supported we can find $Z\Subset X$ so that $Pu=0$ in Y_1 if $u=0$ in Z. Fix $\phi\in C_0^\infty(X)$ with $\phi=1$ in Z. Then we have $Pu=P(\phi u)$ in Y_1 if $u\in\mathscr{D}'(X)$, and $\operatorname{supp}\phi u\subset M=\operatorname{supp}\phi$.

Condition (i) means that we can choose $f_1,\ldots,f_r\in C^\infty(X)$ so that for any $f\in C^\infty(X)$ we have

$$Pu=f+\sum_1^r a_j f_j \quad \text{in } Y_N \tag{26.4.2}$$

for some positive integer N, $a_j\in\mathbb{C}$ and $u\in\mathscr{D}'(X)$. Since u can be replaced by ϕu we can always choose $u\in\mathscr{E}'(M)$, hence $u\in H^c_{(-N)}(M)$ for some N. The union of the sets

$$F_N=\{f\in C^\infty(X);\ (26.4.2)\text{ is valid with } u\in H^c_{(-N)}(M),\quad \|u\|_{(-N)}+\sum|a_j|\leqq N\}$$

is therefore equal to $C^\infty(X)$. The set F_N is convex, symmetric and closed since the set of permitted $(u,a_1,\dots,a_r)$ in the definition is convex, symmetric and weakly compact. Hence it follows from Baire's theorem that F_N has 0 as an interior point when N is large. Thus we can find $\chi\in C_0^\infty(X)$ and N' so that

$$f\in C^\infty(X),\quad \|\chi f\|_{(N')}\leqq 1\Rightarrow f\in F_N.$$

Using the same compactness argument again we conclude that (26.4.2) has a solution $u\in H^c_{(-N)}(M)$, $a_1,\dots,a_r$ with $\|u\|_{(-N)}+\sum|a_j|\leqq N$ for every $f\in H^{\text{loc}}_{(N')}$ with $\|\chi f\|_{(N')}\leqq 1$. This gives (ii) with N replaced by max (N,N').

It remains to prove that (iv) $\Rightarrow$ (v). To do so we now denote by G_ν the set of all $f\in H^c_{(N)}(\overline{Y}_1)=H$ such that

$$Pu=f+g\quad \text{in } Y_\nu$$

for some $g\in H^c_{(N+1)}(\overline{Y}_1)$ and $u\in H^c_{(-\nu)}(M)$ with

$$\|u\|^2_{(-\nu)}+\|g\|^2_{(N+1)}\leqq\nu^2. \tag{26.4.3}$$

Baire's theorem gives as above that G_ν contains the unit ball in H for large ν. The minimum of the left-hand side of (26.4.3) is attained precisely when (u,g) is orthogonal to all (u',g') with $Pu'=g'$ in Y_ν, so g is then a linear function Tf of f. (All norms are taken Hilbertian.) The map T: $H\to H^c_{(N+1)}(\overline{Y}_1)$ has norm $\leqq\nu$. Thus T defines a compact operator in H, which implies that the range of $I+T$ has finite codimension. The equation $Pu=h$ in Y_ν has a solution $u\in H^c_{(-\nu)}(M)$ for every $h\in H$ in the range of $I+T$. This proves (v) with N replaced by max(N,ν).

Remarks. a) If P satisfies (v) and Q is of order $-2N-1$ then it is clear that $P+Q$ satisfies (iv). Thus solvability at K is not destroyed by perturbations of P of sufficiently low order.

b) In view of (iii) it follows from Theorem 26.3.9 that P is not solvable at $\{x\}$ if x is in the projection in X of the set Σ_+ defined by (26.3.27).

c) In proving that (i) $\Rightarrow$ (ii) it would have been sufficient to know that $\bigcup F_N$ is of the second category. If P is not solvable at K it follows therefore that for every finite dimensional subspace W of $C^\infty(X)$ the set

$$\{f+g;\ f\in C^\infty(X),\ Pu=f \text{ in a neighborhood of } K \text{ for some } u\in\mathscr{D}'(X),\ g\in W\}$$

is of the first category in $C^\infty(X)$. For any sequence K_j, W_j with these properties we can thus find $f\in C^\infty(X)$ so that the equation $Pu=f$ cannot be solved modulo W_j in a neighborhood of K_j for any j. In particular, we can choose $f\in C^\infty(X)$ so that the equation $Pu=f$ cannot be solved in a neighborhood of any point in the projection of Σ_+ in X. An example is the Lewy operator $P=D_1+iD_2+i(x_1+ix_2)D_3$ in $\mathbb{R}^3$. Since

$$\Sigma_+=\{(x,\xi);\ \xi_1=x_2\xi_3,\ \xi_2=-x_1\xi_3,\ \xi_3>0\}$$

has surjective projection we can choose $f \in C^\infty(\mathbb{R}^3)$ so that the equation $Pu = f$ does not have a distribution solution in any open set.

The condition (iii) in Theorem 26.4.2 suggests a definition of solvability at a set in the cosphere bundle:

Definition 26.4.3. If K is a compactly based cone $\subset T^*(X) \setminus 0$ we shall say that P is solvable at K if there is an integer N such that for every $f \in H^{loc}_{(N)}(X)$ we have $K \cap WF(Pu - f) = \emptyset$ for some $u \in \mathscr{D}'(X)$.

Solvability at a compact set $M \subset X$ is equivalent to solvability at $T^*(X)|_M \setminus 0$, by condition (iii) in Theorem 26.4.2. Note that solvability at $K \subset T^*(X) \setminus 0$ implies solvability at any smaller closed cone, and that solvability at K only depends on the symbol of P in a conic neighborhood of K. This makes it possible to prove necessary conditions for solvability by local arguments where the following proposition can be used:

Proposition 26.4.4. *Let $K \subset T^*(X) \setminus 0$ and $K' \subset T^*(Y) \setminus 0$ be compactly based cones and let χ be a homogeneous symplectomorphism from a conic neighborhood of K' to one of K such that $\chi(K') = K$. Let $A \in I^{m'}(X \times Y, \Gamma')$ and $B \in I^{m''}(Y \times X, (\Gamma^{-1})')$ where Γ is the graph of χ, and assume that A and B are properly supported and non-characteristic at the restriction of the graphs of χ and χ^{-1} to K' and to K respectively, while $WF'(A)$ and $WF'(B)$ are contained in small conic neighborhoods. Then the pseudo-differential operator P in X is solvable at K if and only if the pseudo-differential operator BPA in Y is solvable at K'.*

Proof. Choose $A_1 \in I^{-m''}(X \times Y, \Gamma')$ and $B_1 \in I^{-m'}(Y \times X, (\Gamma^{-1})')$ properly supported so that

$$K' \cap WF(BA_1 - I) = \emptyset, \quad K \cap WF(A_1 B - I) = \emptyset,$$
$$K' \cap WF(B_1 A - I) = \emptyset, \quad K \cap WF(AB_1 - I) = \emptyset.$$

Assume that P is solvable at K and choose N as in Definition 26.4.3. Let $g \in H^{loc}_{(N-m'')}(Y)$ and set $f = A_1 g \in H^{loc}_{(N)}(X)$. We can then find $u \in \mathscr{D}'(X)$ with $K \cap WF(Pu - f) = \emptyset$. Let $v = B_1 u \in \mathscr{D}'(\mathbb{R}^n)$. Then

$$WF(Av - u) = WF((AB_1 - I)u)$$

does not meet K, so $K \cap WF(PAv - f) = \emptyset$. Hence

$$K' \cap WF(BPAv - Bf) = \emptyset.$$

Since $K' \cap WF((BA_1 - I)g) = \emptyset$ it follows that

$$K' \cap WF(BPAv - g) = \emptyset.$$

Hence BPA is solvable at K'. Conversely, if BPA is solvable at K' it follows that $A_1 BPAB_1$ is solvable at K. Since $K \cap WF(A_1 BPAB_1 - P) = \emptyset$ this means that P is solvable at K, which completes the proof.

As a final analytic preparation for the proof of necessary conditions for solvability we shall show that solvability of P implies an a priori estimate for the adjoint operator P^*.

Lemma 26.4.5. *Let K be a compactly generated cone $\subset T^*(X)\setminus 0$ such that P is solvable at K, and choose $Y\Subset X$ so that $K\subset T^*(Y)$. If N is the integer in Definition 26.4.3 we can find an integer ν and a properly supported pseudo-differential operator A with $WF(A)\cap K=\emptyset$ such that*

$$\|v\|_{(-N)}\leqq C(\|P^*v\|_{(\nu)}+\|v\|_{(-N-n)}+\|Av\|_{(0)}),\quad v\in C_0^\infty(Y). \tag{26.4.4}$$

Proof. Let $Y\Subset Z\Subset X$. We claim that for fixed f in the Hilbert space $H_{(N)}^c(\overline{Z})$ we have for some C, ν and A as in the lemma

$$|(f,v)|\leqq C(\|P^*v\|_{(\nu)}+\|v\|_{(-N-n)}+\|Av\|_{(0)}),\quad v\in C_0^\infty(Y). \tag{26.4.5}$$

In fact, by hypothesis we can find u and g in $\mathscr{E}'(X)$ so that $f=Pu+g$ and $K\cap WF(g)=\emptyset$. Thus

$$(f,v)=(u,P^*v)+(g,v),\quad v\in C_0^\infty(Y).$$

Choose properly supported pseudo-differential operators B_1 and B_2 of order 0 with $I=B_1+B_2$ and $WF(B_1)\cap WF(g)=\emptyset$, $WF(B_2)\cap K=\emptyset$ which is possible since $WF(g)\cap K=\emptyset$. Then $B_1 g\in C^\infty$ so $(B_1 g,v)$ can be estimated by $C\|v\|_{(-N-n)}$. We have for some μ

$$|(B_2 g,v)|\leqq \|B_2^* v\|_{(\mu)}\leqq C(\|BB_2^* v\|_{(0)}+\|v\|_{(-N-n)})$$

if B is elliptic of order μ and properly supported. This gives (26.4.5) with $A=BB_2^*$.

Let V be the space $C_0^\infty(Y)$ equipped with the topology defined by the semi-norms $\|v\|_{(-N-n)}$, $\|P^*v\|_{(\nu)}$, $\nu=1,2,\dots$, and $\|Av\|_{(0)}$ where A is a properly supported pseudo-differential operator with $K\cap WF(A)=\emptyset$. It suffices to use a countable sequence $A_1,A_2,\dots$ where A_ν is noncharacteristic of order ν in a set which increases to $T^*(X)\setminus 0\setminus K$ as $\nu\to\infty$. Thus V is a metrizable space. The sesquilinear form (f,v) in the product of the Hilbert space $H_{(N)}^c(\overline{Z})$ and the metrizable space V is obviously continuous in f for fixed v, and by (26.4.5) it is also continuous in v for fixed f. Hence it is continuous, which means that for some ν and C

$$|(f,v)|\leqq C\|f\|_{(N)}(\|P^*v\|_{(\nu)}+\|A_\nu v\|_{(0)}+\|v\|_{(-N-n)}),$$
$$f\in H_{(N)}^c(\overline{Z}),\quad v\in C_0^\infty(Y).$$

This implies (26.4.4).

Proposition 26.3.8 suggests that an operator $P\in\Psi_{\text{phg}}^m$ with principal symbol p is not solvable at a characteristic point where $\operatorname{Im}p$ changes sign from $-$ to $+$ on the oriented bicharacteristic of $\operatorname{Re}p$. However, from Proposition 26.4.4 we know that a necessary condition for solvability stated in terms of p should be invariant under multiplication by non-vanishing homogeneous functions, so we are led to the following somewhat more complicated looking condition:

Definition 26.4.6. The positively homogeneous function $p \in C^\infty(T^*(X) \setminus 0)$ is said to satisfy condition (Ψ) in the open set $Y \subset X$ if there is no positively homogeneous complex valued function q in $C^\infty(T^*(Y) \setminus 0)$ such that $\operatorname{Im} qp$ changes sign from $-$ to $+$ when one moves in the positive direction on a bicharacteristic of $\operatorname{Re} qp$ over Y on which $q \neq 0$. (Sometimes $\bar{p}$ is then said to satisfy ($\overline{\Psi}$).)

Recall that a bicharacteristic of r is an integral curve of the Hamilton field H_r where $r=0$. We shall say that a bicharacteristic of $\operatorname{Re} qp$ where $q \neq 0$ is a *semi-bicharacteristic* of p. The main purpose of this section is to prove the following theorem.

Theorem 26.4.7. *Suppose that there is a C^∞ positively homogeneous function q in $T^*(X) \setminus 0$ and a bicharacteristic interval $t \mapsto \gamma(t)$, $a \leqq t \leqq b$, for $\operatorname{Re} qp$ such that $q(\gamma(t)) \neq 0$, $a \leqq t \leqq b$, and*

$$\operatorname{Im} qp(\gamma(a)) < 0 < \operatorname{Im} qp(\gamma(b)).$$

Then P is not solvable at the cone generated by $\gamma([a,b])$.

Corollary 26.4.8. *If P is solvable at the compact set $K \subset X$ then K has an open neighborhood Y in X where p satisfies condition* (Ψ).

Proof. By condition (v) in Theorem 26.4.2 we can find a neighborhood Y of K such that P is solvable at any compactly generated cone $M \subset T^*(Y)$. Hence the statement follows from Theorem 26.4.7.

Without using Theorem 26.4.7 but only results already established we can prove that $\operatorname{Im} qp$ cannot change sign from $-$ to $+$ on a bicharacteristic of $\operatorname{Re} qp$ at a point $(x^0, \xi^0) \in T^*(Y) \setminus 0$ where $\operatorname{Im} qp$ vanishes of finite order. In fact, if Q is a pseudo-differential operator with principal symbol q we know from Proposition 26.4.4 that QP must be solvable in a neighborhood of (x^0, ξ^0). On every bicharacteristic of $\operatorname{Re} qp$ nearby there must be a zero (x^1, ξ^1) where the same sign change occurs, and we choose it so that the order of the zero is minimal. Then qp satisfies the hypothesis of Theorem 21.3.5 at (x^1, ξ^1) so using Proposition 26.3.1 we can transform QP microlocally at (x^1, ξ^1) to the operator $D_1 + i x_1^k D_n$ at $(0, \varepsilon_n)$, where it is not solvable by Proposition 26.3.8. In view of Proposition 26.4.4 this is a contradiction proving the weaker form of condition (Ψ).

Before proving Theorem 26.4.7 in complete generality we must study the geometrical situation in some detail; this will also lead to a simpler form of condition (Ψ). Suppose that the hypotheses of Theorem 26.4.7 are fulfilled, and choose a pseudo-differential operator Q with principal symbol q. Then the principal symbol of $P_1 = QP$ is $p_1 = qp$, so $\operatorname{Im} p_1$ changes sign from $-$ to $+$ along a bicharacteristic of $\operatorname{Re} p_1$. We then set $P_2 = Q_1 P_1$ where Q_1 is of degree $1-$degree P_1 and has positive, homogeneous principal symbol. If p_2 is the principal symbol of P_2 then $\operatorname{Im} p_1$ and $\operatorname{Im} p_2$ have the same sign and $\operatorname{Re} p_2$ has the same bicharacteristics as $\operatorname{Re} p_1$ including the orientation. In view of Proposition 26.4.4 it is therefore sufficient to prove Theorem 26.4.7 in the case where $q=1$ and p is of degree 1. The bicharacteristics of $\operatorname{Re} p$ can then be considered as curves on the cosphere bundle. If the curve where $\operatorname{Im} p$ changes sign is closed on $S^*(X)$ we can always pick an arc which is not closed where the change of sign still occurs, and this we assume done in what follows. We can then use Proposition 26.1.6 and Proposition 26.4.4 to reduce the proof further to the case $X = \mathbb{R}^n$, $\operatorname{Re} p = \xi_1$, and the bicharacteristic of $\operatorname{Re} p$ given by

$$a \leqq x_1 \leqq b, \quad x' = (x_2, \ldots, x_n) = 0, \quad \xi = \varepsilon_n. \tag{26.4.6}$$

Global problems might occur in our constructions if $b-a$ is large so we shall examine how small the intervals can be where the crucial sign change occurs. To do so we set

$$L(x', \xi') = \inf\{t-s;\ a<s<t<b,\ \operatorname{Im} p(s, x', 0, \xi') < 0 < \operatorname{Im} p(t, x', 0, \xi')\}$$

when (x', ξ') is close to $(0, \varepsilon'_n)$, and we denote by L_0 the lower limit of $L(x', \xi')$ as $(x', \xi') \to (0, \varepsilon'_n)$. For small $\delta > 0$ we can choose an open neighborhood V_δ of $(0, \varepsilon'_n)$ in $\mathbb{R}^{2n-2}$ with diameter $<\delta$ such that $L(x', \xi') > L_0 - \delta/2$ in V_δ. For some $(x'_\delta, \xi'_\delta) \in V_\delta$ and s_δ, t_δ with $a < s_\delta < t_\delta < b$ we have

$$t_\delta - s_\delta < L_0 + \delta/2, \quad \operatorname{Im} p(s_\delta, x'_\delta, 0, \xi'_\delta) < 0 < \operatorname{Im} p(t_\delta, x'_\delta, 0, \xi'_\delta).$$

It follows that $\operatorname{Im} p(t, x', 0, \xi')$ and all derivatives with respect to x', ξ' must vanish at $(t, x'_\delta, 0, \xi'_\delta)$ if $s_\delta + \delta < t < t_\delta - \delta$, for otherwise we could choose $(x', \xi') \in V_\delta$ so close to (x'_δ, ξ'_δ) that

$$\operatorname{Im} p(t, x', 0, \xi') \neq 0, \quad \operatorname{Im} p(s_\delta, x', 0, \xi') < 0 < \operatorname{Im} p(t_\delta, x', 0, \xi').$$

The required change of sign must then occur in one of the intervals (s_δ, t) and (t, t_δ) which is impossible since they are shorter than $L_0 - \delta/2$.

Choose a sequence $\delta_j \to 0$ such that $\lim s_{\delta_j} = a_0$ and $\lim t_{\delta_j} = b_0$ exist. Then $b_0 - a_0 = L_0$ and $\operatorname{Im} p_{(\beta)}^{(\alpha)}(t, 0, \varepsilon_n) = 0$ for all α, β with $\alpha_1 = 0$ if $a_0 < t < b_0$. If $a_0 < b_0$ it follows in particular that we have a one dimensional bicharacteristic in the following sense:

Definition 26.4.9. A one dimensional bicharacteristic of the pseudo-differential operator with homogeneous principal symbol p is a C^1 map $\gamma: I \to T^*(X) \setminus 0$ where I is an interval on $\mathbb{R}$, such that

(i) $p(\gamma(t)) = 0$, $t \in I$,

(ii) $0 \neq \gamma'(t) = c(t) H_p(\gamma(t))$ if $t \in I$.

In order to achieve a simplification of p similar to that in Theorem 21.3.6 near a one dimensional bicharacteristic we shall now prove that the choice of the function q in Definition 26.4.6 is not very essential there.

Lemma 26.4.10. *Let $\gamma\colon I\to T^*(X)\setminus 0$ be the inclusion of a characteristic point for p or a compact one dimensional bicharacteristic interval and assume that for some $q\in C^\infty$ we have*

(i) *$q\neq 0$ and $\operatorname{Re} H_{qp}\neq 0$ on $\gamma(I)$,*

(ii) *there is a neighborhood U of $\gamma(I)$ where $\operatorname{Im} qp$ never changes sign from $-$ to $+$ along a bicharacteristic of $\operatorname{Re} qp$.*

Then (ii) is valid for every q satisfying (i).

Note that no homogeneity is assumed here so we could in fact have an arbitrary symplectic manifold. This will be allowed in the following more general statement of the result which is actually easier to prove.

Lemma 26.4.10′. *Let I be a point or a compact interval on $\mathbb{R}$, and let $\gamma\colon I\to M$ be an embedding of I in a symplectic manifold M as a one dimensional bicharacteristic of $p=p_1+ip_2$, if I is not reduced to a point, and any characteristic point otherwise. Let*

$$f_j=\sum_1^2 a_{jk}p_k,\quad j=1,2,$$

where $\det(a_{jk})>0$ on $\gamma(I)$. Assume that $H_{p_1}\neq 0$ and that $H_{f_1}\neq 0$ on $\gamma(I)$. If $\gamma(I)$ has a neighborhood U such that p_2 does not change sign from $-$ to $+$ along any bicharacteristic for p_1 in U, then U can be chosen so that f_2 has no such sign change along the bicharacteristics of f_1 in U.

Proof. First note that if $p=0$ at a point in U then

$$\{p_1,p_2\}=H_{p_1}p_2\leqq 0.$$

Hence we have at the same point

$$\begin{aligned}\{f_1,f_2\}&=\{a_{11}p_1+a_{12}p_2,\,a_{21}p_1+a_{22}p_2\}\\&=(a_{11}a_{22}-a_{12}a_{21})\{p_1,p_2\}\leqq 0.\end{aligned}$$

The proof is now divided into two steps, the first of which is quite trivial.

(i) Assume first that $a_{12}=0$. Since $a_{11}a_{22}>0$ either a_{11} and a_{22} are both positive or both negative. Thus the bicharacteristics of $f_1=a_{11}p_1$ are equal to those of p_1 with preserved and reversed orientation respectively, and $f_2=a_{22}p_2$ when $p_1=0$ so f_2 has the same and opposite sign as p_2, respectively. The lemma is therefore true in this case.

(ii) Proposition 26.1.6 obviously has an analogue for a general symplectic manifold where we just drop everything referring to the multiplicative structure in $T^*(X)\setminus 0$. The proof is the same except that we start from Theorem 21.1.6 instead of Theorem 21.3.1. By a canonical change of variables we can therefore make $M=\mathbb{R}^{2n}$, $p_1=\xi_1$ and $\Gamma=\gamma(I)$ equal to an interval on the x_1 axis. Let T be a vector $\in\mathbb{R}^{2n}$ with

$$\langle T, dp_1\rangle=1, \quad \langle T, df_1\rangle\neq 0 \quad \text{on } \Gamma.$$

Since dp_1 and df_1 do not vanish on Γ, the existence of T is obvious if Γ consists of a single point. Otherwise dp_2 is proportional to dp_1 on Γ so df_1 is proportional to dp_1. We can take any T with ξ_1 coordinate equal to 1 then.

Set

$$q_2(x,\xi)=p_2((x,\xi)-\xi_1 T)$$

which means that $p_2=q_2$ when $\xi_1=0$ and that q_2 is constant in the direction T. Then there is a C^∞ function ϕ such that

$$q_2=\phi p_1+p_2$$

so it follows from step (i) that the hypotheses in the lemma are fulfilled for p_1+iq_2. We have

$$f_1=(a_{11}-a_{12}\phi)p_1+a_{12}q_2,$$

hence

$$0\neq\langle T, df_1\rangle=(a_{11}-a_{12}\phi) \quad \text{on } \Gamma.$$

In a neighborhood of Γ we can therefore divide f_1 by $a_{11}-a_{12}\phi$ and set

$$q_1=f_1/(a_{11}-a_{12}\phi)=p_1+a_{12}(a_{11}-a_{12}\phi)^{-1}q_2$$

which implies

$$f_j=\sum_1^2 b_{jk}q_k, \quad j=1,2,$$

where $b_{11}=a_{11}-a_{12}\phi$, $b_{12}=0$ and $\det b=\det a>0$. Thus it follows from step (i) that it is sufficient to prove that (q_1,q_2) satisfies the hypothesis made on (p_1,p_2) in the lemma. The difficulty here is that the surfaces $p_1=0$ and $q_1=0$ are not the same. We shall identify them by projecting in the direction T.

Let U be a neighborhood of Γ where q_2 does not change sign from $-$ to $+$ on the bicharacteristics of p_1. Since T is transversal to the surface $f_1=q_1=0$ we can choose U so small that

$$Y=\{(x,\xi)\in U;\, q_1(x,\xi)=0\}$$

is mapped diffeomorphically by the projection π in the direction T on

$$X=\{(x,\xi)\in U;\, \xi_1=0\}.$$

When $q_1=q_2=0$, thus $p_1=p_2=0$, we have $H_{q_1}q_2=H_{p_1}p_2\leqq 0$. At a point in Y where $q_2=0$ and dq_2 vanishes on the tangent space of Y, we have $dq_2=0$ since $\langle T, dq_2\rangle=0$. Hence $w=H_{q_1}=H_{p_1}$ there so $\pi_* w=H_{p_1}$. If we apply the following lemma to $f=q_2=\pi^* q_2$ and the vector fields $v=(\pi^{-1})_* H_{p_1}$ and $w=H_{q_1}$ in Y, it follows that q_2 cannot change sign from $-$ to $+$ along a bicharacteristic of q_1 in Y, which proves the lemma.

Lemma 26.4.11. *Let $f \in C^1(Y)$ where Y is a C^2 manifold and let v be a Lipschitz continuous vector field in Y such that for any integral curve $t \mapsto y(t)$ of v we have*

$$f(y(0))<0 \Rightarrow f(y(t)) \leqq 0 \quad \text{for } t>0. \tag{26.4.7}$$

Let w be another Lipschitz continuous vector field such that

$$\langle w, df \rangle \leqq 0 \quad \text{when } f=0 \tag{26.4.8}$$

$$w=v \quad \text{when } f=df=0. \tag{26.4.9}$$

Then (26.4.7) *remains valid if $y(t)$ is an integral curve of w.*

Note that (26.4.8) is empty when $f=df=0$ so it is natural that another condition must be imposed then.

Proof. Let F be the closure of the union of all forward orbits for v starting at a point with $f(y)<0$. By (26.4.7) we have $f \leqq 0$ in F, and F contains the closure of the set where $f<0$. Orbits of v which start in F must remain in F. If now $(y,\eta) \in N_e(F)$ (Definition 8.5.7) then y is in the boundary of F so $f(y)=0$. If $df(y) \neq 0$ then F is bounded by the surface $f=0$ in a neighborhood of y, so η must be a positive multiple of $df(y)$ and $\langle w(y), \eta \rangle \leqq 0$ by (26.4.8). If $df(y)=0$ we have $\langle w(y),\eta\rangle = \langle v(y),\eta\rangle$ by (26.4.9), and $\langle v(y),\eta\rangle \leqq 0$ by condition (ii) in Theorem 8.5.11. Hence w satisfies condition (ii) in Theorem 8.5.11 so condition (i) there is also fulfilled, which proves the lemma.

Before proceeding with the proof of Theorem 26.4.7 we digress to give two alternative forms of condition (Ψ).

Theorem 26.4.12. *Each of the following conditions is necessary and sufficient for the homogeneous C^∞ function p in $T^*(Y) \setminus 0$ to satisfy condition (Ψ):*

(Ψ_1) *There is no C^∞ complex valued function q in $T^*(Y) \setminus 0$ such that* $\operatorname{Im} qp$ *changes sign from $-$ to $+$ when one moves in the positive direction on a bicharacteristic of* $\operatorname{Re} qp$ *where $q \neq 0$.*

(Ψ_2) *If Γ is a characteristic point with $H_p \neq 0$ or a compact one dimensional bicharacteristic interval with injective regular projection in $S^*(Y)$ then there exists a C^∞ function q in a neighborhood Ω of Γ such that* $\operatorname{Re} H_{qp} \neq 0$ *in Ω and* $\operatorname{Im} qp$ *does not change sign from $-$ to $+$ when one moves in the positive direction on a bicharacteristic of* $\operatorname{Re} qp$ *in Ω.*

Proof. It is clear that $(\Psi_1) \Rightarrow (\Psi)$; the difference is just that q is not assumed homogeneous in (Ψ_1). To prove that $(\Psi) \Rightarrow (\Psi_2)$ we only have to show that $\operatorname{Re} H_{qp} \neq 0$ on Γ for some homogeneous q. This is clear if Γ is a point. Otherwise Γ has a parametrization $t \mapsto \Gamma(t)$ with $\Gamma'(t) = c(t) H_p(\Gamma(t))$. If the parameter is suitably normalized then $c(t)$ and $\Gamma(t)$ are C^∞ functions. If π: $T^*(Y) \setminus 0 \to S^*(Y)$ is the projection then $t \mapsto \pi\Gamma(t)$ is an embedding of an interval so we can find a C^∞ function q_s on $S^*(X)$ with $q_s(\pi\Gamma(t))=c(t)$. Thus $q=\pi^* q_s$ is homogeneous of degree 0 and $\operatorname{Re} H_{qp}\neq 0$ on Γ.

It remains to prove that $(\Psi_2)\Rightarrow(\Psi_1)$ or equivalently that (Ψ_2) is false if (Ψ_1) is. So let q be any function in $C^\infty(T^*(Y)\setminus 0)$ such that $\operatorname{Im} qp$ changes sign from $-$ to $+$ on a bicharacteristic γ of $\operatorname{Re} qp$ where $q\neq 0$. As above we can find a compact one dimensional bicharacteristic interval $\Gamma\subset\gamma$ or a point $\Gamma\in\gamma$ such that the sign change occurs on bicharacteristics of $\operatorname{Re} qp$ arbitrarily close to Γ. By Lemma 26.4.10 this remains true for any other choice of q with $H_{\operatorname{Re} qp}\neq 0$ on Γ, so (Ψ_2) will be proved false if we show that π is injective and has injective differential on Γ, when Γ is a one dimensional bicharacteristic interval. If H_p has the radial direction at some point on Γ then the whole orbit of $H_{\operatorname{Re} qp}$ starting at Γ, and in particular γ, would just be a ray where $p=0$ identically. This contradicts our assumptions so π restricted to Γ has injective differential. If $\pi\Gamma$ is a closed smooth curve then p would also vanish identically on γ which is again contradictory. Finally it cannot happen that $\pi\circ\Gamma$ returns to the same position with a change of orientation, for a one dimensional bicharacteristic is uniquely determined by its starting point and the choice of orientation there. If $\pi\circ\Gamma(t_1)=\pi\circ\Gamma(t_2)$, $t_1<t_2$, and the orientations are opposed, then we can for any $t_1'>t_1$ close to t_1 find t_2' with $t_1'<t_2'<t_2$ and $\pi\circ\Gamma(t_1')=\pi\circ\Gamma(t_2')$. The supremum t of such t_1' must be equal to the infimum of the corresponding t_2' which contradicts that $\pi\circ\Gamma$ has a nonzero tangent at $\pi\circ\Gamma(t)$. Thus (Ψ_2) is false and the theorem is proved.

The interest of condition (Ψ_2) is of course that it eliminates the need to consider arbitrary functions q. In case Γ is a point it suffices to check it for $q=1$ and for $q=i$.

To simplify the principal symbol near a one dimensional bicharacteristic we need a global version of Theorem 21.3.6.

Proposition 26.4.13. *Let p be a C^∞ homogeneous function on $T^*(X)\setminus 0$, let I be a compact interval on $\mathbb{R}$ not reduced to a point and $I\ni t\mapsto\gamma(t)\in T^*(X)\setminus 0$ a one dimensional bicharacteristic, $\gamma\in C^\infty$. Assume also that the composition of γ and the projection $T^*(X)\setminus 0\to S^*(X)$ is injective, which means in particular that $H_p(\gamma(t))$ never has the radial direction. Then there is a homogeneous C^∞ canonical transformation χ from a conic neighborhood of $\{(x,\varepsilon_n),\, x_1\in I,\, x'=0\}$ in $T^*(\mathbb{R}^n)\setminus 0$ to a conic neighborhood of $\gamma(I)$ in $T^*(X)\setminus 0$ and a C^∞ homogeneous function a of degree $1-m$ with no zero on $\gamma(I)$ such that $\chi(x_1,0,\varepsilon_n)=\gamma(x_1)$, $x_1\in I$, and*

$$\chi^*(ap)=\xi_1+if(x,\xi') \tag{26.4.10}$$

where f is real valued, homogeneous of degree 1 and independent of ξ_1.

Proof. Essentially we just have to inspect the proof of Theorem 21.3.6 to see that it works globally. First choose as in the proof of Theorem 26.4.12 a C^∞

function q, homogeneous of degree $1-m$, such that $q(\gamma(t))=c(t)$ where c is the function in Definition 26.4.9. Then

$$\gamma'(t)=H_{\operatorname{Re} qp}(\gamma(t)), \quad d\operatorname{Im} qp=0 \quad \text{at } \gamma(t),\ t\in I.$$

From Proposition 26.1.6 it follows that we can find a canonical transformation χ satisfying the conditions in the theorem except that

$$\chi^*(qp)=\xi_1+ig(x,\xi)$$

where we only know that $dg=0$ on $I\times(0,\varepsilon_n)$. Using Malgrange's preparation theorem we can find h and r homogeneous of degree 0 and 1 respectively, and C^∞ in a neighborhood of $I\times(0,\varepsilon_n)$, so that

$$\xi_1=h(x,\xi)(\xi_1+ig(x,\xi))+r(x,\xi'). \tag{24.4.11}$$

In fact, it suffices to prove this when $\xi_n=1$ and then extend from there by homogeneity. As in the proof of Theorem 21.3.6 the preparation theorem gives a local solution at any point in $I\times(0,\varepsilon_n)$, and the local solutions can be pieced together by a partition of unity in x_1 to a solution in a neighborhood of $I\times(0,\varepsilon_n)$. Note that $h=1$ and $dr=0$ on $I\times(0,\varepsilon_n)$. Writing $r=r_1+ir_2$ we want to introduce

$$y_1=x_1, \quad \eta_1=\xi_1-r_1(x,\xi')$$

as new canonical variables. We choose

$$y_2=x_2,\ \eta_2=\xi_2,\ \dots,\ \eta_n=\xi_n \quad \text{when } x_1=0$$

and determine these canonical variables so that they are constant along the orbits of H_{η_1}. One of these contains $I\times(0,\varepsilon_n)$, so $y_2,\eta_2,\dots$ will be defined in a neighborhood. The commutation relations are fulfilled by the Jacobi identity since they hold when $x_1=0$. Hence we obtain a canonical transformation χ_1 keeping $I\times(0,\varepsilon_n)$ fixed, such that $h(x,\xi)(\xi_1+ig(x,\xi))$ composed with χ_1 is equal to $\eta_1+if(y,\eta)$ where $f(y,\eta)=-r_2(x,\xi')$. Now

$$\partial f/\partial\eta_1=\{f,y_1\}=-\{r_2,x_1\}=0$$

so $\chi\circ\chi_1$ and $q(\chi^{-1})^*h$ have the desired properties.

If we combine the discussion preceding Definition 26.4.9 with Proposition 24.4.13 or Theorem 21.3.6 we conclude in view of Lemma 26.4.10 that Theorem 26.4.7 follows if we prove

Theorem 26.4.7′. *Suppose that in a conic neighborhood of*

$$\Gamma=\{(x_1,0,0,\xi^0),\ a_0\leqq x_1\leqq b_0\}\subset T^*(\mathbb{R}^n)\setminus 0$$

the principal symbol of P has the form

$$p(x,\xi)=\xi_1+if(x,\xi')$$

where f is real valued and vanishes of infinite order on Γ if $b_0>a_0$. Assume also that in any neighborhood of Γ one can find an interval in the x_1

direction where f changes sign from $-$ to $+$ for increasing x_1. Then P is not solvable at Γ.

In the proof of Theorem 26.4.7′ we may also assume that the lower order terms $p_0, p_{-1}, \dots$ in the symbol of P are independent of ξ_1 near Γ. In fact, Malgrange's preparation theorem implies that

$$p_0(x,\xi)=q(x,\xi)(\xi_1+if(x,\xi'))+r(x,\xi')$$

where q is homogeneous of degree -1 and r homogeneous of degree 0. (See the proof of Proposition 26.4.13.) The term of degree 0 in the symbol of $(I-q(x,D))P$ is equal to $r(x,\xi')$. Repetition of the argument allows us to make the lower order terms successively independent of ξ_1.

To prove Theorem 26.4.7′ we shall use Lemma 26.4.5 which shows that it suffices to construct approximate solutions of the equation $P^*v=0$ concentrated so near Γ that (26.4.4) cannot hold. Let us first show how this can be done in the simple case where $\Gamma=\{(0,\varepsilon_n)\}\in T^*(\mathbb{R}^n)$ and $P=D_1+ix_1D_n$. (In that case we know of course already from Proposition 26.3.8 that there is no solvability.) Set

$$v_\tau(x)=\phi(x)\,e^{i\tau w(x)} \tag{26.4.12}$$

where $\phi\in C_0^\infty(\mathbb{R}^n)$ is equal to 1 in a neighborhood of 0 and

$$w(x)=x_n+i(x_1^2+x_2^2+\dots+x_{n-1}^2+(x_n+ix_1^2/2)^2)/2$$

satisfies the equation $P^*w=0$. If $\operatorname{supp}\phi$ is small enough then

$$\operatorname{Im} w(x)>|x|^2/4,\quad x\in\operatorname{supp}\phi,$$

so $v_\tau\to 0$ in $C^\infty(\mathbb{R}^n\setminus 0)$ and $\tau^N P^*v_\tau=\tau^N(P^*\phi)e^{i\tau w}\to 0$ in $C_0^\infty(\mathbb{R}^n)$ for any N. We have $v_\tau(x)=e^{i\tau x_n}V_\tau(x\sqrt{\tau})$ where $V_\tau(x)\to V(x)=e^{-|x|^2/2}$ in $\mathscr{S}$ as $\tau\to+\infty$. Since $\hat{v}_\tau(\xi)=\tau^{-n/2}\hat{V}_\tau((\xi-\tau\varepsilon_n)/\sqrt{\tau})$ it is clear that $\hat{v}_\tau(\xi)(1+|\xi|)^N\to 0$ uniformly for any N outside any conic neighborhood of ε_n and on any compact set, so $Av_\tau\to 0$ uniformly for any properly supported pseudo-differential operator A such that $(0,\varepsilon_n)\notin WF(A)$. We also have

$$\|v_\tau\|_{(s)}^2\,\tau^{-2s+n/2}\to\|V\|_{L^2}^2\quad\text{as } \tau\to\infty,$$

and these statements together show that (26.4.4) cannot be valid.

Using Theorem 21.3.3 and Proposition 26.3.1 we can adapt the preceding construction to prove that (26.4.4) is not valid if there is a point $(x,\xi)\in T^*(Y)\setminus 0\setminus WF(A)$ where $p(x,\xi)=0$ and $\{\operatorname{Re}p,\operatorname{Im}p\}(x,\xi)>0$. When proving Theorem 26.4.7′ we may therefore assume that

$$f(x,\xi')=0\Rightarrow \partial f(x,\xi')/\partial x_1\le 0 \tag{26.4.13}$$

in a neighborhood of Γ. This will be important for an application of Lemma 26.4.11 later on.

In the general proof of Theorem 26.4.7′ we shall take v_τ of the form

$$v_\tau(x)=e^{i\tau w(x)}\sum_0^M \phi_j(x)\tau^{-j} \tag{26.4.14}$$

where $\operatorname{Im} w \geqq 0$ with equality at some point and strict inequality outside a compact set, which makes v_τ very small and ϕ_j irrelevant there as $\tau\to\infty$. The principal symbol of P^* is $\xi_1 - i f(x,\xi')$ near Γ. To make $P^* v_\tau$ small the first step is therefore to construct a phase function w satisfying the *eiconal equation*

$$\partial w/\partial x_1 - i f(x, \partial w/\partial x')=0 \tag{26.4.15}$$

approximately. When that has been done, which is the main problem, we shall choose appropriate *amplitude functions* $\phi_0, \phi_1, \ldots$ successively. (Roughly speaking these steps correspond in the preceding proof of (26.4.13) to the application of Theorem 21.3.3 and of Proposition 26.3.1 respectively.)

To simplify notation we shall in what follows write t instead of x_1 and x instead of x', so (26.4.15) takes the form

$$\partial w/\partial t - i f(t, x, \partial w/\partial x)=0. \tag{26.4.15′}$$

To keep as much as possible of the qualitative properties of (26.4.12) we shall choose w so that $\operatorname{Im} w$ is strictly convex in x for fixed t and has its minimum on a smooth real curve $x=y(t)$. Thus we shall have

$$\operatorname{Im} \partial w(t,x)/\partial x=0 \quad \text{when } x=y(t),$$

so we are led to looking for a solution of (26.4.15)′ which has the form

$$w(t,x)=w_0(t)+\langle x-y(t), \eta(t)\rangle + \sum_{2\leqq|\alpha|\leqq M} w_\alpha(t)(x-y(t))^\alpha/|\alpha|!. \tag{26.4.16}$$

Here M is a large integer and it is convenient to use, during the present discussion only, the notation $\alpha=(\alpha_1,\ldots,\alpha_s)$ for a sequence of $s=|\alpha|$ indices between 1 and the dimension $n-1$ of the x variable. w_α will be symmetric in these indices. If we make sure that the matrix $(\operatorname{Im} w_{jk})$ is positive definite then $\operatorname{Im} w$ will have a strict minimum when $x=y(t)$ as a function of the x variables, for $\eta(t)$ will be *real valued.*

On the curve $x=y(t)$ the equation (26.4.15)′ reduces to

$$w_0'(t)=\langle y'(t), \eta(t)\rangle + i f(t, y(t), \eta(t)). \tag{0}$$

This is the only equation where w_0 occurs so it can be used to determine w_0 after y and η have been chosen. In particular

$$d \operatorname{Im} w_0(t)/dt = f(t, y(t), \eta(t)). \tag{0′}$$

If $f(t, y(t), \eta(t))$ has a sign change from $-$ to $+$ then $\operatorname{Im} w_0(t)$ will start decreasing and end increasing, so the minimum is attained at an interior point. We can normalize the minimum value to zero and have then for a suitable interval of t that $\operatorname{Im} w_0>0$ at the end points and $\operatorname{Im} w_0=0$ at some interior point. Thus $\operatorname{Im} w \geqq 0$ with equality attained but strict inequality valid outside a compact subinterval of the curve.

Our purpose is to make (26.4.15)′ valid apart from an error of order $M+1$ in $x-y(t)$. Actually $f(t,x,\xi)$ is not defined for complex ξ, but since

$$\partial w(t,x)/\partial x_j - \eta_j(t) = \sum w_{\alpha,j}(t)(x-y(t))^\alpha/|\alpha|!$$

this is given a meaning if $f(t,x,\partial w/\partial x)$ is replaced by the finite Taylor expansion

$$\sum_{|\beta|\leqq M} f^{(\beta)}(t,x,\eta(t))(\partial w(t,x)/\partial x - \eta(t))^\beta/|\beta|!.$$

Note that to compute the coefficient of $(x-y(t))^\alpha$ we just have to consider the terms with $|\beta|\leqq|\alpha|$. We have

$$\begin{aligned}\partial w/\partial t = w_0' - \langle y', \eta\rangle + \langle x-y, \eta'\rangle + \sum w_\alpha'(t)(x-y)^\alpha/|\alpha|!\\ -\sum_k \sum_{1\leqq|\alpha|\leqq M-1} w_{\alpha,k}(t)(x-y)^\alpha\, dy_k/dt/|\alpha|!,\end{aligned}$$

so the first order terms in the equation (26.4.15)′ give

$$(1)\qquad d\eta_j/dt - \sum_k w_{jk}(t)\,dy_k/dt = i\Big(f_{(j)}(t,y,\eta) + \sum_k f^{(k)}(t,y,\eta)\,w_{jk}(t)\Big).$$

Since y and η are real, this is a system of $2n$ equations

$$(1)'\qquad d\eta_j/dt - \sum_k \operatorname{Re} w_{jk}(t)\,dy_k/dt = -\sum_k \operatorname{Im} w_{jk}(t) f^{(k)}(t,y,\eta),$$

$$(1)''\qquad \sum_k \operatorname{Im} w_{jk}(t)\,dy_k/dt = -f_{(j)}(t,y,\eta) - \sum_k \operatorname{Re} w_{jk}(t) f^{(k)}(t,y,\eta).$$

When $\operatorname{Im} w_{jk}$ is positive definite we can solve these equations for dy/dt and $d\eta/dt$. At a point where $f=df=0$ they just mean that $dy/dt = d\eta/dt = 0$.

When $2\leqq|\alpha|\leqq M$ we obtain from (26.4.15)′ a differential equation

$$(\alpha)\qquad dw_\alpha/dt - \sum_k w_{\alpha,k}\,dy_k/dt = F_\alpha(t,y,\eta,\{w_\beta\})$$

where F_α is a linear combination of the derivatives of f of order $\leqq|\alpha|$ multiplied by polynomials in w_β with $2\leqq|\beta|\leqq|\alpha|+1$. (When $|\alpha|=M$ the sum on the left-hand side of (α) should of course be dropped and $|\beta|\leqq|\alpha|$.) Altogether (1)′, (1)″ and (α) form a quasilinear system of differential equations with as many equations as unknowns, so it is clear that we have local solutions with prescribed initial data. As seen above $F_\alpha(t,0,\xi^0,.)=0$ if $a_0<t<b_0$, so when $a_0<t<b_0$ we have the solution $y=0$, $\eta=\xi^0 = (0,\ldots,0,1)$, $w_\alpha=$ constant. Hence we can find $c>0$ such that the equations (1) and (α) with initial data

$$(26.4.17)\qquad w_{jk} = i\delta_{jk},\quad w_\alpha = 0 \quad \text{when } 2<|\alpha|\leqq M,\ t=(a_0+b_0)/2$$

$$(26.4.18)\qquad y=x,\quad \eta=\xi \quad \text{when } t=(a_0+b_0)/2$$

have a unique solution in (a_0-c, b_0+c) for all x,ξ with $|x|+|\xi-\xi^0|<c$. Moreover,

(i) $(\operatorname{Im} w_{jk}-\delta_{jk}/2)$ is positive definite,

(ii) the map

$$(x, \xi, t)\mapsto(y, \eta, t); \quad |x|+|\xi-\xi^0|<c,\ a_0-c<t<b_0+c$$

is a diffeomorphism.

In the range X_c of the map (ii) we denote by v the image of the vector field $\partial/\partial t$ under the map. Thus v is the tangent vector field of the integral curves. Note that $v=\partial/\partial t$ when $f=df=0$. Since we have assumed above that $f=0$ implies $\partial f/\partial t\leqq 0$ in X_c (cf. (26.4.13)), if c is small enough, we can now apply Lemma 26.4.11 with the vector field v just defined and $w=\partial/\partial t$. The conclusion is that f must have a change of sign from $-$ to $+$ along an integral curve of v in X_c, for otherwise there would be no such sign change for increasing t and fixed (x, ξ), and that contradicts the hypothesis in Theorem 26.4.7'. Recalling the discussion of the equation (0) above we have therefore proved

Lemma 26.4.14. *Assume that the hypotheses of Theorem 26.4.7' are fulfilled and that in a neighborhood of Γ we have $\partial f/\partial t\leqq 0$ when $f=0$, the variables being denoted by (t, x) now. Given M one can then find*

(i) *a curve $t\mapsto(t, y(t), 0, \eta(t))\in\mathbb{R}^{2n}$, $a'\leqq t\leqq b'$, as close to Γ as desired,*

(ii) *C^∞ functions $w_\alpha(t)$, $2\leqq|\alpha|\leqq M$, with $(\operatorname{Im} w_{jk}-\delta_{jk}/2)$ positive definite when $a'\leqq t\leqq b'$*

(iii) *a function w_0 with $\operatorname{Im} w_0(t)\geqq 0$, $a'\leqq t\leqq b'$, $\operatorname{Im} w_0(a')>0$,*

$$\operatorname{Im} w_0(b')>0 \quad \text{and} \quad \operatorname{Im} w_0(c')=0 \quad \text{for some } c'\in(a', b')$$

such that (26.4.16) is a formal solution of (26.4.15)' with an error which is $O(|x-y(t)|^{M+1})$.

Before passing to the choice of the functions ϕ_j in (26.4.14) we shall make some general remarks which show what is required to disprove (26.4.4). In doing so we revert to the symmetric notation in (26.4.14) where x denotes all the variables in $\mathbb{R}^n$.

Lemma 26.4.15. *Let v_τ be defined by (26.4.14) where $w\in C^\infty(X)$, $\phi_j\in C_0^\infty(X)$, $\operatorname{Im} w\geqq 0$ in X and $d\operatorname{Re} w\neq 0$. Here X is an open set in $\mathbb{R}^n$. For any positive integer N we have then*

$$\|v_\tau\|_{(-N)}\leqq C\tau^{-N}, \quad \tau>1. \tag{26.4.19}$$

If $\operatorname{Im} w(x_0)=0$ and $\phi_0(x_0)\neq 0$ for some $x_0\in X$ then

$$\|v_\tau\|_{(-N)}\geqq c\tau^{-n/2-N}, \quad \tau>1, \tag{26.4.20}$$

for some $c>0$. If $\tilde\Gamma$ is the cone generated by

$$\{(x, w'(x)),\ x\in\bigcup\operatorname{supp}\phi_j,\ \operatorname{Im} w(x)=0\} \tag{26.4.21}$$

then $\tau^k v_\tau \to 0$ in $\mathscr{D}'_{\tilde\Gamma}$ as $\tau\to\infty$, hence $\tau^k A v_\tau \to 0$ in $C^\infty(\mathbb{R}^n)$, if A is a pseudo-differential operator with $WF(A)\cap\tilde\Gamma=\emptyset$, and k is any real number.

Proof. For every neighborhood U of the projection of (26.4.21) on the second component in $\mathbb{R}^n$ and every positive integer v we have

$$|\hat v_\tau(\xi)| \leqq C_v(1+|\xi|+\tau)^{-v} \quad \text{if } \tau>1,\ \xi/\tau\notin U. \tag{26.4.22}$$

This follows from Theorem 7.7.1 since $x\mapsto(\tau w-\langle x,\xi\rangle)/(\tau+|\xi|)$ is in a compact set of functions with non-negative imaginary part and differential $\neq 0$ at the real points. If we choose U bounded with $0\notin\bar U$ then

$$\int_{\tau U} |\hat v_\tau(\xi)|^2(1+|\xi|^2)^{-N}\,d\xi = O(\tau^{-2N})$$

since v_τ is bounded in L^2. Together with (26.4.22) this gives (26.4.19). If $\chi\in C_0^\infty$ the estimate (26.4.22) is applicable to χv_τ as well. Hence

$$|\widehat{\chi v_\tau}(\xi)| \leqq C_v(1+|\xi|+|\tau|)^{-v}, \qquad \tau>1,\ \xi\in V,$$

if V is any closed cone with $\tilde\Gamma\cap(\operatorname{supp}\chi\times V)=\emptyset$; hence $\tau^k v_\tau\to 0$ in $\mathscr{D}'_{\tilde\Gamma}$ for every k. To prove (26.4.20) finally we assume that $x_0=0$ and observe that when $\psi\in C_0^\infty$ we have if $w(0)=0$

$$\begin{aligned}\tau^n\langle v_\tau,\psi(\tau\,.)\rangle &= \int e^{i\tau w(x/\tau)}\psi(x)\sum \phi_j(x/\tau)\tau^{-j}\,dx\\ &\to \int e^{i\langle x,w'(0)\rangle}\psi(x)\phi_0(0)\,dx,\end{aligned}$$

which is not equal to 0 for a suitable choice of ψ. Since

$$\|\psi(\tau\,.)\|_{(N)} = O(\tau^{N-n/2})$$

it follows that $c\leqq \tau^{N+n/2}\|v_\tau\|_{(-N)}$, which proves (26.4.20).

As already pointed out Theorem 26.4.7′ will be proved by showing that (26.4.4) cannot be valid. To do so we first use Lemma 26.4.14 to choose a function w in a neighborhood Y of $\{(x_1,0);\ a_0\leqq x_1\leqq b_0\}\subset\mathbb{R}^n$ such that $\operatorname{Im} w>0$ in Y except on a compact non-empty subset K of a curve $x'=y(x_1)$. In addition

$$\Gamma_0=\{(x,w'(x)),\ x\in K\}$$

is in a small conic neighborhood of Γ which does not meet $WF(A)$ and where the symbol of P^* is of the form $\xi_1+iF(x,\xi')$, $-f$ being the principal symbol of F. If we apply (26.4.4) to a function of the form (26.4.14) where $\phi_0\neq 0$ at some point in K, it follows from Lemma 26.4.15 that the left-hand side has a lower bound $c\tau^{-n/2-N}$ and that there is a smaller bound for the last two terms in the right-hand side. If we can prove that

$$\|P^* v_\tau\|_{(v)} = O(\tau^{-N-(n+1)/2}) \tag{26.4.23}$$

it will follow that (26.4.4) is not valid.

If B is a pseudo-differential operator of order 0 with symbol 1 in a conic neighborhood of Γ_0 then $(I-B)P^*\tau^k v_\tau\to 0$ in C^∞ for any k. We can choose

B with $WF(B)$ so small that $\xi' \neq 0$ and the symbol of P^* is $\xi_1 + iF(x, \xi')$ in a conic neighborhood. Then it follows from Theorem 18.1.35 that the product $B(P^* - D_{x_1} - iF(x, D'))$ is a pseudo-differential operator with wave front set disjoint with Γ_0. Here $F(x, D')$ is a pseudo-differential operator in $n-1$ variables depending on x_1 as parameter. Hence (26.4.23) will follow if

$$\|(D_{x_1} + iF(x, D'))v_\tau\|_{(v)} = O(\tau^{-N-(n+1)/2}). \tag{26.4.24}$$

Since $F(x, D')$ is a pseudo-differential operator in the variables $(x_2, \ldots, x_n)$ it is convenient to change notation again so that x_1 is denoted by t and the other variables are denoted by x. Thus (26.4.24) is written now

$$\|(D_t + iF(t, x, D))v_\tau\|_{(v)} = O(\tau^{-N-(n+1)/2}). \tag{26.4.24)'$$

In our construction we shall actually aim for the estimate

$$|(D_t + iF(t, x, D))v_\tau| \leq C\tau^{-N-(n+2)/2-v}, \tag{26.4.24)''$$

and we shall see afterwards that (26.4.24)' is obtained at the same time.

To make the left-hand side of (26.4.24)' small we need a formula for how $F(t, x, D)$ acts on functions of the form (26.4.14). This could be obtained from the work in Section 25.3, but we prefer a direct elementary proof. To simplify notation we suppress the parameter t in the proof.

Lemma 26.4.16. *Let* $q(x, \xi) \in S^\mu(\mathbb{R}^{n-1} \times \mathbb{R}^{n-1})$, *let* $\phi \in C_0^\infty(\mathbb{R}^{n-1})$, $w \in C^\infty(\mathbb{R}^{n-1})$, *and assume that* $\operatorname{Im} w > 0$ *except at a point* y *where* $w'(y) = \eta \in \mathbb{R}^{n-1} \setminus 0$ *and* $\operatorname{Im} w''$ *is positive definite. Then*

$$|q(x, D)(\phi e^{i\tau w}) - \sum_{|\alpha|<k} q^{(\alpha)}(x, \tau\eta)(D - \tau\eta)^\alpha(\phi e^{i\tau w})/\alpha!| \leq C_k \tau^{\mu-k/2}; \tag{26.4.25}$$

$$\tau > 1, \; k = 1, 2, \ldots.$$

Proof. Let us first observe that

$$|(D - \tau\eta)^\alpha \phi e^{i\tau w}| = |D^\alpha \phi e^{i\tau(w - \langle \cdot, \eta\rangle)}| \leq C\tau^{|\alpha|/2}. \tag{26.4.26}$$

In fact, if j of the $|\alpha|$ derivatives fall on the exponential they bring out a factor τ^j but also j factors $\partial w/\partial x_i - \eta_i$ vanishing at y. If $j > |\alpha|/2$ the remaining $|\alpha| - j$ derivatives can only reduce the order of the zero to $j - (|\alpha| - j) = 2j - |\alpha|$, so the term is bounded by a constant times

$$\tau^j |x - y|^{2j - |\alpha|} e^{-c\tau|x-y|^2} \leq C\tau^{|\alpha|/2}.$$

Note that (26.4.26) explains the power of τ in the right-hand side of (26.4.25).

To prove (26.4.25) we set $u_\tau(x) = \phi(x) e^{i\tau w(x)}$ and study

$$\hat{u}_\tau(\xi) = \int \phi(x) e^{i(\tau w(x) - \langle x, \xi\rangle)} dx.$$

As in the proof of (26.4.22) it is clear that for every v

$$|\hat{u}_\tau(\xi)| \leq C(|\xi| + \tau)^{-v}, \quad \tau > 1, \tag{26.4.27}$$

if $|\xi/\tau-\eta|\geqq|\eta|/2$, say. On the other hand, if $|\xi/\tau-\eta|<|\eta|/2$ we can also apply Theorem 7.7.1 with $f(x)=w(x)-\langle x,\xi/\tau\rangle$ noting that

$$|\eta-\xi/\tau|^2+|x-y|^2\leqq C(|\operatorname{grad} f|^2+\operatorname{Im} f)$$

since $f'(x)=w'(x)-\xi/\tau=\eta-\xi/\tau+O(|x-y|)$. Hence

$$|\hat{u}_\tau(\xi)|\leqq C_k\tau^{-k}|\eta-\xi/\tau|^{-2k}=C_k\tau^k|\tau\eta-\xi|^{-2k}. \tag{26.4.28}$$

In the integral

$$q(x,D)(\phi e^{i\tau w})=(2\pi)^{-n}\int e^{i\langle x,\xi\rangle}q(x,\xi)\hat{u}_\tau(\xi)\,d\xi$$

the contributions when $|\xi/\tau-\eta|>|\eta|/2$ are rapidly decreasing by (26.4.27). When $|\xi/\tau-\eta|<|\eta|/2$ we replace q by the Taylor expansion at $\tau\eta$ of order k. This gives the sum

$$\sum_{|\alpha|<k}(2\pi)^{-n}\int e^{i\langle x,\xi\rangle}q^{(\alpha)}(x,\tau\eta)(\xi-\tau\eta)^\alpha\hat{u}_\tau(\xi)\,d\xi/\alpha!.$$

Extending the integration to the whole space will only mean a change by a rapidly decreasing function, again by (26.4.27), and the sum is then equal to that in (26.4.25). The error term in Taylor's formula can be estimated by $C\tau^{\mu-k}|\tau\eta-\xi|^k$. If we use (26.4.28) with k replaced by $k/2$ it follows that the contribution from the error term to the integral when $|\xi/\tau-\eta|<|\eta|/2$ is $O(\tau^{\mu+n-k/2})$. This proves (26.4.25) apart from an extra factor τ^n in the right-hand side. If we apply this weaker result with k replaced by $k+2n$ and recall (26.4.26), we obtain (26.4.25).

If q is homogeneous of degree μ, then the sum in (26.4.25) consists apart from the factor $e^{i\tau w}$ of terms which are homogeneous in τ of degree μ, $\mu-1,\dots$. The terms of degree μ are those in

$$\phi\sum q^{(\alpha)}(x,\tau\eta)(\tau w'(x)-\tau\eta)^\alpha/\alpha!$$

which is the Taylor expansion at $\tau\eta$ of the possibly undefined quantity $q(x,\tau w')$, just as in the discussion of (26.4.15)' above. The terms of degree $\mu-1$ where ϕ is differentiated are similarly

$$\sum_1^{n-1}q^{(k)}(x,\tau w'(x))D_k\phi$$

where $q^{(k)}$ should be replaced by the Taylor expansion at $\tau\eta$ representing the value at $\tau w'(x)$.

End of Proof of Theorem 26.4.7'. With v_τ defined by (26.4.14) we have now proved that

$$(D_t+iF(t,x,D))v_\tau=e^{i\tau w}\left(\sum_0^M\psi_j\tau^{-j}\right)+O(\tau^{(1-M)/2}),$$

where

$$\psi_j=D_t\phi_j+\sum_2^n c_k(t,x)D_k\phi_j+c_0\phi_j+R_j \tag{26.4.29}$$

with $R_0=0$ and R_j otherwise determined by $\phi_0, \dots, \phi_{j-1}$. Here c_k is a partial sum of the Taylor series at $\eta(t)$ for $-if^{(k)}(t,x,w'_x(t,x))$ but this is of no importance. Set

$$\phi_0(t,x)=\sum_{|\alpha|<M} \phi_{0\alpha}(t)(x-y(t))^\alpha$$

with $y(t)$ as in Lemma 26.4.14. Then

$$\psi_0=\left(D_t+\sum_2^n c_k(t,x)D_k+c_0\right)\phi_0=O((x-y(t))^M)$$

if $\phi_{0\alpha}$ satisfy a certain linear system of ordinary differential equations

$$D_t\phi_{0\alpha}+\sum_{|\beta|<M} a_{\alpha\beta}\phi_{0\beta}=0.$$

We can solve these equations so that $\phi_0(0)=1$ at a chosen point in K. In the same way we can successively choose ϕ_j so that

$$\psi_j(t,x)=O((x-y(t))^{M-2j}), \quad \text{when } j<M/2.$$

If M is chosen so that $(1-M)/2\leqq -N-(n+1)/2-\nu$, we obtain (26.4.24)″.

The asymptotic series in (26.4.25) remains valid if we differentiate with respect to x or a parameter t, though a factor τ may be lost in the estimate. In proving (26.4.24)″ we have also used that a function of the form $\chi(t,x)e^{i\tau w}$ can be estimated by $\tau^{-k/2}$ if χ vanishes of order k when $x=y(t)$. A differentiation can lead to a decrease of the order of the zero by one unit or to a factor τ when the exponential is differentiated, so the estimate may deteriorate by a factor τ. In any case it is clear that we obtain (26.4.24)′ when we compute the derivatives of order $\leqq\nu$, so (26.4.4) is not valid. This completes the proof of Theorem 26.4.7′.

26.5. Geometrical Aspects of Condition (P)

Unfortunately it is not yet known if the converse of Corollary 26.4.8 is valid. However, if P is a differential operator one can strengthen condition (Ψ) since the principal symbol $p(x,\xi)$ then has the symmetry property

$$p(x,-\xi)=(-1)^m p(x,\xi).$$

If $t\mapsto(x(t),\xi(t))$ is a bicharacteristic of $\operatorname{Re}p(x,\xi)$ it follows that $t\mapsto(x(t), -\xi(t))$ is also a bicharacteristic curve with the correct (reversed) orientation if m is odd (even). If the condition (Ψ) is fulfilled in X by p it follows that $\operatorname{Im}p(x,\xi)$ cannot take both positive and negative values on the bicharacteristic, that is, $\bar{p}$ also satisfies condition (Ψ).

Definition 26.5.1. A C^∞ homogeneous function p in $T^*(X)\setminus 0$ is said to satisfy condition (P) if p and $\bar{p}$ both satisfy condition (Ψ), that is, there is no

C^∞ complex valued function q in $T^*(X)\setminus 0$ such that $\operatorname{Im} qp$ takes both positive and negative values on a bicharacteristic of $\operatorname{Re} qp$ where $q \neq 0$.

Here we have chosen to use the equivalent form (Ψ_1) of (Ψ) in Theorem 26.4.12, which is applicable in any symplectic manifold. Of course we could have used (Ψ_2) or (Ψ) as well.

Thus condition (P) is necessary for solvability in the case of differential operators although not for general pseudo-differential operators. At the end of this chapter we shall prove that, conversely, an analogue of Theorem 26.1.7 is valid for every pseudo-differential operator P satisfying condition (P). The proof will be based on theorems concerning propagation of singularities which extend Theorems 26.1.4 and 26.2.1. These will be the main topic of the following sections. As a preparation we shall discuss in this section some geometrical properties of the characteristic set

$$N=\{(x,\xi)\in T^*(X)\setminus 0,\ p(x,\xi)=0\}$$

which follow from condition (P), *which we assume fulfilled throughout.*

As in Section 26.2 we set

$$N_2=\{(x,\xi)\in N;\ H_{\operatorname{Re} p} \text{ and } H_{\operatorname{Im} p} \text{ are linearly independent at } (x,\xi)\}.$$

This is a conic manifold of codimension 2. By condition (P) it is involutive, that is,

$$\{\operatorname{Re} p, \operatorname{Im} p\}(x,\xi)=0, \quad (x,\xi)\in N_2,$$

for $\{\operatorname{Re} p, \operatorname{Im} p\}=H_{\operatorname{Re} p}\operatorname{Im} p$ must vanish at the zeros of $\operatorname{Im} p$ on a bicharacteristic of $\operatorname{Re} p$ since the sign would otherwise change. Thus $H_{\operatorname{Re} p}$ and $H_{\operatorname{Im} p}$ are tangents to N_2.

In Section 26.4 we introduced the term semi-bicharacteristic of p for a bicharacteristic of $\operatorname{Re} qp$ where $q\neq 0$. The advantage of this notion is that through every point in N with $dp\neq 0$ there is at least one semi-bicharacteristic curve. We shall now examine to what extent semi-bicharacteristics are one dimensional bicharacteristics in the sense of Definition 26.4.9. In doing so we shall use the following lemma, which is obtained by applying to $\operatorname{Re} qp$ the non-homogeneous version of Proposition 26.1.6. (This was also used in the proof of Lemma 26.4.10′.)

Lemma 26.5.2. *Let $I=[a,b]$ be a compact interval on $\mathbb{R}$ not reduced to a point, and let $I\ni t\mapsto\gamma(t)$ be a bicharacteristic arc for $\operatorname{Re} qp$ where $0\neq q\in C^\infty$. Then there is a symplectomorphism χ from a neighborhood V of $J=\{(x_1,0,\dots,0),\ x_1\in I\}\subset T^*(\mathbb{R}^n)$ to a neighborhood of $\gamma(I)\subset T^*(X)\setminus 0$ such that $\chi(x_1,0,\dots,0)=\gamma(x_1)$ and*

$$\chi^*(qp)(x,\xi)=\xi_1+if(x,\xi)$$

where f is real valued.

Assume now that $\gamma(a)\in N_2$. In a neighborhood of $a_0=(a,0,\dots,0)\in T^*(\mathbb{R}^n)$ the manifold $\chi^{-1}N_2$ is invariant under the vector field $\partial/\partial x_1$ so it is defined by $\xi_1=0$, $g(x',\xi')=f(a,x',0,\xi')=0$ where $x'=(x_2,\dots,x_n)$, $\xi'=(\xi_2,\dots,\xi_n)$. Since $\partial f/\partial x_1=\{\xi_1,f\}=0$ at a_0 we know that $dg\neq 0$ at 0. The parallels of the x_1 axis in the plane $\xi_1=0$ are semi-bicharacteristics so (P) gives in a neighborhood of J that $f(x,\xi)\geqq 0$ (resp. $f(x,\xi)\leqq 0$) when $\xi_1=0$ and $g(x',\xi')>0$ (resp. $g(x',\xi')<0$). Hence $f(x,\xi)=0$ when $\xi_1=0$ and $g(x',\xi')=0$, and for some $\varepsilon>0$

$$f(x,0,\xi')=g(x',\xi')h(x,\xi'),\quad \text{if } a-\varepsilon<x_1<b+\varepsilon,\ |x'|+|\xi'|<\varepsilon.$$

Here $h\geqq 0$, $h(a,x',\xi')>0$ if $|x'|+|\xi'|<\varepsilon$, $g(0)=0$ and $dg(x',\xi')\neq 0$ when $|x'|+|\xi'|<\varepsilon$. Thus $f=0$ on J and

$$df=\partial f/\partial\xi_1\, d\xi_1+h\,dg \quad \text{on } J,$$

so df is proportional to $d\xi_1$ except at points where $h\neq 0$, and they are in N_2. A semi-bicharacteristic starting in N_2 is therefore a one-dimensional bicharacteristic when it is not in N_2. (If an isolated point is not in N_2 the tangent is still proportional to H_p there.) Also note that if $f(x,0,\xi')=0$ and $g(x',\xi')\neq 0$ then $h(x,\xi')=0$ which implies $dh(x,\xi')=0$ since $h\geqq 0$, so $df=\partial f/\partial\xi_1\, d\xi_1$. If a semi-bicharacteristic starting in N_2 contains some point $\chi(x,\xi)$ with (x,ξ) in

$$V=\{(x,\xi);\ \xi_1=0,\ a-\varepsilon<x_1<b+\varepsilon,\ |x'|+|\xi'|<\varepsilon\}$$

and $g(x',\xi')\neq 0$ it must therefore run in the x_1 direction as a one dimensional bicharacteristic until it leaves V. Extending it if necessary so that it contains a point with $x_1=a$ we obtain a contradiction since $f(a,x',\xi')\neq 0$ when $g(x',\xi')\neq 0$. It follows that in $\chi(V)$ the union of all semi-bicharacteristics emanating from a point anywhere in N_2 is defined by $\xi_1=g(x',\xi')=0$, so it is an involutive manifold of codimension 2. Hence we have proved

Proposition 26.5.3. *The union N_2^e of all semi-bicharacteristics of p which meet N_2 is a locally closed conic involutive submanifold of $T^*(X)\smallsetminus 0$ of codimension 2 on which $p=0$. Thus $H_{\operatorname{Re}p}$ and $H_{\operatorname{Im}p}$ are tangents to N_2^e, and N_2 is an open subset of N_2^e.*

From Theorem 21.2.7 and the remark following it we know that as involutive manifold N_2^e has a natural foliation with 2 dimensional leaves, having the symplectically orthogonal plane of the tangent plane of N_2^e as tangent planes. It is natural to extend the terminology used in Section 26.2 as follows:

Definition 26.5.4. The leaves of the natural foliation of N_2^e are called two dimensional bicharacteristics.

We recall from Theorem 21.2.7 that a leaf B is either conic or else the radial vector field is never tangent to it. In any case H_p is a complex tangent vector field of B. In Section 26.2 we used H_p to define a complex structure in the leaves of the foliation of N_2 such that the analytic functions are the solutions of the equation $H_p w=0$. In a leaf B of the foliation of N_2^e this equation implies that w is constant on the one dimensional bicharacteristics which may be embedded in B. Let B_0 be the subset of B consisting of semi-bicharacteristics with both end points in N_2, and let $\tilde{B}_0$ be the reduced two dimensional bicharacteristic obtained by identifying points in B_0 which are connected by a one dimensional bicharacteristic. We shall prove in Section 26.7 that $\tilde{B}_0$ has a natural structure as Riemann surface such that the analytic functions lifted to B_0 are precisely the solutions of the equation $H_p w=0$. In Section 26.9 we shall show that Theorem 26.2.1 can be extended to N_2^e with superharmonicity defined in terms of this analytic structure. In the proofs we shall use the following supplement to Proposition 26.4.13 to simplify the principal symbol.

Proposition 26.5.5. *Suppose with the notation in Proposition 26.4.13 that $\gamma(I)$ is a one dimensional bicharacteristic contained in N_2^e which cannot be extended at both end points as a one dimensional bicharacteristic. Then Proposition 26.4.13 is applicable, and in a neighborhood of $I\times\{0\}\times\{\varepsilon_n\}$ we have*

$$f(x,\xi')=g(x',\xi')\,h(x,\xi')$$

where $h\geqq 0$, $dg\neq 0$, both factors are in C^∞ and h is homogeneous of degree 0, g homogeneous of degree 1.

Proof. The tangent of $\gamma(I)$ cannot be radial and the projection of $\gamma(I)$ on $S^*(X)$ cannot be a closed curve since $\gamma(I)$ is maximal at one end. (In the proof that $(\Psi_2)\Rightarrow(\Psi_1)$ in Theorem 26.4.12 we saw that the projection cannot return to the same point with reversed orientation.) Hence Proposition 26.4.13 is applicable, and since $f(x_1,0,\varepsilon_n)=0$ for x_1 in a neighborhood of I and $df\neq 0$ at some point there, the factorization follows from the discussion of N_2^e following Lemma 26.5.2.

Remark. Proposition 26.5.5 and its proof remain valid if I is a point and $\gamma(I)\in N_2^e$ but does not belong to any one dimensional bicharacteristic. Proposition 26.4.13 is just replaced by Theorem 21.3.6.

Set $N_1=N\smallsetminus N_2^e$. We shall now discuss semi-bicharacteristics $I_0\ni t\mapsto\gamma(t)$ such that $\gamma(t_0)\in N_1$ for some $t_0\in I_0$. By the definition of N_2^e we know that $\gamma(I_0)$ cannot intersect N_2, so $H_{\mathrm{Re}\,p}$ and $H_{\mathrm{Im}\,p}$ are linearly dependent at every point in $\gamma(I_0)\cap N$. If $\gamma(I_0)\subset N$ it follows that γ is a one dimensional bicharacteristic. Let I be the largest subinterval of I_0 containing t_0 such that $\gamma(I)\subset N$. If I has an end point contained in the interior of I_0 it is clear that $\gamma(I)$ cannot be continued there as a one dimensional bicharacteristic, for it would have to coincide with $\gamma(I_0)$ then.

Proposition 26.5.6. *With the notation in Proposition 26.4.13 assume that $\gamma(I)\cap N_2^e=\emptyset$ and that $\gamma(I)$ cannot be extended in both directions as a one* *dimensional bicharacteristic. Then Proposition* 26.4.13 *is applicable and* $f \geqq 0$, *or* $f \leqq 0$, *in a neighborhood of* $I \times \{0\} \times \{\varepsilon_n\}$.

Proof. The maximality of $\gamma(I)$ shows as in the proof of Proposition 26.5.5 that Proposition 26.4.13 can be applied. If $f(x_1, 0, \varepsilon_n)$ vanishes for all x_1 in a neighborhood of I then $df=0$ since $\gamma(I) \cap N_2^e = \emptyset$. Hence we would have a one dimensional bicharacteristic with $\gamma(I)$ in its interior. This is a contradiction proving that $f(x_1, 0, \varepsilon_n) \neq 0$ somewhere. By the continuity of f and condition (P) this sign is kept in the wide sense in a neighborhood of $I \times \{0\} \times \{\varepsilon_n\}$, since f does not depend on ξ_1.

We shall denote by N_{11} the set of all points in N_1 which lie on a semi-bicharacteristic with one non-characteristic end point. In Section 26.6 we shall show that Theorem 26.1.4 can be extended to control the singularities in N_{11} with the slight modification that regularity only propagates in one direction, determined by the sign of f in Proposition 26.5.6.

A semi-bicharacteristic starting at a point in $N_1 \setminus N_{11}$ is always a one dimensional bicharacteristic, for it never leaves the characteristic set by the definition of N_{11} and it cannot enter N_2 by the definition of N_2^e. Thus $N_1 \setminus N_{11}$ is the set through which one dimensional bicharacteristics can be prolonged indefinitely.

Let us now consider Proposition 26.4.13 when $\gamma(I) \subset N_1 \setminus N_{11}$. Condition (P) then states that the sign of f is independent of x_1, and $f = df = 0$ on $I \times \{0\} \times \{\varepsilon_n\}$. However, it is not always possible to factor f as in Proposition 26.5.5 into a product of a non-negative function and one which does not depend on x_1. (This can be done in the real analytic case.) When f just vanishes of second order, this is nearly possible though:

Proposition 26.5.7. *Suppose with the hypotheses and notation of Proposition* 26.4.13 *that for some* $t_0 \in I$ *and complex number* c *the quotient* dp/c *is real and the Hessian of* $\operatorname{Im} p/c$ *is not identically* 0 *at* $\gamma(t_0)$, *in the plane defined by* $dp=0$. *Then we have near* $I \times \{0\} \times \{\varepsilon_n\}$

$$f(x, \xi') = g(x', \xi')\, h(x, \xi') + r(x, \xi')$$

where $g(x', \xi') = f(t_0, x', \xi')$, h *and* r *are in* C^∞, r *is homogeneous of degree* 1, $h \geqq 0$. *Moreover,* $r \equiv 0$ *unless the Hessian of* g *is positive (resp. negative) semi-definite at* 0, *and then* r *must still vanish when* $g < 0$ (*resp.* $g > 0$).

If f has constant sign near $I \times \{0\} \times \{\varepsilon_n\}$, we can of course take $r = f$ so the assertion is trivial then. Note that the Hessian of $\operatorname{Im} p/c$ is invariantly defined since $\operatorname{Im} p/c$ vanishes of the second order at $\gamma(t_0)$. If p is multiplied by a function $q_1 + iq_2$ which is real there, then $\operatorname{Im} p/c$ is replaced by $q_1 \operatorname{Im} p/c + q_2 \operatorname{Re} p/c$, and the Hessian of the second term is zero in the plane $dp=0$. Thus the hypothesis is invariant under coordinate changes and multiplication by non-vanishing functions as well.

Proof of Proposition 26.5.7. It is enough to make the decomposition when $\xi_n=1$ and extend by homogeneity to $\xi_n>0$ afterwards. Let us write $t=x_1$ and $y=(x_2, \dots, x_n, \xi_2, \dots, \xi_{n-1})$. That $\gamma(I)$ is a one dimensional bicharacteristic means that $f=df=0$ on $I\times\{0\}$, and by hypothesis the Hessian of $g(y)=f(t_0, y, 1)$ with respect to y is not 0 when $y=0$, for the second derivatives containing some t or ξ_n must vanish. Assume for example that $\partial^2 g(0)/\partial y_1^2 \neq 0$. By Malgrange's preparation theorem (Theorem 7.5.5)

$$g(y)=k(y)(y_1^2+a_1(y')y_1+a_0(y'))$$

where $k(0)\neq 0$ and $y'=(y_2, \dots)$. We can take $y_1+a_1(y')/2$ as a new variable and divide by $k(y)$ so we may assume that with $a=a_0-a_1^2/4$

$$g(y)=y_1^2+a(y').$$

Now Malgrange's preparation theorem (Theorem 7.5.6) gives, near $I\times\{0\}$,

$$f(t, y, 1)=h(t, y)g(y)+r(t, y); \quad r(t, y)=b(t, y')y_1+c(t, y'),$$

where $h, b, c\in C^\infty$. When $a(y')<0$ we have two simple zeros $y_1=\pm(-a(y'))^{\frac{1}{2}}$ and obtain $b(t, y')=c(t, y')=0$ by condition (*P*), hence $r(t, y)=0$ when $g<0$. This completes the proof if the Hessian of g is semi-definite. Otherwise we can apply the same argument to prove that $f(t, y, 1)/g(y)$ is also equal to a C^∞ function when $g(y)\geqq 0$. When $g(y)=0$ the two quotients must have the same Taylor expansion which proves that f is divisible by g.

An exact factorization is not always possible even if $n=2$. An example is given by

$$f(x_1, x_2, \xi_2)=(\xi_2-x_1\exp(-1/x_2))^2 \quad \text{if } x_2>0,$$

$$f(x_1, x_2, \xi_2)=\xi_2(\xi_2-\exp(1/x_2)) \quad \text{if } x_2<0; \quad f(x_1, 0, \xi_2)=\xi_2^2.$$

The proof can be found in the references or may be supplied by the reader.

We shall denote by N_{12} the set of points in $N_1\setminus N_{11}$ such that there is some complex c for which dp/c is real and the Hessian of $\operatorname{Im} p/c$ is not identically zero in the plane $dp=0$, and we write N_{12}^i for the subset where the Hessian is not semi-definite. By N_{12}^e and N_{12}^{ie} we denote the union of the one dimensional bicharacteristics intersecting these sets. Proposition 26.5.7 gives a simple representation of the principal symbol, particularly in the set N_{12}^i. In the remaining part N_{13} of the characteristic set we can apply Proposition 26.4.13 and obtain a function f vanishing of third order on $I\times\{0\}\times\{\varepsilon_n\}$.

On all one dimensional bicharacteristics we shall prove a result on singularities which is considerably weaker than Theorem 26.1.4; roughly speaking it states when $Pu\in C^\infty$ that s_u^* is either monotonic or rises to a maximum value monotonically and then falls monotonically again. As a byproduct of the study of N_2^e we shall be able to prove more in N_{12}^{ie}; on one dimensional bicharacteristics there s_u^* is concave with respect to an affine structure which we shall now define. Under the hypotheses in Proposition 26.5.7 it will be defined by the differential $h(x_1, 0, \varepsilon_n)\,dx_1$ on $\gamma(I)$. We shall prove that this is invariantly defined apart from a constant factor, and this means that we have a unique affine structure if we identify points on a subinterval where $h=0$ identically.

Let us first note that if p is real and γ, γ' are two points on the same integral curve of H_p in the surface $p=0$, then the Hamilton flow gives a symplectic map between the tangent spaces of $p=0$ modulo H_p at γ and at γ'. (This is obvious if p is taken as a symplectic coordinate ξ_1 for example.) The observation remains true when p is complex valued provided that γ and γ' lie on a one dimensional bicharacteristic and we consider the complexified tangent planes, for analytically the same computations will be involved. When p has the special form in Proposition 26.5.7 and we consider two points on $\gamma(I)$, the map is obtained from the Hamilton equations for the Hamiltonian

$$\xi_1 + i h(x_1, 0, \varepsilon_n') Q(x', \xi')$$

where Q is the second order part of the Taylor expansion of g at $(0, \varepsilon_n)$. These equations

$$dx_1/dt = 1, \quad dx'/dt = i h(x_1, 0, \varepsilon_n')\partial Q/\partial \xi', \quad d\xi'/dt = -i h(x_1, 0, \varepsilon_n')\partial Q/\partial x'$$

are easy to integrate but we only need the obvious fact that $Q(x', \xi')$ is constant along the orbits. This leads to the following general definition. Let $\gamma(t)$ be a one dimensional bicharacteristic of p,

$$H_p(\gamma(t)) = c(t)\,d\gamma/dt,$$

and let Q_t be the Hessian of $\operatorname{Im} p/c(t)$ in the plane $dp=0$ modulo H_p, at $\gamma(t)$. Note that if p is replaced by qp where $q \neq 0$ then $c(t)$ must be replaced by $q(\gamma(t))\,c(t)$ and Q_t remains unchanged. When p has the special form in Proposition 26.5.7 we have just seen that the pullback of $Q_t = h(t, 0, \varepsilon_n')Q$ to the tangent space at t_0 is equal to

$$h(t, 0, \varepsilon_n')Q = (h(t, 0, \varepsilon_n')/h(0, 0, \varepsilon_n'))\,Q_0.$$

In general, if $\gamma(t_0) \in N_{12}$ and $Q_{t_0} \neq 0$ it follows that the pullback of Q_t to the tangent space at t_0 is of the form $h_0(t) Q_0$. The differential $h_0(t)\,dt$ is well defined on the bicharacteristic and changes only by a constant factor if the base point t_0 is changed. Now assume that we introduce another parameter s on the bicharacteristic. Then

$$H_p(\gamma(t)) = c(t)\,ds/dt\,d\gamma/ds$$

so Q_t is multiplied by dt/ds which makes the differential invariant apart from the normalization which depends on the choice of base point. We have now proved

Proposition 26.5.8. *There is a natural affine structure in every bicharacteristic in N_{12}^e if one identifies points in a subinterval which does not meet N_{12}.*

Let us sum up the notation introduced in this section and which will be referred to in all the rest of the chapter:

N is the characteristic set consisting of all zeros of p;

N_2 is the subset where $H_{\operatorname{Re} p}$ and $H_{\operatorname{Im} p}$ are linearly independent;

N_2^e is the union of semi-bicharacteristics starting in N_2; it is an involutive manifold obtained by attaching one dimensional bicharacteristics to N_2;

N_{11} is the set of points in N which can be joined by a semi-bicharacteristic to a non-characteristic point;

N_{12} is the set of points in $N \setminus (N_2^e \cup N_{11})$ such that for some complex c with dp/c real the Hessian of $\operatorname{Im} p/c$ is not identically zero in the plane $dp = 0$;

N_{12}^i is the subset where the Hessian is not semi-definite;

N_{12}^e and N_{12}^{ie} are the unions of the one dimensional bicharacteristics intersecting these sets;

$N_{13} = N \setminus (N_2^e \cup N_{11} \cup N_{12}^e)$ is the rest of the characteristic set.

It should be kept in mind that this classification of the characteristic points has been made under the assumption that p satisfies (P).

26.6. The Singularities in N_{11}

We shall start with studying an operator for which the principal symbol has the special form which by Proposition 26.5.6 can be achieved by a homogeneous canonical transformation at any one dimensional bicharacteristic (or point) in N_{11}. Afterwards we shall put the result in a general invariant form.

The proof may seem a bit technical so it might be useful to see the simple idea first for the ordinary differential equation

$$du/dt + fu = g$$

in (a, b). If $f \geq 0$ we obtain by multiplying with $\bar{u} e^{-2\lambda t}$, integrating and taking the real part

$$\operatorname{Re} \int_a^b g \bar{u} e^{-2\lambda t} dt = \int_a^b e^{-2\lambda t} d|u|^2/2 + \int_a^b f |u|^2 e^{-2\lambda t} dt$$
$$\geq \lambda \int_a^b |u|^2 e^{-2\lambda t} dt - |u(a)|^2 e^{-2\lambda a}/2.$$

An application of Cauchy-Schwarz' inequality now gives

$$(2\lambda - 1) \int_a^b |u|^2 e^{-2\lambda t} dt \leq \int_a^b |g|^2 e^{-2\lambda t} dt + e^{-2\lambda a} |u(a)|^2.$$

When $\lambda > \frac{1}{2}$ we get control of the L^2 norm of u in terms of that of g and $u(a)$.

Proposition 26.6.1. *Let $P \in \Psi^1_{phg}(\mathbb{R}^n)$ be properly supported and have principal symbol p satisfying*

$$p(x,\xi) = \xi_1 + i f(x,\xi), \quad f(x,\xi) \leqq 0, \tag{26.6.1}$$

in a conic neighborhood of $\Gamma = \{(x_1, 0, \varepsilon_n), \ a \leqq x_1 \leqq b\} \subset T^*(\mathbb{R}^n) \smallsetminus 0$. If $u \in \mathscr{D}'(\mathbb{R}^n)$ and $Pu \in H_{(s)}$ at Γ, $u \in H_{(s)}$ at $(a, 0, \varepsilon_n)$ for some $s \in \mathbb{R}$, then $u \in H_{(s)}$ at Γ. Moreover, if $\Gamma \cap WF(Pu) = \emptyset$ and $(a, 0, \varepsilon_n) \notin WF(u)$, then $\Gamma \cap WF(u) = \emptyset$.

Proof. It suffices to prove the first statement for it implies the result on wave front sets when applied to all bicharacteristics of ξ_1 near Γ and all $s \in \mathbb{R}$. In the proof we may also make the additional hypothesis that $u \in H_{(s-\frac{1}{2})}$ at Γ. For suppose that the theorem is proved under that hypothesis. Since $u \in H_{(s-k/2)}$ at Γ for some positive integer k we can then conclude successively that $u \in H_{(s-(k-1)/2)}$ at $\Gamma, \dots, u \in H_{(s)}$ at Γ. We may even assume that

$$u \in H^{comp}_{(s-\frac{1}{2})}, \text{ and } u \in H^{loc}_{(s)} \text{ in a neighborhood of } \{x;\, x_1 = a\}. \tag{26.6.2}$$

In fact, we can choose $T \in \Psi^0$ with symbol 0 outside such a small conic neighborhood of Γ that Tu satisfies (26.6.2) but $WF(I-T) \cap \Gamma = \emptyset$, hence $PTu \in H_{(s)}$ at Γ since $PTu - Pu$ is in C^∞ at Γ. An application of Proposition 26.6.1 to Tu will then give $Tu \in H_{(s)}$ at Γ, hence $u \in H_{(s)}$ at Γ.

Now we come to the heart of the proof. Choose a compactly generated conic neighborhood V of Γ in $T^*(\mathbb{R}^n) \smallsetminus 0$ such that (26.6.1) is valid in a neighborhood of V and $Pu \in H_{(s)}$ at V. Then choose $\chi \in C_0^\infty(\mathbb{R})$ and a real valued $C(x', \xi) \in S^s(\mathbb{R}^{n-1} \times \mathbb{R}^n)$ so that

(i) $\chi = 1$ on $[a, b]$, and C is non-characteristic at $(0, \varepsilon_n)$;

(ii) $\chi'(x_1) = \chi^-(x_1) - \chi^+(x_1)$ where $0 \leqq \chi^+$, $0 \leqq \chi^-$, $x_1 \leqq a$ in $\operatorname{supp} \chi^-$, $x_1 \geqq b$ in $\operatorname{supp} \chi^+$, and $\chi^-(x_1) u \in H_{(s)}$.

(iii) $\chi(x_1) C(x', \xi)$ vanishes outside V.

Set for $\lambda, \delta > 0$

$$Q_{\lambda,\delta}(x,\xi) = e^{-\lambda x_1} \chi(x_1) C(x', \xi)(1 + |\delta\xi|^2)^{-1}.$$

Then $Q_{\lambda,\delta} \in S^{s-2}$ and is uniformly bounded in S^s when $\delta \to 0$ for fixed λ. Writing $Q = Q_{\lambda,\delta}(x, D)$ for the sake of brevity we now form

$$(QPu, Qu) = (PQu, Qu) + ([Q, P]u, Qu)$$

which is legitimate since $Qu \in H_{(1)}$ and $QPu \in H_{(0)}$. Write $P = A + iB$ where A and B are self adjoint; thus the principal symbols are ξ_1 and f in a neighborhood of V. Taking the imaginary part we then obtain

$$\operatorname{Im}(QPu, Qu) = (BQu, Qu) + \operatorname{Re}([Q, B]u, Qu) + \operatorname{Im}([Q, A]u, Qu). \tag{26.6.3}$$

We shall discuss the terms in order from right to left.

The symbol of $[Q, A]$ is equal to the symbol $i\partial Q_{\lambda,\delta}/\partial x_1$ of $[Q, D_{x_1}]$ with an error which is bounded in S^{s-1} for fixed λ. Since

$$\partial Q_{\lambda,\delta}/\partial x_1 = -\lambda Q_{\lambda,\delta} + e^{-\lambda x_1}\chi'(x_1)\, C(x', \xi)(1+|\delta\xi|^2)^{-1}$$

and $\chi\chi' \leqq \chi\chi^-$, because $\chi \geqq 0$, we obtain

(26.6.4) $\operatorname{Im}([Q, A]u, Qu) \leqq -\lambda\|Qu\|^2 + \|Q^- u\|\,\|Qu\| + K_\lambda\|u\|_{(s-1)}\|Qu\|$

where $Q^- = Q^-_{\lambda,\delta}(x, D)$ with $Q^-_{\lambda,\delta}$ defined as $Q_{\lambda,\delta}$ with χ replaced by χ^-.

The symbol of $Q^*[Q, B]$ is $-iQ_{\lambda,\delta}\{Q_{\lambda,\delta}, f\}$ with an error which is bounded in S^{2s-1}, so the symbol of the self adjoint part

$$\tfrac{1}{2}(Q^*[Q, B] + [Q, B]^* Q)$$

is bounded in S^{2s-1}, when $\delta \to 0$. Hence

(26.6.5) $$|\operatorname{Re}([Q, B]u, Qu)| \leqq K_\lambda\|u\|^2_{(s-\frac{1}{2})}.$$

Choose $C_1 \in S^0(\mathbb{R}^n \times \mathbb{R}^n)$ equal to 1 in $\{(x, \xi) \in V;\ |\xi| > 1\}$ and 0 outside such a small conic neighborhood of V that $0 \geqq F = C_1 f \in S^1$. By Theorem 18.1.14 we have

(26.6.6) $$\operatorname{Re}(F(x, D)v, v) \leqq K\|v\|^2, \quad v \in H_{(1)}.$$

If b_0 is the term of order 0 in the symbol of B then the symbol of $(B - F(x, D) - (C_1 b_0)(x, D))Q$ is bounded in S^{s-1} for fixed λ. Since $(C_1 b_0)(x, D)$ is bounded in L^2 we obtain with K' independent of λ

(26.6.7) $$(BQu, Qu) \leqq K'\|Qu\|^2 + K_\lambda\|Qu\|\,\|u\|_{(s-1)}.$$

Summing up (26.6.3), (26.6.4), (26.6.5), (26.6.7) we have

$$(\lambda - K')\|Qu\|^2 \leqq \|Qu\|(\|Q^- u\| + \|QPu\| + K_\lambda\|u\|_{(s-1)}) + K_\lambda\|u\|^2_{(s-\frac{1}{2})}.$$

Using Cauchy-Schwarz' inequality we obtain

(26.6.3)′ $$(\lambda - K' - 1)\|Qu\|^2 \leqq \|Q^- u\|^2 + \|QPu\|^2 + K'_\lambda\|u\|^2_{(s-\frac{1}{2})}.$$

The symbol of

$$Q_{\lambda,\delta}(x, D) - \sum_{|\alpha| \leqq 1} \operatorname{Op}\partial_\xi^\alpha(1+\delta|\xi|^2)^{-1}\operatorname{Op}(-D_x^\alpha)Q_{\lambda,0}(x, \xi)$$

is bounded in S^{s-2} when $\delta \to 0$, and $\operatorname{Op}(-D_x^\alpha Q_{\lambda,0})Pu \in L^2$ since $Pu \in H_{(s)}$ at V. Hence $\|Q_{\lambda,\delta}Pu\|$ is bounded when $\delta \to 0$. The same is true of $Q^-_{\lambda,\delta}u$ since $\chi^-(x_1)u \in H_{(s)}$. If we take $\lambda = K' + 2$ in (26.6.3)′ and let $\delta \to 0$, it follows now that $Q_{\lambda,0}(x, D)u \in L^2$. Hence $u \in H_{(s)}$ at Γ and the proposition is proved.

In the preceding proof the crucial point is the semi-boundedness (26.6.6). Now Theorem 18.6.8 shows that (26.6.6) is also valid if $0 \leqq F \in S^1_{1-\varepsilon,\varepsilon}$ and $0 < \varepsilon \leqq \frac{1}{4}$. This leads to an extension of Proposition 26.6.1 which will be important in Section 26.9.

Proposition 26.6.1′. *Let* $P \in \Psi^1_{\text{phg}}(\mathbb{R}^n)$, $C_1 \in S^0_{1-\varepsilon,\varepsilon}(\mathbb{R}^n)$; *assume that* $C_1(x,\xi)=0$ *for large* $|x|$ *and that the principal symbol* p *of* P *satisfies*

$$p(x,\xi)=\xi_1+if(x,\xi), \quad f(x,\xi)\leqq 0 \quad \text{in } \operatorname{supp} C_1. \tag{26.6.1)$'$}$$

Let $\chi \in C_0^\infty(\mathbb{R})$ *be equal to* 1 *in* $[a,b]$ *and* $\chi'(t)=\chi^-(t)-\chi^+(t)$ *where* χ^- *and* χ^+ *are non-negative with support to the left and right of* a *and* b *respectively. Finally let* $C(x',\xi)\in S^s_{1-\varepsilon,\varepsilon}$ *and assume that*

$$C_1(x,\xi)=1 \quad \text{in a neighborhood of } \operatorname{supp}\chi(x_1)C(x',\xi). \tag{26.6.8}$$

If $0\leqq\varepsilon\leqq\frac{1}{4}$ *and*

$$u\in H^{\text{comp}}_{(s+(3\varepsilon-1)/2)}, \quad \chi(x_1)C(x',D)Pu\in L^2, \quad \chi^-(x_1)C(x',D)u\in L^2 \tag{26.6.9}$$

it follows then that

$$\chi(x_1)C(x',D)u\in L^2. \tag{26.6.10}$$

Proof. We just have to inspect the proof of Proposition 26.6.1. The identity (26.6.3) is of course unchanged. In (26.6.4) we just have to replace $\|u\|_{(s-1)}$ by $\|u\|_{(s-1+\varepsilon)}$, and $s-1+\varepsilon<s+(3\varepsilon-1)/2$. The self-adjoint part of $Q^*[Q,B]$ is now bounded of order $2s-1+3\varepsilon$ so (26.6.5) remains valid with $\|u\|_{(s+(3\varepsilon-1)/2)}$ in the right-hand side. There is no change at all in (26.6.7), but the proof now relies on Theorem 18.6.8. The proof is then completed as before.

The important point in Proposition 26.6.1′ is that f does not have to be of constant sign in a conic neighborhood of $\operatorname{supp}\chi(x_1)C(x',\xi)$. In fact, we can choose $C_1\in S^0_{1-\varepsilon,\varepsilon}$ satisfying (26.6.8) and (26.6.1)′ if f is $\leqq 0$ at all points with a fixed distance to $\operatorname{supp}\chi(x_1)C(x',\xi)$ in the metric

$$|dx_1|^2+(1+|\xi'|^2)^\varepsilon|dx'|^2+(1+|\xi'|^2)^{\varepsilon-1}|d\xi|^2$$

(see Section 18.4). However, we shall not continue in this direction now but turn instead to more invariant formulations of Proposition 26.6.1.

Theorem 26.6.2. *Let* $P\in\Psi^m_{\text{phg}}(X)$ *be properly supported and have principal symbol* p *satisfying* (P), *and let*

$$[a,b]\ni t\mapsto\gamma(t), \quad a<b,$$

be a bicharacteristic of $\operatorname{Re} qp$ *where* $q\neq 0$ *and* $\operatorname{Im}(qp)(\gamma(a))<0$. *If* $u\in\mathscr{D}'(X)$ *and* $Pu\in H_{(s)}$ *at* $\gamma([a,b])$ *it follows that* $u\in H_{(s+m-1)}$ *at* $\gamma([a,b])$.

Proof. If $\gamma(t_0)\notin N$ we have $u\in H_{(s+m)}$ at $\gamma(t_0)$ so it suffices to prove that $u\in H_{(s+m-1)}$ at $\gamma(t_0)$ if $\gamma(t_0)\in N$. Let I be the maximal interval $\subset[a,b]$ containing t_0 such that $\gamma(I)\subset N$. Choose an interval $[a',b']$ with $I\subset[a',b']\subset[a,b]$ such that $\gamma(a')\notin N$ and $[a',b']$ is so close to I that Proposition 26.4.13 is applicable. Then we can find a C^∞ function $\tilde q$ which is homogeneous of degree $1-m$ and a homogeneous canonical transformation χ such that $\chi^*(\tilde q p)=\xi_1+if(x,\xi')$ in a neighborhood of $[a',b']\times\{0\}\times\{\varepsilon_n\}$.

We have $\chi(x_1, 0, \varepsilon_n)=\gamma(x_1)$ when $x_1 \in [a', b']$ and $\tilde{q}=q$ on this arc, for these properties are obvious after the first step of the proof of Proposition 26.4.13 and they are not affected by the second step. Now $\operatorname{Im}(qp)(\gamma(t)) \leqq 0$ when $t \in [a, b]$ by condition (P) since $\operatorname{Im}(qp)(\gamma(a))<0$, so we conclude that $f(x, \xi') \leqq 0$ in a neighborhood of $[a', b'] \times \{0\} \times \{\varepsilon_n\}$. If we now transform P as in Proposition 26.4.4 with Fourier integral operators A and B belonging to the graphs of χ and χ^{-1} respectively, and the principal symbol of AB is $\tilde{q}$ near $\gamma(I)$, the theorem follows from Proposition 26.6.1 since $u \in H_{(s+m)}$ at $\gamma(a')$.

We shall now study the propagation of singularities on one dimensional bicharacteristic arcs $\Gamma \subset N_{11}$. We can extend Γ as a one dimensional bicharacteristic so that it is maximal at one end point Γ_0 and choose $q \in C^\infty$ so that $q \neq 0$ and $H_{\operatorname{Re} qp} \neq 0$ on Γ. Then $\operatorname{Im} qp$ must be non-zero at some points arbitrarily close to Γ_0 on the bicharacteristic of $\operatorname{Re} qp$ extending Γ there, and by condition (P) it follows that either $\operatorname{Im} qp \geqq 0$ or else $\operatorname{Im} qp \leqq 0$ in a neighborhood of Γ in the surface $\operatorname{Re} qp=0$. If $\tilde{q}$ is another function with the same properties as assumed for q, and $H_{\operatorname{Re} \tilde{q}p}$ has the same direction as $H_{\operatorname{Re} qp}$ on Γ, then we conclude that for each $t \in [0, 1]$ the imaginary part of $(tq+(1-t)\tilde{q})p$ has a fixed sign and is not identically 0 in a neighborhood of Γ in the surface $\operatorname{Re}(tq+(1-t)\tilde{q})p=0$. Clearly the sign must then be independent of t. Put differently, if we choose q so that $\operatorname{Im} qp \leqq 0$ in a neighborhood of Γ when $\operatorname{Re} qp=0$, then $H_{\operatorname{Re} qp}$ gives Γ an orientation which does not depend on the choice of q.

Definition 26.6.3. If p satisfies condition (P) and Γ is a one dimensional bicharacteristic arc $\subset N_{11}$ which is maximal at one end, then Γ is given the orientation of $H_{\operatorname{Re} qp}$ when q is chosen so that $q \neq 0$, $H_{\operatorname{Re} qp} \neq 0$ on Γ, and $\operatorname{Im} qp \leqq 0$ in a neighborhood of Γ (in the surface defined by $\operatorname{Re} qp=0$).

Exactly as in the proof of Theorem 26.6.2 we now obtain from Proposition 26.6.1:

Theorem 26.6.4. *Let $P \in \Psi^m_{\mathrm{phg}}(X)$ be properly supported and have a principal symbol satisfying condition (P). If Γ is a compact one dimensional bicharacteristic interval $\subset N_{11}$ and $u \in \mathscr{D}'(X)$, $Pu \in H_{(s)}$ at Γ, $u \in H_{(s+m-1)}$ at the starting point on Γ, then $u \in H_{(s+m-1)}$ at Γ.*

The difference between this result and Theorem 26.1.4 is that regularity only propagates in the direction of the orientation, which of course agrees with the direction in which regularity enters from the non-characteristic set by Theorem 26.6.2. In general one cannot expect propagation in the opposite direction. We give an example.

Example 26.6.5. Let $P=D_1 D_n+iQ(D'')$ where Q is a real quadratic form in $D''=(D_2, \ldots, D_{n-1})$ which is not negative semi-definite. Then one can find a solution u of the equation $Pu=0$ such that

$$WF(u)=\{(x_1, 0, s\varepsilon_n), \quad s>0, \ x_1\in\mathbb{R}\}$$

and $s_u^*(x_1, 0, \varepsilon_n)=h(x_1)$ where h is any given decreasing concave function on $\mathbb{R}$. If Q takes both positive and negative values then one can obtain an arbitrary concave function h. Since every (decreasing) concave function is the infimum of countably many (decreasing) linear functions the assertion follows from standard category arguments (cf. the proof of Theorem 26.2.3) if it is verified when h is linear.

Fourier transformation of the equation $Pu=0$ with respect to $x'=(x_2, \dots, x_n)$ gives formally the equation

$$(\xi_n\partial/\partial x_1-Q(\xi''))\,\hat{u}(x_1, \xi'')=0$$

with the solution $\hat{u}(x_1, \xi'')=c(\xi')\exp(x_1 Q(\xi'')/\xi_n)$. Choose $\theta\in\mathbb{R}^{n-2}$ with $Q(\theta)>0$, a function $\psi\in C_0^\infty(\mathbb{R}^{n-2})$ with $\psi(0)=1$ and a function $\phi\in C^\infty(\mathbb{R})$ with $\phi=0$ in $(-\infty, 1)$, $\phi=1$ in $(2, \infty)$, and set with some real number b

$$\hat{u}(x_1, \xi')=\xi_n^b\psi(\eta'')\,\phi(\xi_n)\exp(x_1 Q(\xi'')/\xi_n),$$

where

$$\eta''=\xi''((\log\xi_n)/\xi_n)^{\frac{1}{2}}-\theta\log\xi_n.$$

Note that

$$\begin{aligned} Q(\xi'')/\xi_n &= Q(\theta(\log\xi_n)^{\frac{1}{2}}+\eta''(\log\xi_n)^{-\frac{1}{2}}) \\ &= (\log\xi_n)Q(\theta)+L(\eta'')+Q(\eta'')/\log\xi_n \end{aligned}$$

where L is a linear function of η''. Since $|\eta''|<C$, $|\xi''-\theta(\xi_n\log\xi_n)^{\frac{1}{2}}|<C(\xi_n/\log\xi_n)^{\frac{1}{2}}$ in supp $\hat{u}$, and

$$\begin{aligned} &\partial\eta_j/\partial\xi_k=\delta_{jk}((\log\xi_n)/\xi_n)^{\frac{1}{2}} \quad \text{if } 1<j, k<n, \\ &\partial\eta_j/\partial\xi_n=\xi_j((\log\xi_n)/\xi_n)^{-\frac{1}{2}}\xi_n^{-2}(1-\log\xi_n)/2-\theta/\xi_n, \end{aligned}$$

it is easy to see that

$$|D_{\xi'}^\alpha\hat{u}(x_1, \xi')|\le C_\alpha(1+|\xi'|)^{b+x_1 Q(\theta)-\rho|\alpha|} \tag{26.6.11}$$

if $0<\rho<\frac{1}{2}$ is fixed. From (26.6.11) it follows that $\hat{u}$ is indeed the partial Fourier transform of a distribution u satisfying the equation $Pu=0$, and $x'^\alpha u(x)$ has N bounded derivatives where

$$b+x_1 Q(\theta)-\rho|\alpha|+N<-n.$$

Hence $x'=0$ in sing supp$u$. If $\chi\in C_0^\infty(\mathbb{R})$ then the Fourier transform of $\chi(x_1)u$ is

$$(\widehat{\chi u})(\xi)=\xi_n^b\psi(\eta'')\,\phi(\xi_n)\,\hat{\chi}(\xi_1+iQ(\xi'')/\xi_n).$$

We have $\xi_n>1$ and $|\xi''|=o(\xi_n)$ in the support. Since $Q(\xi'')/\xi_n=O(\log\xi_n)$ there we can also estimate $|\hat{\chi}(\xi_1+iQ(\xi'')/\xi_n)|$ by $C_N\xi_n^K(1+|\xi_1|)^{-N}$ for any N, with K independent of N. This proves that $\widehat{\chi u}$ is rapidly decreasing in any cone where $\xi_n<C|\xi_1|$, so $\widehat{\chi u}$ is rapidly decreasing outside any conic neigh-

borhood of $\varepsilon_n=(0, \dots, 0, 1)$. We have therefore proved that

$$WF(u) \subset \{(x_1, 0, s\varepsilon_n),\ s>0,\ x_1 \in \mathbb{R}\}.$$

In view of the rapid decrease of u and all its derivatives as $x' \to \infty$ we can determine the microlocal $H_{(s)}$ class by just examining when $\chi(x_1)u \in H_{(s)}$, that is,

$$\int |\widehat{\chi u}(\xi)|^2 (1+|\xi|^2)^s \, d\xi < \infty.$$

We have just seen that $\widehat{\chi u}$ is rapidly decreasing when $|\xi_1|>|\xi'|$, so this is equivalent to

$$\int |\widehat{\chi u}(\xi)|^2 (1+|\xi'|^2)^s \, d\xi < \infty.$$

The integral with respect to ξ_1 can then be calculated by Parseval's formula, so an equivalent condition is that

$$\iint |\xi_n^b \psi(\eta'') \phi(\xi_n) \exp(x_1 Q(\xi'')/\xi_n) \chi(x_1)|^2 (1+|\xi'|^2)^s \, dx_1 \, d\xi' < \infty.$$

Here the exponential can be replaced by $\xi_n^{x_1 Q(\theta)}$ and $|\xi'|^2$ can be replaced by ξ_n^2. The integral with respect to ξ'' can then be worked out, so we obtain the simpler equivalent condition

$$\iint_{\xi_n>2} (\xi_n/\log \xi_n)^{(n-2)/2} \xi_n^{2(b+x_1 Q(\theta)+s)} |\chi(x_1)|^2 \, dx_1 \, d\xi_n < \infty. \tag{26.6.12}$$

(26.6.12) implies that $n+4(b+x_1 Q(\theta)+s)<0$ if $\chi(x_1) \neq 0$, and conversely (26.6.12) follows if this is true in $\operatorname{supp}\chi$. Hence

$$s_u^*(x_1, 0, 0, \varepsilon_n) = -x_1 Q(\theta) - b - n/4$$

which is an arbitrary decreasing linear function. If we take θ with $Q(\theta)=0$ or $Q(\theta)<0$ we get instead a constant or an increasing linear function, which completes the verification.

That $s_u^*(x_1, 0, 0, \varepsilon_n)$ must be decreasing if $Q \geqq 0$ follows from Theorem 26.6.4. We shall see later (Theorem 26.9.6) that the concavity is also a necessary condition so the construction above is optimal.

26.7. Degenerate Cauchy-Riemann Operators

In Proposition 26.5.5 we have seen that the principal symbol of an operator satisfying condition (P) can be reduced to the form $\xi_1 + ig(x', \xi')h(x, \xi')$ in a neighborhood of a one dimensional bicharacteristic (or a single point) embedded in a two dimensional bicharacteristic. Here $h \geqq 0$ and $dg \neq 0$. By a possibly non-homogeneous canonical transformation the function g can be taken as the ξ_2 variable. Then the two dimensional bicharacteristics are the leaves of the foliation of the plane $\xi_1=\xi_2=0$ by the planes parallel to the $x_1 x_2$ plane, and the Hamilton field is

$$\partial/\partial x_1 + ih(x, \xi')\partial/\partial x_2. \tag{26.7.1}$$

The purpose of this section is to discuss such first order differential operators. (An example is the model equation $D_1+ix_1^k D_2$ with even k studied in Section 26.3.) This will lead to a complex structure in reduced two dimensional bicharacteristics as indicated in Section 26.5, and this will be the basis of the extension of Theorem 26.2.1 in the next section. The study of singularities in N_{13} will require information on solutions of families of operators of the form (26.7.1) where there is only uniform control of the derivatives of h which do not depend on x_1. We shall therefore state the results in the generality which will be required then. The special role of the x_1 variable motivates a change of notation so that the x_1 variable in (26.7.1) becomes t and all other variables are called x.

Let $B^\infty(\mathbb{R}^{1+k})$ be the set of continuous function $u(t,x)$, where $t\in\mathbb{R}$ and $x=(x_1,\dots,x_k)\in\mathbb{R}^k$, which have continuous bounded derivatives of all orders with respect to x. It is a Fréchet space with the semi-norms

$$u\mapsto \sup|D_x^\alpha u(t,x)|.$$

We shall study first order differential operators of the form

$$P=D_t+ia(t,x)D_{x_1}+ib(t,x) \tag{26.7.2}$$

where $a,b\in B^\infty$. At first we take $k=1$ and prove a lemma closely related to Proposition 26.6.1′. The norms are L^2 norms unless otherwise indicated.

Lemma 26.7.1. *For every bounded subset B of $B^\infty(\mathbb{R}^2)$ there is a constant C such that if P is defined by* (26.7.2) *with $a,b\in B$ and $a\geqq 0$ then*

$$\|u\|+\|a^{\frac12}\Lambda^{\frac12}u\|\leqq C(\|Pu\|+\|\Lambda^{-1}u\|), \tag{26.7.3}$$

if $u\in\mathscr{S}$ and $u=0$ when $|t|>1$. Here $\Lambda^s u=(1+|D_x|^2)^{s/2}u$.

Proof. Choose $h\in C^\infty(\mathbb{R})$ decreasing, equal to 1 in $(-\infty,-2)$ and 0 in $(-1,\infty)$. We shall apply (26.6.3), that is,

$$\operatorname{Im}(QPu,Qu)=\operatorname{Re}((aD_x+b)Qu,Qu)+\operatorname{Im}([Q,P]u,Qu)$$

with $Q=e^{-\lambda t}h(D_x)$. The commutator is

$$[Q,P]=-i\lambda Q+ie^{-\lambda t}[h(D_x),aD_x+b].$$

Here aD_x+b can be regarded as a pseudo-differential operator in x with symbol bounded in $S^1(\mathbb{R}\times\mathbb{R})$, with t as parameter, and from the calculus it follows then that $[h(D_x),aD_x+b]$ has a symbol uniformly bounded in S^{-1} (in fact, bounded in S^{-N} for any N). Hence

$$\operatorname{Im}([Q,P]u,Qu)\leqq -\lambda\|Qu\|^2+K_\lambda\|\Lambda^{-1}u\|\,\|Qu\|.$$

We shall compare the term $\operatorname{Re}((aD_x+b)Qu,Qu)$ with the positive quantity

$$\|e^{-\lambda t}a^{\frac12}\Lambda^{\frac12}h(D_x)u\|^2=(e^{-\lambda t}\Lambda^{\frac12}a\Lambda^{\frac12}h(D_x)u,e^{-\lambda t}h(D_x)u).$$

The principal symbol of $\Lambda^{\frac12}a\Lambda^{\frac12}+aD_x+b$ is $a(|\xi|+\xi)=0$, $\xi<0$, so the symbol is the sum of one which is uniformly bounded in S^0 and one which is uniformly bounded in S^1 and supported where $\xi>0$, so the product by $h(D_x)$ is bounded in S^{-1}. Hence we obtain, with K independent of λ,

$$\begin{aligned}\operatorname{Re}((aD_x+b)Qu, Qu)&+\|e^{-\lambda t}a^{\frac{1}{2}}\Lambda^{\frac{1}{2}}h(D_x)u\|^2\\ &\leqq K\|Qu\|^2+K_\lambda\|\Lambda^{-1}u\|\,\|Qu\|.\end{aligned}$$

Summing up, we obtain

$$\begin{aligned}(\lambda-K)\|Qu\|^2&+\|e^{-\lambda t}a^{\frac{1}{2}}\Lambda^{\frac{1}{2}}h(D_x)u\|^2\\ &\leqq\|Qu\|(\|QPu\|+K_\lambda\|\Lambda^{-1}u\|)+K_\lambda\|\Lambda^{-1}u\|^2.\end{aligned}$$

We fix $\lambda=K+2$ now and obtain

$$\|Qu\|^2+\|e^{-\lambda t}a^{\frac{1}{2}}\Lambda^{\frac{1}{2}}h(D_x)u\|^2\leqq\|QPu\|^2+K'_\lambda\|\Lambda^{-1}u\|^2.$$

Now the same argument can be applied with $Q=e^{\lambda t}h(-D_x)$, which just amounts to changing the signs of both t and x. Since $h_0(\xi)=1-h(\xi)-h(-\xi)$ has compact support, we have

$$\|h_0(D)u\|\leqq C\|\Lambda^{-1}u\|,\qquad \|a^{\frac{1}{2}}\Lambda^{\frac{1}{2}}h_0(D)u\|\leqq C\|\Lambda^{-1}u\|.$$

The triangle inequality now gives (26.7.3).

When the support is small in the x direction we can eliminate the second term on the right-hand side of (26.7.3):

Lemma 26.7.2. *Under the hypotheses in Lemma* 26.7.1 *one can find positive constants c and C such that*

$$\|u\|\leqq C\|Pu\|,\quad \text{if } u\in C_0^\infty(\{(t,x);\,|t|<1,\,|x|<c\}). \tag{26.7.3$'$}$$

Proof. If

$$\hat{u}(t,\xi)=\int e^{-ix\xi}u(t,x)\,dx$$

is the partial Fourier transform of u with respect to x, then

$$|\hat{u}(t,\xi)|^2\leqq c\int|u(t,x)|^2\,dx$$

by Cauchy-Schwarz' inequality. Hence

$$(2\pi)^{-1}\int(1+|\xi|^2)^{-1}|\hat{u}(t,\xi)|^2\,d\xi\leqq c/2\int|u(t,x)|^2\,dx,$$

so $\|\Lambda^{-1}u\|\leqq(c/2)^{\frac{1}{2}}\|u\|$. If we use this estimate in (26.7.3) and choose c so small that $C(c/2)^{\frac{1}{2}}<\frac{1}{2}$, we obtain (26.7.3)$'$ with twice the constant in (26.7.3).

Lemma 26.7.2 leads immediately to a local existence theorem for the adjoint operator. (Thanks to the lower order term in (26.7.2) the class of operators we consider does not change if we take adjoints and change the sign of x_1.) However, we want to find C^∞ solutions so we must also have estimates in $H_{(s)}$ norms for large negative s. We must then work directly with the parameters, so we allow an arbitrary dimension $1+k$ again now. The set of functions in $B^\infty(\mathbb{R}^2)$ obtained by fixing the variables $x_2,\dots,x_k$ in the functions belonging to a bounded subset of $B^\infty(\mathbb{R}^{1+k})$ is of course bounded, so (26.7.3)′ remains valid with norms in $L^2(\mathbb{R}^{k+1})$.

We shall actually work with the $H_{(s)}$ norm in the x variables only,

$$\|u\|'_{(s)}=((2\pi)^{-k}\int|\hat{u}(t,\xi)|^2(1+|\xi|^2)^s\,d\xi\,dt)^{\frac{1}{2}},\quad u\in\mathscr{S}, \tag{26.7.4}$$

where $\hat{u}(t,\xi)$ is the partial Fourier transform in the x variables. For technical reasons we must use equivalent norms defined as follows. Let $\varepsilon_1>\varepsilon_2>\dots$ be a decreasing positive sequence and set

$$E_N(\xi)=\prod_1^N(1+|\varepsilon_j\xi|^2)^{-1}.$$

Then

$$|||u|||_N=((2\pi)^{-k}\int|\hat{u}(t,\xi)|^2|E_N(\xi)|^2\,d\xi\,dt)^{\frac{1}{2}},\quad u\in\mathscr{S},$$

is equivalent to $\|u\|'_{(-2N)}$. We shall show that if ε_j are successively chosen small enough then (26.7.3)′ remains valid with a slight change of the constants for all the norms $|||\ |||_N$. The problem is of course that P does not commute with the operator $E_N(D_x)$. To be able to estimate commutators we must extract more information from the second term in the left-hand side of (26.7.3). In doing so we write

$$F_\varepsilon=(1+|\varepsilon D_x|^2)^{-1}.$$

With this notation we have

$$|||u|||_{N+1}=|||F_{\varepsilon_{N+1}}u|||_N=\|E_N(D_x)F_{\varepsilon_{N+1}}u\|.$$

Lemma 26.7.3. *If P is defined by (26.7.2) with a and b in a fixed bounded subset of $B^\infty(\mathbb{R}^{1+k})$, $a\geqq 0$, then*

$$\|[P,F_\varepsilon]v\|\leqq C\varepsilon^{\frac{1}{2}}(\|PF_\varepsilon v\|+\|F_\varepsilon v\|),\quad 0<\varepsilon<1, \tag{26.7.5}$$

if $v\in\mathscr{S}(\mathbb{R}^{1+k})$ and $v=0$ when $|t|>1$.

Proof. If a denotes multiplication by the function a then

$$[a,1+|\varepsilon D_x|^2]=\varepsilon^2\Big(-\Delta_x a+2i\sum_1^k D_j\partial a/\partial x_j\Big).$$

Multiplication left and right by F_ε gives (cf. the resolvent equation)

$$[a,F_\varepsilon]=\varepsilon^2F_\varepsilon\Big((\Delta_x a)-2i\sum_1^k D_j\partial a/\partial x_j\Big)F_\varepsilon. \tag{26.7.6}$$

Since $a\geqq 0$ and $a''_{x_jx_j}$ is uniformly bounded we have by Lemma 7.7.2

$$|\partial a/\partial x_j|\leqq Ca^{\frac{1}{2}}. \tag{26.7.7}$$

With the notation $\Lambda^{\frac{1}{2}}=(1+D_{x_1}^2)^{\frac{1}{2}}$ from Lemma 26.7.1 we now obtain

$$\begin{aligned}[aD_1,F_\varepsilon]=&-2i\varepsilon^2F_\varepsilon\sum D_jD_1\Lambda^{-\frac{1}{2}}\partial a/\partial x_j\Lambda^{\frac{1}{2}}F_\varepsilon+\varepsilon^2F_\varepsilon D_1(\Delta_x a)F_\varepsilon\\&-\varepsilon^2F_\varepsilon((D_1\Delta_x a)+2i\sum D_j[\partial a/\partial x_j,D_1\Lambda^{-\frac{1}{2}}]\Lambda^{\frac{1}{2}})F_\varepsilon.\end{aligned}$$

Here $\varepsilon^{\frac{1}{2}} F_\varepsilon D_j D_1 \Lambda^{-\frac{1}{2}}$ is bounded in L^2, and so are F_ε, $\varepsilon F_\varepsilon D_j$ and the commutator $[\partial a/\partial x_j, D_1 \Lambda^{-\frac{1}{2}}]\Lambda^{\frac{1}{2}}$. By (26.7.3) applied to $F_\varepsilon v$ and (26.7.7) this gives

$$\|[aD_1, F_\varepsilon]v\| \leq C\varepsilon^{\frac{1}{2}}(\|PF_\varepsilon v\| + \|F_\varepsilon v\|),$$

and a similar estimate for $[b, F_\varepsilon]v$ follows from (26.7.6) with a replaced by b. The lemma is proved.

In the following lemma we collect information on some other commutators which will occur. We use the notation $E_N = E_N(D_x)$.

Lemma 26.7.4. *For fixed E_N and ϕ, ψ in a fixed bounded subset of B^∞, regarded as multiplication operators, we have for $0<\varepsilon<1$, $u \in \mathscr{S}$*

(i) $\|E_N F_\varepsilon \phi u\| \leq C\|E_N F_\varepsilon u\|$,
(ii) $\|E_N[\phi, F_\varepsilon]u\| \leq C\varepsilon\|E_N F_\varepsilon u\|$,
(iii) $\|E_N[D_j\phi, F_\varepsilon]u\| \leq C\|E_N F_\varepsilon u\|$,
(iv) $\|E_N[D_j\phi, [\psi, F_\varepsilon]]u\| \leq C\varepsilon\|E_N F_\varepsilon u\|$,
(v) $\|[D_j\phi, E_N]F_\varepsilon u\| \leq C\|E_N F_\varepsilon u\|$,
(vi) $\|[F_\varepsilon, [D_j\phi, E_N]]u\| \leq C\varepsilon\|E_N F_\varepsilon u\|$.

In (iii)–(vi) *j runs from* 1 *to k and D_j may be omitted.*

Proof. To prove (ii) we replace a by ϕ in (26.7.6) and multiply by E_N to the left and by $E_N^{-1}E_N$ to the right. The operators $E_N(\Delta_x\phi)E_N^{-1}$ and $E_N\partial\phi/\partial x_j E_N^{-1}$ have uniformly bounded symbols in S^0 so the L^2 norms are uniformly bounded. Since $\varepsilon F_\varepsilon D_j$ and F_ε have norm ≤ 1 we obtain (ii). If ϕ is multiplied by D_j to the left we just get another factor D_j to the left, and since the norm of $\varepsilon^2 F_\varepsilon D_j D_i$ is at most 1, this proves (iii). The estimate (i) follows from (ii), for

$$\|E_N\phi F_\varepsilon u\| = \|E_N\phi E_N^{-1}E_N F_\varepsilon u\| \leq C\|E_N F_\varepsilon u\|.$$

To prove (iv) we observe that taking the commutator of $D_j\phi$ and (26.7.6) with $a=\psi$ gives three terms,

$$[D_j\phi, [\psi, F_\varepsilon]] = \varepsilon^4 F_\varepsilon A_1 F_\varepsilon B_1 F_\varepsilon + \varepsilon^2 F_\varepsilon B_2 F_\varepsilon + \varepsilon^4 F_\varepsilon B_1 F_\varepsilon A_2 F_\varepsilon$$

where A_j are second order differential operators and B_j are first order differential operators with coefficients in a bounded subset of B^∞. If we multiply by E_N to the left and insert factors $E_N^{-1}E_N$ at appropriate places, the estimate (iv) follows, for $\varepsilon^2 F_\varepsilon E_N A_j E_N^{-1}$ and $\varepsilon F_\varepsilon E_N B_j E_N^{-1}$ have uniformly bounded norms since differentiations can be moved through E_N to F_ε. (v) follows since

$$[D_j\phi, E_N]F_\varepsilon = [D_j\phi, E_N]E_N^{-1}E_N F_\varepsilon$$

and $[D_j\phi, E_N]E_N^{-1}$ is bounded. Now

$$[F_\varepsilon, [D_j\phi, E_N]]F_\varepsilon^{-1}E_N^{-1} = F_\varepsilon[[D_j\phi, E_N], F_\varepsilon^{-1}]E_N^{-1} = \varepsilon^2 F_\varepsilon(A_0 + \sum D_i A_i)$$

where $A_0, \dots, A_k$ are uniformly bounded in L^2. The norm is therefore $O(\varepsilon)$, which completes the proof.

We can now give the main inductive step in the extension of (26.7.3)′ to the norms $|||\ |||_N$.

Lemma 26.7.5. *Assume that for a set of operators of the form (26.7.2) with coefficients in a bounded subset of B^∞ we have*

$$|||u|||_N \leqq C_N |||Pu|||_N \quad \text{if } u \in \mathscr{S}(\mathbb{R}^{k+1}) \text{ and } u=0 \text{ when } |t|>1 \text{ or } |x_1|>c_N. \tag{26.7.8}$$

If $C_{N+1} > C_N$ and $0 < c_{N+1} < c_N$ it follows that (26.7.8) remains valid with N replaced by $N+1$ provided that ε_{N+1} is small enough.

Proof. Choose $\psi \in C_0^\infty(-c_N, c_N)$ equal to 1 in $(-c_{N+1}, c_{N+1})$. To prove (26.7.8) with the next norm we shall apply (26.7.8) to $v = \psi F_\varepsilon u$ where $u=0$ when $|t|>1$ or $|x_1|>c_{N+1}$. Since $u=\psi u$ we have

$$F_\varepsilon u = F_\varepsilon \psi u = v - [\psi, F_\varepsilon] u,$$

so (ii) in Lemma 26.7.4 and (26.7.8) give

$$|||F_\varepsilon u|||_N \leqq |||v|||_N + C\varepsilon |||F_\varepsilon u|||_N \leqq C_N |||Pv|||_N + C\varepsilon |||F_\varepsilon u|||_N.$$

We have

$$Pv = P\psi F_\varepsilon u = PF_\varepsilon u + P[\psi, F_\varepsilon] u = F_\varepsilon Pu + [P, F_\varepsilon] u + [\psi, F_\varepsilon] Pu + [P, [\psi, F_\varepsilon]] u.$$

By (ii) and (iv) in Lemma 26.7.4

$$|||[\psi, F_\varepsilon] Pu|||_N + |||[P, [\psi, F_\varepsilon]] u|||_N \leqq C\varepsilon (|||F_\varepsilon Pu|||_N + |||F_\varepsilon u|||_N).$$

Using the Jacobi identity and Lemma 26.7.3 we obtain

$$\begin{aligned} |||[P, F_\varepsilon] u|||_N = \|E_N [P, F_\varepsilon] u\| &\leqq \|[E_N, [P, F_\varepsilon]] u\| + \|[P, F_\varepsilon] E_N u\| \\ &\leqq \|[F_\varepsilon, [P, E_N]] u\| + \varepsilon^{\frac{1}{2}} (\|PF_\varepsilon E_N u\| + \|F_\varepsilon E_N u\|). \end{aligned}$$

The first term can be estimated by $C\varepsilon |||F_\varepsilon u|||_N$ by (vi) in Lemma 26.7.4, and

$$PF_\varepsilon E_N = E_N PF_\varepsilon + [P, E_N] F_\varepsilon = E_N F_\varepsilon P + E_N [P, F_\varepsilon] + [P, E_N] F_\varepsilon,$$

so by (iii) and (v) in Lemma 26.7.4 we have

$$\|PF_\varepsilon E_N u\| \leqq C(|||F_\varepsilon Pu|||_N + |||F_\varepsilon u|||_N).$$

Summing up, we have proved that

$$|||F_\varepsilon u|||_N \leqq C_N (1 + C\varepsilon^{\frac{1}{2}}) |||F_\varepsilon Pu|||_N + C\varepsilon^{\frac{1}{2}} |||F_\varepsilon u|||_N.$$

Choosing ε so small that $C_N(1 + C\varepsilon^{\frac{1}{2}})/(1 - C\varepsilon^{\frac{1}{2}}) < C_{N+1}$ we obtain (26.7.8) with N replaced by $N+1$ if $\varepsilon_{N+1} < \varepsilon$. The proof is complete.

We can now prove an existence theorem by a standard duality argument.

Theorem 26.7.6. *Let M be a set of operators of the form (26.7.2) with coefficients in a bounded subset of $B^\infty(\mathbb{R}^{1+k})$ and $a\geqq 0$. Then there is a constant $c_1>0$ such that for every f in a bounded subset F of $B^\infty(\mathbb{R}^{1+k})$ one can find u with u and $\partial u/\partial t$ in another bounded subset U of $B^\infty(\mathbb{R}^{1+k})$, and satisfying the equation $Pu=f$ in $\{(t,x);|t|<1,|x_1|<c_1\}$.*

Proof. Application of Lemma 26.7.2 to the adjoint of P with x_2 replaced by $-x_2$ shows that for suitable positive constants C and c

$$\|v\|\leqq C\|P^*v\| \quad \text{if } v\in C_0^\infty(\{(t,x);|t|<1,|x_1|<c\}),\ P\in M.$$

Assume for the moment that all $f\in F$ have support in a set of fixed measure. Then there is a constant C' such that

$$\|f\|\leqq C',\quad f\in F.$$

Let $0<c_0<c$. We shall choose a sequence $\varepsilon_1>\varepsilon_2>\ldots>0$ such that for $N=0,1,\ldots$

$$|||v|||_N\leqq C(2N+1)/(N+1)|||P^*v|||_N \tag{26.7.9}$$
$$\text{if } v\in C_0^\infty(\{(t,x);|t|<1,|x_1|<(Nc_0+c)/(N+1)\}),\ P\in M,$$

$$\|E_N^{-1}f\|\leqq C'(2N+1)/(N+1),\quad f\in F. \tag{26.7.10}$$

We know this already when $N=0$, and Lemma 26.7.5 states that if (26.7.9) is valid for one value of N then it is valid with N replaced by $N+1$ if ε_{N+1} is small enough. This is also true for (26.7.10), for

$$\|E_{N+1}^{-1}f\|^2=\|E_N^{-1}f\|^2+\varepsilon_{N+1}^2\|E_N^{-1}\Delta_x f\|^2$$

where the last term tends to 0 uniformly for $f\in F$ when $\varepsilon_{N+1}\to 0$. If

$$v\in C_0^\infty(\Omega),\quad \Omega=\{(t,x);|t|<1,|x_1|<c_0\},$$

and if $f\in F$, we obtain

$$|(f,v)|=|(E_N^{-1}f,E_N v)|\leqq 4CC'|||P^*v|||_N\to 4CC'|||P^*v|||_\infty$$

as $N\to\infty$. Hence the Hahn-Banach theorem shows that we can find g with

$$(2\pi)^{-n}\int|g(t,\xi)|^2\prod_1^\infty(1+|\varepsilon_j\xi|^2)\,d\xi\leqq(4CC')^2,$$

$$(v,f)=(2\pi)^{-n}\int\widehat{P^*v}(t,\xi)\overline{g(t,\xi)}\,d\xi,\quad v\in C_0^\infty(\Omega).$$

If we set

$$u(t,x)=(2\pi)^{-n}\int g(t,\xi)e^{i\langle x,\xi\rangle}\,d\xi$$

we obtain $Pu=f$ in Ω, and there is a fixed bound for $\|D_x^\alpha u\|$ for every α. Since $D_t u=f-iaD_{x_1}u-ibu$ in Ω we also have a uniform bound for $\|D_t D_x^\alpha u\|_{L^2(\Omega)}$. Hence $D_x^\alpha u$ is continuous and uniformly bounded in Ω for all α, and so is $D_t D_x^\alpha u$. After multiplying by a function $\psi\in C_0^\infty(-c_0,c_0)$ of x_1, which is 1 in a somewhat smaller interval $(-c_1,c_1)$ we can extend u to a function in $B^\infty(\mathbb{R}^{k+1})$ with fixed bounds for all x derivatives.

To remove the support condition on f we choose a function $\chi\in C_0^\infty(\mathbb{R}^{k-1})$ such that $\sum\chi(x'-g)^2=1$ in Ω if g runs through the lattice points in the variables $x'=(x_2,\dots,x_k)$. After solving the equations

$$Pu_g=\chi(x'-g)f$$

as above we just have to set $u(t,x)=\sum\chi(x'-g)u_g(t,x)$ to get the desired solution of $Pu=f$ for we may assume that $|t|<2$ and $|x_1|<2C_0$ in $\operatorname{supp} f$.

Remark. By repeated differentiation of the equation $Pu=f$ it follows that $u\in C^\infty$ in any open set where a, b and f are in C^∞. Bounds for the derivatives of u follow from bounds for those of a, b and f.

The substitution $u=ve^w$ where $D_t w+iaD_{x_1}w+ib=0$ can now be used to reduce the equation $Pu=f$ to the form $D_t v+iaD_{x_1}v=fe^{-w}$. In the following corollary we construct non-trivial solutions of the corresponding homogeneous equation.

Corollary 26.7.7. *Let $A\subset B^\infty(\mathbb{R}^{k+1})$ be a bounded set of non-negative functions. Then there exists a constant $c_1>0$ and another bounded set $U\subset B^\infty(\mathbb{R}^{k+1})$ such that for every $a\in A$ there is a solution of the equation*

$$D_t u+iaD_{x_1}u=0 \quad \text{in } \Omega=\{(t,x);|t|<1,|x_1|<c_1\}, \tag{26.7.11}$$

such that u and $\partial u/\partial t$ are in U, $\partial u/\partial x_1=e^w$ in Ω, and $w,\partial w/\partial t\in U$.

Proof. If u satisfies (26.7.11) and $v=\partial u/\partial x_1$, then the equation

$$D_t v+iaD_{x_1}v+\partial a/\partial x_1 v=0 \tag{26.7.12}$$

follows by differentiation of (26.7.11). We solve this equation first by writing $v=e^w$, which gives the inhomogeneous equation

$$D_t w+iaD_{x_1}w+\partial a/\partial x_1=0. \tag{26.7.13}$$

By Theorem 26.7.6 there is a solution of this equation when $|t|<1$, $|x_1|<c_1$ such that w and $D_t w$ are in a bounded subset of B^∞. Now

$$u(t,x)=\int_0^{x_1} v(t,s,x')\,ds-u_1(t,x'),\qquad x'=(x_2,\dots,x_n),$$

satisfies (26.7.11) if

$$D_t u_1(t,x')=(av)(t,0,x').$$

We choose

$$u_1(t,x')=i\int_0^t (av)(s,0,x')\,ds.$$

It is then clear that u and $\partial u/\partial t$ are in a bounded subset of B after a cutoff for large $|t|+|x_1|$, and this proves the corollary.

In Section 26.10 we shall need solutions which are small at the boundary in the x direction:

Corollary 26.7.8. *Let $A \subset B^\infty(\mathbb{R}^{k+1})$ be a bounded set of non-negative functions. For every $\varepsilon>0$ one can then find arbitrarily small neighborhoods $V_0 \Subset V_1 \Subset V_2$ of the origin in $\mathbb{R}^k$ and $T>0$ such that the equation $D_t U + iaD_{x_1} U = 0$ for every $a \in A$ has a solution in $(-T, T) \times V_2$ with uniform bounds*

$$|D_x^\alpha U(t,x)| \leqq C_\alpha; \quad |t|<T, \ x \in V_2; \tag{26.7.14}$$

independent of a, and

$$\operatorname{Re} U(t,x) \geqq 0 \quad \text{if } |t|<T, \ x\in V_2, \tag{26.7.15}$$

$$\operatorname{Re} U(t,x) > 1 \quad \text{if } |t|<T, \ x\in V_2 \setminus V_1, \tag{26.7.16}$$

$$\operatorname{Re} U(t,x) < \varepsilon \quad \text{if } |t|<T, \ x\in V_0. \tag{26.7.17}$$

Proof. We start from the solution u in Corollary 26.7.7 where we may assume that $w(0)=0$, since this can be achieved by multiplication with a suitable constant. Decreasing c_1 if necessary we may then assume that $|w(0, x_1, 0, \dots, 0)| < \pi/4$ when $|x_1| \leqq c_1$. The curve

$$\Gamma = \{u(0, x_1, 0, \dots, 0); \ |x_1| \leqq c_1\}$$

then has slope $<\pi/4$ and the element of arc is $\geqq e^{-\pi/4}|dx_1|$. The distance between the end points $z_\pm = u(0, \pm c_1, 0, \dots, 0)$ is therefore at least $2^{\frac{1}{2}} c_1 e^{-\pi/4}$. The function

$$F_\delta(z) = 2\delta((\delta + z - z_-))^{-1} + (\delta + z_+ - z)^{-1})$$

is analytic and $\operatorname{Re} F_\delta > 0$ in the δ neighborhood of Γ, since $|\arg(z-z_-)| < \pi/4$ and $|\arg(z_+ - z)| < \pi/4$ on Γ. We have $\operatorname{Re} F_\delta(z) < \varepsilon/2$ near $\{u(0, x_1, 0, \dots, 0); |x_1| \leqq c_1/2\}$ if δ is small, and $\operatorname{Re} F_\delta(z) > \frac{4}{3}$ when $|z - z_-| < \delta/2$ or $|z - z_+| < \delta/2$. Since u is uniformly Lipschitz continuous it follows that the function

$$U(t,z) = F_\delta(u(t,x)) + |x''|^2/\delta^4$$

has the desired properties for small δ if

$$V_0 = \{x; |x_1| < c_1/2, |x''| < \delta^2 \varepsilon/2\},$$
$$V_1 = \{x; |x_1| < c_1 - \delta^2, |x''| < \delta^2\},$$
$$V_2 = \{x; |x_1| < c, |x''| < 2\delta^2\}.$$

In the following results we have no need to insist on uniformity with respect to the coefficients so we consider a single operator.

Corollary 26.7.9. *Let $0 \leqq a \in B^\infty(\mathbb{R}^{1+k})$, and let u_0 be a solution of the equation*

$$D_t u_0 + ia(t, x_1, 0) D_{x_1} u_0 = 0$$

when $|t|<1$ and $|x_1|<c_1$, with u_0 and $\partial u_0/\partial t \in B^\infty(\mathbb{R}^2)$. Then one can find $u \in B^\infty(\mathbb{R}^{k+1})$ with $\partial u/\partial t \in B^\infty(\mathbb{R}^{k+1})$ satisfying (26.7.11) such that

$$u(t, x_1, 0) = u_0(t, x_1).$$

Proof. Since $f(t,x)=D_t u_0+ia(t,x)D_{x_1}u_0\in B^\infty$ vanishes in Ω when $x''=0$ we can find $f_j\in B^\infty(\mathbb{R}^{1+k})$ such that

$$f(t,x)=\sum_2^k x_j f_j(t,x)(1+|x''|^2)^{-\frac12} \quad \text{in } \Omega;$$

we first define f_j in Ω and then extend to the whole space. By Theorem 26.7.6 we can find $v_j\in B^\infty$ with $\partial v_j/\partial t\in B^\infty$ and

$$D_t v_j+iaD_{x_1}v_j=f_j \quad \text{in } \Omega.$$

Now we just have to take

$$u(t,x)=u_0(t,x_1)-\sum_2^k x_j v_j(t,x)(1+|x''|^2)^{-\frac12}.$$

We shall simplify the following discussion by dropping the parameters x'', so x will denote a real variable from now on. The characteristics of the operator

$$D_t+ia(t,x)D_x,$$

where we assume $0\leq a\in B^\infty(\mathbb{R}^2)$, are defined by $a(t,x)=0$ and $\tau=0$. Since $a=0$ implies $da=0$ the corresponding direction of the Hamilton field is that of the t axis, so the base projection of a one dimensional bicharacteristic is precisely an interval $I\times\{x_0\}$ in the direction of the t axis where a is identically 0. It is clear that every solution of the homogeneous equation

$$D_t u+ia(t,x)D_x u=0 \tag{26.7.11$'$}$$

is constant on $I\times\{x_0\}$. Differentiation with respect to x (cf. (26.7.12)) shows that also $\partial u/\partial x_1$ is constant in $I\times\{x_0\}$. For the solution given by Corollary 26.7.7 it follows then that w is also constant in $I\times\{x_0\}$. Now assume that $I\subset(-1,1)$ is a maximal compact interval such that a vanishes on $I\times\{x_0\}$. Choose a closed rectangle R with axes parallel to the coordinate axes so that $I\times\{x_0\}\subset R\subset\Omega$ and $a\neq 0$ on the sides parallel to the x axis. We choose R so small that w varies by less than $\pi/4$ in R. Then it is clear that intervals in R parallel to the t or x axis are mapped by u to C^1 curves with tangent direction differing by less than $\pi/4$ from $\operatorname{Im} w-\pi/2$ resp. $\operatorname{Im} w$, evaluated at a point in $I\times\{x_0\}$. If we join two points in R by a curve consisting of two line segments parallel to the coordinate axes, it follows that they have different images under u unless they lie on a line $x=$constant and a vanishes between them. In particular, the boundary of R is mapped to a Jordan curve Γ which has the value of u at $I\times\{x_0\}$ in its interior. If z is in the interior (exterior) of Γ then the winding number of $u(t,x)-z$ when (t,x) goes around Γ is 1 (resp. 0). Hence $u(R)$ contains the interior of Γ but no point in the exterior since regular values there would have to be taken an even number of times and we know that they can only be taken once. It follows that u is a homeomorphism from $\tilde{R}$ to $u(R)$ if $\tilde{R}$ is obtained from R by identifying points in R which lie on a line segment $x=$constant where a vanishes identically.

Now assume that v is any other solution of (26.7.11)′ in a neighborhood of R. Then $v=f(u)$ where f is a continuous function from $u(R)$ to $v(R)$. We claim that f is analytic in the interior of $u(R)$. This is obvious if $a \neq 0$ everywhere, for u is then a C^1 diffeomorphism which transforms (26.7.11)′ to the Cauchy-Riemann equation. In general we can obtain the same conclusion by using Corollary 26.7.9 to extend u to a solution $U(t,x,\varepsilon)$ of the equation

$$D_t U + i(a(t,x)+\varepsilon^2) D_x U = 0$$

with $U(t,x,0)=u(t,x)$. We extend v similarly to a solution V and obtain

$$V(t,x,\varepsilon)=f_\varepsilon(U(t,x,\varepsilon)), \quad \varepsilon \neq 0$$

where f_ε is a continuous map from $U(R,\varepsilon)$ to $V(R,\varepsilon)$ which is analytic in the interior of $U(R,\varepsilon)$. In particular, f_ε is uniformly bounded so there is a uniform limit f_0 of f_ε in the interior of $u(R)$. Thus f_0 is analytic and $f_0(u) = v$, so f_0 is the function we wanted to prove analytic. Thus we have proved:

Theorem 26.7.10. *Let $0 \leqq a \in B^\infty(\mathbb{R}^2)$ and let u be the solution of the equation* (26.7.11)′ *given in Corollary* 26.7.7. *Set*

$$\Omega_0 = \{(t,x)\in\Omega;\ a(t_1,x)\,a(t_2,x)\neq 0 \text{ for some } t_1,t_2 \text{ with } -1<t_1<t<t_2<1\},$$

and let $\tilde\Omega_0$ be the quotient of Ω_0 by the equivalence relation identifying (t,x) with (t',x) if $a(s,x)=0$ when $s\in[t,t']$. Then u defines a local homeomorphism $\tilde\Omega_0 \to \mathbb{C}$ giving $\tilde\Omega_0$ an analytic structure such that the analytic functions in $\tilde\Omega_0$ lifted to Ω_0 are precisely the solutions of the equation (26.7.11)′.

Theorem 26.7.10 can be applied to the Hamilton field in a two dimensional bicharacteristic of an operator satisfying condition (P), for with the coordinates in Proposition 26.5.5 it is of the form

$$\partial/\partial x_1 + ih H_g$$

where $h \geqq 0$. The two dimensional bicharacteristic is generated by the x_1 axis and a bicharacteristic of g in the x', ξ' variables. Hence we have a natural analytic structure in the reduced two dimensional bicharacteristics defined after Definition 26.5.4. The special case of a conic two dimensional bicharacteristic deserves a special discussion. Again with the coordinates in Proposition 26.5.5 the bicharacteristic through $I \times \{0\} \times \{\varepsilon_n\}$ is the product of the x_1 axis and the positive ξ_n axis, and the Hamilton field is

$$H_p = \partial/\partial x_1 + i b(x_1)\xi_n \partial/\partial\xi_n$$

where $b(x_1) = -\partial f(x_1,0,\varepsilon'_n)/\partial x_n$ is different from 0 at some points close to the end points of I. It follows that

$$u = \int b(x_1)\,dx_1 + i\log\xi_n$$

is a solution of the equation. $H_p u=0$ in B with $\operatorname{Re} u$ constant in the radial direction. Any other such solution is of the form $au+c$ for some real a. In fact, an analytic function f with $\operatorname{Re} f(z)$ depending only on $\operatorname{Re} z$ must be of the form $az+c$ with a real, for the harmonic function $\operatorname{Re} f(z)$ must be a linear function of $\operatorname{Re} z$. Thus we obtain:

Theorem 26.7.11. *Let p satisfy condition (P) and let B be a two dimensional bicharacteristic, $\tilde{B}_0$ the corresponding reduced bicharacteristic. Then $\tilde{B}_0$ has a natural analytic structure such that the (local) analytic functions on $\tilde{B}_0$ lifted to B are precisely the (local) solutions of the equation $H_p u=0$. If B is conic, then the set $\tilde{B}_0'$ obtained by identification of points on $\tilde{B}_0$ on the same ray has a natural affine structure such that the linear functions lifted to B are precisely the real parts of solutions of $H_p u=0$ which are constant in the radial direction.*

26.8. The Nirenberg-Treves Estimate

In this section we shall prove a general version of Lemma 26.7.1 where (26.7.2) is replaced by an ordinary differential equation

$$du/dt - A(t)Bu = f \tag{26.8.1}$$

in a Hilbert space H. We make the following assumptions:

(i) $A(t)$ is a bounded non-negative self adjoint operator which is uniformly continuous as a function of t.

(ii) B is bounded and self adjoint.

The boundedness condition on B could be dropped but it is convenient in the statement and proof of the following theorem, and it is quite harmless in our applications.

Theorem 26.8.1. *Assume that the conditions* (i) *and* (ii) *above are fulfilled and that when $|t|<T$*

$$10\|A(t)\|^{\frac14}\|[B,A(t)]\|^{\frac14}\|[B,[B,A(t)]]\|^{\frac14} \leq M. \tag{26.8.2}$$

If u is a continuously differentiable function of t with values in H which satisfies (26.8.1) *and vanishes for $|t|>T$ and if $TM<\frac12$, then*

$$\int \|u(t)\|^2\,dt \leq (4T/(1-2TM))^2 \int \|f(t)\|^2\,dt. \tag{26.8.3}$$

Proof. Let E_λ be the spectral projections of B and write $E_-=E_0$, $E_+=I-E_0$ for the projections corresponding to the half axes. They will replace the operators $h(\pm D_x)$ in the proof of Proposition 26.7.1. Set $u_\pm=E_\pm u$ and form

$$\begin{aligned}\operatorname{Re}(u_-,f_-)&=\operatorname{Re}(u_-,f)=\operatorname{Re}(u_-,\partial u/\partial t)-\operatorname{Re}(u_-,A(t)Bu)\\ &=\tfrac12 d\|u_-\|^2/dt-\operatorname{Re}(u_-,A(t)(B_+u_+-B_-u_-)).\end{aligned}$$

Here $B_+ = E_+B$ and $B_- = -E_-B$ are positive operators and we have used that $\partial u_+/\partial t$ is orthogonal to u_-. Now

$$\begin{aligned}\operatorname{Re}(u_-, A(t)B_-u_-) &= \operatorname{Re}(u_-, [A(t), B_-^{\frac12}]B_-^{\frac12}u_-) + (B_-^{\frac12}u_-, A(t)B_-^{\frac12}u_-)\\ &\geq (u_-, [[A(t), B_-^{\frac12}], B_-^{\frac12}]u_-)/2\end{aligned}$$

for the adjoint of $[A(t), B_-^{\frac12}]B_-^{\frac12}$ is $-B_-^{\frac12}[A(t), B_-^{\frac12}]$. Similarly

$$\begin{aligned}\operatorname{Re}(u_-, A(t)B_+u_+) &= \operatorname{Re}(u_-, [A(t), B_+^{\frac12}]B_+^{\frac12}u_+)\\ &= \operatorname{Re}(u_-, [[A(t), B_+^{\frac12}], B_+^{\frac12}]u_+)\end{aligned}$$

since $B_+^{\frac12}u_- = 0$. In Lemma 26.8.2 we shall show that the norms of these commutators are $\leq M/3$ by (26.8.2). If we now multiply by $T-t$ and integrate, we obtain

$$\tfrac12\int \|u_-\|^2\,dt \leq 2T\int\Big(\|u_-\|\,\|f_-\| + \frac{M}{3}(\|u_-\|^2/2 + \|u_-\|\,\|u_+\|)\Big)\,dt.$$

Taking scalar product with u_+ instead and multiplying by $-T-t$ we obtain

$$\tfrac12\int \|u_+\|^2\,dt \leq 2T\int\Big(\|u_+\|\,\|f_+\| + \frac{M}{3}(\|u_+\|^2/2 + \|u_-\|\,\|u_+\|)\Big)\,dt.$$

If we add and use Cauchy-Schwarz' inequality it follows that

$$\tfrac12\int \|u\|^2\,dt \leq 2T\Big(\int\|u\|^2\,dt\Big)^{\frac12}\Big(\int\|f\|^2\,dt\Big)^{\frac12} + TM\int\|u\|^2\,dt,$$

and this gives (26.8.3) when a factor $(\int\|u\|^2\,dt)^{\frac12}$ is cancelled.

Lemma 26.8.2. *Let A and B be bounded operators in a Hilbert space H, and assume that B is self-adjoint. If $B_\pm = (|B| \pm B)/2$ it follows that*

$$\|[B_\pm^{\frac12}, [B_\pm^{\frac12}, A]]\| \leq \tfrac{10}{3}\|A\|^{\frac14}\,\|[B,A]\|^{\frac14}\,\|[B,[B,A]]\|^{\frac12}. \tag{26.8.4}$$

Proof. If $R(z) = (B-z)^{-1}$ is the resolvent of B and $\varepsilon > 0$, then

$$(1+\varepsilon|B|)^{-1}B_+^{\frac12} = (2\pi i)^{-1}\int_{-i\infty}^{i\infty} z^{\frac12}(1+\varepsilon z)^{-1}R(z)\,dz. \tag{26.8.5}$$

Here $\sqrt{z}$ is analytic when $\operatorname{Re} z > 0$. In fact,

$$\begin{aligned}(2\pi i)^{-1}\int_{-i\infty}^{i\infty} z^{\frac12}(z-\lambda)^{-1}(1+\varepsilon z)^{-1}\,dz &= -\lambda^{\frac12}(1+\varepsilon\lambda)^{-1} \quad \text{if } \lambda > 0,\\ &= 0 \quad \text{if } \lambda \leq 0,\end{aligned}$$

so (26.8.5) follows if we write $R(z) = \int(\lambda - z)^{-1}\,dE_\lambda$ by the spectral theorem. The integral in (26.8.5) is absolutely convergent since $\|R(z)\| \leq 1/|\operatorname{Im} z|$. The resolvent equation

$$[A, R(z)] = R(z)\,[B, A]\,R(z)$$

follows by multiplying the equation $[B-z, A] = [B, A]$ left and right by $R(z)$, and it gives the estimate

$$\|[A, R(z)]\| \leq \min(2\|A\|\,|\operatorname{Im} z|^{-1}, \|[B,A]\|\,|\operatorname{Im} z|^{-2}).$$

Hence we have for every $T>0$

$$\|[A,(1+\varepsilon|B|)^{-1}B_+^{\frac{1}{2}}]\| \leq \frac{2}{\pi}\int_0^T t^{-\frac{1}{2}}\|A\|\,dt + 1/\pi\int_T^\infty t^{-\frac{3}{2}}\|[B,A]\|\,dt$$
$$= 2/\pi(2T^{\frac{1}{2}}\|A\| + T^{-\frac{1}{2}}\|[B,A]\|).$$

We minimize the right-hand side and conclude when $\varepsilon\to 0$ that

(26.8.6) $$\|[A,B_+^{\frac{1}{2}}]\| \leq 4\sqrt{2}/\pi\,\|A\|^{\frac{1}{2}}\|[B,A]\|^{\frac{1}{2}}.$$

To prove (26.8.4) we shall now estimate $[[A,B_+^{\frac{1}{2}}],B]$. We have

$$[[A,R(z)],B] = R(z)[[B,A],B]\,R(z),\ [[A,R(z)],B] = -[[B,A],R(z)]$$

since B and $R(z)$ commute, so $\|[[A,R(z)],B]\|$ can be estimated by

$$\min(2\|[B,A]\|\,|\operatorname{Im} z|^{-1},\ \|[[B,A],B]\|\,|\operatorname{Im} z|^{-2}).$$

Hence it follows from (26.8.5) as in the proof of (26.8.6) that

(26.8.7) $$\|[[A,B_+^{\frac{1}{2}}],B]\| \leq 4\sqrt{2}/\pi\,\|[B,A]\|^{\frac{1}{2}}\|[B,[B,A]]\|^{\frac{1}{2}}.$$

The preceding estimates are of course valid for $B_-^{\frac{1}{2}}$ also. If we now apply (26.8.6) with A replaced by $[A,B_\pm^{\frac{1}{2}}]$ we obtain by (26.8.7)

$$\|[[A,B_\pm^{\frac{1}{2}}],B_\pm^{\frac{1}{2}}]\|^2 \leq 32/\pi^2\,\|[A,B_\pm^{\frac{1}{2}}]\|\,4\sqrt{2}/\pi\,\|[B,A]\|^{\frac{1}{2}}\|[B,[B,A]]\|^{\frac{1}{2}}$$
$$\leq (32/\pi^2)^2\,\|A\|^{\frac{1}{2}}\|[B,A]\|\,\|[B,[B,A]]\|^{\frac{1}{2}}.$$

Since $32/\pi^2 = 3.24\ldots < \frac{10}{3}$, the estimate (26.8.4) follows.

In our application of Theorem 26.8.1 A and B will be pseudo-differential operators with symbols bounded in S^0 and in S^1 respectively, which makes (26.8.2) valid for some M which we can estimate. It would have been possible to avoid the abstract operator theory in Theorem 26.8.1 by using Fourier integral operators corresponding to non-homogeneous canonical transformations to reduce to a situation where $B=D_{x_2}$ and pseudo-differential operators can be used as in the proof of Lemma 26.7.1. However, Theorem 26.8.1 may serve as a useful reminder that abstract operator theory may sometimes be more efficient than pseudo-differential operator theory; the operators $B_\pm^{\frac{1}{2}}$ are not pseudo-differential operators unless B is of a very special form.

In the proof of Theorem 26.8.1 we discarded a positive quantity corresponding to the one which gave the second term in the left-hand side of (26.7.3), which was essential for the commutator estimate in Lemma 26.7.3. We shall now prove an analogous estimate in the abstract context of Theorem 26.8.1 which will also allow us to control some commutators which occur in Sections 26.9 and 26.10.

Theorem 26.8.3. *If* $A(t)$ *and* B *satisfy* (i), (ii) *and* (26.8.2) *above, and if* $u\in C^1(\mathbb{R},H)$ *vanishes for* $|t|>T$ *and satisfies* (26.8.1), *then*

$$\int (Bu, A(t)Bu)\,dt \leqq T\|B\| \int (18\|f\|^2+2M^2\|u\|^2)\,dt. \tag{26.8.8}$$

Proof. With the notation in the proof of Theorem 26.8.1 we have

$$\begin{aligned}\operatorname{Re}(B_-u_-,f) &= \operatorname{Re}(B_-u_-, du/dt - A(t)Bu)\\ &= \tfrac{1}{2}d\|B_-^{\frac{1}{2}}u_-\|^2/dt + (B_-u_-, A(t)B_-u_-) - \operatorname{Re}(B_-u_-, A(t)B_+u_+).\end{aligned}$$

Since $B_+^{\frac{1}{2}}B_-=0$ the last term can be estimated by

$$|(B_-u_-, [[A(t), B_+^{\frac{1}{2}}], B_+^{\frac{1}{2}}]u_+)| \leqq \|B_-u_-\|\, M\|u_+\|/3$$

where we have used (26.8.4) and (26.8.2). If we integrate after multiplication by $2T-t$ we obtain

$$\begin{aligned}\tfrac{1}{2}\int \|B_-^{\frac{1}{2}}u_-\|^2\,dt + T\int (B_-u_-, A(t)B_-u_-)\,dt\\ \leqq T\int (3\|B_-u_-\|\,\|f_-\| + M\|B_-u_-\|\,\|u_+\|)\,dt.\end{aligned}$$

We have a similar estimate for $B_+^{\frac{1}{2}}u_+$, and since $Bu=B_+u_+-B_-u_-$ we have

$$(Bu, A(t)Bu) \leqq 2(B_+u_+, A(t)B_+u_+) + 2(B_-u_-, A(t)B_-u_-),$$

for $A(t)$ is positive. Hence we obtain by adding

$$\int \||B|^{\frac{1}{2}}u\|^2\,dt + T\int (Bu, A(t)Bu)\,dt \leqq T\int (6\|Bu\|\,\|f\| + 2M\|Bu\|\,\|u\|)\,dt.$$

Here $\|Bu\| \leqq \|B\|^{\frac{1}{2}}\,\||B|^{\frac{1}{2}}u\|$ so the right-hand side is bounded by

$$\int \||B|^{\frac{1}{2}}u\|^2\,dt + T^2\|B\|\int (18\|f\|^2 + 2M^2\|u\|^2)\,dt.$$

Hence

$$T\int (Bu, A(t)Bu)\,dt \leqq T^2\|B\|\int (18\|f\|^2+2M^2\|u\|^2)\,dt,$$

which proves (26.8.8).

26.9. The Singularities in N_2^e and in N_{12}^e

We are now ready to extend Theorem 26.2.1 to a general two dimensional bicharacteristic B, that is, a leaf of the foliation of the involutive manifold N_2^e. More precisely we shall consider the corresponding reduced bicharacteristic $\tilde{B}_0$. (See Proposition 26.5.3, Definition 26.5.4 and the discussion after it, as well as Theorems 26.7.10 and 26.7.11.) If $u\in\mathscr{D}'(X)$ we define $\tilde{s}_u(\tilde{\gamma})$ for $\tilde{\gamma}\in\tilde{B}_0$ as $\inf s_u^*(x,\xi)$ for (x,ξ) in the inverse image γ of $\tilde{\gamma}$ in B, which is a compact maximal embedded one dimensional bicharacteristic. Since s_u^* is semi-continuous from below it is clear that $\tilde{s}_u(\tilde{\gamma})$ is the supremum of all $s\in\mathbb{R}$ such that $u\in H_{(s)}$ at every point in γ. The central result in this section is the following one.

Theorem 26.9.1. *Let $P\in\Psi^m_{\rm phg}(X)$ be properly supported and satisfy condition (P). Let $u\in\mathscr{D}'(X)$, let B be a two dimensional bicharacteristic of P and $\tilde{s}$ a superharmonic function in an open subset ω of the corresponding reduced bicharacteristic $\tilde{B}_0$ such that $\tilde{s}_{Pu}\geqq\tilde{s}$ in ω. Then it follows that*

$$\min(\tilde{s}_u, \tilde{s}+m-1) \tag{26.9.1}$$

is superharmonic in ω. In the special case where B is conic, this means that (26.9.1) is a concave function on $\tilde{B}'_0$ if $\tilde{s}$ is concave; here $\tilde{B}'_0$ is obtained from $\tilde{B}_0$ by identifying points on the same ray.

Superhamonicity is a local property so it suffices to prove that Theorem 26.9.1 is valid for small neighborhoods ω of any point in $\tilde{B}_0$. Using Propositions 26.4.13 and 26.5.5 with the remark following the proof of the latter we can then transform P as in Proposition 26.4.4 (see also the proof of Theorem 26.6.2). Thus we may assume that $P\in\Psi^1_{\rm phg}(\mathbb{R}^n)$ and that the principal symbol is of the form

$$p(x,\xi)=\xi_1+ig(x',\xi')h(x,\xi'); \quad x'=(x_2,\dots,x_n),\ \xi'=(\xi_2,\dots,\xi_n) \tag{26.9.2}$$

in a conic neighborhood of $I'=I\times\{0\}\times\varepsilon_n\subset T^*(\mathbb{R}^n)\smallsetminus 0$. Here $h\geqq 0$ is homogeneous of degree 0, g is homogeneous of degree 1, and $g(0,\varepsilon'_n)=0$, $dg(0,\varepsilon'_n)\neq 0$. The interval $I\subset\mathbb{R}$ is compact and $h(x_1,0,\varepsilon'_n)$ vanishes in I but not in any strictly larger interval. The first step in the proof is to derive estimates from Theorems 26.8.1 and 26.8.3.

Lemma 26.9.2. *Let $\psi\in C_0^\infty(\mathbb{R}^{2n-2})$ be equal to 1 in a neighborhood of 0, $0\leqq\psi\leqq 1$ everywhere, and set*

$$\psi_{\delta,\lambda}(x',\xi')=\psi(x'/\delta,(\lambda\xi'-\varepsilon'_n)/\delta); \quad \lambda,\ \delta>0;$$

$I_\delta=\{t+t';\ t\in I,\ |t'|\leqq\delta\}$. Then there is a constant C such that

$$\|v\|\leqq C\|D_1v+i(\psi^4_{\delta,\lambda}gh)(x,D')v\|, \tag{26.9.3}$$

$$\sum_{|\alpha+\beta|=1}\lambda^{|\beta|}\|(\psi^4_{\delta,\lambda}h^{(\alpha)}_{(\beta)}g)(x,D')v\|\leqq C\lambda^{\frac12}\|D_1v+i(\psi^4_{\delta,\lambda}gh)(x,D')v\|, \tag{26.9.4}$$

if $v\in C_0^\infty(I_\delta\times\mathbb{R}^{n-1})$, δ is sufficiently small and $0<\lambda<\lambda_\delta$. The norms are L^2 norms.

Proof. After multiplying g and h by a cut off function which is equal to 1 in a conic neighborhood of $(0,\varepsilon'_n)$ in $T^*(\mathbb{R}^{n-1})$ when $|\xi'|>\frac12$ we may assume that $g\in S^1(\mathbb{R}^{n-1}\times\mathbb{R}^{n-1})$ and that $0\leqq h\in S^0(\mathbb{R}^n\times\mathbb{R}^{n-1})$ are homogeneous of degree 1 and 0 for $|\xi'|>\frac12$. We shall apply Theorems 26.8.1 and 26.8.3 to the self-adjoint operators in $L^2(\mathbb{R}^{n-1})$ defined by

$$A(x_1)=\psi_{\delta,\lambda}(x',D')^*(h(x,D')+h(x,D')^*)\psi_{\delta,\lambda}(x',D')/2+C_0\lambda,$$
$$B=\psi_{\delta,\lambda}(x',D')^*(g(x',D')+g(x',D')^*)\psi_{\delta,\lambda}(x',D')/2,$$

where the constant C_0 will be determined so that $A(x_1)\geqq 0$. By Theorem 18.1.13 we have

$$(A(x_1)v, v)=C_0\lambda\|v\|^2+\operatorname{Re}(h(x,D')\psi_{\delta,\lambda}(x',D')v, \psi_{\delta,\lambda}(x',D')v)$$
$$\geqq C_0\lambda\|v\|^2-C\|(1+|D'|^2)^{-\frac{1}{4}}\psi_{\delta,\lambda}(x',D')v\|^2.$$

If δ is small we have $\frac{1}{2}<|\xi'\lambda|<2$ in the support of the symbol of $(1+|D'|^2)^{-\frac{1}{2}}\psi_{\delta,\lambda}(x',D')$, the product of the symbol by $\lambda^{-\frac{1}{2}}$ is uniformly bounded in S^0 for fixed δ, and the maximum has a bound independent of δ for small λ. Hence it follows from Theorem 18.1.15 that

$$A(x_1)\geqq C_0\lambda-\lambda(C_1+C_\delta\lambda^{\frac{1}{2}})>0$$

for small λ if $C_1<C_0$. From Theorem 18.1.15 we also obtain

$$\|A(x_1)\|\leqq|\psi_{\delta,\lambda}^2 h|_\infty+C_\delta\lambda^{\frac{1}{2}},$$
$$\|[B, A(x_1)]\|\leqq|\{\psi_{\delta,\lambda}^2 g, \psi_{\delta,\lambda}^2 h\}|_\infty+C_\delta\lambda^{\frac{1}{2}}$$
$$\|[B,[B, A(x_1)]]\|\leqq|\{\psi_{\delta,\lambda}^2 g, \{\psi_{\delta,\lambda}^2 g, \psi_{\delta,\lambda}^2 h\}\}|_\infty+C_\delta\lambda^{\frac{1}{2}},$$

for all symbols have support where $\frac{1}{2}<|\xi'\lambda|<2$. The maximum norms on the right-hand side are independent of λ. Since g vanishes and h vanishes of second order on $I\times\{0\}\times\{\varepsilon_n'\}$ we have $D^\alpha(\psi_\delta, h/\delta^2, g/\delta)=O(\delta^{-|\alpha|})$ in $I_\delta\times\operatorname{supp}\psi_\delta$ so the maximum norms are $O(\delta^2)$, $O(\delta)$ and $O(1)$ respectively if $x_1\in I_\delta$. For $x_1\in I_\delta$ and $0<\lambda<\lambda_\delta$ we therefore have $\|B\|\leqq C/\lambda$ and

$$\|A(x_1)\|\leqq C\delta^2, \quad \|[B, A(x_1)]\|\leqq C\delta, \quad \|[B,[B, A(x_1)]]\|\leqq C.$$

When δ is small enough it follows from (26.8.3) and (26.8.8) that

$$\|v\|\leqq C\|D_1v+iABv\|, \quad v\in C_0^\infty(I_\delta\times\mathbf{R}^{n-1}), \tag{26.9.5}$$

$$\int(A(x_1)Bv, Bv)\,dx_1\leqq C/\lambda(\|D_1v+iABv\|^2+\|v\|^2), \quad v\in C_0^\infty(I_\delta\times\mathbf{R}^{n-1}). \tag{26.9.6}$$

The symbol of $A(x_1)B$ is

$$\psi_{\delta,\lambda}^4 hg+C_0\lambda\psi_{\delta,\lambda}^2 g+S_{\delta,\lambda}+R_{\delta,\lambda}$$

where $R_{\delta,\lambda}$ is bounded in S^{-1} for fixed δ and $S_{\delta,\lambda}$ is a finite linear combination of terms obtained from $\psi^4 gh$ by applying $\partial/\partial\xi_j$ and $\partial/\partial x_j$ to two factors, possibly the same. Thus $S_{\delta,\lambda}$ is bounded in S^0 for fixed δ and $|S_{\delta,\lambda}|_\infty\leqq C\delta$ with C independent of λ, by the arguments above. Hence Theorem 18.1.15 gives

$$\|A(x_1)B-(\psi_{\delta,\lambda}^4 gh)(x,D')\|\leqq C\delta, \quad x_1\in I_\delta,$$

if λ is small enough. Hence (26.9.5) implies

$$\|v\|\leqq C\|D_1v+iABv\|\leqq C\|D_1v+i(\psi_{\delta,\lambda}^4 gh)(x,D')v\|+C'\delta\|v\|,$$
$$v\in C_0^\infty(I_\delta\times\mathbf{R}^{n-1}),$$

which gives (26.9.3) if δ is small enough and also shows that we may replace ABv by $(\psi_{\delta,\lambda}^4 gh)(x,D')v$ in the right-hand side of (26.9.6). To prove (26.9.4) we first observe that Lemma 7.7.2 implies

$$|h'_x|^2+(1+|\xi|^2)|h'_\xi|^2 \leq Ch.$$

Hence it follows from Theorem 18.1.14 that

$$\sum \|h_{(j)}(x,D')v\|^2+\sum \|(1+|D'|^2)^{\frac12} h^{(j)}(x,D')v\|^2 \leq C\operatorname{Re}(h(x,D')v,v)+C_1\|v\|^2_{(-\frac12)}. \tag{26.9.7}$$

To combine (26.9.7) with (26.9.6) where the leading term in the symbol of BAB is $\psi^6_{\delta,\lambda}g^2h$ we replace v by $(\psi^3_{\delta,\lambda}g)(x',D')v$ in (26.9.7). The symbol of

$$(\psi^3_{\delta,\lambda}g)(x',D')^* h(x,D')(\psi^3_{\delta,\lambda}g)(x',D')-BA(x_1)B$$

is uniformly bounded in S^0 for fixed δ and as above it follows that the maximum of the symbol has a bound independent of δ when λ is small. Hence the norm has a bound independent of δ when $\lambda<\lambda_\delta$, again by Theorem 18.1.15. Combining (26.9.6) and (26.9.7) we now obtain

$$\lambda\sum \|h_{(j)}(x,D')(\psi^3_{\delta,\lambda}g)(x',D')v\|^2 \leq C(\|D_1v+i(\psi^4_{\delta,\lambda}gh)(x,D')v\|^2+\|v\|^2), \quad v\in C_0^\infty(I_\delta\times\mathbb{R}^{n-1}),$$

if δ is small and $\lambda<\lambda_\delta$. The symbol of

$$h_{(j)}(x,D')(\psi^3_{\delta,\lambda}g)(x',D')-(\psi^3_{\delta,\lambda}h_{(j)}g)(x',D')$$

is bounded in S^0 for fixed δ and the maximum has a bound independent of δ when $\lambda<\lambda_\delta$. Hence Theorem 18.1.13′ again gives a fixed bound for the norm, and (26.9.4) follows when we sum for $\alpha=0$, $|\beta|=1$. The estimate of the other terms follows in the same way by means of the second sum in (26.9.7), and this completes the proof, if we use (26.9.3) to estimate $\|v\|$.

In the proof of Theorem 26.9.1 we shall also need a description in terms of L^2 norms for the regularity function s^*_u defined in (18.1.41).

Lemma 26.9.3. *Let χ and ϕ be functions in $C_0^\infty(T^*(\mathbb{R}^n)\smallsetminus 0)$ and set*

$$q_\lambda(x,\xi)=\chi(x,\lambda\xi)\lambda^{-\phi(x,\lambda\xi)}. \tag{26.9.8}$$

If $u\in\mathscr{E}'(\mathbb{R}^n)$ and $s^*_u>\operatorname{Re}\phi$ in $\operatorname{supp}\chi$, then $\|q_\lambda(x,D)u\|_{L^2}$ is bounded as $\lambda\to 0$. Conversely, if $\|q_\lambda(x,D)u\|_{L^2}$ is bounded as $\lambda\to 0$, then

$$s^*_u(x,\xi)\geq \operatorname{Re}\phi(x,\xi) \quad \text{if } \chi(x,\xi)\neq 0.$$

Proof. First note that if $\operatorname{Re}\phi<\mu$ in $\operatorname{supp}\chi$, then q_λ is bounded in S^μ as $\lambda\to 0$, that is

$$|D^\alpha_\xi D^\beta_x q_\lambda(x,\xi)|\leq C_{\alpha\beta}(1+|\xi|)^{\mu-|\alpha|}, \quad \lambda<1. \tag{26.9.9}$$

In fact, $|\lambda\xi|$ lies between two fixed positive bounds when $(x,\lambda\xi)\in\operatorname{supp}\chi$, and we have

$$\partial q_\lambda/\partial x=((\partial\chi/\partial x-\chi\log\lambda\,\partial\phi/\partial x)\lambda^{-\phi})(x,\lambda\xi)$$

and a similar formula with another factor λ for $\partial q_\lambda/\partial\xi$. This proves (26.9.9), and it follows that $\|q_\lambda(x,D)u\|_{L^2}$ is bounded as $\lambda\to 0$ if $s_u^*>\mu>\operatorname{Re}\phi$ in $\operatorname{supp}\chi$. If we just have $s_u^*>\operatorname{Re}\phi$ in $\operatorname{supp}\chi$, we note that since s_u^* is semi-continuous from below we can write $\chi=\sum\chi_j$ where $\operatorname{Re}\phi<\mu_j<s_u^*$ in $\operatorname{supp}\chi_j$ for some μ_j. This proves the first part of the lemma.

It remains to prove the second part of the lemma. Let $\chi(y,\eta)\neq 0$ and choose for given $\mu<\operatorname{Re}\phi(y,\eta)$ a non-negative function $\chi_0\in C_0^\infty$ such that $\operatorname{Re}\phi>\mu$ and $\chi\neq 0$ in $\operatorname{supp}\chi_0$, $\chi_0(y,\eta)>0$. Let $u\in H_{(-M)}^{\text{comp}}$. One can find $a_\lambda(x,\xi)$ bounded in S^0 and R_λ bounded in S^{-M} so that

$$\chi_0(x,\lambda\xi)\lambda^{-\mu}=a_\lambda(x,D)q_\lambda(x,D)+R_\lambda(x,D). \tag{26.9.10}$$

As a first approximation to a_λ we take

$$a_\lambda^0(x,\xi)=\chi_0(x,\lambda\xi)/\chi(x,\lambda\xi)\lambda^{-\mu+\phi(x,\lambda\xi)}$$

which gives an error of the desired form apart from a finite number of terms in the series

$$-\sum_{\alpha\neq 0}(iD_\xi)^\alpha a_\lambda^0(x,\xi)D_x^\alpha q_\lambda(x,\xi)/\alpha!.$$

These can again be handled in the same way, and after a finite number of iterations we obtain the desired function a_λ. It follows from (26.9.10) that

$$\|\chi_0(x,\lambda D)\lambda^{-\mu}u\|_{L^2}\leq C,\quad \lambda<1.$$

If we multiply by $\lambda^{\varepsilon-1}$ where $\varepsilon>0$ and integrate from 0 to 1, we obtain $\|r(x,D)u\|_{L^2}<\infty$ where

$$r(x,\xi)=\int_0^1\chi_0(x,\lambda\xi)\lambda^{\varepsilon-\mu-1}\,d\lambda.$$

When $|\xi|$ is large enough the upper bound in the integration may be replaced by ∞ which shows that $r(x,\xi)$ is homogeneous of degree $\mu-\varepsilon$ at ∞. It is clear that r is positive in the direction (y,η). Hence $s_u^*(y,\eta)\geq\mu-\varepsilon$ which completes the proof.

Proof of Theorem 26.9.1. As already pointed out we may assume that $P\in\Psi_{\text{phg}}^1(\mathbb{R}^n)$, that the principal symbol p satisfies (26.9.2) in a conic neighborhood of the inverse image $I'=I\times\{0\}\times\{\varepsilon_n\}\subset T^*(\mathbb{R}^n)\setminus 0$ of a point in $\tilde{B}_0$, and that B is the leaf containing $(0,\varepsilon_n)$ generated by the vector fields $\partial/\partial x_1$ and H_g. We may also assume that the term p_0 of order 0 in the symbol of p vanishes in I'. In fact, the equation $Pu=f$ implies $SPS^{-1}(Su)-Sf\in C^\infty$ if $S\in\Psi_{\text{phg}}^0$ is elliptic with parametrix S^{-1}. We have $s_u^*=s_{Su}^*$, $s_f^*=s_{Sf}^*$, and the term of order 0 in the symbol of SPS^{-1} vanishes in I' if for the principal symbol S_0 of S we have

$$S_0D_1S_0^{-1}+p_0=0,\quad\text{that is, } i\,\partial S_0/\partial x_1+S_0p_0=0 \text{ in } I'.$$

This ordinary differential equation is easy to solve, and S_0 can be extended to an elliptic symbol of order 0. Replacing P by SPS^{-1} we can thus assume that p_0 vanishes in I'. In view of Theorem 18.1.15 we can then choose $R\in S^0$ so that p_0-R is of order $-\infty$ in a conic neighborhood V of I' in $T^*(\mathbb{R}^n)\setminus 0$ and $C\|R(x,D)\|<\frac{1}{2}$ where C is the constant in Lemma 26.9.2.

We can find a neighborhood $V_0\subset V$ of I' and functions $w, w_0\in C^\infty(V_0)$ such that

$$(\partial/\partial x_1+ihH_g)w=(\partial/\partial x_1+ihH_g)w_0=0 \quad\text{and}\quad H_g w\neq 0 \quad\text{in } V_0, \tag{26.9.11}$$

$$C_1 d_B(x,\xi)^2\leqq \operatorname{Re} w_0(x,\xi)\leqq C_2 d_B(x,\xi)^2, \quad (x,\xi)\in V_0, \tag{26.9.12}$$

where d_B is the distance to B. Note that (26.9.11) implies $H_p w=H_p w_0=0$ when $g=0$. When verifying this we may assume that $g=\xi_2$, for this can be achieved by a possibly non-homogeneous canonical transformation. Then we have the equations

$$(\partial/\partial x_1+ih\,\partial/\partial x_2)w=(\partial/\partial x_1+ih\,\partial/\partial x_2)w_0=0.$$

The existence of w is therefore an immediate consequence of Corollary 26.7.7. Recalling that w must be constant on I' we may assume that $w=0$ on I' and can then take

$$w_0(x,\xi)=(x_3^2+\dots+x_n^2+\xi_1^2+\dots+\xi_{n-1}^2+(\xi_n-1)^2)\exp w(x,\xi).$$

Now we return to the original coordinates. Let $\tilde{K}$ be a compact subset of ω such that the inverse image K in B is contained in V_0, $x_1\in I_{\delta/2}$ in K and the function $\psi_{\delta,1}$ in Lemma 26.9.2 is equal to 1 in a neighborhood of K for some δ such that (26.9.3), (26.9.4) are valid. We fix δ. Let H be a harmonic polynomial in $\mathbb{C}$ such that, with s denoting the lifting of $\tilde{s}$ to B,

$$H(w)<\min(s_u^*, s) \qquad \text{on the boundary of } K \text{ in } B_0.$$

The theorem will be proved if we show that the same inequality is then valid in K, for the restriction of w to B is the lifting to B of a local analytic coordinate in $\tilde{B}_0$. Since $H(w)$ is harmonic in $\tilde{B}_0$ and $\tilde{s}$ is superharmonic, we have $H(w)<s$ in K. Furthermore, the boundary ∂K of K in B is in the inverse image of the boundary of $\tilde{K}$ in $\tilde{B}_0$, for the inverse image of the interior is open. Hence we have if $f=Pu$

$$H(w)<s_u^* \quad \text{on } \partial K, \tag{26.9.13}$$

$$H(w)<s_f^* \quad \text{in } K. \tag{26.9.14}$$

We may also assume in the proof that $u\in\mathscr{E}'$ and that

$$H(w)-1<s_u^* \quad \text{in } K, \tag{26.9.15}$$

for if the assertion is proved under that additional hypothesis we can just start from the fact that $s_u^*>H(w)-k$ for some positive integer k and decrease k successively until $k=0$ and the theorem is proved.

Choose $\chi\in C_0^\infty(V_0)$ equal to 1 in a neighborhood of K but with $\operatorname{supp}\chi$ so close to K that $x_1\in I_\delta$ and $\psi_{\delta,1}=1$ in $\operatorname{supp}\chi$ and (26.9.13)–(26.9.15) imply

$$s_u^*>H(w) \quad \text{in } B\cap\operatorname{supp} d\chi;$$
$$s_f^*>H(w) \quad \text{and} \quad s_u^*>H(w)-1 \quad \text{in } B\cap\operatorname{supp}\chi.$$

We can write $H=\operatorname{Re}F$ where F is an analytic polynomial in $\mathbb{C}$. In view of (26.9.12) we can choose a constant τ so large that if

$$\phi=F(w)-\tau w_0$$

we even have

$$s_u^*>\operatorname{Re}\phi \quad \text{in } \operatorname{supp} d\chi \tag{26.9.16}$$
$$s_f^*>\operatorname{Re}\phi \quad \text{in } \operatorname{supp}\chi \tag{26.9.17}$$
$$s_u^*>\operatorname{Re}\phi-1 \quad \text{in } \operatorname{supp}\chi. \tag{26.9.18}$$

Note that $\operatorname{Re}\phi=H(w)$ in K and that $(\partial/\partial x_1+ihH_g)\phi=0$.

From (26.9.3) and the fact that $C\|R\|<\frac{1}{2}$ it follows that

$$\|v\|\leqq 2C\|D_1 v+i(\psi_{\delta,\lambda}^4 gh)(x,D')v+R(x,D)v\| \tag{26.9.3$'$}$$

if $v\in\mathscr{S}$ and $x_1\in I_\delta$ in $\operatorname{supp} v$. We shall apply (26.9.3)$'$ to

$$v=q_\lambda(x,D)u, \quad q_\lambda(x,\xi)=\chi(x,\lambda\xi)\lambda^{-\phi(x,\lambda\xi)}.$$

To estimate

$$M=\|(D_1+i(\psi_{\delta,\lambda}^4 gh)(x,D')+R(x,D))\,q_\lambda(x,D)u\|$$

we want to commute $q_\lambda(x,D)$ through the operator in front. Choose μ so that $u\in H_{(-\mu)}$. Since

$$\partial q_\lambda(x,\xi)/\partial x=(\chi_x'(x,\lambda\xi)-\log\lambda\,\phi_x'(x,\lambda\xi))\lambda^{-\phi(x,\lambda\xi)}$$

and similarly for $\partial q_\lambda/\partial\xi$, the symbol of $[R(x,D),q_\lambda(x,D)]$ is, apart from an error which is bounded in $S^{-\mu}$, a finite sum of functions of the same form as q_λ but with ϕ replaced by $\phi-j$ where j is a positive integer, and multiplied by a power of $\log\lambda$. Hence it follows from (26.9.18) and Lemma 26.9.3 that

$$\|[R(x,D),q_\lambda(x,D)]u\|\leqq C_u, \quad 0<\lambda<1.$$

The symbol of

$$[D_1+i(\psi_{\delta,\lambda}^4 gh)(x,D'),\,q_\lambda(x,D)]$$

is similar apart from the first term in the symbol which is

$$-i(\partial/\partial x_1+igH_h+ihH_g)q_\lambda.$$

(Here we have used that $\psi_{\delta,\lambda}=1$ in $\operatorname{supp} q_\lambda$.) In view of the differential equation $(\partial/\partial x_1+ihH_g)\phi=0$ we obtain

$$-i(\partial/\partial x_1+igH_h+ihH_g)q_\lambda=\tilde q_\lambda(x,\xi)+i\log\lambda\, g\{h,\phi\}(x,\lambda\xi)\,q_\lambda. \tag{26.9.19}$$

Here

$$\tilde q_\lambda(x,\xi)=-i\tilde\chi(x,\lambda\xi)\lambda^{-\phi(x,\lambda\xi)}, \quad \tilde\chi=H_p\chi,$$

is of the same form as q_λ but with $\operatorname{supp}\tilde{\chi}\subset\operatorname{supp} d\chi$. Hence it follows from (26.9.16) and Lemma 26.9.3 that $\|\tilde{q}_\lambda(x,D)u\|$ is bounded when $\lambda\to 0$. The main term in the symbol of

$$(26.9.20)\quad i\log\lambda\big(\sum\phi_{(j)}(x,\lambda D)(\psi^3_{\delta,\lambda}h^{(j)}g)(x,D')q_\lambda(x,D)\\ -\lambda\sum\phi^{(j)}(x,\lambda D)(\psi^3_{\delta,\lambda}h_{(j)}g)(x,D')q_\lambda(x,D)\big)$$

is equal to the last term in (26.9.19), and the others are of the same form but smaller by at least a factor $\lambda\log\lambda$. Estimating (26.9.20) by means of (26.9.4) we obtain in view of (26.9.16)

$$M\leqq\|q_\lambda(x,D)(D_1+i(\psi^4_{\delta,\lambda}gh)(x,D')+R(x,D))u\|+C\lambda^{\frac{1}{2}}|\log\lambda|M+C_u.$$

When λ is so small that $C\lambda^{\frac{1}{2}}|\log\lambda|<\frac{1}{2}$ we can cancel the middle term on the right-hand side against half the left hand side. The symbol of

$$q_\lambda(x,D)(D_1+i(\psi^4_{\delta,\lambda}gh)(x,D')+R(x,D)-P)$$

is bounded in $S^{-\infty}$ since $\psi_{\delta,\lambda}=1$ in $\operatorname{supp} q_\lambda$ and the complete symbol of P is $\xi_1+igh(x,\xi')+R(x,\xi)$ in V. Hence we obtain for small λ

$$M/2\leqq\|q_\lambda(x,D)Pu\|+C_u.$$

By (26.9.17) the right-hand side is bounded when $\lambda\to 0$, so using (26.9.3)′ with $v=q_\lambda(x,D)u$ we conclude that $\|q_\lambda(x,D)u\|$ is bounded as $\lambda\to 0$. By Lemma 26.9.3 it follows that $s^*_u\geqq\phi$ in K. In view of Theorem 26.7.11 the proof is now complete.

By the method of descent from an operator in a higher dimensional space we can derive from Theorem 26.9.1 a similar result on the singularities in N^{ie}_{12} (see the summary at the end of Section 26.5).

Theorem 26.9.4. *Let $P\in\Psi^m_{\text{phg}}(X)$ be properly supported and satisfy condition (P), let I be a compact interval on a one dimensional bicharacteristic with end points in N^i_{12}, and let $\tilde{I}$ be the affine interval obtained by Proposition 26.5.8 when subintervals not meeting N^i_{12} are collapsed to points. If $u\in\mathscr{D}'(X)$, s is a concave function on $\tilde{I}$ and $\tilde{s}_{Pu}\geqq s$ on I, it follows that*

$$\min(\tilde{s}_u, s+m-1)$$

is a concave function on $\tilde{I}$.

Here the definition of $\tilde{s}_u$ and $\tilde{s}_{Pu}$ on $\tilde{I}$ is completely analogous to the definitions used in Theorem 26.9.1.

Proof. We may assume that $m=1$ and that the principal symbol p has the form ξ_1+igh in Propositions 26.4.13 and 26.5.7, with $h>0$ at the end points of I. Let $H(x,\xi)$ be a homogeneous function of degree 0 which is equal to $h(x,\xi')$ in a neighborhood of I. If we regard u and f as distributions U and

F in $\mathbb{R}^{n+1}$ independent of x_{n+1}, then

$$(P+iH(x,D)D_{n+1})U=F. \tag{26.9.21}$$

Strictly speaking the operator in (26.9.21) is not a pseudo-differential operator but it becomes one if we multiply by an operator with symbol

$$\chi(\xi_{n+1}/|\xi|) \quad \text{where } \chi\in C_0^\infty(\mathbb{R}), \quad \chi(0)=1 \quad \text{and} \quad |\xi|^2=\xi_1^2+\ldots+\xi_{n+1}^2.$$

Since $D_{n+1}U=0$ it is clear that $\xi_{n+1}=0$ in $WF(U)$. By Theorem 8.2.9

$$WF(U)=\{(x,x_{n+1},\xi,0);\,(x,\xi)\in WF(u)\},$$

and the same obvious proof gives the more precise statement

$$s_u^*(x,\xi)=s_U^*(x,x_{n+1},\xi,0).$$

Now the principal symbol of the operator in (26.9.21) is

$$\xi_1+ih(x,\xi')(\xi_{n+1}+g(x',\xi')).$$

Near $I\times(\mathbb{R}\times\{0\})$ it satisfies condition (P) and defines a two dimensional bicharacteristic which is the product of the x_1 axis, the x_{n+1} axis and $\varepsilon_n=(0,\ldots,0,1,0)$, for $H_g=0$ on I. The Hamilton field is

$$\partial/\partial x_1+ih(x_1,0,\varepsilon_n')\partial/\partial x_{n+1},$$

so introducing $\int h(x_1,0,\varepsilon_n')\,dx_1$ as a variable in $\tilde{I}$ we obtain the standard Cauchy-Riemann operator in $\tilde{I}\times\mathbb{R}$. A superharmonic function independent of x_{n+1} is then the same as a concave function on $\tilde{I}$, so Theorem 26.9.4 follows from Theorem 26.9.1.

The preceding argument does not fully use the hypothesis that the end points of I are in N_{12}^i; the conclusion is always valid when a factorization is available with $h\geq 0$, $h>0$ at the end points of I, and g independent of x_1. In particular this is always true by Weierstrass' preparation theorem in the analytic case if the end points are just in N_{12}. However, it seems necessary to weaken the notion of concavity in general. We shall now discuss two such weaker concepts here, where the weakest will be useful in Section 26.10.

Definition 26.9.5. A function s defined on an interval $I\subset\mathbb{R}$ with values in $(-\infty,+\infty]$ will be called semi-concave if it is semi-continuous from below and for every compact interval $J\subset I$ and linear decreasing function L with $s\geq L$ on ∂J we have $s\geq L$ in J. We shall say that s is quasi-concave if this is true for all constants L.

Semi-concavity is well defined on a semi-bicharacteristic I with end points in $N_{12}\setminus N_{12}^i$. In fact, putting the principal symbol in the standard form $\xi_1+ig(x',\xi')h(x,\xi')+r(x,\xi')$ with $r(x,\xi')$ vanishing of infinite order on I, Hess $g\leq 0$ and $h\geq 0$, we obtain a natural orientation from the orientation of the x_1 axis; the form $h(x_1,0,\varepsilon_n')\,dx_1$ defines a natural oriented affine structure in $\tilde{I}$.

Semi-concavity is clearly invariant under linear increasing changes of variable whereas quasi-concavity is invariant under strictly monotonic changes of variable. Thus an affine structure and an orientation are required for the definition of semi-concavity. The meaning of the conditions is further clarified in the following

Proposition 26.9.6. *s is semi-concave in $I \subset \mathbb{R}$ if and only if either*

(i) *s is increasing and continuous to the left, or*

(ii) *s is decreasing and concave, or*

(iii) *there is a point $a \in I$ such that s satisfies* (i) *to the left of a and* (ii) *to the right of a, and $s(a) = s(a-0) \leqq s(a+0)$.*

s is quasi-concave if and only if either (i) *is valid or*

(ii)′ *s is decreasing and continuous to the right, or*

(iii)′ *there is a point $a \in I$ such that s satisfies* (i) *to the left of a and* (ii)′ *to the right of a, and $s(a) = \min(s(a+0), s(a-0))$.*

Proof. Assume first that s is quasi-concave. If s is monotonic we must of course have (i) or (ii)′. Otherwise s is not decreasing so we can choose $t_1 < t_2$ with $s(t_1) < s(t_2)$. Then $s(t) \leqq s(t_1)$ for $t < t_1$ since $s(t) > s(t_1) < s(t_2)$ would contradict the definition. Moreover, if $t' \leqq t \leqq t_1$ we have $s(t') \leqq s(t)$ for otherwise $s(t') > s(t) \leqq s(t_1) < s(t_2)$ which is also a contradiction. If a is the supremum of all $t_1 \in I$ with $s(t_1) < s(t_2)$ for some $t_2 > t_1$ in I, it follows that s is increasing to the left of a and decreasing to the right of a. The lower semi-continuity gives

$$s(a) \leqq \min(s(a+0), s(a-0)).$$

Inequality here would imply that $s(a) < s(t_1)$ and $s(a) < s(t_2)$ for suitable $t_1 < a$ and $t_2 > a$, which is also impossible. This proves that (iii)′ is valid. If s is even semi-concave it is obvious that s is concave to the right of a since the linear interpolation between two values there is decreasing. If $s(a-0) > s(a+0)$ we have for $a < t \in I$ and small $\varepsilon > 0$

$$s(a) \geqq (\varepsilon s(t) + (t-a)s(a-\varepsilon))/(t-a+\varepsilon) \to s(a-0), \quad \varepsilon \to 0,$$

so $s(a) = s(a-0)$, which is also obvious from (iii)′ if $s(a-0) \leqq s(a+0)$. This proves the necessity of (i), (ii) or (iii) in the semi-concave case. Conversely, assume that s satisfies (iii) and let L be a linear decreasing function, J a compact interval $\subset I$ with $L \leqq s$ at ∂J. Then $s - L$ is increasing to the left of a so $L(t) \leqq s(t)$ if $a \geqq t \in J$. If $a \in J$ we obtain $s(a+0) \geqq L(a)$ by (iii). The concavity now gives $s \geqq L$ if $a < t \in I$ also. The sufficiency of the conditions in Proposition 26.9.6 is trivial in all other cases.

We can now give an exact analogue of Theorem 26.9.4:

Theorem 26.9.7. *Let $P \in \Psi^m_{\mathrm{phg}}(X)$ be properly supported and satisfy condition* (P), *let I be a compact interval on a one dimensional bicharacteristic with end points in $N_{12} \setminus N^i_{12}$, and let $\tilde{I}$ be the affine interval obtained by Proposition*

26.5.8 when subintervals not meeting N_{12} are identified to points, with the orientation just defined. If $u\in\mathscr{D}'(X)$, s is a semi-concave function on $\tilde{I}$ and $\tilde{s}_{Pu}\geqq s$ on $\tilde{I}$, it follows that

$$\min(\tilde{s}_u, s+m-1)$$

is a semi-concave function on $\tilde{I}$. Here $\tilde{s}_u$ and $\tilde{s}_{Pu}$ are defined as in Theorem 26.9.1.

Proof. We may assume that $P\in\Psi^1_{\text{phg}}(\mathbb{R}^n)$, that $x'=0$, $\xi=\varepsilon_n$ on I and that

$$p(x,\xi)=\xi_1+if(x,\xi'),\qquad (26.9.22)$$
$$f(x,\xi')=g(x',\xi')\,h(x,\xi')+r(x,\xi')$$

in a neighborhood of I. Here Hess $g\leqq 0$ at $(0,\varepsilon'_n)$, $h\geqq 0$, and $r=0$ when $g>0$, by Proposition 26.5.7. Since g is strictly concave in some variable the set where $g\leqq 0$ is the closure of the set where $g<0$ so $g(x',\xi')\leqq 0$ implies $f(x,\xi')\leqq 0$ for all x_1 by condition (P), for $g(x',\xi')=f(x,\xi')$ for a certain x_1. Let L be a linear decreasing function of $\int h(x_1,0,\varepsilon'_n)\,dx_1$ and let J be a compact subinterval of I with end points in N_{12} such that

$$L<\min(s_u^*, s)\quad \text{at } \partial J. \qquad (26.9.23)$$

(Here s denotes the function s lifted from $\tilde{I}$ to I.) The theorem will be proved if we show that the same estimate is valid in J. Since s is semi-concave we have $L<s$ in J, hence

$$L<s_{Pu}^*\quad \text{in } J. \qquad (26.9.24)$$

If $g\leqq 0$ then $f\leqq 0$ in a neighborhood of I, so we can apply Proposition 26.6.1. Let γ_0 be the first point in J and γ an arbitrary point in J. Then $s_{Pu}^*\geqq L(\gamma)$ on $[\gamma_0,\gamma]\subset J$ and $s_u^*(\gamma_0)>L(\gamma_0)\geqq L(\gamma)$ so Proposition 26.6.1 gives $s_u^*(\gamma)>L(\gamma)$ which proves the statement.

We now allow g to change sign. To prove the theorem we shall then first use Proposition 26.6.1′ to get hold of u when $g<0$. We may then assume that we already know that

$$L-\tfrac{1}{8}<s_u^*\quad \text{in } J. \qquad (26.9.25)$$

Let $0<\varepsilon\leqq\frac{1}{4}$ and choose χ, $\chi_1\in S^0_{1-\varepsilon,\varepsilon}(\mathbb{R}^{n-1}\times\mathbb{R}^{n-1})$ non-negative so that $\chi_1=1$ in supp χ, $\chi_1=0$ (resp. $\chi=1$) at all points in a conic neighborhood of $(0,\varepsilon'_n)$ with distance <1 (resp. >2) to $\{(x',\xi');\, g(x',\xi')>0\}$ with respect to the metric

$$(1+|\xi'|^2)^{\varepsilon}(|dx'|^2+(1+|\xi'|^2)^{-1}|d\xi'|^2). \qquad (26.9.26)$$

This is possible by Corollary 1.4.11 (or by using the closely related partitions of unity in Section 18.4). Choose ψ, $\psi_1\in S^0(\mathbb{R}^{n-1}\times\mathbb{R}^n)$ non-negative so that $\psi_1=1$ in supp ψ and $\psi=1$ at infinity in a conic neighborhood of $(0,\varepsilon_n)$. If we set

$$C^s(x',\xi)=(1+|\xi|^2)^{s/2}\,\psi(x',\xi)\,\chi(x',\xi'),$$
$$C_1(x',\xi)=\psi_1(x',\xi)\,\chi_1(x',\xi')$$

the hypotheses of Proposition 26.6.1′ are fulfilled in $[\gamma_0, \gamma]$ if supp ψ is contained in a sufficiently small conic neighborhood of $(0, \varepsilon_n)$ and $s = L(\gamma)$ for some $\gamma \in J$. (Note that $(3\varepsilon - 1)/2 \leqq -\frac{1}{8}$ so the first part of (26.6.9) follows from (26.9.25) in $[\gamma_0, \gamma]$.) Hence

$$C^s(x', D) u \in H_{(0)} \quad \text{at } \gamma \text{ if } \gamma \in J \quad \text{and} \quad s = L(\gamma).$$

Since $C^s(x', D) - (1 + |D|^2)^{s/2} C^0(x', D)$ is of order $s + \varepsilon - 1$ and $\varepsilon - 1 < -\frac{1}{8}$, it follows in view of (26.9.25) that $(1 + |D|^2)^{s/2} C^0(x', D) u \in H_{(0)}$ at γ, if $s = L(\gamma)$, that is,

$$C^0(x', D) u \in H_{(L)} \quad \text{at } J. \tag{26.9.27}$$

Now let $v = u - C^0(x', D) u$. From (26.9.23) and (26.9.27) it follows that

$$v \in H_{(L(\gamma))} \quad \text{if } \gamma \in \partial J.$$

We have

$$Pv = Pu - C^0(x', D) Pu - [P, C^0(x', D)] u.$$

The symbol of $[P, C^0(x', D)]$ is in $S^{2\varepsilon - 1}_{1-\varepsilon, \varepsilon}$ apart from the term

$$\{f, C^0\} = (\{f, C^0\}/C_1) C_1$$

which is in $S^{\varepsilon}_{1-\varepsilon}$. Thus we have

$$[P, C^0(x', D)] = Q_1(x, D) C_1(x, D) + Q_2(x, D)$$

where $Q_1 \in S^{\varepsilon}_{1-\varepsilon, \varepsilon}$ and $Q_2 \in S^{3\varepsilon - 1}_{1-\varepsilon, \varepsilon}$. The argument which led to (26.9.27) also gives

$$C_1(x', D) u \in H_{(L)} \quad \text{at } J,$$

so it follows that

$$Pv \in H_{(L-\varepsilon)} \quad \text{at } J.$$

Since $r = 0$ when $g > 0$ we can estimate $r(x, \xi')/|\xi'|$ by any power of the distance to this set in the norm (26.9.26) with $\varepsilon = 0$. In a conic neighborhood of I we therefore have $r(x, \xi') = O(|\xi|^{1 - N\varepsilon})$ for any N in supp $(1 - C^0)$, and all derivatives of r are also rapidly decreasing there. Hence the symbol of $r(x, D')(1 - C^0(x', D))$ is of order $-\infty$ in a conic neighborhood of J. If P_1 is obtained from P by changing the principal symbol to $\xi_1 + igh$ in a conic neighborhood of J, it follows that $P_1 v \in H_{(L-\varepsilon)}$ at J. The proof of Theorem 26.9.4 works for the operator P_1 since an exact factorization is available, so we may conclude that $v \in H_{(L-\varepsilon)}$ at J since this is true at ∂J. Hence

$$u \in H_{(L-\varepsilon)} \quad \text{at } J$$

for every $\varepsilon > 0$ and every L satisfying (26.9.23), which completes the proof.

26.10. The Singularities on One Dimensional Bicharacteristics

On a general one dimensional bicharacteristic we have no affine structure which permits us to define (semi-)concavity. However, quasi-concavity is meaningful (see Definition 26.9.5), and we shall prove

Theorem 26.10.1. *Let $P\in\Psi^m_{\text{phg}}(X)$ be properly supported and satisfy condition (P), and let I be a one dimensional bicharacteristic interval. If $u\in\mathscr{D}'(X)$, s is a quasi-concave function on I and $s^*_{Pu}\geqq s$ on I, then*

$$\min(s^*_u, s+m-1)$$

is a quasi-concave function on I.

We have here chosen a statement which is analogous to Theorems 26.9.1, 26.9.4 and 26.9.7. However, it is useful to rephrase the result. Explicitly it means that if J is a compact interval $\subset I$ and $\min(s^*_u, s+m-1)\geqq t+m-1\in\mathbb{R}$ at ∂J, then this is also true in J. Since s is quasi-concave we have $s^*_{Pu}\geqq s\geqq t$ in J, so the assertion is that $s^*_u\geqq t+m-1$ in J. Theorem 26.10.1 is therefore equivalent to the following apparently weaker statement:

Theorem 26.10.1′. *Let $P\in\Psi^m_{\text{phg}}(X)$ be properly supported and satisfy condition (P), and let I be a compact one dimensional bicharacteristic interval. If $u\in\mathscr{D}'(X)$, $s\in\mathbb{R}$ and $s^*_{Pu}\geqq s$ on I, $s^*_u\geqq s+m-1$ at ∂I, then $s^*_u\geqq s+m-1$ in I.*

In the proof we may by Proposition 26.4.13 assume that $P\in\Psi^1_{\text{phg}}(\mathbb{R}^n)$, that $x'=(x_2,\dots,x_n)=0$ and $\xi=(0,\dots,0,1)=\varepsilon_n$ on I, and that

$$p(x,\xi)=\xi_1+if(x,\xi') \tag{26.10.1}$$

in a conic neighborhood of I, where f does not change sign for fixed (x',ξ'). Theorem 26.10.1′ is a consequence of Theorem 26.6.4 if $I\subset N_{11}$. If $I\subset N^e_{12}$ then Theorem 26.10.1′ follows from Theorems 26.9.4 and 26.9.7. In fact, if Γ is the one dimensional bicharacteristic containing I and $\varepsilon>0$ we can then choose $\phi\in S^0$ equal to 1 at infinity in a conic neighborhood of I and equal to 0 outside another neighborhood which is so small that

$$P\phi(x,D)u=\phi(x,D)Pu+[P,\phi(x,D)]u\in H_{(s-\varepsilon)}\quad\text{in }\Gamma,$$

and $\phi(x,D)u\in H_{(s-\varepsilon+m-1)}$ at every point in $\Gamma\smallsetminus I$. If $\Gamma\smallsetminus I$ contains points in N_{12} on both sides of I it follows from Theorem 26.9.4 or Theorem 26.9.7 that $\phi(x,D)u\in H_{(s-\varepsilon+m-1)}$ in I, hence that $u\in H_{(s-\varepsilon+m-1)}$ in I. Otherwise we can modify the definition of p outside $WF(\phi(x,D))$ so that such points occur, for example by changing the principal symbol to $p(y_1(x_1),x',\xi)$ where $y_1(x_1)=x_1$ in a neighborhood of I but $(y_1(x_1),0,\varepsilon_n)\in N_{12}$ for large $|x_1|$. This does not affect condition (P) or the condition that $P\phi(x,D)u\in H_{(s-\varepsilon)}$ on Γ.

If $I\subset N^e_2$ then f vanishes of third order on I since $f=gh$ and $g=h=0$ on I, $h\geqq 0$ in a neighborhood. In the proof of Theorem 26.10.1′ we may therefore always assume that

(26.10.2) $\qquad f$ vanishes of third order in I.

As in the proof of Theorem 26.4.7′ we may also assume that the term of order 0 in the symbol of P is of the form $r(x,\xi')$, and the argument at the beginning of the proof of Theorem 26.9.1 shows that by conjugation with an elliptic operator we can achieve that

(26.10.3) $$r=0 \quad \text{in } I.$$

As in Lemma 26.9.2 we shall cut f and r off outside a small neighborhood of $x'=0$, $\xi'=\varepsilon'_n/\lambda$ by introducing

(26.10.4) $$\begin{aligned} f_{\delta,\lambda}(x,\xi') &= \psi(x'/\delta,(\lambda\xi'-\varepsilon'_n)/\delta)\, f(x,\xi'),\\ r_{\delta,\lambda}(x,\xi') &= \psi(x'/\delta,(\lambda\xi'-\varepsilon'_n)/\delta)\, r(x,\xi') \end{aligned}$$

where $\psi\in C_0^\infty$. Let $I=J\times\{0\}\times\{\varepsilon_n\}$ and $J_\delta=\{t+t';\ t\in J,\ |t'|\leqq\delta\}$. For $\lambda>0$ and small $\delta>0$ we have if $x_1\in J_\delta$

(26.10.5) $$|D^\alpha_{\xi'}D^\beta_{x'} f_{\delta,\lambda}(x,\xi')|\,\lambda/\delta^2+|D^\alpha_{\xi'}D^\beta_{x'} r_{\delta,\lambda}(x,\xi')|\leqq C_{\alpha\beta}\delta^{1-|\alpha|-|\beta|}\lambda^{|\alpha|}.$$

For reasons of homogeneity it suffices to prove (26.10.5) when $\lambda=1$ and then it follows from the fact that by Taylor's formula

$$D^\alpha_{\xi'}D^\beta_{x'} f(x,\xi')=O(\delta^{3-|\alpha|-|\beta|}),\qquad D^\alpha_{\xi'}D^\beta_{x'} r(x,\xi')=O(\delta^{1-|\alpha|-|\beta|})$$

when $|x'|^2+|\xi'-\varepsilon'_n|^2<C\delta^2$ and $x_1\in J_\delta$.

To simplify notation we shall write F and R instead of $f_{\delta,\lambda}$ and $r_{\delta,\lambda}$, taking $x'/\sqrt{\lambda}$, $\sqrt{\lambda}\,\xi'$ as variables instead of x', ξ'. Then the right-hand side of (26.10.5) becomes $C_{\alpha\beta}\delta a^{|\alpha|+|\beta|}$ where $a^2=\lambda/\delta^2$. (Note that the change of variables is the symplectic change of variables in $T^*(\mathbb{R}^{n-1})$ induced by the change of x' variables.) Dropping the primes and writing t instead of x_1 we have to study the operator

$$D_t+iF(t,x,D)+R(t,x,D)$$

where all x, ξ derivatives of F and R are continuous in t, x, ξ and

(i) $\qquad F$ is real valued and $F(t,x,\xi)F(s,x,\xi)\geqq 0$,

(ii) $$a^2|F^{(j)}_{x,\xi}(t,x,\xi)|+|R^{(j)}_{x,\xi}(t,x,\xi)|\leqq\delta C_j a^j,\qquad j=0,1,2,\dots.$$

Actually F and R are only defined in an interval, say $|t|\leqq T$, but since no differentiability with respect to t is assumed we can extend them to all $t\in\mathbb{R}$ so that they are independent of t when $t\geqq T$ or $t\leqq -T$. We can also introduce $t'=\delta t$ as a new variable instead of t. This gives the operator

$$\delta(D_{t'}+i\delta^{-1}F(t'/\delta,x,D)+\delta^{-1}R(t'/\delta,x,D)).$$

The interval $|t|<T$ has now become the interval $|t'|<\delta T$, and the operators involve functions satisfying (ii) with $\delta=1$. This is finally the situation in which we are going to work. It is no restriction to assume that $C_j=1$ for $j\leqq 2$. In the first lemma we write $X=(x,\xi)$.

Lemma 26.10.2. *Let $F(t,X)$ be a real valued function in $\mathbb{R}^{1+k}$ such that $F(t,X)F(s,X)\geqq 0$ for all t,s,X, all X derivatives are continuous and*

$$|F_X^{(j)}(t,X)|\leqq C_j a^{j-2},\quad j=0,1,\ldots;\ (t,X)\in\mathbb{R}^{1+k}. \tag{26.10.6}$$

Let $1\leqq\rho\leqq 1/a$ and set

$$\tilde a(X)^{-2}=\max(\rho^2,\sup_t|F(t,X)|,\sup_t|F'_X(t,X)|^2). \tag{26.10.7}$$

Then

$$a\leqq\tilde a(X)\leqq 1/\rho,\quad \tilde a(X+Y)\leqq 2\tilde a(X)\quad \text{if } |Y|\,\tilde a(X)<\tfrac12, \tag{26.10.8}$$

$$\tilde a(X)\leqq\tilde a(X+Y)(1+|Y|\,\tilde a(X)), \tag{26.10.9}$$

$$|F_X^{(j)}(t,X)|\leqq C_j\tilde a(X)^{j-2},\quad j=0,1,\ldots;\ (t,X)\in\mathbb{R}^{1+k}; \tag{26.10.10}$$

and for every X one of the following cases occurs:

I) $\tilde a(X)=1/\rho$; *then* $\frac12\leqq\rho\tilde a(X+Y)\leqq 1$ *and* $|F_Y^{(j)}(t,X+Y)|\leqq 4C_j\rho^{2-j}$, $j=0,1,\ldots$ *if* $\tilde a(X)|Y|\leqq\frac12$.

II$_+$) $\tilde a(X)^{-2}=\sup_t F(t,X)$; *then* $F(s,X+Y)\geqq 0$ *if* $\tilde a(X)|Y|\leqq\frac12$.

II$_-$) $\tilde a(X)^{-2}=\sup_t -F(t,X)$; *then* $F(s,X+Y)\leqq 0$ *if* $\tilde a(X)|Y|\leqq\frac12$.

III) $\tilde a(X)^{-1}=\sup|F'_X(t,X)|$; *then* $F(t,Y)=G(Y)H(t,Y)$ *if* $\tilde a(X)|Y-X|<\frac12$, *and then we have* $H\geqq 0$, $|G'(X)|=1/\tilde a(X)$,

$$|G^{(j)}(Y)|\leqq C_j\tilde a(Y)^{j-2},\quad |H_Y^{(j)}(t,Y)|\leqq C'_j\tilde a(Y)^j \tag{26.10.11}$$

where the constants C'_j only depend on $C_0,\ldots,C_{j+1}$.

Proof. That $a\leqq\tilde a(X)\leqq 1/\rho$ follows at once from the definition and the hypothesis. Since

$$|F(t,X+Y)-F(t,X)|\leqq|Y|\,|F'_X(t,X)|+|Y|^2/2\leqq|Y|/\tilde a(X)+|Y|^2/2,$$
$$|F'(t,X+Y)-F'(t,X)|\leqq|Y|,$$

we obtain if $\tilde a(X)|Y|\leqq\frac12$ that

$$|F(t,X+Y)|\geqq|F(t,X)|-3\tilde a(X)^{-2}/4,$$
$$|F'(t,X+Y)|\geqq|F'(t,X)|-\tfrac12\tilde a(X)^{-1}$$

which proves the second inequality in (26.10.8). Also (26.10.9) follows for

$$|F(t,X+Y)|\leqq\tilde a(X)^{-2}(1+|Y|\,\tilde a(X))^2,$$
$$|F'(t,X+Y)|\leqq\tilde a(X)^{-1}(1+|Y|\,\tilde a(X)).$$

The estimate (26.10.10) follows from (26.10.6) if $j\geqq 2$, since $a\leqq\tilde a$, and from the definition of $\tilde a$ if $j=0$ or $j=1$. In case I) we have $\frac12\leqq\rho\tilde a(X+Y)\leqq 1$ when $\tilde a(X)|Y|\leqq 1$ in view of (26.10.9), hence $\tilde a(X+Y)^{j-2}\leqq 4\rho^{2-j}$, $j\geqq 0$. In case II$_+$) we have

$$F(t,X+Y)\geqq F(t,X)-3\tilde a(X)^{-2}/4>0,\quad \tilde a(X)|Y|<\tfrac12,$$

provided that $F(t,X)>3\tilde{a}(X)^{-2}/4$. By hypothesis it follows that we have $F(s,X+Y)\geqq 0$ for all s then. Case II$_-$) is of course similar.

In case III) we choose t_j so that $|F'_X(t_j,X)|\to\tilde{a}(X)^{-1}$ and $F(t_j,Y)\to G(Y)$. Then the first estimate in (26.10.11) follows from (26.10.6), and $|G'(X)|=\tilde{a}(X)^{-1}$. Hence $G'(Y)\neq 0$ if $|X-Y|\tilde{a}(X)<1$. Since the zeros of G are simple then, they are limits of simple zeros of $F(t_j,Y)$. At such zeros we have $F(s,Y)=0$ for all s since Y is in the closure of the set where $F(t_j,.)>0$ as well as the set where $F(t_j,.)<0$. It follows that $F=0$ when $G=0$ and that $H(t,Y)=F(t,Y)/G(Y)$ is a non-negative C^∞ function when $\tilde{a}(X)|Y-X|<1$. To estimate H we assume for example that $\partial G(X)/\partial X_1>0$, $\partial G(X)/\partial X_j=0$, $j>1$. Then

$$\partial G(X+Y)/\partial Y_1>\tilde{a}(X)^{-1}-|Y|.$$

If $\tilde{a}(X)|Y|<\frac{1}{2}$ and $G(X+Y+se_1)\neq 0$ for $|s|<1/(2\tilde{a}(X))$, where $e_1=(1,0,\dots,0)$, then $G(X+Y)>\int_0^{1/(2\tilde{a}(X))} s\,ds=1/(8\tilde{a}(X)^2)$. Since $|F(t,X+Y)|\leqq\tilde{a}(X)^{-2}(1+\frac{1}{2}+\frac{1}{8})$ we obtain $|H(t,X+Y)|\leqq 13$ and of course similar estimates for the derivatives with respect to Y. On the other hand, if $G(X+Y+se_1)=0$ for some s with $\tilde{a}(X)|s|<\frac{1}{2}$ then $F(X+Y+se_1)/s$ and $G(X+Y+se_1)/s$ are equal to averages of $\partial_1 F$ and of $\partial_1 G$ on the intervals between $X+Y$ and $X+Y+se_1$ so the ratio is bounded by

$$\int_{\frac{1}{2}}^{1}(1+s)\,ds\Big/\int_{\frac{1}{2}}^{1}(1-s)\,ds=7.$$

Hence $|H(t,X+Y)|\leqq 7$, and we have similar estimates for the derivatives of H since this is true of the averages of $\partial_1 G$ and $\partial_1 F$. The proof is complete.

Note that the proof also shows in the third case that if $G(Y)$ has no zero with $|Y-X|\tilde{a}(X)<\frac{1}{2}$ then $|G(X)|\geqq\frac{3}{8}\tilde{a}(X)^{-2}$, so we could essentially have classified this situation as case II. There are similar borderline cases between the others, but this will not be important. What is crucial is that the lemma cleanly separates areas where F is bounded and therefore controlled, or F is of constant sign so that the methods of Section 26.6 are applicable, or finally where F can be factored so that we have essentially the situation studied in Section 26.9.

Since the proof of the following proposition will depend on the advanced calculus of pseudo-differential operators in Sections 18.5.6, we shall use the Weyl calculus already in the statement. Note that the hypotheses (26.10.2) and (26.10.3) remain valid for the Weyl symbols.

Proposition 26.10.3. *Let $P=D_t+iF^w(t,x,D)+R^w(t,x,D)$ where F is real valued, $F(t,x,\xi)\,F(s,x,\xi)\geqq 0$ for all $(s,t,x,\xi)\in\mathbb{R}^{2n}$ and*

$$a^2|F^{(j)}_{x,\xi}(t,x,\xi)|+|R^{(j)}_{x,\xi}(t,x,\xi)|\leqq\delta C_j a^j,\quad j=0,1,\dots. \tag{26.10.12}$$

If a and δT are smaller than positive constants depending only on $C_0, C_1, \dots$ then

(26.10.13) $\|u\| \leqq 16T\|Pu\|$, *if* $u \in \mathscr{S}(\mathbb{R}^n)$ *and* $u(t,x)=0$ *when* $|t|>T$.

Proof. As already observed we may assume in the proof that $\delta=1$ and that $C_j=1$ when $j\leqq 2$. At first we also assume that $R=0$. From Lemma 26.10.2 with ρ still to be chosen we obtain a metric

$$g=\tilde{a}(x,\xi)^2(|dx|^2+|d\xi|^2)$$

which is slowly varying and σ temperate (cf. (18.5.11)′) by (26.10.8) and (26.10.9), for $1+|Y|\tilde{a}(X)\leqq 1+|Y|\leqq 1+|Y|/\tilde{a}(X)$. Choose $X_\nu=(x_\nu,\xi_\nu)$ so that the balls

$$B_\nu(r)=\{X;\tilde{a}(X_\nu)|X-X_\nu|<r\}$$

cover $\mathbb{R}^{2(n-1)}$ when $r=\frac{1}{5}$ and there is a fixed bound for the number of balls $B_\nu(\frac{1}{2})$ with a non-empty intersection. Choose $\phi_\nu \in C_0^\infty(B_\nu(\frac{1}{4}))$ real valued so that $\sum \phi_\nu^2=1$ and $\{\phi_\nu\}$ is uniformly bounded as a symbol in $S(1,g)$ with values in l^2. (See Lemma 18.4.4.) We have

$$\sum \|\phi_\nu^w u\|^2=\sum(\phi_\nu^w\phi_\nu^w u,u)=(u,u)+(\phi^w u,u)$$

where ϕ is uniformly bounded in $S(\tilde{a}^2,g)$, hence in $S(\rho^{-2},g)$. It follows that $\|\phi^w\|\leqq C\rho^{-2}$. Fixing ρ now so that $\rho^2=\max(1,2C)$ we obtain

(26.10.14) $$\|u\|^2\leqq 2\sum\|\phi_\nu^w u\|^2\leqq 3\|u\|^2.$$

Choose $\psi_\nu \in C_0^\infty(B_\nu(\frac{1}{2}))$ non-negative and equal to 1 in $B_\nu(\frac{1}{3})$ so that $\{\psi_\nu\}$ is uniformly bounded in $S(1,g)$, and set

$$F_\nu(t,x,\xi)=\psi_\nu(x,\xi)^2 F(t,x,\xi).$$

F_ν is uniformly bounded in $S(\tilde{a}(X_\nu)^{-2}, \tilde{a}(X_\nu)^2(|dx|^2+|d\xi|^2))$. If X is in case I) of Lemma 26.10.2 then $\tilde{a}(X_\nu)^{-2}=\rho^2$, so Theorem 26.8.1 applied with $A=B=0$ gives

$$\|u\|\leqq 4T\|D_t u\|\leqq 4T\|D_t u+iF_\nu^w(t,x,D)u\|+CT\rho^2\|u\|.$$

When $CT\rho^2<\frac{1}{2}$ it follows that

(26.10.15) $$\|u\|\leqq 8T\|(D_t+iF_\nu^w(t,x,D))u\|.$$

In case II$_+$) we observe that by Theorem 18.6.7 (in fact, by Theorem 18.1.14) there is a constant c such that

$$F_\nu^w(t,x,D)+cI\geqq 0.$$

Applying Theorem 26.8.1 with $B=I$ and $A(t)=F_\nu^w(t,x,D)+cI$ we obtain

$$\|u\|\leqq 4T\|(D_t+iF_\nu^w+c)u\|\leqq 4T\|(D_t+iF_\nu^w)u\|+4Tc\|u\|,$$

so (26.10.15) is valid if $4Tc<\frac{1}{2}$. The same is true in case II$_-$; we just have to change the sign of t. In case III we set $G_\nu=\psi_\nu G$ and $H_\nu=\psi_\nu H$ where $F=GH$ is the local factorization in Lemma 26.10.2. We can apply Theorem 26.8.1

with $B=G_\nu^w(x,D)$ and $A(t)=H_\nu^w(t,x,D)+c\tilde{a}(X_\nu)^2$ with c chosen so large that $A(t)\geqq 0$. Then $B\in S(\tilde{a}(X_\nu)^{-2}, \tilde{a}(X_\nu)^2(|dx|^2+|d\xi|^2))$ and $A(t)\in S(1, \tilde{a}(X_\nu)^2(|dx|^2+|d\xi|^2))$ uniformly, so $\|A\|$, $\|[B,A]\|$ and $\|[B,[B,A]]\|$ are uniformly bounded; later we shall need that $\tilde{a}(X_\nu)^2\|B\|$ is also uniformly bounded. Hence (26.8.3) gives if T is small enough

$$\|u\|\leqq 6T\|(D_t+iA(t)B)u\|\leqq 6T\|(D_t+iF_\nu^w(t,x,D))u\|+CT\|u\|$$

since the symbol of $A(t)B-F_\nu^w$ is uniformly bounded in $S(1,g)$. When $CT<\frac{1}{4}$ we obtain (26.10.15), which is now established in all cases.

Now we apply (26.10.15) to $\phi_\nu^w u$ and obtain using (26.10.14)

$$\|u\|^2\leqq 2\sum\|\phi_\nu^w u\|^2\leqq 2^7T^2\sum\|(D_t+iF_\nu^w(t,x,D))\phi_\nu^w u\|^2.$$

Regarding $\{F_\nu\}$ and $\{\phi_\nu\}$ as symbols with values in diagonal matrices in $\mathscr{L}(l^2,l^2)$ or in $l^2=\mathscr{L}(\mathbb{C},l^2)$, we obtain from the calculus that

$$(D_t+iF_\nu^w(t,x,D))\phi_\nu^w u=\phi_\nu^w(D_t+iF^w(t,x,D))u+K_\nu^w(t,x,D)$$

where $\{K_\nu\}$ is uniformly bounded in $S(1,g)$ (with values in l^2). Since

$$\sum\|\phi_\nu^w(x,D)(D_t+iF^w(t,x,D))u\|^2\leqq\tfrac{3}{2}\|(D_t+iF^w(t,x,D))u\|^2$$

by (26.10.14) and

$$\sum\|K_\nu^w(t,x,D)u\|^2\leqq C\|u\|^2,$$

we obtain for small T

$$\|u\|\leqq 14T\|(D_t+iF^w(t,x,D))u\|.$$

This proves (26.10.13) if $R=0$ with the constant 14 instead of 16. When T is so small that

$$14T\|R^w(t,x,D)u\|\leqq\|u\|/8,$$

the estimate (26.10.13) follows.

In case III) the estimate (26.10.15) is analogous to (26.9.3). We shall also need the analogue of (26.9.4) obtained when Theorem 26.8.3 is applied to the operators $A(t)=H_\nu^w(t,x,D)+c\tilde{a}(X_\nu)^2$ and $B=G_\nu^w(x,D)$ in the preceding proof. Using (26.10.15) we obtain for small T

$$\int(B^w u, H_\nu^w(t,x,D)B^w u)\,dt\leqq CT\tilde{a}(X_\nu)^{-2}\|P_\nu u\|^2;\quad P_\nu=D_t+iF_\nu^w(t,x,D).$$

Since $H_\nu\geqq 0$ and the second derivatives of $\tilde{a}(X_\nu)^{-2}H_\nu$ have uniform bounds, it follows from Lemma 7.7.2 that

$$|\tilde{a}(X_\nu)^{-2}H'_\nu|^2\leqq C\tilde{a}(X_\nu)^{-2}H_\nu$$

if $H'_\nu=\partial H_\nu/\partial x_j$ or $\partial H_\nu/\partial\xi_j$ for some j. Hence

$$0\leqq C\tilde{a}(X_\nu)^2H_\nu-(H'_\nu)^2\in S(\tilde{a}(X_\nu)^2,\tilde{a}(X_\nu)^2(|dx|^2+|d\xi|^2))$$

so we can find another constant C' such that the Weyl operator with symbol

$$C\tilde{a}(X_\nu)^2H_\nu-H_\nu'^2+C'\tilde{a}(X_\nu)^4$$

is non-negative. Since the symbol of $(H_v'^2)^w - ((H_v')^w)^2$ belongs to $S(\tilde{a}(X_v)^4, \tilde{a}(X_v)^2(|dx|^2+|d\xi|^2))$ we obtain

$$\|H_v'^w v\|^2 \leqq C\tilde{a}(X_v)^2(H_v^w v, v) + C'\tilde{a}(X_v)^4(v,v).$$

Taking $v = G_v^w(x,D)u$ we obtain with another constant C

$$\|H_v'^w G_v^w(x,D)u\|^2 \leqq CT\|P_v u\|^2 + C\|u\|^2.$$

The symbol of $H_v'^w G_v^w - (H_v' G_v)^w$ is bounded in $S(1, \tilde{a}(X_v)^2(|dx|^2+|d\xi|^2))$. In view of (26.10.15) it follows for small T that with still another C

$$\|(H_v' G_v)^w(t,x,D)u\|^2 \leqq CT\|P_v u\|^2 \tag{26.10.16}$$

if $u \in \mathscr{S}$, $u=0$ for $|t|>T$.

Proposition 26.10.3 is all one needs to prove local existence theorems. However, the proof of Theorem 26.10.1′ requires a localized form of (26.10.13) which we shall now prove.

Proposition 26.10.4. *Let the hypotheses of Proposition* 26.10.3 *be fulfilled, let* χ_0 *and* χ_1 *be uniformly bounded in* $S(1, a^2(|dx|^2+|d\xi|^2))$, *and assume that* $\chi_1 = 1$ *in* supp χ_0. *For every* $\varepsilon>0$ *we have then if* a *and* δT *are sufficiently small*

$$a^{\varepsilon}\|\chi_0^w(x,D)u\| \leqq CT(\|\chi_1^w(x,D)Pu\| + a^3\|Pu\|) + a^{\frac{1}{2}}\|u\|, \tag{26.10.17}$$

for all $u \in \mathscr{S}(\mathbb{R}^n)$ *vanishing when* $|t|>T$.

Proof. We may assume in the proof that $|\chi_1| > \frac{1}{2}$ at all points with distance $\leqq 1/a$ from supp χ_0, for this is true at distance $\leqq \gamma/a$ for some fixed $\gamma \in (0,1)$, and changing the constants in (26.10.12) we may replace a by a/γ in the hypothesis. As in the proof of Proposition 26.10.3 we may also assume that $\delta=1$ so that a and T are small. Now we apply Lemma 26.10.2 with $\rho = a^{-\frac{1}{2}}$, define ϕ_v, ψ_v and F_v as in the proof of Proposition 26.10.3 and set $R_v(t,x,\xi) = \psi_v(x,\xi)^2 R(t,x,\xi)$, $P_v = D_t + iF_v^w(t,x,D) + R_v^w(t,x,D)$. If a is small enough we have (26.10.14), hence

$$\|\chi_0^w(x,D)u\|^2 \leqq 2\sum \|\phi_v^w(x,D)\chi_0^w(x,D)u\|^2. \tag{26.10.18}$$

Application of Proposition 26.10.3 to $\phi_v^w(x,D)\chi_0^w(x,D)u$ and P_v gives

$$\|\phi_v^w(x,D)\chi_0^w(x,D)u\| \leqq 16T\|P_v\phi_v^w(x,D)\chi_0^w(x,D)u\|.$$

We shall estimate the right-hand side by writing

$$P_v\phi_v^w\chi_0^w = [P_v, \phi_v^w]\chi_0^w + \phi_v^w[P_v, \chi_0^w] + \phi_v^w(P_v - P) + \phi_v^w\chi_0^w P.$$

When computing $[P_v, \phi_v^w] = (iF_v^w + R_v^w)\phi_v^w - \phi_v^w(iF_v^w + R_v^w)$ we regard the right-hand factor as a symbol with values in $l^2 = \mathscr{L}(\mathbb{C}, l^2)$ and the left factor as a symbol with values in diagonal matrices in $\mathscr{L}(l^2, l^2)$. (All later calculations are made similarly.) With the notation $g = \tilde{a}^2(|dx|^2+|d\xi|^2)$ it follows that the symbol is $\{F_v - iR_v, \phi_v\} \in S(1,g)$ (with l^2 values) apart from an error in $S(\tilde{a}^2, g)$

$\subset S(a,g)$. Hence

$$\sum \|[P,\phi_\nu^w]\chi_0^w u\|^2 \leqq C\|\chi_0^w u\|^2$$

which is harmless to have multiplied by T^2 in the right hand side of our estimates since $\|\chi_0^w u\|^2$ will occur on the left. The symbol of $\phi_\nu^w[P_\nu,\chi_0^w]$ is $\phi_\nu\{F_\nu - iR_\nu, \chi_0\}$ with an error in $S(\tilde{a}^2, g)$, for $\{F_\nu - iR_\nu\} \in S(\tilde{a}^{-2}, g)$ and $\chi_0 \in S(1,g)$. Since $\chi_0 \in S(1, a^2(|dx|^2+|d\xi|^2))$ the symbol is bounded in $S(a/\tilde{a}, g)$ $\subset S(a^{\frac{1}{2}}, g)$ if we restrict ν to the set N_I of indices such that X_ν is in case I) of Lemma 26.10.2. Thus

$$\sum_{N_\mathrm{I}} \|\phi_\nu^w[P_\nu,\chi_0^w]u\|^2 \leqq Ca\|u\|^2.$$

The symbol of $\phi_\nu^w\chi_0^w(P_\nu - P)$ is bounded in $S(a,g)$ for all terms in the composition series vanish because $\psi_\nu = 1$ in $\operatorname{supp}\phi_\nu$. These terms have therefore an even better estimate. Finally the symbol of $\chi_0^w - \chi_0^w\chi_1^w$ belongs to $S(a^4, a^2(|dx|^2+|d\xi|^2))$ so

$$\|\chi_0^w Pu\|^2 \leqq C(\|\chi_1^w Pu\|^2 + a^8\|Pu\|^2).$$

Using (26.10.14) again we therefore obtain

$$\sum_{N_\mathrm{I}} \|\phi_\nu^w(x,D)\chi_0^w(x,D)u\|^2 \leqq CT^2(\|\chi_1^w(x,D)Pu\|^2 + a^8\|Pu\|^2 + a\|u\|^2 + \|\chi_0^w(x,D)u\|^2). \tag{26.10.19}$$

In cases II) and III) the commutator of P_ν and χ_0^w is too large to make the preceding estimates useful. There is no point in keeping the factor χ_0^w then. Since $\phi_\nu^w\chi_0^w = \chi_0^w\phi_\nu^w + [\phi_\nu^w, \chi_0^w]$ and $[\phi_\nu^w,\chi_0^w]$ has a vector valued symbol in $S(a^{\frac{1}{2}}, g)$ (recall that $\tilde{a} \leqq a^{\frac{1}{2}}$), we have

$$\sum_{\nu\notin N_\mathrm{I}} \|\phi_\nu^w\chi_0^w u\|^2 \leqq C\Big(\sum_{N_\mathrm{II}\cup N_\mathrm{III}} \|\phi_\nu^w u\|^2 + a^3\|u\|^2\Big), \tag{26.10.20}$$

where N_II and N_III denote the sets of indices ν such that X_ν is in case $\mathrm{II}_\pm$ resp. III of Lemma 26.10.2 and $\phi_\nu\chi_0 \neq 0$. We recall that this implies $|\chi_1| > \frac{1}{2}$ in $B_\nu(\frac{1}{2})$. (See the beginning of the proof.)

If $\nu \in N_\mathrm{II}$ we go back to the proofs of Proposition 26.6.1 and of Theorem 26.8.1. Thus we start from the fact that for fixed t

$$\operatorname{Im}(\phi_\nu^w Pu, \phi_\nu^w u) = \operatorname{Im}(P\phi_\nu^w u, \phi_\nu^w u) + \operatorname{Im}([\phi_\nu^w, P]u, \phi_\nu^w u). \tag{26.10.21}$$

We have in case II_+

$$2\operatorname{Im}(P\phi_\nu^w u, \phi_\nu^w u) = -\frac{\partial}{\partial t}\|\phi_\nu^w u\|^2 + 2(F^w\phi_\nu^w u, \phi_\nu^w u) + 2\operatorname{Im}(R^w\phi_\nu^w u, \phi_\nu^w u)$$

$$\geqq -\frac{\partial}{\partial t}\|\phi_\nu^w u\|^2 - C\|\phi_\nu^w u\|^2,$$

since F^w is bounded from below and R^w is bounded. Furthermore

$$2\operatorname{Im}([\phi_\nu^w, P]u, \phi_\nu^w u) = ([\phi_\nu^w,[\phi_\nu^w, F^w]]u, u) + 2\operatorname{Im}(\phi_\nu^w[\phi_\nu^w, R^w]u, u),$$

and the symbols of

$$\sum_{N_{\mathrm{II}+}} [\phi_\nu^w, [\phi_\nu^w, F^w]], \quad \sum_{N_{\mathrm{II}+}} \phi_\nu^w[\phi_\nu^w, R^w]$$

are bounded in $S(\tilde{a}^2, g) \subset S(a, g)$. If we multiply (26.10.21) by $T+t$, sum and integrate with respect to t also, we obtain with L^2 norms in (t, x)

$$\sum_{N_{\mathrm{II}}} \|\phi_\nu^w u\|^2 \leqq CT\Big(\sum_{N_{\mathrm{II}}} \|\phi_\nu^w u\|^2 + a\|u\|^2 + \sum_{N_{\mathrm{II}}} \|\phi_\nu^w Pu\| \, \|\phi_\nu^w u\|\Big).$$

(The case II_- is reduced to II_+ by changing the sign of t.) Since

$$CT\|\phi_\nu^w Pu\| \, \|\phi_\nu^w u\| \leqq (C^2T^2\|\phi_\nu^w Pu\|^2 + \|\phi_\nu^w u\|^2)/2$$

we obtain for small T and another C

$$\sum_{N_{\mathrm{II}}} \|\phi_\nu^w u\|^2 \leqq CT^2 \sum_{N_{\mathrm{II}}} \|\phi_\nu^w Pu\|^2 + CTa\|u\|^2. \tag{26.10.22}$$

If $\tilde{\phi}_\nu = \phi_\nu/\chi_1 + \{\phi_\nu, 1/\chi_1\}/2i$ then $\{\tilde{\phi}_\nu\}_{\nu\in N_{\mathrm{II}}}$ is bounded in $S(1, g)$ since $|\chi_1| > \frac{1}{2}$ in $\operatorname{supp}\phi_\nu$ when $\nu \in N_{\mathrm{II}}$, and the symbol of $\phi_\nu^w - \tilde{\phi}_\nu^w \chi_1^w$ is bounded in $S(a^3, g)$ since $\phi_\nu - \chi_1\tilde{\phi}_\nu - \{\tilde{\phi}_\nu, \chi_1\}/2i = \{\{\phi_\nu, 1/\chi_1\}, \chi_1\}/4$. Hence

$$\sum_{N_{\mathrm{II}}} \|\phi_\nu^w Pu\|^2 \leqq C(\|\chi_1^w Pu\|^2 + a^6\|Pu\|^2). \tag{26.10.23}$$

In case III) we shall use operators commuting approximately with P which are similar to the operators with symbol (26.9.8) used in the proof of Theorem 26.9.1. They are constructed in the following lemma, which will be proved after completion of the proof of Proposition 26.10.4.

Lemma 26.10.5. *One can find an integer J and for every $\nu \in N_{\mathrm{III}}$ and $j = 1, \ldots, J$ functions $\phi_{\nu j}, \tilde{\phi}_{\nu j} \in C_0^\infty(B_\nu(\frac{1}{3}))$, $v_{\nu j} \in C_0^\infty(B_\nu(\frac{1}{2}))$ such that if $|t| < T$ and T is small*

(i) *$\{\phi_{\nu j}\}_{\nu\in N_{\mathrm{III}}}$, $\{\tilde{\phi}_{\nu j}\}_{\nu\in N_{\mathrm{III}}}$, and $\{v_{\nu j}\}_{\nu\in N_{\mathrm{III}}}$ are uniformly bounded in $S(1, g)$,*

(ii) *$\phi_{\nu j} = 1$ in $\operatorname{supp}\tilde{\phi}_{\nu j}$, $\sum_j \tilde{\phi}_{\nu j} = 1$ in $B_\nu(\frac{1}{4})$,*

(iii) *$\partial v_{\nu j}/\partial t - iH_\nu\{v_{\nu j}, G_\nu\} = 0$ in $\operatorname{supp}\phi_{\nu j}$,*

(iv) *$v_{\nu j} > \varepsilon/3$ in $\operatorname{supp}\phi_{\nu j}$, $v_{\nu j} < 2\varepsilon/3$ in $\operatorname{supp}\tilde{\phi}_{\nu j}$, $v_{\nu j} > 1 + \varepsilon/3$ in $\operatorname{supp} d\phi_{\nu j}$.*

End of proof of Proposition 26.10.4. With

$$m_{\nu j} = \phi_{\nu j} a^{v_{\nu j}}$$

we have by Proposition 26.10.3 applied to P_ν and $m_{\nu j}^w u$

$$\|m_{\nu j}^w u\| \leqq 16T\|P_\nu m_{\nu j}^w u\|. \tag{26.10.24}$$

Here we want to commute P_ν and $m_{\nu j}^w$. The first part of (iv) shows that $\{m_{\nu j}\}_{\nu\in N_{\mathrm{III}}}$ is bounded in $S(1, g)$, for the powers of $\log a$ which occur when $m_{\nu j}$ is differentiated can be estimated by $a^{-\varepsilon/3}$. The symbol of $[m_{\nu j}, P_\nu]$ is therefore in $S(a, g)$ (with values in l^2) apart from the term

$$-i\{m_{\nu j}, \tau + iF_\nu\} = -i\{\phi_{\nu j}, \tau + iF_\nu\}a^{v_{\nu j}} + (\log a)G_\nu\{v_{\nu j}, H_\nu\}m_{\nu j}. \tag{26.10.25}$$

Here we have used (iii) when calculating $\{v_{\nu j}, \tau+iF_\nu\}$. Since $\{\phi_{\nu j}, \tau+iF_\nu\}$ is in $S(1,g)$ with values in l^2, the last part of (iv) gives

$$\{\phi_{\nu j}, \tau+iF_\nu\}a^{v_{\nu j}}\in S(a,g)$$

with values in l^2, and with a uniform bound. The second term in (26.10.25) differs from the symbol of

$$\log a(\sum_k((\partial_{\xi_k}v_{\nu j})^w(G_\nu\partial_{x_k}H_\nu)^w-(\partial_{x_k}v_{\nu j})^w(G_\nu\partial_{\xi_k}H_\nu)^w))m^w_{\nu j}$$

by a symbol in $S(a,g)$ with values in l^2. Since $\partial_{\xi_k}v_{\nu j}$ and $\partial_{x_k}v_{\nu j}$ are bounded in $S(a^{\frac{1}{2}},g)$, it follows from (26.10.16) that

$$\|P_\nu m^w_{\nu j}u\|\leqq\|m^w_{\nu j}Pu\|+Ca^{\frac{1}{2}}\log a\|P_\nu m^w_{\nu j}u\|+\|\rho^w_{\nu j}u\|$$

where $\{\rho_{\nu j}\}$ is bounded in $S(a,g)$. (Note that the composition series of $m^w_{\nu j}(P-P_\nu)$ has only zero terms.) Hence

$$\|P_\nu m^w_{\nu j}u\|\leqq 2\|m^w_{\nu j}Pu\|+2\|\rho^w_{\nu j}u\|$$

if a is so small that $2Ca^{\frac{1}{2}}\log a<1$, so we obtain using (26.10.24)

$$\sum\|m^w_{\nu j}u\|^2\leqq CT^2(\sum\|m^w_{\nu j}Pu\|^2+a^2\|u\|^2). \tag{26.10.26}$$

The proof of (26.10.23) also gives

$$\sum\|m^w_{\nu j}Pu\|^2\leqq C(\|\chi^w_1 Pu\|^2+a^6\|Pu\|^2). \tag{26.10.27}$$

Since (ii) in Lemma 26.10.5 implies

$$\phi_\nu=\sum_j\phi_\nu\tilde{\phi}_{\nu j}=a^{-\varepsilon}\sum_j\psi_{\nu j}m_{\nu j},\qquad \psi_{\nu j}=\phi_\nu\tilde{\phi}_{\nu j}a^{\varepsilon-v_{\nu j}},$$

and $\{\psi_{\nu j}\}$ is bounded in $S(1,g)$ by the second part of (iv), it follows that the symbol of $a^\varepsilon\phi^w_\nu-\sum_j\psi^w_{\nu j}m^w_{\nu j}$ is bounded in $S(a,g)$ with values in l^2, hence

$$\sum_{N_{III}}\|a^\varepsilon\phi^w_\nu u\|^2\leqq C(\sum\|m^w_{\nu j}u\|^2+a^2\|u\|^2).$$

If we combine this estimate with (26.10.26), (26.10.27), and recall the estimates (26.10.18), (26.10.19), (26.10.20), (26.10.22) and (26.10.23), we have proved that

$$a^{2\varepsilon}\|\chi^w_0(x,D)u\|^2\leqq CT^2(\|\chi^w_1(x,D)Pu\|^2+a^6\|Pu\|^2+a^{2\varepsilon}\|\chi^w_0(x,D)u\|^2)+Ca\|u\|^2.$$

When T is so small that $CT^2<\frac{1}{2}$, the estimate (26.10.17) follows.

Proof of Lemma 26.10.5. The essential point is to use Corollary 26.7.8 to construct a solution of the equation

$$\partial v/\partial t-iH_\nu\{v,G_\nu\}=0$$

for $|t|<T$, where T is small, and for (x,ξ) in a neighborhood of an arbitrary $Y\in B_\nu(\frac{1}{4})$ with diameter proportional to $1/a_\nu$ where $a_\nu=\tilde{a}(X_\nu)$. To do so we

set $\kappa_\nu(y,\eta)=(x_\nu+y/a_\nu, \xi_\nu+\eta/a_\nu)$ and obtain the equation

$$\partial(\kappa_\nu^* v)/\partial t + i\kappa_\nu^* H_\nu\{a_\nu^2\kappa_\nu^* G_\nu, \kappa_\nu^* v\}=0.$$

(Note that κ_ν is not symplectic but multiplies the symplectic form by a constant factor.) Here $\kappa_\nu^* H_\nu$ and $a_\nu^2\kappa_\nu^* G_\nu$ have uniformly bounded $y\eta$ derivatives. There is a fixed positive lower bound for $a_\nu^2|d\kappa_\nu^* G_\nu(0)|$, so Theorem 21.1.6 and its proof show that there is a canonical transformation $\tilde\kappa_\nu$ from a fixed neighborhood of 0 to a neighborhood of $\chi_\nu^{-1}(Y)$ with uniform bounds for all derivatives of $\tilde\kappa_\nu$ and $\tilde\kappa_\nu^{-1}$ such that

$$a_\nu^2\tilde\kappa_\nu^*\kappa_\nu^* G_\nu(y,\eta)=\eta_1.$$

Thus we obtain the equation

$$\frac{\partial}{\partial t}((\kappa_\nu\tilde\kappa_\nu)^* v)+i(\kappa_\nu\tilde\kappa_\nu)^* H_\nu\partial((\kappa_\nu\kappa_\nu)^* v)/\partial y_1=0$$

which we solve using Corollary 26.7.8 with ε replaced by $\varepsilon/3$. Since $\kappa_\nu\tilde\kappa_\nu(0)=Y\in B_\nu(\frac14)$ the neighborhoods can be chosen so small that $\kappa_\nu\tilde\kappa_\nu V_2\subset B_\nu(\frac13)$. Choose $\Phi\in C_0^\infty(V_2)$ equal to 1 in V_1 and $\Psi\in C_0^\infty(V_2)$ equal to 1 in $\operatorname{supp}\Phi$. Then

$$v=(\tilde\kappa_\nu^{-1}\kappa_\nu^{-1})^*(\Psi(U+\varepsilon/3)), \qquad \phi=(\tilde\kappa_\nu^{-1}\kappa_\nu^{-1})^*\Phi$$

are in $C_0^\infty(B_\nu(\frac13))$ and satisfy for small T the conditions on $\phi_{\nu j}$, $v_{\nu j}$ in (i), (iii), (iv). In addition we have $\phi=1$ and $v<2\varepsilon/3$ in $\{X; a_\nu|X-Y|<c\}$ where c is a fixed constant. Now we can cover $B_\nu(\frac14)$ by a fixed number J of such neighborhoods so that there is a subordinate partition of unity $\tilde\phi_{\nu j}$ with uniform bounds in $S(1, a_\nu^2(|dx|^2+|d\xi|^2)$. The corresponding functions v and ϕ are denoted by $v_{\nu j}$ and $\phi_{\nu j}$. This completes the proof of the lemma.

Proof of Theorem 26.10.1'. We must show that if $u\in\mathscr{E}'(\mathbb{R}^n)$, $s_u^*\geqq s$ at ∂I and $s_{Pu}^*\geqq s$ in I then $s_u^*\geqq s$ in I. In doing so we may assume that $s_u^*\geqq s-\frac14$ in I, for if the theorem is known then we can start from the fact that $s_u^*\geqq s-k/4$ in I for some integer k and deduce that $s_u^*\geqq s-(k-1)/4,\dots, s_u^*\geqq s$ in I. The hypothesis $s_{Pu}^*\geqq s$ in I is then preserved if we change the terms of order $\leqq -1$ in the symbol of P, so we may assume that the Weyl symbol of P is equal to $\xi_1+if(x,\xi')+r(x,\xi')$ in a conic neighborhood V of I, when $|\xi|>1$. We can take V so small that $s_u^*>s-\frac14-\varepsilon$ and that $s_{Pu}^*>s-\varepsilon$ in V where ε is an arbitrary positive number kept fixed in the following discussion. The conditions (26.10.2), (26.10.3) are also assumed valid. Choose χ_0, χ_1, $\psi\in C_0^\infty(\mathbb{R}^{2n-2})$ so that

$$\chi_0(0)=1, \quad \chi_1=1 \text{ in } \operatorname{supp}\chi_0, \quad \psi=1 \text{ in } \operatorname{supp}\chi_1,$$

and define $f_{\delta,\lambda}$, $r_{\delta,\lambda}$ by (26.10.4) and similarly

$$\chi_{j,\delta,\lambda}(x',\xi')=\chi_j(x'/\delta, (\lambda\xi'-\varepsilon_n')/\delta).$$

After a symplectic dilation we can then, as observed after (26.10.5), apply Proposition 26.10.4 with $a^2=\lambda/\delta^2$ and obtain for sufficiently small δ and λ

$$(26.10.28)\quad (\lambda/\delta^2)^{\varepsilon/2}\|\chi^w_{0,\delta,\lambda}(x',D')v\| \le C(\|\chi^w_{1,\delta,\lambda}(x',D')P_{\delta,\lambda}v\|+(\lambda/\delta^2)^{\frac{1}{2}}\|P_{\delta,\lambda}v\|+(\lambda/\delta^2)^{\frac{1}{2}}\|v\|),$$

if $v\in\mathscr{S}$ and $v=0$ when $x_1\notin J$. (Recall that $I=J\times\{0\}\times\{\varepsilon_n\}$.) Here

$$P_{\delta,\lambda}=D_1+if^w_{\delta,\lambda}(x,D')+r^w_{\delta,\lambda}(x,D').$$

Choose a compact interval I_0 in the interior of I such that $s^*_u>s-\varepsilon$ in $I\setminus I_0$ and then a function $\chi\in C_0^\infty(V)$ with $x_1\in J$ in $\operatorname{supp}\chi$ and $\chi=1$ in a neighborhood of I_0. We shall apply (26.10.28) to $v=q_\lambda(x,D)u$ where

$$q_\lambda(x,\xi)=\chi(x,\lambda\xi)\,\lambda^{\varepsilon-s}.$$

(As this point we prefer not to use the Weyl calculus to be sure that $x_1\in J$ in supp v.) By Lemma 26.9.3 there is a bound for $\|\lambda^{\frac{1}{2}}q_\lambda(x,D)u\|$ as $\lambda\to 0$. Since $\lambda D_1 q_\lambda$ is a sum of two operators of the same form with $\chi(x,\xi)$ replaced by $\xi_1\chi(x,\xi)$ or by $\lambda D_1\chi(x,\xi)$ and since $\lambda(f^w_{\delta,\lambda}(x,D')+r^w_{\delta,\lambda}(x,D'))$ is uniformly bounded, it follows that $\lambda^{\frac{1}{2}}\|P_{\delta,\lambda}q_\lambda u\|\to 0$ as $\lambda\to 0$.

When computing the symbol of

$$\chi^w_{1,\delta,\lambda}P_{\delta,\lambda}q_\lambda(x,D)$$

we note that the symbol of $P_{\delta,\lambda}$ is equal to the symbol of P in the intersection of $\operatorname{supp}q_\lambda$ and $\operatorname{supp}\chi_{1,\delta,\lambda}$. By Theorem 18.5.4 the symbol of

$$\chi^w_{1,\delta,\lambda}(x,D)\,P_{\delta,\lambda}q_\lambda(x,D)-\chi^w_{1,\delta,\lambda}(x,D)\,q_\lambda(x,D)P$$

is therefore equal to $\lambda^{\varepsilon-s}\tilde\chi(x,\lambda\xi)+\rho_{\delta,\lambda}$ where

$$\tilde\chi(x,\xi)=\chi_{1,\delta,1}\{p,q_1\}/i$$

and $\rho_{\delta,\lambda}$ is uniformly bounded in S^{s-1} as $\lambda\to 0$, for q_λ is uniformly bounded in $S^{s-\varepsilon}$. Hence $\|\rho_{\delta,\lambda}u\|$ is bounded as $\lambda\to 0$. If δ is small we have $s^*_u>s-\varepsilon$ in $\operatorname{supp}\tilde\chi\subset\operatorname{supp}\chi_{1,\delta}\cap\operatorname{supp}d\chi$, for $s^*_u>s+1-\varepsilon$ when $\xi_1\neq 0$ since P is non-characteristic then, and $s^*_u>s-\varepsilon$ in $\operatorname{supp}d\chi$ when $x'=0$, $\xi=\varepsilon_n$, by the choice of χ. Hence

$$\|\lambda^{\varepsilon-s}\tilde\chi(x,\lambda D)u\|$$

is bounded as $\lambda\to 0$. This is also true for $\|q_\lambda(x,D)Pu\|$, so (26.10.28) shows that

$$\lambda^{\varepsilon/2}\|\chi^w_{0,\delta,\lambda}(x',D')\,q_\lambda(x,D)u\|$$

is bounded as $\lambda\to 0$. The symbol of $\chi^w_{0,\delta,\lambda}q_\lambda(x,D)$ is $\chi_{0,\delta,\lambda}q_\lambda$ apart from a term which is bounded in S^{s-1}, so it follows that

$$\lambda^{\varepsilon/2}\|(\chi_{0,\delta,\lambda}q_\lambda)(x,D)u\|$$

is bounded as $\lambda\to 0$. Hence $s^*_u\ge s-3\varepsilon/2$ on I_0 by Lemma 26.9.3, and this proves the theorem since ε is an arbitrary positive number.

26.11. A Semi-Global Existence Theorem

For arbitrary operators satisfying condition (P) we have now proved substitutes for Theorem 26.1.4 which permit us to prove an analogue of Theorem 26.1.7 with essentially the same arguments. Before stating it we shall examine the geometrical conditions involved. The notation is that in the summary at the end of Section 26.5.

Theorem 26.11.1. *Let P be a pseudo-differential operator in $\Psi^m_{\text{phg}}(X)$ satisfying condition (P), and let K be a compact subset of X. Then the following two conditions are equivalent:*

(i) *Every characteristic point over K lies on a compact semi-bicharacteristic interval with no characteristic endpoint over K.*

(ii) *No two dimensional bicharacteristic and no complete one dimensional bicharacteristic in $N\setminus(N_{11}\cup N_2^e)$ lies entirely over K.*

Proof. It is clear that (i) ⇒ (ii). Assume now that (ii) is fulfilled. We can also assume that the order of P is 1. The Hamilton field H_p of the principal symbol p can then be regarded as a vector field v on the cosphere bundle $S^*(X)$. It follows from (ii) that v cannot vanish anywhere over K in the characteristic set for then there would exist a radial bicharacteristic curve which contradicts (ii). If $\gamma_0\in N\setminus N_2^e$ then a semi-bicharacteristic through γ_0 is a one dimensional bicharacteristic until it leaves the characteristic set. If (i) is false for some $\gamma_0\in N\setminus N_2^e$ we can therefore find a C^1 map $\mathbb{R}_+\ni t\mapsto\gamma(t)\in S^*(X)|_K$ with

$$p(\gamma(t))=0,\quad \gamma'(t)=c(t)\,v(\gamma(t)),\quad |c(t)|=1,\quad \gamma(0)=\pi\gamma_0$$

where π is the projection $T^*(X)\setminus 0\to S^*(X)$. Now choose a sequence $t_j\to\infty$ such that $\gamma(t_j)$ converges. Then it follows that

$$\tilde\gamma(t)=\lim_{j\to\infty}\gamma(t+t_j)$$

exists and is a complete one dimensional bicharacteristic curve, which contradicts (ii). Assume now that $\gamma_0\in N_2^e$ and let B be the two dimensional bicharacteristic containing γ_0. We may assume that B contains some point $\hat\gamma\in N_2$ over the complement of K, for Proposition 26.5.5 shows that without violating condition (P) one can modify the symbol at a point in $N_2^e\setminus N_2$ to make it lie in N_2. If $\gamma_0\in B_0$ (see the discussion after Definition 26.5.4) then the assertion (i) follows since in the Riemann surface $\tilde B_0$ we can obviously choose a smooth curve through the class of γ_0 with endpoints near $\hat\gamma$. If $\gamma_0\in B\setminus B_0$ one can still find a semi-bicharacteristic from $\hat\gamma$ to γ_0 by the definition of N_2^e, and (i) follows again unless it continues indefinitely in the opposite direction as a one dimensional bicharacteristic over K. But we saw in the first part of the proof that this would contradict (ii), so the proof is now complete.

Our microlocal regularity theorems have the following consequence:

Theorem 26.11.2. *Let P be a pseudo-differential operator in $\Psi^m_{phg}(X)$ where X is a manifold. Assume that P satisfies condition (P), and let K be a compact subset of X such that the equivalent conditions in Theorem 26.11.1 are fulfilled. If $u\in\mathscr{E}'(K)$ and $s^*_{Pu}\geqq s$ where s is a real number or $+\infty$, it follows then that $s^*_u\geqq s+m-1$.*

Proof. Assume that the assertion is false so that

$$s_0=\inf s^*_u<s+m-1. \tag{26.11.1}$$

We shall prove that this leads to a contradiction. Since s^*_u is lower semi-continuous, there is some $\gamma\in T^*(X)\setminus 0$ over K such that $s^*_u(\gamma)=s_0$. We have $s^*_u\geqq s+m$ outside N so it is clear that $\gamma\in N$. Choose a semi-bicharacteristic interval Γ containing γ with no characteristic end point over K. Then $s^*_u\geqq s+m$ at the end points of Γ. If Γ is not contained in N it follows from Theorems 26.6.2 and 26.6.4 that $s^*_u\geqq s+m-1$ at Γ, which contradicts (26.11.1). Thus $\Gamma\subset N$. If Γ is a one dimensional bicharacteristic we also obtain a contradiction in view of Theorem 26.10.1′. The remaining possibility is that $\Gamma\subset N_2^e$. Without violating condition (P) we can then as in the proof of Theorem 26.11.1 change the principal symbol at the end points of Γ so that they are in N_2. This does not affect the condition $s^*_{Pu}\geqq s$ if the change is made in a small enough set, for $u\in\mathscr{E}'(K)$. Let B be the leaf of the foliation of N_2^e containing γ. Then $\gamma\in B_0$ and the function

$$S=\min(\tilde{s}_u, s+m-1)$$

which is superharmonic by Theorem 26.9.1 is $\geqq s_0$ in $\tilde{B}_0$ with equality in the class of γ. Hence S is identically equal to s_0 which contradicts the fact that $S=s+m-1$ at the class of any end point of Γ. This completes the proof.

We can now prove a slightly weakened analogue of Theorem 26.1.7.

Theorem 26.11.3. *Assume that $P\in\Psi^m_{phg}(X)$ is properly supported and satisfies condition (P). Let K be a compact subset of X such that the equivalent conditions in Theorem 26.11.1 are fulfilled. Then it follows that*

$$N(K)=\{v\in\mathscr{E}'(K), P^*v=0\}$$

is a finite dimensional subspace of $C_0^\infty(K)$ orthogonal to $P\mathscr{D}'(X)$. For every $f\in H^{loc}_{(s)}(X)$ with $(f,N(K))=0$ and every $t<s+m-1$ one can find $u\in H^{loc}_{(t)}(X)$ satisfying the equation $Pu=f$ in a neighborhood of K. (If $s=\infty$ one can take $t=\infty$.)

Proof. That $N(K)$ is a finite dimensional subspace of $C_0^\infty(K)$ follows from Theorem 26.11.2 exactly as in the proof of Theorem 26.1.7. By condition (i) in Theorem 26.11.1 we can choose a compact neighborhood K' of K for which the hypotheses are still fulfilled and $N(K')=N(K)$, so it suffices to prove that the equation can be satisfied in the interior of K. The proof then proceeds as that of Theorem 26.1.7 except that in (26.1.5) and (26.1.6) we must replace t by a larger number in $\|P^*v\|_{(t)}$, so we obtain (26.1.7) for any $t>1-m-s$. The existence of a solution then follows as before.

It is now natural to extend Definition 26.1.8 and end the chapter by defining the terminology used in the title:

Definition 26.11.4. Let $P\in\Psi^m_{\mathrm{phg}}(X)$ be properly supported and satisfy condition (*P*) in X. We shall then say that P is of principal type in X if the conditions in Theorem 26.11.1 are satisfied for every K.

When P is of principal type we have proved in this chapter that the equation $Pu=f$ can be solved on an arbitrary compact set when f satisfies a finite number of compatibility conditions there.

Notes

For operators of real principal type a local existence theorem was proved in Hörmander [1]. The example $D_1+iD_2+i(x_1+ix_2)D_3$ due to Lewy [1] showed that the result was not true in general for complex coefficients. This led to the proof in Hörmander [11] of a necessary condition for solvability. Solvability was proved in Hörmander [10] under a stronger form of this condition, and the results were made semi-global in "Linear partial differential operators". (See also Calderón [2].) Mizohata [4] observed that the same methods are applicable in some other cases such as the "Mizohata operators" $D_1+ix_1^kD_2$. The importance of this became clear when Nirenberg-Treves [1] showed that the local solvability properties of arbitrary first order differential operators with analytic coefficients could be analysed by means of closely related examples. A few years later Nirenberg-Treves [2] extended their results to the higher order case and even to pseudo-differential operators. They proved that P is not solvable if with the notation in Theorem 26.4.7 $\operatorname{Im} qp$ changes sign from $-$ to $+$ at a zero of finite order on a bicharacteristic for $\operatorname{Re} qp$. For first order zeros this was known from Hörmander [11, 17]. The same necessary condition was found by Egorov [2]. A decisive point in this work is the theorem of Egorov [1] which allows a simplification of the principal symbol by conjugation with a Fourier integral operator. Nirenberg and Treves [2] also conjectured the necessity of condition (Ψ) for local solvability and proved its invariance. The idea of the full proof given here is due to Moyer [1]. It contains the invariance proof for condition (Ψ) as an essential component. The proof was previously presented in Hörmander [40].

Nirenberg and Treves [2] proved the sufficiency of condition (*P*) for local solvability in the analytic case. The analyticity assumption was removed by Beals-Fefferman [1] but the result remained local. Indeed, it did not even give local existence of C^∞ solutions for C^∞ right-hand sides. That such solutions exist was proved in Hörmander [37] where a semi-global existence theory was also added. The key to this is the proof of theorems on propagation of singularities. In the real constant coefficient case such results go back to Grušin [1] (see the notes to Chapter VIII) and were proved in general by Hörmander [25]. The detailed discussion of operators of real principal type in Section 26.1 is taken from Duistermaat-Hörmander [1] where it was given as an application of the theory of Fourier integral operators. The results for the involutive case in Section 26.2 were also proved there. Normal forms in the symplectic case were first given by Sato-Kawai-Kashiwara [1] in the analytic (hyperfunction) case. The C^∞ results in Section 26.3 are due to Duistermaat and Sjöstrand [1]. The geometrical arguments in Section 26.5 come from Hörmander [37]. Section 26.7 is an improvement of results there due to Dencker [1], and the key estimates in Section 26.8 are due to Nirenberg-Treves [2]. They are first used in Section 26.9 to prove the extension of the superharmonicity theorem of Duistermaat-Hörmander [1] given in Hörmander [37], and they are also essential in Section 26.10. The main result there is due to Dencker [1]. The methods of Beals-Fefferman [1] are also very essential in the proof. The standard conclusions in Section 26.11 are taken from Hörmander [37].

Chapter XXVII. Subelliptic Operators

Summary

If P is an elliptic operator of order m in a C^∞ manifold X then $Pu\in H^{loc}_{(s)}$ implies $u\in H^{loc}_{(s+m)}$ (Theorem 18.1.29). This result can be microlocalized (Theorem 18.1.31): If $Pu\in H^{loc}_{(s)}$ at a point in the cotangent bundle where P is non-characteristic then $u\in H^{loc}_{(s+m)}$ there. This is the strongest possible result on (micro-)hypoellipticity.

The purpose of this chapter is to give a complete study of the next simplest case where $Pu\in H^{loc}_{(s)}$ implies $u\in H^{loc}_{(s+m-\delta)}$ for some fixed $\delta\in(0,1)$. One calls P subelliptic with loss of δ derivatives then. The condition $\delta<1$ guarantees that subellipticity is only a condition on the principal symbol.

In Section 26.4 we have already seen that condition $(\overline{\Psi})$ is necessary for hypoellipticity. In Section 27.1 another necessary condition on the principal symbol of a subelliptic operator is obtained by a scaling argument. These results together suggest the necessary and sufficient condition for subellipticity stated as Theorem 27.1.11. However, to prove the necessity completely we also need a symplectic study of the Taylor expansion of the symbol given in Section 27.2. The general proof of sufficiency is long so we give a short proof for operators satisfying condition (P) in Section 27.3. Section 27.4 is devoted to a detailed discussion of the local properties of a general subelliptic symbol. The proof of Theorem 27.1.11 is then completed in Sections 27.5 and 27.6 by means of a localization argument in several steps.

27.1. Definitions and Main Results

Let P be a properly supported pseudo-differential operator of order m in a C^∞ manifold X, which has a homogeneous principal symbol p. Let δ be a number with $0<\delta<1$.

Definition 27.1.1. *P is microsubelliptic at $\gamma\in T^*(X)\setminus 0$ with loss of δ derivatives if for every $s\in\mathbb{R}$*

$$(27.1.1)\qquad u\in\mathscr{D}'(X),\quad Pu\in H^{loc}_{(s)} \text{ at } \gamma \Rightarrow u\in H^{loc}_{(s+m-\delta)} \text{ at } \gamma.$$

L. Hörmander, *Classics in Mathematics*
The Analysis of Linear Partial Differential Operators IV,
DOI: 10.1007/978-3-642-00117-8_3, © Springer-Verlag Berlin Heidelberg 2009

It is often useful to know that (27.1.1) may be apparently weakened:

Lemma 27.1.2. *If for some* $s\in\mathbb{R}$

$$u\in H^{loc}_{(s+m-1)}(X),\quad Pu\in H^{loc}_{(s)}(X)\Rightarrow u\in H^{loc}_{(s+m-\delta)} \text{ at } \gamma \tag{27.1.1$'$}$$

then P is microsubelliptic at γ *with loss of* δ *derivatives.*

Proof. First we shall extend (27.1.1)′ to arbitrary s. To do so we choose A properly supported and elliptic of order μ. If

$$u\in H^{loc}_{(s+\mu+m-1)}(X),\quad Pu\in H^{loc}_{(s+\mu)}(X)$$

then $v=Au\in H^{loc}_{(s+m-1)}(X)$ and

$$Pv=APu+[P,A]u\in H_{(s)}(X)$$

since $[P,A]\in\Psi^{m+\mu-1}$. Hence (27.1.1)′ gives $Au\in H^{loc}_{(s+m-\delta)}$ at γ which means that $u\in H^{loc}_{(s+m+\mu-\delta)}$ at γ. Thus (27.1.1)′ remains valid with s replaced by any real number $s+\mu$, which will be used from now on.

If $u\in\mathscr{D}'(X)$, $Pu\in H^{loc}_{(s)}$ at γ, we can choose some $t\leqq s$ such that $u\in H^{loc}_{(t+m-1)}$ at γ. Choose $B\in\Psi^0$ properly supported and non-characteristic at γ so that $Pu\in H^{loc}_{(s)}$ and $u\in H^{loc}_{(t+m-1)}$ in $WF(B)$. Then we obtain $Bu\in H^{loc}_{(t+m-1)}(X)$, $PBu\in H^{loc}_{(t)}$ as in the first part of the proof. Hence $Bu\in H^{loc}_{(t+m-\delta)}$ at γ by the extended version of (27.1.1)′ which means that $u\in H^{loc}_{(t+m-\delta)}$ at γ. Thus we have replaced t by min $(s,t+1-\delta)$ in the hypothesis. After a finite number of iterations of the argument we obtain $t=s$ and then $u\in H^{loc}_{(s+m-\delta)}$ at γ. The proof is complete.

If P is microsubelliptic at every $\gamma\in T^*(X)\setminus 0$ with loss of δ derivatives, it follows that

$$u\in\mathscr{D}'(X),\quad Pu\in H^{loc}_{(s)}(X)\Rightarrow u\in H^{loc}_{(s+m-\delta)}(X). \tag{27.1.2}$$

Conversely, it follows from Lemma 27.1.2 that (27.1.2) implies microsubellipticity at every point in $T^*(X)\setminus 0$. Thus we obtain complete information on subellipticity in the sense of (27.1.2) if we master microsubellipticity. From now on we shall only deal with the latter property and use the term subellipticity as an abbreviation of microsubellipticity.

Lemma 27.1.3. *Subellipticity for P at* γ, *with loss of* δ *derivatives, implies the same property for every operator with principal symbol ap in a conic neighborhood of* γ, *if a is homogeneous and* $a(\gamma)\neq 0$.

Proof. The characterization (27.1.1)′ shows that lower order terms in P are irrelevant, and Definition 27.1.1 shows that only the restriction of the symbol to a conic neighborhood of γ matters. If A is a pseudo-differential operator with principal symbol a, then it follows from Definition 27.1.1 and Theorem 18.1.31 that AP is subelliptic at γ with loss of δ derivatives. Since the principal symbol is ap, this completes the proof.

Lemma 27.1.4. *Let χ be a homogeneous canonical transformation from a conic neighborhood of $(x_0,\xi_0)\in T^*(\mathbb{R}^n)\setminus 0$ to a conic neighborhood of $\gamma\in T^*(X)\setminus 0$. If P is subelliptic at γ with loss of δ derivatives, it follows that operators with principal symbol $p\circ\chi$ in a conic neighborhood of (x_0,ξ_0) are also subelliptic with loss of δ derivatives.*

Proof. As in the proof of Proposition 26.1.3 we choose $A\in I^0(X\times\mathbb{R}^n,\Gamma')$ and $B\in I^0(\mathbb{R}^n\times X,(\Gamma^{-1})')$, where Γ is the graph of χ, so that the corresponding operators are properly supported and $(\gamma,\gamma)\notin WF'(AB-I)$, $(x_0,\xi_0,x_0,\xi_0)\notin WF'(BA-I)$. Then $Q=BPA$ is a pseudodifferential operator with principal symbol $p\circ\chi$ in a conic neighborhood of (x_0,ξ_0). If $v\in\mathscr{D}'(\mathbb{R}^n)$ and $Qv\in H^{loc}_{(s)}$ at (x_0,ξ_0), that is, $BP(Av)\in H^{loc}_{(s)}$ at (x_0,ξ_0), it follows that $ABP(Av)\in H^{loc}_{(s)}$ at γ, hence $P(Av)\in H^{loc}_{(s)}$ at γ. From (27.1.1) it follows now that $Av\in H^{loc}_{(s+m-\delta)}$ at γ, hence $BAv\in H_{(s+m-\delta)}$ at (x_0,ξ_0) and $v\in H_{(s+m-\delta)}$ at (x_0,ξ_0). This completes the proof.

From now on we assume that $X\subset\mathbb{R}^n$. To prove necessary conditions for subellipticity we must convert (27.1.1) to an estimate.

Lemma 27.1.5. *Let $K\Subset X$ be a compact neighborhood of x_0 and assume that P is subelliptic at (x_0,ξ_0) with loss of δ derivatives. Then one can find $a\in S^0$ non-characteristic at (x_0,ξ_0) such that*

$$\|a(x,D)u\|_{(m-\delta)}\leq C(\|Pu\|_{(0)}+\|u\|_{(m-1)}),\quad u\in C_0^\infty(K). \tag{27.1.3}$$

Conversely, (27.1.3) *implies that P is subelliptic with loss of δ derivatives at every $(x,\xi)\in T^*(X)\setminus 0$ such that x is in the interior of K and a is non-characteristic at (x,ξ).*

Proof. $H=\{u\in\mathscr{E}'(K)\cap H_{(m-1)};\ Pu\in H_{(0)}\}$ is a Hilbert space with the norm $\|Pu\|_{(0)}+\|u\|_{(m-1)}$. The hypothesis means that every $u\in H$ is in $H^{loc}_{(m-\delta)}$ at (x_0,ξ_0). This means that $a(x,D)u\in H_{(m-\delta)}$ if $a\in S^0$ vanishes outside a sufficiently small conic neighborhood of (x_0,ξ_0). Choose $a_j\in S^0$ non-characteristic at (x_0,ξ_0) such that every conic neighborhood of (x_0,ξ_0) contains supp a_j for some $j=1,2,\dots$. Then we have $\|a_j(x,D)u\|_{(m-\delta)}<\infty$ for some j if $u\in H$. Set

$$M_{j,N}=\{u\in H;\ \|a_j(x,D)u\|_{(m-\delta)}\leq N\}.$$

This is a closed convex symmetric set, and $\bigcup M_{j,N}=H$. By Baire's theorem it follows that some $M_{j,N}$ has an interior point. The origin must then be an interior point which means that

$$\|a_j(x,D)u\|_{(m-\delta)}\leq C(\|Pu\|_{(0)}+\|u\|_{(m-1)})$$

for some C. This proves (27.1.3).

Suppose now that (27.1.3) is fulfilled and let Y be the interior of K. If $u\in\mathscr{E}'(Y)\cap H_{(m-1)}$ and $Pu\in H_{(0)}$ it follows then that $a(x,D)u\in H_{(m-\delta)}$. In fact, if we choose $\chi\in C_0^\infty$ with $\int\chi\,dx=1$ and set $\chi_\varepsilon(x)=\chi(x/\varepsilon)/\varepsilon^n$ we have

$u_\varepsilon = u*\chi_\varepsilon \to u$ in $H_{(m-1)}$. Since $u*\chi_\varepsilon = \hat{\chi}_\varepsilon(D)u$ and $\hat{\chi}_\varepsilon(\xi)=\hat{\chi}(\varepsilon\xi)$ is bounded in S^0, it follows that

$$Pu_\varepsilon = \hat{\chi}_\varepsilon(D)Pu + [P, \hat{\chi}_\varepsilon(D)]u$$

is bounded in $H_{(0)}$ as $\varepsilon \to 0$. Hence (27.1.3) shows that $a(x,D)u_\varepsilon$ is bounded in $H_{(m-\delta)}$ so $a(x,D)u \in H_{(m-\delta)}$.

If we just have $u \in H^{\mathrm{loc}}_{(m-1)}(X)$ and $Pu \in H^{\mathrm{loc}}_{(0)}(X)$ it follows if $\phi \in C_0^\infty(Y)$ that $\phi u \in H^{\mathrm{loc}}_{(m-1)}(X)$, $P(\phi u) \in H^{\mathrm{loc}}_{(0)}(X)$, so $a(x,D)(\phi u) \in H_{(m-\delta)}$. Hence $u \in H^{\mathrm{loc}}_{(m-\delta)}$ at every point in $T^*(Y) \setminus 0$ which is non-characteristic for a, so subellipticity follows from Lemma 27.1.2. The proof is complete.

An important consequence of Lemma 27.1.4 is

Theorem 27.1.6. *The set of all $\gamma \in T^*(X)\setminus 0$ such that P is subelliptic with loss of δ derivatives at γ is an open cone.*

Proposition 27.1.7. *If p is the principal symbol of P and for some odd integer k*

$$p(\gamma)=0, \quad H^j_{\mathrm{Re}\,p}\,\mathrm{Im}\,p(\gamma)=0, \quad j<k; \quad H^k_{\mathrm{Re}\,p}\,\mathrm{Im}\,p(\gamma)<0 \tag{27.1.4}$$

then P is not for any δ subelliptic at γ with loss of δ derivatives.

Proof. The hypothesis means that $\mathrm{Im}\,p$ has a zero of order k on the bicharacteristic for $\mathrm{Re}\,p$ through γ and that the sign changes from $+$ to $-$. Let κ be the smallest odd integer such that in any neighborhood of γ there is a characteristic γ' where (27.1.4) is valid with k replaced by κ. On any bicharacteristic of $\mathrm{Re}\,p$ starting near such a point γ' it is then clear that $\mathrm{Im}\,p$ has a zero of order exactly κ, so P is not even microhypoelliptic at γ' by Theorem 26.3.6. In view of Theorem 27.1.6 it follows that P is not subelliptic at γ either, which proves the corollary.

We shall now prove a necessary condition for subellipticity with loss of δ derivatives which actually depends on δ.

Proposition 27.1.8. *Let k be a positive integer and let $m_1, \dots, m_n$, $\mu_1, \dots, \mu_n$ be positive numbers with*

$$m_j + \mu_j = k+1, \quad j=1,\dots,n. \tag{27.1.5}$$

Assume that P is subelliptic at (x_0, ξ_0) with loss of δ derivatives and that the principal symbol p vanishes of weight k there, that is

$$D_\xi^\alpha D_x^\beta p(x_0,\xi_0)=0 \quad \text{when } \langle\alpha,\mu\rangle + \langle\beta, m\rangle < k. \tag{27.1.6}$$

Then it follows that $\delta \geq k/(k+1)$.

Proof. To simplify notation we assume $x_0=0$ in the proof. By Lemma 27.1.3 we may also assume that P is of order 1 and that $P=P(x,D)$ where $P \in S^1$

and $P(x,\xi)=p(x,\xi)$ when $|\xi|>1$. We can then extend (27.1.3) to all $u\in\mathscr{S}$. In fact, if $\chi\in C_0^\infty(K)$ we can apply (27.1.3) to χu for every $u\in\mathscr{S}$. This gives

$$\|\chi a(x,D)u\|_{(1-\delta)}\leq C(\|Pu\|_{(0)}+\|u\|_{(0)})$$

for some new constant C. Choosing χ with $\chi(x_0)\neq 0$ and replacing χa by a we have an estimate of the form (27.1.3) valid in $\mathscr{S}$.

Choose u so that $\hat{u}\in C_0^\infty(\mathbb{R}^n)$ and set for $\varepsilon\in(0,1)$

$$M_\varepsilon x=(\varepsilon^{m_1/(k+1)}x_1,\ldots,\varepsilon^{m_n/(k+1)}x_n).$$

Then

$$P(x,D)(e^{i\langle x,\xi_0\rangle/\varepsilon}u(M_\varepsilon^{-1}x))=e^{i\langle x,\xi_0\rangle/\varepsilon}f_\varepsilon(M_\varepsilon^{-1}x),$$

where

$$f_\varepsilon(x)=(2\pi)^{-n}\int e^{i\langle x,\xi\rangle}p(M_\varepsilon x,\xi_0/\varepsilon+M_\varepsilon^{-1}\xi)\hat{u}(\xi)\,d\xi. \tag{27.1.7}$$

By Taylor's formula the hypothesis (27.1.6) means that

$$|p(x,\xi_0+\xi)|\leq C(\sum|x_j|^{k/m_j}+\sum|\xi_j|^{k/\mu_j}). \tag{27.1.6$'$}$$

Hence

$$\begin{aligned}\varepsilon p(M_\varepsilon x,\xi_0/\varepsilon+M_\varepsilon^{-1}\xi)&=p(M_\varepsilon x,\xi_0+\varepsilon M_\varepsilon^{-1}\xi)\\&=O(\varepsilon^{k/(k+1)})(\sum|x_j|^{k/m_j}+\sum|\xi_j|^{k/\mu_j})\end{aligned}$$

in $\operatorname{supp}\hat{u}$, which proves that

$$|f_\varepsilon(x)|\leq C\varepsilon^{-1/(k+1)}(1+\sum|x_j|^{k/m_j}).$$

For any β we obtain a similar estimate for $x^\beta f_\varepsilon(x)$ if we multiply (27.1.7) by x^β and integrate by parts, thus replacing x^β by $(-D_\xi)^\beta$. Hence it follows that

$$\|f_\varepsilon\|_{(0)}=O(\varepsilon^{-1/(k+1)})$$

as $\varepsilon\to 0$.

Let $b(x,D)=(1+|D|^2)^{(1-\delta)/2}a(x,D)$ and set

$$\begin{aligned}b(x,D)(e^{i\langle x,\xi_0\rangle/\varepsilon}u(M_\varepsilon^{-1}x))&=e^{i\langle x,\xi_0\rangle/\varepsilon}g_\varepsilon(M_\varepsilon^{-1}x);\\g_\varepsilon(x)&=(2\pi)^{-n}\int e^{i\langle x,\xi\rangle}b(M_\varepsilon x,\xi_0/\varepsilon+M_\varepsilon^{-1}\xi)\hat{u}(\xi)\,d\xi.\end{aligned}$$

The principal symbol of b is $|\xi|^{1-\delta}a_0(x,\xi)$ where a_0 is the principal symbol of a, so we have

$$g_\varepsilon(x)\varepsilon^{1-\delta}\to|\xi_0|^{1-\delta}a_0(0,\xi_0)u(x)$$

uniformly in any compact set as $\varepsilon\to 0$. Now the extension of (27.1.3) made at the beginning of the proof, applied to $e^{i\langle x,\xi_0\rangle/\varepsilon}u(M_\varepsilon^{-1}x)$, gives

$$\|g_\varepsilon\|_{(0)}\leq C(\|f_\varepsilon\|_{(0)}+\|u\|_{(0)})$$

so it follows that $\delta-1\geq -1/(k+1)$. The proof is complete.

Corollary 27.1.9. *If P is subelliptic with loss of δ derivatives at the characteristic point γ and if p is the principal symbol of P, then $H_p(\gamma)$ does not have the radial direction.*

Proof. Assume that $\gamma=(0,\varepsilon_1)$ is a characteristic point where H_p has the radial direction $\partial/\partial\xi_1$; $\varepsilon_1=(1,0,\dots,0)$. This means that dp is proportional to dx_1 at $(0,\varepsilon_1)$. In view of the homogeneity it follows that

$$D_\xi^\alpha D_x^\beta p(0,\varepsilon_1)=0 \quad \text{if } \beta_1=0 \text{ and } \alpha_2+\dots+\alpha_n+\beta_2+\dots+\beta_n\leqq 1.$$

The hypothesis of Proposition 27.1.8 is therefore fulfilled if $m_j=\mu_j=(k+1)/2$ when $j>1$ and $m_1=k$, $\mu_1=1$, where k is any positive integer. Hence $\delta\geqq k/(k+1)$ which is a contradiction for large k since $\delta<1$. The proof is complete.

The interest of Corollary 27.1.9 is that in view of Lemmas 27.1.3 and 27.1.4 it allows us to use Theorem 21.3.6 to reduce the principal symbol to the special form $\xi_1+if(x,\xi')$, where f is independent of ξ_1. Further symplectic reductions can be made, and this will be the main topic in Sections 27.2 and 27.4. The following simple lemma suggests an invariant interpretation of the condition in Proposition 27.1.8.

Lemma 27.1.10. *If $p=p_1+ip_2$ vanishes of weight k at (x_0,ξ_0) as in Proposition 27.1.8, then any Poisson bracket*

$$\{p_{i_1},\{p_{i_2},\{\dots,p_{i_j}\}\}\}$$

of $j<k+1$ factors p_1, p_2 vanishes of weight $k+1-j$ at (x_0,ξ_0). If $j=k+1$ the Poisson bracket depends at (x_0,ξ_0) only on the derivatives $D_\xi^\alpha D_x^\beta p(x_0,\xi_0)$ with $\langle\alpha,\mu\rangle+\langle\beta,m\rangle=k$.

Proof. From the definition of the Poisson bracket

$$\{f,g\}=\sum(\partial f/\partial\xi_j\,\partial g/\partial x_j-\partial f/\partial x_j\,\partial g/\partial\xi_j)$$

it follows that if f and g vanish of weights w_f and w_g respectively then each term vanishes of weight $w_f+w_g-m_j-\mu_j=w_f+w_g-k-1$. If $w_f=k$ the weight is w_g-1, so the lemma follows at once by induction with respect to j.

In Section 27.2 we shall show that if all Poisson brackets with at most k factors $\operatorname{Re}p$, $\operatorname{Im}p$ vanish at (x_0,ξ_0) then the hypothesis of Proposition 27.1.8 is fulfilled for suitable m_j,μ_j and some homogeneous symplectic coordinates. Hence $\delta\geqq k/(k+1)$, that is, $k\leqq\delta/(1-\delta)$. If k is as large as possible we must have a non-vanishing Poisson bracket with at most $1+\delta/(1-\delta)=1/(1-\delta)$ factors. Together with Proposition 27.1.7 this will give the necessity in the following main theorem of this chapter.

Theorem 27.1.11. *The pseudo-differential operator P with principal symbol p is subelliptic at $\gamma_0\in T^*(X)\setminus 0$ with loss of $\delta<1$ derivatives if and only if there is a neighborhood V of γ_0 such that*

(i) *For every $\gamma\in V$ the repeated Poisson bracket*

$$(H_{\operatorname{Re}zp})^j\operatorname{Im}zp(\gamma) \tag{27.1.8}$$

is different from 0 for some $z\in\mathbb{C}$ and some $j\leqq\delta/(1-\delta)$.

(ii) (27.1.8) *is non-negative if j is odd and j is the smallest integer such that* (27.1.8) *does not vanish for all* $z\in\mathbb{C}$.

It is of course sufficient to take $\gamma=\gamma_0$ in condition (i). In Section 27.2 we shall see that it is fulfilled if (and only if) some Poisson bracket with at most $1/(1-\delta)$ factors $\operatorname{Re}p$, $\operatorname{Im}p$ is different from 0 at γ_0. The necessity of (ii) follows from Proposition 27.1.7. Condition (ii) follows if condition $(\overline{\Psi})$ in Definition 26.4.6 is valid in V. On the other hand, if (i) and (ii) are valid, then the proof of Proposition 27.1.7 shows that $\operatorname{Im} zp$ does not change sign from + to − along a bicharacteristic of $\operatorname{Re} zp$ in V if z has the property (i). Thus $(\overline{\Psi})$ is a consequence of (i) and (ii).

27.2. The Taylor Expansion of the Symbol

If $p=p_1+ip_2$ is the principal symbol of an operator which is subelliptic at $\gamma\in T^*(X)\setminus 0$ then Proposition 27.1.7 gives in particular that

$$\{p_1,p_2\}\geqq 0 \quad \text{when } p_1=p_2=0, \tag{27.2.1}$$

in a neighborhood of γ. By Corollary 27.1.9 we also know that

$$H_p\neq 0 \quad \text{when } p=0. \tag{27.2.2}$$

In this section we shall study the Taylor expansion of p at γ under these conditions. At first we shall ignore the homogeneity so p may be any C^∞ function satisfying (27.2.1) and (27.2.2) in a symplectic manifold S.

We shall use the notation $I=(i_1,\dots,i_j)$ for a sequence of $j=|I|$ elements which are either 1 or 2, and we shall write

$$H^I=H_{i_1}\dots H_{i_j},\qquad (\operatorname{ad}H)^I=\operatorname{ad}H_{i_1}\dots\operatorname{ad}H_{i_j},\qquad p_I=H_{i_1}\dots H_{i_{j-1}}p_{i_j}.$$

Here H_i is the Hamilton field H_{p_i} of p_i and we have used the standard notation

$$(\operatorname{ad} v)w=[v,w]$$

for the commutator of two vector fields v and w. By the Jacobi identity (21.1.3)′

$$H_{\{f,g\}}=[H_f,H_g]=\operatorname{ad}H_f H_g,$$

and since $\{f,g\}=H_f g$ it follows that the Hamilton field of the repeated Poisson bracket $H^I p_2$ is the repeated commutator $(\operatorname{ad}H)^I H_2$ of the Hamilton fields.

Before introducing specially adapted symplectic coordinates we introduce two integers which are important for the understanding of the local behavior of p:

$$k(\gamma)=\sup\{j\in\mathbb{Z};\ p_I(\gamma)=0 \text{ when } |I|\leqq j\}, \tag{27.2.3}$$

(27.2.4) $s(\gamma)=\sup\{j\in\mathbf{Z};\, j\leqq k(\gamma)$ and all commutators of H_1 and H_2 with at most j factors are linearly dependent at $\gamma\}$.

Here H_1 and H_2 are regarded as commutators with 1 factor. Thus $k(\gamma)=0$ means that $p(\gamma)\neq 0$ while $k(\gamma)=1$ means that $p(\gamma)=0$ but $\{p_1,p_2\}(\gamma)\neq 0$. If $k(\gamma)\neq 0$ then $s(\gamma)=0$ means that $H_1(\gamma)$ and $H_2(\gamma)$ are linearly independent.

The definitions are obviously invariant under canonical transformations. Moreover, if we multiply p by a zero free complex valued function a, then the Poisson brackets of $\operatorname{Re} ap$ and $\operatorname{Im} ap$ of order $\leqq j$ are linear combinations with smooth coefficients of those of p_1 and p_2 and vice versa. This shows that the numbers $k(\gamma)$ and $s(\gamma)$ are unchanged if p is multiplied by a zero free function.

When $k(\gamma)=0$ we can reduce p to the constant 1 locally by multiplication with p^{-1}. If $k(\gamma)=1$ then $s(\gamma)=0$ and the product with a suitable non-zero function is equal to ξ_1+ix_1 for appropriate symplectic coordinates (Theorem 21.3.3). We shall therefore usually assume $k(\gamma)>1$ in what follows.

Replacing p by the product with some non-zero function we can by the non-homogeneous version of Theorem 21.3.6 choose local symplectic coordinates vanishing at γ such that

$$p(x,\xi)=\xi_1+ip_2(x,\xi') \tag{27.2.5}$$

where $\xi'=(\xi_2,\dots,\xi_n)$. Thus p_2 is independent of ξ_1. Since $H_{p_1}=\partial/\partial x_1$ the definitions (27.2.3) and (27.2.4) imply that $(\partial/\partial x_1)^j p_2(0)=0$ if $j<k(0)$ and that $d(\partial/\partial x_1)^j p_2(0)=0$ if $j<s(0)$. In the opposite direction we shall prove

Proposition 27.2.1. *Assume that p is of the form* (27.2.5) *and that*

$$\begin{aligned}(\partial/\partial x_1)^j p_2(0)&=0 \quad \text{if} \quad j<k;\\ d(\partial/\partial x_1)^j p_2(0)&=0 \quad \text{if} \quad 2j<k.\end{aligned} \tag{27.2.6}$$

Then it follows that p vanishes of weight k at 0 if the coordinates are given the weights (cf. Proposition 27.1.8)

$$\mu_1=k,\quad m_1=1,\quad m_j=\mu_j=(k+1)/2 \quad \text{if } j>1. \tag{27.2.7}$$

The only terms of weight k are $\xi_1+icx_1^k/k!$ where $c=\partial^k p_2(0)/\partial x_1^k$. We have $k(0)\geqq k$ (with strict inequality if $c=0$), $s(0)\geqq k/2$, and for arbitrary $f_1, f_2, g_1, g_2\in C^\infty$

$$(H_{f_1p_1+f_2p_2})^k(g_1p_1+g_2p_2)(0)=c(f_1g_2-f_2g_1)f_1^{k-1}(0). \tag{27.2.8}$$

Proof. If $\langle\alpha,\mu\rangle+\langle\beta,m\rangle\leqq k$ then $|\alpha'+\beta'|\leqq 1$. When $\alpha'=\beta'=0$ it follows that $\beta_1\leqq k$, and if $|\alpha'+\beta'|=1$ it follows that $\beta_1\leqq(k-1)/2$. Condition (27.2.6) means that $D_\xi^\alpha D_x^\beta p_2(0)$ is then equal to 0 except when $\beta_1=k$ and $\alpha'=\beta'=0$, so $p-\xi_1-icx_1^k/k!$ vanishes of weight $k+\frac{1}{2}$. By Lemma 27.1.10 it follows that $k(0)\geqq k$ with strict inequality if $c=0$, and that we can replace p_2 by $cx_1^k/k!$

and assume $f_1,\dots,g_2$ constant when verifying (27.2.8). Since

$$H_{f_1p_1+f_2p_2}=f_1\,\partial/\partial x_1-f_2cx_1^{k-1}/(k-1)!\,\partial/\partial\xi_1,$$
$$\{f_1p_1+f_2p_2,g_1p_1+g_2p_2\}=(f_1g_2-f_2g_1)cx_1^{k-1}/(k-1)!$$

we obtain (27.2.8). The Poisson brackets with $j+1$ factors vanish at 0 of weight $k-j+\frac{1}{2}$ apart from multiples of x_1^{k-j}. When $k-j+\frac{1}{2}>(k+1)/2$, that is, $j<k/2$, it follows that the Taylor expansion contains no first order term so the Hamilton field is equal to 0 at 0. Hence $s(0)\geqq k/2$ as claimed.

If $c\neq 0$ then $k(0)=k$ and we have the information required in the proof of Theorem 27.1.11. However, in general the second part of (27.2.6) may fail for values of k which are much smaller than $k(0)$. This can only happen if $k(0)>2s(0)$, for if $k\leqq k(0)\leqq 2s(0)$ and $2j<k$ then $j<k/2\leqq s(0)$ so $j+1\leqq s(0)$. On the other hand, if $k(0)\geqq k>2s(0)$ then $d(\partial/\partial x_1)^s p_2(0)\neq 0$ for some s with $2s<k$, for Lemma 27.2.1 would otherwise give $s(0)\geqq k/2$. This is the situation which we shall consider now, so we assume that for some integers s and $k\geqq 2$

$$k>2s, \tag{27.2.9}$$

$$p_I(0)=0 \quad \text{for } |I|\leqq k, \tag{27.2.10}$$

$$d(\partial/\partial x_1)^j p_2(0)=0 \quad \text{when } j<s, \tag{27.2.11}$$

$$d(\partial/\partial x_1)^s p_2(0)\neq 0. \tag{27.2.12}$$

Note that $(\partial/\partial x_1)^{s+1}p_2(0)=0$ by (27.2.10) since $s+2\leqq(k-1)/2+2=(k+3)/2\leqq k$ if $k\geqq 3$ and $s=0$ if $k=2$. Hence (27.2.12) means in fact that the differential with respect to x',ξ' of $(\partial/\partial x_1)^s p_2$ does not vanish at 0. It is therefore possible to take this function with x_1 fixed at 0 as a coordinate in a new symplectic coordinate system. To simplify the notation we keep the same notation for the new coordinates, that is, we assume that

$$\partial^s p_2(0,x',\xi')/\partial x_1^s=\xi_2. \tag{27.2.13}$$

Proposition 27.2.2. *Assume that p is of the form* (27.2.5) *and that* (27.2.1), (27.2.9), (27.2.10), (27.2.11), (27.2.13) *are fulfilled,* $k\geqq 2$. *Then it follows that p vanishes of weight k at 0 if the weights of the* ξ, x *coordinates are*

$$\mu_1=k,\quad m_1=1,\quad \mu_2=k-s,\quad m_2=s+1, \tag{27.2.14}$$
$$m_j=\mu_j=(k+1)/2 \quad \text{if } j>2.$$

The only terms of weight k are $\xi_1+i(b(x_1,x_2)+x_1^s\xi_2/s!)$ *where b is a polynomial in* x_1,x_2 *with all terms of weight k, and* $B(x_1,x_2)=\partial b(x_1,x_2)/\partial x_1-sb(x_1,x_2)/x_1$ *is a non-negative polynomial. We have* $k(0)\geqq k$ *(with strict inequality if* $B\equiv 0$), $s(0)=s$, *all commutators of* $\leqq(k+1)/2$ *factors* H_1, H_2 *are linear combinations of* $\partial/\partial x_1$ *and* $\partial/\partial x_2$ *at 0, and for arbitrary* f_1, f_2, g_1,

$g_2 \in C^\infty$

$$(27.2.15)\qquad (H_{f_1p_1+f_2p_2})^k(g_1p_1+g_2p_2)(0) = (f_1g_2-f_2g_1)k!\,(k-s)^{-1}\,B(f_1, f_1^s f_2/(s+1)!).$$

For the proof we need an elementary lemma:

Lemma 27.2.3. *If Y is an open subset of $\mathbb{R}^{1+N}$ then the set of all real valued $F \in C^1(Y)$ such that*

$$(27.2.16)\qquad F(t,y)=0 \Rightarrow \partial F(t,y)/\partial t \geqq 0$$

is closed in $C^1(Y)$. If $y'=(y_1,\dots,y_M)$ and

$$F(t,y)=F_0(t,y')+\sum_{M+1}^{N} F_j(t,y')y_j$$

satisfies (27.2.16) then $\partial(F_j/F_k)/\partial t=0$ where $F_k \neq 0$, if $j,k \neq 0$.

Proof. If F does not satisfy (27.2.16) then $F(t,y)=0$ and $\partial F(t,y)/\partial t<0$ for some $(t,y)\in Y$. If G is sufficiently close to F in the C^1 topology it follows that $G(s,y)=0$, $\partial G(s,y)/\partial s<0$ for some s close to t. Hence the complement of the set of all F satisfying (27.2.16) is open. To prove the last statement we observe that

$$\sum_{M+1}^{N} F_j(t,y')y_j \quad\text{and}\quad \sum_{M+1}^{N} \partial F_j(t,y')/\partial t\; y_j$$

must be linearly dependent if (27.2.16) is fulfilled. The proof is complete.

Proof of Proposition 27.2.2. We shall first prove that $p_2(x_1,x_2,0)$ vanishes of weight k at 0, that is,

$$(27.2.17)\qquad (\partial/\partial x_2)^j(\partial/\partial x_1)^i\, p_2(0)=0 \quad \text{if } i+(s+1)j<k.$$

To do so we observe that (27.2.10) implies

$$((\operatorname{ad} H_1)^s H_2)^j H_1^i\, p_2(0)=0 \quad \text{if } i+(s+1)j<k,$$

and that $(\operatorname{ad} H_1)^s H_2$ is the Hamilton field of $H_1^s p_2$. We may replace it by the Hamilton field of $H_1^s p_2(0,x',\xi')=\xi_2$ for both can be considered as vector fields in the plane $x_1=0$ differing only by a multiple of $\partial/\partial\xi_1$. Thus we obtain (27.2.17). (From now on we shall make no further use of (27.2.10).)

From (27.2.17) and (27.2.11) it follows that p_2 vanishes of weight k at 0 if we replace (27.2.14) by

$$(27.2.14)'\qquad \mu_1=k,\quad m_1=1,\quad \mu_2=k-s,\quad m_2=s+1,$$
$$m_j=\mu_j=\kappa \quad \text{if } j>2$$

and take $\kappa=k-s$. Let κ be the smallest positive number $\geqq k/2$ for which this is true, that is,

$$(k-s)\alpha_2+\beta_1+(s+1)\beta_2+\kappa|\alpha''+\beta''|\geqq k \quad \text{if } D_\xi^\alpha D_x^\beta p_2(0)\neq 0.$$

Here $\alpha''=(\alpha_3,\dots,\alpha_n)$, $\beta''=(\beta_3,\dots,\beta_n)$. If $\kappa>k/2$ we must have equality for some α, β with $|\alpha''+\beta''|=1$, hence $\alpha_2=0$ and $\beta_1\neq s$, by (27.2.13). Now (27.2.1) means that $\partial p_2(x,\xi')/\partial x_1\geqq 0$ when $p_2(x,\xi')=0$. By Lemma 27.2.3 it follows that

$$\lim_{\varepsilon\to 0}\varepsilon^{-k}p_2(\varepsilon x_1,\varepsilon^{s+1}x_2,\varepsilon^{\kappa}x'',\varepsilon^{k-s}\xi_2,\varepsilon^{\kappa}\xi'')=\sum(\partial_\xi^\alpha\partial_x^\beta p_2(0))\xi^\alpha x^\beta/\alpha!\beta!$$

has the same property. The sum is taken over all α and β with

$$(k-s)\alpha_2+\beta_1+(s+1)\beta_2+\kappa|\alpha''+\beta''|=k$$

so it is linear in ξ_2, x'', ξ''. The coefficient of ξ_2 is $x_1^s/s!$ and the coefficients of x'',ξ'' have no term $x_1^ix_2^j$ so they must be equal to 0 by Lemma 27.2.3. This is a contradiction proving that $\kappa=k/2$. With the weights (27.2.14) all terms involving x'',ξ'' are therefore of weight at least $k+\frac{1}{2}$, so $p_2(x,\xi')-(b(x_1,x_2)+x_1^s\xi_2/s!)$ vanishes of weight $k+\frac{1}{2}$ at 0 for some polynomial b of homogeneous weight k. By Lemma 27.2.3 we know that

$$b(x_1,x_2)+x_1^s\xi_2/s!=0\Rightarrow \partial b(x_1,x_2)/\partial x_1+x_1^{s-1}\xi_2/(s-1)!\geqq 0,$$

that is, $B(x_1,x_2)=\partial b(x_1,x_2)/\partial x_1-sb(x_1,x_2)/x_1$ is non-negative when $x_1\neq 0$ and therefore a polynomial also in x_1. If we write

$$b(x_1,x_2)=\sum_{i+(s+1)j=k}b_{ij}x_1^ix_2^j$$

then

$$B(x_1,x_2)=\sum_{i+(s+1)j=k}b_{ij}(i-s)x_1^{i-1}x_2^j\geqq 0$$

and since b contains no term with $i=s$ it follows that $b\equiv 0$ if $B\equiv 0$. The homogeneity

$$B(-x_1,(-1)^{s+1}x_2)=(-1)^{k-1}B(x_1,x_2)$$

shows that k must be odd if b is not identically 0.

By Lemma 27.1.10 we know that H^Ip_2 vanishes of weight $\geqq k-|I|$ at 0. Replacing p_2 by $b(x_1,x_2)+x_1^s\xi_2/s!$ changes H^Ip_2 by a term vanishing of order at least $k-|I|+\frac{1}{2}$ at 0. Since all Poisson brackets of ξ_1 and $x_1^s\xi_2/s!$ vanish at 0 because of the factor ξ_2, it follows that $k(0)>k$ if $b\equiv 0$. On the other hand, it will follow from (27.2.15) that $k(0)=k$ if $B\not\equiv 0$, and this is equivalent to $b\not\equiv 0$. Before proving (27.2.15) we observe that apart from the terms obtained when p_2 is replaced by $b(x_1,x_2)+x_1^s\xi_2/s!$ the Poisson bracket H^Ip_2 vanishes at 0 of weight $\geqq k+\frac{1}{2}-|I|>(k+1)/2$ if $|I|<k/2$. It follows that dx'' and $d\xi''$ cannot occur in $dH^Ip_2(0)$ then. Nor can dx_1 for the weight 1 of x_1 is $\leqq k/2$ since we have assumed $k\geqq 2$. Nor can dx_2 occur unless b depends on x_2, and then the positivity of B shows that x_2^2 occurs in b so $k\geqq 2(s+1)+1$, thus $k/2>s+1$. Hence the differentials of p_1 and H^Ip_2 are linear combinations of $d\xi_1$ and $d\xi_2$ at 0 if $|I|<k/2$. If $|I|<s$ then H^Ip_2 vanishes of weight $k-|I|>k-s$ at 0 so $d\xi_2$ cannot occur either. However, $H_1^sp_2-\xi_2-\partial^sb/\partial x_1^s$ vanishes of weight $>k-s$ so $dH_1^sp_2=d\xi_2$ at 0. It follows that $s(0)=s$.

To prove (27.2.15) finally we may assume that f_1, f_2, g_1, g_2 are constant and that $p_2 = b(x_1, x_2) + x_1^s \xi_2/s!$. Then

$$(27.2.18) \qquad (f_1 H_1 + f_2 H_2)(g_1 p_1 + g_2 p_2) = (f_1 g_2 - f_2 g_1)(\partial b/\partial x_1 + x_1^{s-1}\xi_2/(s-1)!),$$
$$f_1 H_1 + f_2 H_2 = f_1 \partial/\partial x_1 + f_2(x_1^s/s!\,\partial/\partial x_2 - \partial b/\partial x_2 \partial/\partial \xi_2) \bmod \partial/\partial \xi_1.$$

The integral curve of this vector field starting at 0 is

$$x_1 = f_1 t, \quad x_2 = f_1^s f_2' t^{s+1}, \quad \xi_2 = -t^{k-s} f_2 (k-s)^{-1} b_2(f_1, f_1^s f_2')$$

where $f_2' = f_2/(s+1)!$ and $b_j = \partial b/\partial x_j$. On this curve the right-hand side of (27.2.18) is equal to

$$(27.2.19) \quad (f_1 g_2 - f_2 g_1) t^{k-1}(b_1(f_1, f_1^s f_2') - s(s+1)(k-s)^{-1} f_1^{s-1} f_2' b_2(f_1, f_1^s f_2')).$$

To compute the second paranthesis we observe that

$$b_1(x_1, x_2) - s(s+1)(k-s)^{-1} x_1^{-1} x_2 b_2(x_1, x_2) = \sum (i - s(s+1)j/(k-s)) b_{ij} x_1^{i-1} x_2^j$$

is equal to $kB(x)/(k-s)$ because $i(k-s) - js(s+1) = k(i-s)$. Thus the right-hand side of (27.2.18) is equal to

$$t^{k-1}(f_1 g_2 - f_2 g_1) B(f_1, f_1^s f_2') k/(k-s).$$

Differentiation $k-1$ times with respect to t gives (27.2.15) and completes the proof.

Remark. It is elementary to see that in Proposition 27.2.2 we have $s \neq 1$, and that x_1^3 is a factcr of b if $s \neq 0$. If $\bar{p}$ satisfies (Ψ) then

$$\xi_1 - i(b(x_1, x_2) + x_1^s \xi_2/s!)$$

also satisfies (Ψ) which means that s is even and that $b(x_1, x_2)/x_1^s$ is a polynomial which increases with x_1 and vanishes when $x_1 = 0$. There are no further restrictions on the polynomials b which can occur. These observations will not be needed, and the proof is therefore left for the reader.

The following consequence of Propositions 27.2.1 and 27.2.2 was mentioned after Theorem 27.1.11.

Corollary 27.2.4. *If k is an integer $\geqq 0$ and $p = p_1 + ip_2$ satisfies (27.2.1) then the following conditions are equivalent*

(i) $k(\gamma) > k$;

(ii) $(H_{\mathrm{Re} fp})^j \operatorname{Im} fp(\gamma) = 0$ *if* $f \in C^\infty$ *and* $j \leqq k$;

(iii) $(H_{\mathrm{Re} zp})^j \operatorname{Im} zp(\gamma) = 0$ *if* $z \in \mathbb{C}$ *and* $j \leqq k$.

Proof. The three conditions are independent of k if $H_p(\gamma) = 0$ so we may assume that (27.2.2) is valid. If $k = 0$ the conditions all mean that $p(\gamma) = 0$, and when $k = 1$ they mean that $p(\gamma) = \{\mathrm{Re}\, p, \operatorname{Im} p\}(\gamma) = 0$. Since (i) $\Rightarrow$ (ii)

$\Rightarrow$(iii) is obvious, we just have to prove that if (i)–(iii) are valid for some $k \geqq 1$ and in addition (iii) holds with k replaced by $k+1$ then $k(\gamma) > k+1$. To do so we first choose $a \in C^\infty$ so that with suitable symplectic coordinates

$$\tilde{p} = p/a = \xi_1 + iq(x, \xi').$$

For this symbol the conditions (i)–(iii) remain valid. Hence $\tilde{p}$ satisfies the hypotheses of Proposition 27.2.1 or those of Proposition 27.2.2 for some s, with k replaced by $k+1$ and after a suitable symplectic change of variables. In both cases it follows that if

$$(H_{\operatorname{Re} zp})^{k+1} \operatorname{Im} zp(\gamma) = (H_{\operatorname{Re} za\tilde{p}})^{k+1} \operatorname{Im} za\tilde{p}(\gamma)$$

vanishes for all $z \in \mathbb{C}$ then all Poisson brackets of at most $k+2$ factors $\operatorname{Re} \tilde{p}$, $\operatorname{Im} \tilde{p}$ vanish at γ. This means that $k(\gamma) > k+1$ which completes the proof.

Proof of the necessity in Theorem 27.1.11. **Assume that P is subelliptic at γ_0 with loss of $\delta < 1$ derivatives, and that**

$$(H_{\operatorname{Re} zp})^j \operatorname{Im} zp(\gamma_0) = 0, \quad j \leqq \delta/(1-\delta).$$

By Theorem 27.1.6 and Proposition 27.1.7 we know that $\{\operatorname{Re} p, \operatorname{Im} p\} \geqq 0$ when $p = 0$, in a neighborhood of γ_0. Hence it follows from Corollary 27.2.4 that $k(\gamma_0) \geqq k$ if k is the largest integer $\leqq 1/(1-\delta)$. This condition is invariant under multiplication of p by a non-vanishing homogeneous function and under homogeneous canonical transformations. By Corollary 27.1.9 and Lemma 27.1.4 we may therefore assume that $p = \xi_1 + ip_2(x, \xi')$ and that $\gamma_0 = (0, \varepsilon_2)$. If the hypotheses of Proposition 27.2.1 are fulfilled then Proposition 27.1.8 shows that $\delta \geqq k/(k+1)$, that is, $k+1 \leqq 1/(1-\delta)$. This contradicts the definition of k. If the hypotheses of Proposition 27.2.1 are not fulfilled we can apply Proposition 27.2.2 in the same way if $\eta_2 = \partial^s p_2(0, x', \xi')/\partial x_1^s$ can be taken as a new homogeneous symplectic coordinate, that is, if the Hamilton field is not radial at $(0, \varepsilon_2)$. In that case we consider $q(x, \xi'') = p_2(x, 1, \xi'')$, $\xi'' = (\xi_3, \ldots, \xi_n)$. Then

$$d(\partial/\partial x_1)^j q(0) = 0 \quad \text{when } j < s, \quad d(\partial/\partial x_1)^s q(0) = c\,dx_2$$

for some $c \neq 0$. Thus

$$q(x, \xi'') = bx_1^k + cx_2 x_1^s/s! + O(|x'| + |\xi''|)x_1^{s+1} + O(|x'|^2 + |\xi''|^2),$$

and it follows as in the proof of Proposition 27.2.2 that $q(x, \xi'')$ vanishes of weight k at 0 if

$$m_1 = 1, \quad m_2 = k - s, \quad m_j = \mu_j = (k+1)/2 \quad \text{if } j > 2.$$

Since $p_2(x, \xi) = \xi_2 q(x, \xi''/\xi_2)$ we conclude that p vanishes of weight k at $(0, \varepsilon_2)$ if we take $\mu_1 = k$ and $\mu_2 = s+1$ also. Again Proposition 27.1.8 gives $\delta \geqq k/(k+1)$ which is a contradiction completing the proof.

The last complication of the proof is actually artificial. It can be eliminated by first proving that the problem can be localized to a neighborhood of $(0, \varepsilon_2)$ in such a way that arbitrary symplectic changes of variables can be used. We shall have to do so in the general proof of the sufficiency in Theorem 27.1.11.

27.3. Subelliptic Operators Satisfying (P)

The general proof of sufficiency in Theorem 27.1.11 is quite complicated. We shall therefore discuss in this section the far simpler case when p satisfies not only $(\bar{\Psi})$ but also (P). Then the number $k(\gamma)$ defined by (27.2.3) must be even by Proposition 27.2.2 so only the Taylor expansions discussed in Proposition 27.2.1 can occur. If $k(\gamma)\neq 0$ we have $s(\gamma)\geqq 1$ then. Assuming as we may that the principal symbol is of the form $\xi_1+ip_2(x,\xi')$ in a conic neighborhood of $(0,\varepsilon_n)$, $\varepsilon_n=(0,\ldots,0,1)$, we must have

$$(\partial/\partial x_1)^k p_2(x,\xi')\neq 0 \quad \text{at } (0,\varepsilon_n) \tag{27.3.1}$$

if $k=k((0,\varepsilon_n))$. In fact, since $H_{\operatorname{Im} p}=0$ at $(0,\varepsilon_n)$ we have

$$(H_{\operatorname{Re} zp})^k \operatorname{Im} zp(0,\varepsilon_n)=0 \quad \text{when } \operatorname{Re} z=0,$$

so it follows from (27.2.8) that this is not 0 when $\operatorname{Re} z\neq 0$. Using condition (P) we now conclude that p_2 has a constant sign in a conic neighborhood of $(0,\varepsilon_n)$. Changing the sign of x_1 if necessary we may assume that $p_2\geqq 0$.

Let $\chi(x,\xi')$ be homogeneous of degree 0 with support in a small conic neighborhood of $(0,\varepsilon_n)$ and equal to 1 in another neighborhood. Then

$$q(x,\xi')=\chi(x,\xi')p_2(x,\xi')+(1-\chi(x,\xi'))|\xi'|$$

is non-negative in $\mathbb{R}^n\times(\mathbb{R}^{n-1}\setminus 0)$ and for some $c>0$

$$\sum_{j\leqq k}|\partial^j q(x,\xi')/\partial x_1^j|\geqq c|\xi'|. \tag{27.3.2}$$

The following result extending Proposition 26.3.3 will easily prove subellipticity of P.

Proposition 27.3.1. *Let $Q\in S^1(\mathbb{R}^n\times\mathbb{R}^{n-1})$ have a non-negative principal part q satisfying* (27.3.2). *Then it follows that with L^2 norms*

$$\|(1+|D'|^2)^{1/(2k+2)}u\|\leqq C(\|(D_1+iQ(x,D'))u\|+\|u\|), \quad u\in\mathscr{S}(\mathbb{R}^n). \tag{27.3.3}$$

We shall reduce the proof of (27.3.3) to estimates for ordinary differential operators. These will be proved first, in greater generality than required now and with proofs which can be adapted to more difficult situations in Section 27.5.

Lemma 27.3.2. *Let G be a real valued C^∞ function in $I=(-1,1)$ such that*

$$|G^{(k+1)}|\leqq 1 \quad \textit{in } I, \tag{27.3.4}$$

$$\max_{j\leqq k}|G^{(j)}(0)|=\rho. \tag{27.3.5}$$

With $\varepsilon=2^{-k}$ it follows if ρ is large enough that

$$\rho^{\varepsilon}\|u\|\leqq C(\|Du\|+\|Gu\|), \quad u\in C_0^\infty(I). \tag{27.3.6}$$

Proof. First note that (27.3.4), (27.3.5) imply

$$\rho/e\leqq\max_{j\leqq k}|G^{(j)}(x)|\leqq e\rho, \quad x\in I, \tag{27.3.5$'$}$$

provided that $\rho>e$. In fact, the second inequality follows from Taylor's formula when $\rho\geqq 1$. If $|G^{(j)}(x)|\leqq a$ when $j\leqq k$, for some x, and $a\geqq 1$, it follows that $\rho\leqq ea$, hence $a\geqq\rho/e$ which proves the first inequality. To prove (27.3.6) we shall follow the proof of Lemma 22.2.3. Thus we note that for $0\leqq j<k$

$$\begin{aligned}\|G^{(j+1)}u\|^2&=(i[D,G^{(j)}]u,G^{(j+1)}u)\\&=(iG^{(j)}u,G^{(j+1)}Du)-(G^{(j)}u,G^{(j+2)}u)-(iDu,G^{(j+1)}G^{(j)}u)\\&\leqq\|G^{(j)}u\|\,e\rho(2\|Du\|+\|u\|).\end{aligned}$$

Hence it follows inductively that for $0\leqq j\leqq k$

$$\begin{aligned}\|G^{(j)}u\|&\leqq\|Gu\|^{2^{-j}}(e\rho(2\|Du\|+\|u\|))^{1-2^{-j}}\\&\leqq(e\rho)^{1-\varepsilon}(2\|Du\|+\|Gu\|+\|u\|).\end{aligned}$$

Using (27.3.5)′ we conclude that

$$\rho e^{-1}\|u\|\leqq(k+1)(e\rho)^{1-\varepsilon}(2\|Du\|+\|Gu\|+\|u\|),$$

that is,

$$(\rho^{\varepsilon}e^{\varepsilon-2}-k-1)\|u\|\leqq(k+1)(2\|Du\|+\|Gu\|).$$

For large ρ the estimate (27.3.6) follows, which completes the proof.

Remark. The exponent ε in (27.3.6) can be replaced by $1/(k+1)$. This will follow from the proof of Lemma 27.3.4. The localization argument used in the proof of that result will work equally well no matter what positive exponent ε we have in (27.3.6).

In the next lemma we shall try to replace the estimate by $\|Du\|+\|Gu\|$ in (27.3.6) with an estimate by $\|Du+iGu\|$ only.

Lemma 27.3.3. *Assume, in addition to the hypotheses in Lemma 27.3.2, that G does not change sign from + to − for increasing x. Then we have with $\varepsilon=2^{-k}$*

$$\rho^{2\varepsilon}\int_{-\frac12}^{\frac12}|v|^2\,dx\leqq C\left(\rho^2\int_{-1}^{1}|Dv+iGv|^2\,dx+\int_{-1}^{1}|v|^2\,dx\right), \tag{27.3.8}$$

if $v\in C^\infty(I)$ and ρ is large enough.

Proof. If $\delta e<1$ it follows from (27.3.5)′ that $\{x\in I;\,|G|=\delta\rho\}$ is discrete, for at a limit point we would have $G^{(j)}=0$ for $j\neq 0$, hence $|G|e\geqq\rho$. If $v\in C_0^\infty(I)$ and $f=Dv+iGv$, $g=G/\rho$ then

$$\int_{|g|>\delta}|f|^2/|g|\,dx=\int_{|g|>\delta}(|Dv|^2+|\rho g v|^2)/|g|\,dx-\rho\int_{|g|>\delta}\operatorname{sgn} g\,d|v|^2.$$

The last term here can be integrated. At the boundary of an interval where $|g|<\delta$ and g changes sign from $-$ to $+$ we obtain two positive terms which we drop. At the others we obtain the difference between $|v|^2$ at the end points. The sum of these terms is in absolute value at most

$$2\rho\int_{|g|<\delta}|vDv|\,dx\leqq\delta^{-1}\int_{|g|<\delta}|Dv|^2\,dx+\delta\rho^2\int_{|g|<\delta}|v|^2\,dx.$$

Adding the integral when $|g|<\delta$ of

$$2\delta^{-1}|Dv|^2+\rho^2\delta^{-1}|gv|^2\leqq\delta^{-1}(3|f|^2+7\rho^2\delta^2|v|^2)$$

we obtain

$$\begin{aligned}&\int(|Dv|^2+\rho^2|g|^2|v|^2)/\max(|g|,\delta)\,dx\\ &\quad\leqq 3\int|f|^2/\max(|g|,\delta)\,dx+8\rho^2\delta\int|v|^2\,dx.\end{aligned}$$

We choose $\delta=\rho^{-2}$ and apply (27.3.6), which gives for large ρ

$$\rho^{2\varepsilon}\int|v|^2\,dx\leqq C\int|f|^2/\max(|g|,\rho^{-2})\,dx,\qquad v\in C_0^\infty(I). \tag{27.3.8$'$}$$

To prove (27.3.8) we observe that g is very close to a normalized polynomial so $|g|$ has a fixed lower bound on intervals of fixed length $\subset(\pm\frac12,\pm1)$. Thus we can choose a cutoff function $\phi\in C_0^\infty(I)$ such that $0\leqq\phi\leqq 1$, $\phi=1$ in $(-\frac12,\frac12)$ and $|g|>1/C_1$ in $\operatorname{supp}\phi'$. Since

$$(D+i\rho g)(\phi v)=\phi(D+i\rho g)v+vD\phi,\qquad v\in C^\infty(I),$$

and $|g|>1/C_1$ in $\operatorname{supp} D\phi$, we obtain (27.3.8) if (27.3.8)′ is applied to ϕv. The proof is complete.

Lemma 27.3.4. *Assume that G satisfies* (27.3.4) *in an interval I of length at least* 1, *that G has no sign change from* $+$ *to* $-$ *for increasing x, and that*

$$M_1(x)=\max_{j\leqq k}|G^{(j)}(x)|^{1/(j+1)}$$

is large for every $x\in I$. *Then we have with C independent of G, u and I*

$$\|M_1u\|\leqq C\|(D+iG)u\|,\qquad u\in C_0^\infty(I). \tag{27.3.9}$$

Proof. Let ρ be a large positive number which will be fixed later on in the proof, and set

$$M_\rho(x)=\max_{j\leqq k}|G^{(j)}(x)/\rho|^{1/(j+1)}.$$

We assume that $M_1(x)>\rho$ for every $x\in I$, which implies $M_\rho(x)>1$. Now consider

$$v(y)=u(x+y/M_\rho(x)),\qquad h(y)=f(x+y/M_\rho(x))$$

where $f=(D+iG)u$. Then

$$(D+iG(x+y/M_\rho(x))/M_\rho(x))\,v(y)=h(y)/M_\rho(x),$$

and an application of Lemma 27.3.3 gives

$$\rho^{2\varepsilon}\int_{I_x}|u(y)|^2 M_\rho(x)^3\,dy \le C\int_{J_x}(\rho^2|f(y)|^2+M_\rho(x)^2|u(y)|^2)\,M_\rho(x)\,dy.$$

Here $I_x=\{y;\,|x-y|\,M_\rho(x)<\frac{1}{2}\}$, $J_x=\{y;\,|x-y|\,M_\rho(x)<1\}$. If we apply (27.3.5)′ to $G(x+y/M_\rho(x))/M_\rho(x)$ it follows that the ratio $M_\rho(x)/M_\rho(y)$ has fixed upper and lower bounds when $y\in J_x$, so we may replace $M_\rho(x)$ by $M_\rho(y)$ in the integrands. If C is large enough then we decrease the left-hand side by integrating for $|x-y|\,M_\rho(y)<C^{-1}$ and increase the right-hand side by integrating there for $|x-y|\,M_\rho(y)<C$. After integration with respect to x also we obtain

$$\rho^{2\varepsilon}\|M_\rho u\|^2\le C(\rho^2\|f\|^2+\|M_\rho u\|^2).$$

When ρ is fixed so large that $\rho^{2\varepsilon}>2C$, we obtain (27.3.9). The proof is complete.

Proof of Proposition 27.3.1. First we shall prove that for large $|\xi'|$

$$\int|q(x,\xi')\,v(x_1)|^2\,dx_1+\int|\xi'|^{2/(k+1)}|v(x_1)|^2\,dx_1 \le C\int|D_1 v(x_1)+iq(x,\xi')\,v(x_1)|^2\,dx_1,\quad v\in C_0^\infty(\mathbb{R}). \tag{27.3.10}$$

To do so we rewrite (27.3.10) in terms of a new variable $t=x_1/A$,

$$\int|Aq(tA,x',\xi')\,v(t)|^2\,dt+\int|\xi'|^{2/(k+1)}|Av(t)|^2\,dt \le C\int|D_t v(t)+iAq(tA,x',\xi')\,v(t)|^2\,dt,\quad v\in C_0^\infty(\mathbb{R}). \tag{27.3.10'}$$

We apply Lemma 27.3.4 with $G(t)=Aq(tA,x',\xi')$ choosing A so that

$$C_0A^{k+2}|\xi'|=1\quad\text{ if } C_0=\sup|D_1^{k+1}q(x,\xi')|/|\xi'|.$$

Then (27.3.4) follows, and from (27.3.2) we obtain

$$M_1(t)=\max_{j\le k}|G^{(j)}(t)|^{1/(j+1)}\ge C_1A|\xi'|^{1/(k+1)}\gg 1$$

if $|\xi'|\gg 1$. For large $|\xi'|$ the estimate (27.3.10)′ follows from (27.3.9), for $|G|\le M_1$ also.

To localize the estimate (27.3.3) we first recall that

$$|\partial q(x,\xi')/\partial\xi'|^2|\xi'|+|\partial q(x,\xi')/\partial x|^2/|\xi'|\le Cq(x,\xi') \tag{27.3.11}$$

by Lemma 7.7.2 and the positivity of q. (This was used frequently in Chapter XXII.) Choose ε with $0<\varepsilon<1/(2k+2)$, and set

$$m(x,\xi')=(1+|\xi'|)^{2\varepsilon}+q(x,\xi'). \tag{27.3.12}$$

With respect to the metric of type $\frac{1}{2}+\varepsilon,\ \frac{1}{2}-\varepsilon$

$$g=|dx'|^2(1+|\xi'|)^{1-2\varepsilon}+|d\xi'|^2(1+|\xi'|)^{-1-2\varepsilon}$$

the weight m is uniformly g continuous and σ, g temperate. In fact, if $g_{x',\xi'}(y',\eta')<1$ then $|\eta'|<(1+|\xi'|)^{\frac{1}{2}+\varepsilon}$, $|y'|<(1+|\xi'|)^{\varepsilon-\frac{1}{2}}$ so it follows from (27.3.11) and Taylor's formula that for large $|\xi'|$

$$|q(x,\xi')-q(x_1,x'+y',\xi'+\eta')|<(8\,Cq)^{\frac{1}{2}}(1+|\xi'|)^{\varepsilon}+C'(1+|\xi'|)^{2\varepsilon}$$
$$\leqq C''m(x,\xi'),$$

and $(1+|\xi'+\eta'|)<2(1+|\xi'|)$. When $g_{x',\xi'}(y',\eta')>1$ then $g^{\sigma}_{x',\xi'}(y',\eta')>(1+|\xi'|)^{4\varepsilon}$ which makes (18.5.12) obvious.

We may assume that $Q(x,\xi')=q(x,\xi')$ when $|\xi'|>1$. Then we have for large ξ' since $Q\in S^1$ and (27.3.11) holds

$$|D^{\alpha}_{\xi}D^{\beta}_{x}Q(x,\xi')|\leqq C_{\alpha\beta}(1+|\xi'|)^{1-|\alpha|} \tag{27.3.13}$$
$$\leqq C'_{\alpha\beta}(1+|\xi'|)^{2\varepsilon+|\beta|(\frac{1}{2}-\varepsilon)-|\alpha|(\frac{1}{2}+\varepsilon)};\quad |\alpha+\beta|\geqq 2;$$

$$|D^{\alpha}_{\xi}D^{\beta}_{x}Q(x,\xi')|\leqq C_{\alpha\beta}\,q(x,\xi')^{\frac{1}{2}}(1+|\xi'|)^{(|\beta|-|\alpha|)/2} \tag{27.3.14}$$
$$\leqq C'_{\alpha\beta}\,q(x,\xi')^{\frac{1}{2}}(1+|\xi'|)^{\varepsilon+|\beta|(\frac{1}{2}-\varepsilon)-|\alpha|(\frac{1}{2}+\varepsilon)};\quad |\alpha+\beta|=1.$$

Hence $Q\in S(m,g)$.

Now choose a standard partition of unity $\{\chi_j\}$ for the g metric and $(x'_j,\xi'_j)\in\operatorname{supp}\chi_j$. Thus $\chi_j\in C_0^\infty(\mathbb{R}^{2n-2})$ is real valued, $\sum\chi_j^2=1$, $\{\chi_j\}$ is uniformly bounded in $S(1,g)$, and there is a fixed bound for the number of overlapping supports and for the g distance from (x'_j,ξ'_j) to points in $\operatorname{supp}\chi_j$. Then $\{\chi_j\}$ is a symbol in $S(1,g)$, with values in l^2. With

$$f=(D_1+iQ(x,D'))u,\qquad u\in\mathscr{S},$$

we have

$$\chi_j(x',D')f=(D_1+iq(x_1,x'_j,\xi'_j))\chi_j(x',D')u+R_j(x,D')u.$$

Here $\{R_j\}\in S((1+|\xi'|)^{\varepsilon}m^{\frac{1}{2}},g)$, for this is true for the error term in the symbol of the composition after the main term and also for the error $i(Q(x,\xi')-q(x_1,x'_j,\xi'_j))\chi_j(x',\xi')$. This follows from Taylor's formula since by (27.3.13), (27.3.14) the derivatives of Q although not Q itself have the bounds required in this symbol space. Hence

$$\sum\|(D_j+iq(x_1,x'_j,\xi'_j))\chi_j(x',D')u\|^2\leqq 2((\chi(x',D')f,f)+(R(x',D')u,u)) \tag{27.3.15}$$

where

$$\chi(x',D')=\sum\chi_j(x',D')^*\chi_j(x',D')=1+\psi(x',D'),\qquad \psi\in S^{-2\varepsilon}_{\frac{1}{2}+\varepsilon,\frac{1}{2}-\varepsilon},$$
$$R(x,D')=\sum R_j(x,D')^*R_j(x,D').$$

The symbol of R is bounded in $S((1+|\xi'|)^{2\varepsilon}m,g)$. (Here we have just summed for such j that ξ'_j is sufficiently large for the preceding estimates to hold.)

Using (27.3.10) we can estimate the left-hand side of (27.3.15) from below and obtain

$$\sum\|q(x_1,x'_j,\xi'_j)\chi_j(x',D')u\|^2+\sum(1+|\xi'_j|^2)^{1/(k+1)}\|\chi_j(x',D')u\|^2$$
$$\leqq C(\|f\|^2+(Ru,u)).$$

The proof of (27.3.15) allows us to switch back to Q in the left-hand side, for

$$(\chi(x',D')Qu,Qu)\leqq 2(\sum \|q(x_1,x'_j,\xi'_j)\chi_j(x',D')u\|^2+(Ru,u)).$$

By the logarithmic convexity in s of $\|g\|_{(s)}=\|(1+|D'|^2)^{s/2}u\|$ we have

$$\|g\|^2\leqq(\chi(x',D')g,g)+C\|g\|^2_{(-\varepsilon)}\leqq(\chi(x',D')g,g)+\tfrac{1}{2}\|g\|^2+C'\|g\|^2_{(-1)}.$$

Hence

$$\|Qu\|^2\leqq 2(\chi(x',D')Qu,Qu)+C''\|u\|^2.$$

We can write

$$R=(1+|D'|^2)^{\varepsilon}T_0(x,D')(Q(x,D')+(1+|D'|^2)^{\varepsilon})+T_1$$

where $T_j\in S(1,g)$. This follows from the standard parametrix construction since $(Q+(1+|\xi'|^2)^{\varepsilon})^{-1}\in S(1/m,g)$ for large $|\xi'|$ by Lemma 18.4.3. Hence

$$(Ru,u)\leqq C(\|Qu\|+\|u\|_{(2\varepsilon)})\|u\|_{(2\varepsilon)}.$$

Summing up the preceding estimates we have

$$\begin{aligned}\|Qu\|^2+\|u\|^2_{(1/(k+1))}&\leqq C(\|f\|^2+(\|Qu\|+\|u\|_{(2\varepsilon)})\|u\|_{(2\varepsilon)})\\&\leqq C\|f\|^2+\tfrac{1}{2}\|Qu\|^2+(C+\tfrac{1}{2}C^2)\|u\|^2_{(2\varepsilon)}.\end{aligned}$$

(27.3.3) follows if we estimate the last term by $\tfrac{1}{2}\|u\|^2_{(1/(k+1))}+C'\|u\|^2$. The proof of Proposition 27.3.1 is now complete.

Proof of Theorem 27.1.11 when P Satisfies Condition (*P*). As already observed at the beginning of this section we may assume that $\gamma_0=(0,\varepsilon_n)\in T^*(\mathbb{R}^n)$ and that the principal symbol of P is equal to $\xi_1+iq(x,\xi')$ in a conic neighborhood of γ_0 for some q satisfying the hypotheses of Proposition 27.3.1. Choose $\chi(\xi)\in C^\infty(\mathbb{R}^n)$ homogeneous of degree 0 when $|\xi|>1$ and equal to 0 when $|\xi'|<|\xi_1|$, equal to 1 when $|\xi_1|<\tfrac{1}{2}|\xi'|$ and $|\xi|>1$. Then (27.3.3) applied to $\chi(D)u$ shows by Lemma 27.1.5 that the operator $(D_1+iQ(x,D'))\chi(D)$ is subelliptic with loss of $k/(k+1)$ derivatives at $(0,\varepsilon_n)$. By Lemma 27.1.3 this is also true for P, which completes the proof.

The proof just given is applicable to more general operators, such as that in Proposition 26.3.4, for which there is a satisfactory substitute for (27.3.11). However, in general it is not possible to use such a simple localization argument and keep control of the localization error R in the proof of Proposition 27.3.1. This will require a carefully chosen localization in the x_1, x' and ξ' variables at the same time. It will occupy the next three sections.

27.4. Local Properties of the Symbol

Let $p(x,\xi)=\xi_1+ip_2(x,\xi')$, $\xi'=(\xi_2,\dots,\xi_n)$, be homogeneous and satisfy the conditions (i), (ii) in Theorem 27.1.11 at the characteristic point

$(x_0, \xi_0) \in T^*(\mathbb{R}^n) \setminus 0$. We may assume that $\delta = k/(k+1)$ where k is an integer. Since we shall have to use canonical transformations as in the proof of Proposition 27.2.2 it is useful to make a preliminary localization giving x' and ξ' symmetric roles. Thus we consider for small positive λ

$$p(x+x_0, \xi+\lambda^{-2}\xi_0) = \xi_1 + i\lambda^{-2} p_2(x+x_0, \lambda^2\xi' + \xi_0').$$

By the symplectic dilation $(x', \xi') \mapsto (\lambda x', \xi'/\lambda)$ we obtain the symbol $\xi_1 + iq(x, \xi')$ where

$$q(x, \xi') = \lambda^{-2} p_2(x_1 + x_{01}, \lambda x' + x_0', \lambda\xi' + \xi_0').$$

Assuming as we may that p_2 is smooth when $|x - x_0| \leqq 2$, $|\xi' - \xi_0'| \leqq 1$, we conclude that

$$|D_\xi^\alpha D_x^\beta q(x, \xi')| \leqq C_{\alpha\beta} \lambda^{|\alpha'+\beta'|-2}; \quad |x_1| < 1, \ |(x', \xi')| < 1/\lambda. \tag{27.4.1}$$

Here $C_{\alpha\beta}$ are fixed constants, λ is small but q and λ are allowed to vary. Sometimes we shall write $q_1 = \xi_1$ and $q_2 = q$ to have the symmetric notation of Section 27.2 available. Since $\bar{p}$ satisfies condition (Ψ) we have

(27.4.2) $q(x, \xi')$ does not change sign from $+$ to $-$ for increasing x_1.

Finally, condition (i) is by Corollary 27.2.4 equivalent to

$$\lambda^{-2} \leqq C \sum_{|I| \leqq k+1} |q_I(x, \xi)|. \tag{27.4.3}$$

Here it is of course advantageous to put $\xi_1 = 0$, that is, drop the term ξ_1 on the right-hand side. In what follows we shall study functions satisfying (27.4.1)–(27.4.3) with fixed constants, assuming that λ is small but disregarding the origin of q in the preceding discussion.

We shall introduce a precise measure for the size of the commutators, related to the definition (27.2.3) of $k(x, \xi)$

$$M_\rho(x, \xi') = \max_{|I| \leqq k+1} |q_I(x, \xi)/\rho|^{1/|I|}, \quad \xi_1 = 0. \tag{27.4.4}$$

Here ρ is a large parameter. The purpose of ρ is to guarantee a large constant in the left-hand side of an estimate which makes it possible to cancel some error terms on the right-hand side. (See the proof of Lemma 27.3.4 for a simple case of the argument.) We shall ultimately fix ρ large but independent of λ so that the desired estimates follow for small λ.

By (27.4.1) and (27.4.3) we have if $\lambda^2 \rho \leqq 1$

$$C_1(\lambda^{-2}/\rho)^{1/(k+1)} \leqq M_\rho(x, \xi') \leqq C_2 \lambda^{-2}/\rho. \tag{27.4.5}$$

Here and it what follows the constants are independent of λ and ρ. By (27.4.5) M_ρ would not change very much if k were increased in (27.4.4) so the definition of M_ρ is essentially independent of k if k is only large enough to guarantee (27.4.3).

The following analogue of Lemma 27.1.10 is useful when estimating repeated Poisson brackets.

Lemma 27.4.1. *If* $q=q_1+iq_2$ *and for some positive numbers* $A_1, \dots, A_n$, $B_1, \dots, B_n$ *we have*

$$|D_\xi^\alpha D_x^\beta q(\gamma)| \leqq C_{\alpha\beta}\rho K A^\alpha B^\beta, \tag{27.4.6}$$

and if $A_j B_j \rho \leqq 1$ *for every j, then*

$$|q_I(\gamma)| \leqq C_I \rho K^{|I|}, \tag{27.4.7}$$

where each constant only depends on a finite number of the constants $C_{\alpha\beta}$ *in* (27.4.6). *The terms in* q_I *involving some derivative with respect to* x_j, ξ_j *can be estimated by* $A_j B_j \rho C_I \rho K^{|I|}$.

Proof. By induction with respect to $|I|$ it follows that q_I has estimates of the form (27.4.6) with K replaced by $K^{|I|}$. In fact,

$$\{q_k, q_I\} = \sum (\partial q_k/\partial \xi_j \, \partial q_I/\partial x_j - \partial q_k/\partial x_j \, \partial q_I/\partial \xi_j),$$

and for $D_\xi^\alpha D_x^\beta$ applied to the general term we obtain the bound $C_{I\alpha\beta} A_j B_j \rho^2 K^{|I|+1} A^\alpha B^\beta$. Since $A_j B_j \rho \leqq 1$ the induction goes through and we can also pick out a factor $A_j B_j \rho$ as in the last statement. The proof is complete.

We shall analyze the behavior of q at an arbitrary point, but to simplify notation it is convenient to take it to be the origin. The definition of $M = M_\rho(0)$ means in particular that

$$|D_{x_1}^j q(0)| \leqq \rho M^{j+1}, \quad j \leqq k. \tag{27.4.8}$$

Where (27.4.1) is valid we have by (27.4.5)

$$|D_{x_1}^j q| \leqq C_j \rho M^{k+1} \tag{27.4.9}$$

which is a much better estimate when $j>k$.

If q were non-negative we could apply Lemma 7.7.2 to $q(x, \xi')$ for fixed $x_1 \in (-1/M, 1/M)$ and conclude that $|d_{x',\xi'} q(x_1, 0)| \leqq C(\rho M)^{\frac{1}{2}}$ which easily gives

$$|d_{x',\xi'} D_{x_1}^j q(0)| \leqq C M^j (\rho M)^{\frac{1}{2}}.$$

This corresponds to (27.3.11) with our present notation. However, as in Section 27.2 our discussion of q must be split into two major cases. The first is when a slightly weaker form of this estimate holds.

Let R be a large positive number such that

$$R^2 \leqq M\rho. \tag{27.4.10}$$

In view of (27.4.5) this is true with a wide margin for small λ if we take

$$R = \lambda^{-\kappa}, \quad 0 < \kappa < 1/(k+1). \tag{27.4.11}$$

This choice will be made in the following division in cases.

Case I. Assume first that

(27.4.12) $$|d_{x',\xi'} D^j_{x_1} q(0)| \le \rho R^{-1} M^{j+1}, \quad j \le k/2.$$

Since (27.4.1), (27.4.5) give

(27.4.13) $$|d_{x',\xi'} D^j_{x_1} q| \le C_j (M^{k+1}\rho)^{\frac{1}{2}}$$

where (27.4.1) holds, we have a better estimate when $j > k/2$. Using Taylor's formula and (27.4.8)–(27.4.13) we obtain

(27.4.14) $$|D^\alpha_\xi D^\beta_x q(x,\xi')| \le C'_{\alpha\beta} \rho M^{\beta_1+1} R^{-|\alpha'+\beta'|}, \quad \text{if } |x_1 M| < 1,\ |(x',\xi')| < R.$$

Proposition 27.4.2. *If* (27.4.12) *is valid then* (27.4.14) *follows and for small* λ

(27.4.15) $$M_\rho(x,\xi') \le C M_\rho(0) = C \max_{j \le k} |D^j_{x_1} q(0)/\rho|^{1/(j+1)},$$

if $|x_1| < 1/M$, $|(x',\xi')| < R$. *When* $|(x',\xi')|/R$ *is sufficiently small then*

(27.4.16) $$M_\rho(0) \le C' \max_{j \le k} |D^j_{x_1} q(x,\xi')/\rho|^{1/(j+1)} \le C' M_\rho(x,\xi'), \quad |M x_1| < 1.$$

Proof. That $M_\rho(x,\xi') \le CM$ follows from Lemma 27.4.1 and (27.4.14) if we take $K = M$, $A_1 = 1/M\rho$, $B_1 = M$, $A_j = B_j = 1/R$ when $j \ne 1$. The Poisson brackets containing some x_j, ξ_j derivative with $j \ne 1$ are smaller by a factor $O(\rho/R^2)$ so for small λ we conclude that the second part of (27.4.15) also holds. Thus

$$|D^j_{x_1} q(0)| \ge \rho M^{j+1}$$

for some $j \le k$. By Taylor's formula and (27.4.9) we conclude (cf. (27.3.5)′) that

$$1 \le C \max_{j \le k} |D^j_{x_1} q(x_1, 0)|/\rho M^{j+1}, \quad |x_1| < 1/M,$$

for $q(t/M, 0, 0)/\rho M$ is essentially a polynomial normalized at 0 and therefore essentially normalized at any other point in $(-1,1)$. From (27.4.14) it follows now that when $|x_1| < 1/M$

$$C \max_{j \le k} |D^j_{x_1} q(x,\xi')/\rho M^{j+1}| \ge 1 - C''|(x',\xi')|/R > \tfrac{1}{2}$$

if $|(x',\xi')|/R < \frac{1}{2}C''$. The proof is complete.

Choose $\phi \in C_0^\infty(-1,1)$ equal to 1 in $(-\frac{3}{4},\frac{3}{4})$, with $0 \le \phi \le 1$ everywhere, and set

(27.4.17) $$\phi_0(x,\xi') = \phi(Mx_1)\,\phi((|x'|^2 + |\xi'|^2)/R_1^2)$$

where

(27.4.18) $$R_1 = \lambda^{-\kappa_1}, \quad 0 < \kappa_1 < \kappa.$$

Then $R_1 \ll R$ for small λ so $M_\rho(x,\xi')$ is equivalent to $M_\rho(0)$ in

(27.4.19) $$\omega_0 = \{(x,\xi');\ |x_1| < 1/M,\ |x'|^2 + |\xi'|^2 < R_1^2\}$$

which contains supp ϕ_0. The function ϕ_0 will be used when we construct a partition of unity at the end of the section. Note that the size of ω_0 in the x', ξ' direction only depends on λ. We have in fact an element in the partition of unity in the proof of Proposition 27.3.1 combined with that occurring implicitly in the proof of Lemma 27.3.4. New problems will only occur in case II. Before embarking on that case we shall prove a lemma replacing Lemma 7.7.2 to some extent for functions satisfying (27.4.2).

Lemma 27.4.3. *Let $F(t, y)$ be a real valued C^2 function in $\Omega=\{(t,y)\in\mathbf{R}^{1+N};\ |t|<1,\ |y|<1\}$. Assume that $|F''_{yy}(t,y)|\leqq 1$ in Ω, that $|F(t,0)|\leqq 1$ when $|t|<1$, and that*

$$F(s,y)>0 \Rightarrow F(t,y)\geqq 0 \quad \text{if } (s,y)\in\Omega,\ (t,y)\in\Omega \text{ and } s<t. \tag{27.4.2)'}$$

Then there is a unit vector $\omega\in\mathbf{R}^N$ such that

$$|F'_y(t,0)-\omega\langle F'_y(t,0),\omega\rangle|\leqq 3, \qquad \langle F'_y(t,0),\omega\rangle\geqq -3 \quad \text{if } |t|<1. \tag{27.4.20}$$

For the proof we need an elementary geometrical lemma:

Lemma 27.4.4. *Let $Y_1, Y_2\in\mathbf{R}^N$ and let $a_1, a_2\in\mathbf{R}$. Assume that there is no $y\in\mathbf{R}^N$ with $|y|<1$ and $\langle y, Y_j\rangle>a_j$, $j=1,2$. If $|Y_1|>a_j$ for $j=1,2$ it follows that the distance from Y_2 to $\mathbf{R}_-Y_1$ is at most $(1+|Y_2|/|Y_1|)\max(a_1,a_2)$.*

Proof. By hypothesis the open unit ball and the quarter space defined by $\langle y, Y_j\rangle>a_j$, $j=1,2$, are disjoint. Hence there is a separating hyperplane. This means that for some $\lambda_j\geqq 0$ with $\lambda_1+\lambda_2=1$

$$\langle y, \lambda_1 Y_1+\lambda_2 Y_2\rangle>\lambda_1 a_1+\lambda_2 a_2 \Rightarrow |y|\geqq 1.$$

Thus $|\lambda_1 Y_1+\lambda_2 Y_2|\leqq\lambda_1 a_1+\lambda_2 a_2\leqq\max(a_1,a_2)=R$. There is nothing to prove unless $R<|Y_2|$. Since the segment between Y_1 and Y_2 intersects the ball $\{Y; |Y|\leqq R\}$, the area of the triangle $0, Y_1, Y_2$ is at most $|Y_1-Y_2|R/2$ which proves that the height from Y_2 to $\mathbf{R}Y_1$ is at most equal to $R|Y_1-Y_2|/|Y_1|\leqq R(1+|Y_2|/|Y_1|)$. This proves the statement when $\langle Y_1, Y_2\rangle\leqq 0$. When $\langle Y_1, Y_2\rangle\geqq 0$ the distance in questions is $|Y_2|$ so the asserted inequality is linear in $|Y_2|$ and valid for the smallest possible value R as well as for the largest possible value on the tangent cone from Y_1. The lemma is proved.

Proof of Lemma 27.4.3. By Taylor's formula it follows from (27.4.2)$'$ that when $-1<s<t<1$

$$\langle F'_y(s,0), y\rangle>\tfrac{3}{2}, \quad |y|<1 \Rightarrow \langle F'_y(t,0), y\rangle\geqq -\tfrac{3}{2}.$$

Lemma 27.4.4 can therefore be applied with $a_j=\frac{3}{2}$ and Y_1, Y_2 equal to $F'_y(s,0)$ and $-F'_y(t,0)$. If say $|F'_y(s,0)|>\frac{3}{2}$ and $|F'_y(s,0)|\geqq|F'_y(t,0)|$ we obtain (27.4.20) if ω is the unit vector in the direction $F'_y(s,0)$. Thus (27.4.20) follows if we let ω be a limit of the direction of $F'_y(s,0)$ when $|F'_y(s,0)|$ tends to its supremum M and $M>\frac{3}{2}$. Since (27.4.20) is trivial if $|F'_y(t,0)|\leqq\frac{3}{2}$ for every t, the proof is complete.

The unit vector ω in (27.4.20) is not unique. To estimate how it may be modified we shall need

Lemma 27.4.5. *If ω' is a unit vector in $\mathbb{R}^N$ then* (27.4.20) *implies*

(27.4.20)′ $\quad |F'_y(t,0)-\omega'\langle F'_y(t,0),\omega'\rangle|\leqq 3+4|F'_y(t,0)|\,|\omega'-\omega\langle\omega',\omega\rangle|.$

Proof. Set $\varepsilon=|\omega'-\omega\langle\omega,\omega'\rangle|$. Replacing ω' by $-\omega'$ if necessary we may assume that $\langle\omega',\omega\rangle\geqq 0$. Then $\varepsilon^2=1-\langle\omega',\omega\rangle^2\geqq 1-\langle\omega',\omega\rangle\geqq 0$, hence $|\omega-\omega'|\leqq 2\varepsilon$, which gives (27.4.20)′.

Remark. If $\omega'=F'_y(s,0)/|F'_y(s,0)|$ then it follows from (27.4.20) that $|\omega'-\omega\langle\omega',\omega\rangle|\leqq 3/|F'_y(s,0)|$ so the right-hand side of (27.4.20)′ becomes $3+12|F'_y(t,0)|/|F'_y(s,0)|$. Thus we only get a larger constant if ω is replaced by the direction of $F'_y(s,0)$ for some s for which the norm is not accidentally small.

Case II. Assume now that (27.4.12) fails, that is,

$$A_2=\max_{j\leqq k/2}|d_{x'\xi'}D^j_{x_1}q(0)|/(\rho M^{j+1})>R^{-1}. \tag{27.4.21}$$

Choose $s\leqq k/2$ so that the maximum is attained when $j=s$, that is, with $q^{(j)}=\partial^j q/\partial x_1^j$,

$$|d_{x'\xi'}q^{(s)}(0)|=a,\quad a=A_2\rho M^{s+1}\gg M^s(\rho M)^{\frac{1}{2}}, \tag{27.4.22}$$

$$|d_{x'\xi'}q^{(j)}(0)|\leqq aM^{j-s},\quad j\leqq k/2. \tag{27.4.23}$$

The situation is now analogous to that in the proof of Proposition 27.2.2. Then we introduced $\partial^s p_2(0,x',\xi')/\partial x_1^s$ as a new ξ_2 variable before using an argument related to Lemma 27.4.3 to prove vanishing of a certain weight. We shall use a similar approach here to prove estimates of the form (27.4.6). However, since the canonical transformation used will have to be implemented analytically later on, we want it to be of a very special and explicit form and shall therefore just choose it so that the x_2 axis becomes an orbit of the Hamilton field of $q^{(s)}(0,x',\xi')$. This was actually enough also in the proof of Proposition 27.2.2.

With $\psi(x',\xi')=q^{(s)}(0,x',\xi')$ we have by (27.4.22) and (27.4.1)

$$|\psi'(0,0)|=a,\quad |\psi''|\leqq C.$$

To study the orbit through 0 of the Hamilton field it is convenient to introduce $\Psi(y,\eta)=\psi(ay,a\eta)/a^2$; $(y,\eta)\in T^*(\mathbb{R}^{n-1})$. Then we have $|\Psi'(0,0)|=1$, and $|\Psi''(y,\eta)|\leqq C$ in a fixed ball with center at the origin. (Note that $a\lambda$

is uniformly bounded.) Assume for example that $|\partial\Psi(0,0)/\partial\eta_2|>\frac{1}{2}n$; this can be achieved by a change of labels for the coordinates and if necessary composition with the symplectic map $(y,\eta)\mapsto(\eta,-y)$. Then the Hamilton equations

$$dy/dt=\partial\Psi/\partial\eta, \quad d\eta/dt=-\partial\Psi/\partial y; \qquad y=\eta=0 \text{ when } t=0;$$

have a solution for small t, and $|(y,\eta)|\leq R$ if $|dR/dt|=1+CR$ and $R(0)=0$, that is, $R=(e^{C|t|}-1)/C$. When $|t|$ is smaller than some constant it follows that the solution exists and that $1/(4n)<dy_2/dt<2$, so y_2 can be taken as a new variable and the orbit is defined by the equations

$$y_3=f_3(y_2),\ldots,y_n=f_n(y_2), \quad \eta_2=g_2(y_2),\ldots,\eta_n=g_n(y_2),$$

where $f_3,\ldots,g_n$ belong to a bounded set in $C^\infty([-c,c])$ for some constant $c>0$. Now we return to the original variables and define

$$\begin{aligned}\chi(x',\xi')=(x_2,x_3+af_3(x_2/a),\ldots,\\ \xi_2+ag_2(x_2/a)+G(x',\xi'),\ldots,\xi_n+ag_n(x_2/a))\end{aligned} \tag{27.4.24}$$

where the additional term G given by

$$G(x',\xi')=\sum_3^n(g_j'(x_2/a)x_j-f_j'(x_2/a)\xi_j) \tag{27.4.25}$$

is included to make χ symplectic. For χ we have uniform estimates

$$|D^\alpha\chi(x',\xi')|\leq C_\alpha a^{1-|\alpha|} \quad \text{if } |(x',\xi')|<ca. \tag{27.4.26}$$

Since $H_\Psi=\Psi_2(y,\eta)\partial/\partial y_2+\ldots$ where $\Psi_2=\partial\Psi/\partial\eta_2$, we have

$$H_\psi=a\Psi_2(x'/a,\xi'/a)\partial/\partial x_2+\ldots$$

and conclude that the Hamilton field of $\psi\circ\chi$ is $a\Psi_2(\chi(x',\xi')/a)\partial/\partial x_2$ on the x_2 axis which is an orbit since χ maps it to an orbit of H_ψ. An estimate of the form (27.4.26) is of course also valid for the inverse of χ.

Now we transform q by χ, introducing

$$Q(x,\xi')=q(x_1,\chi(x',\xi')); \quad |Mx_1|<1, \; |(x',\xi')|<ca. \tag{27.4.27}$$

The symplectic invariance of Poisson brackets and the proof of (27.2.17) give

$$H^j_{\psi\circ\chi}(\partial/\partial x_1)^i Q(0)=H^j_\psi(\partial/\partial x_1)^i q(0)=((\operatorname{ad}\partial/\partial x_1)^s H_q)^j(\partial/\partial x_1)^i q(0).$$

By (27.4.4) and (27.4.5) the right-hand side can be estimated by a constant times $\rho M^{(s+1)j+i+1}$. If $c_1(x_2)=\Psi_2(\chi(x_2,0)/a)$ we have here $H_{\psi\circ\chi}=ac_1(x_2)\partial/\partial x_2$, hence $a\partial/\partial x_2=c_1(x_2)^{-1}H_{\psi\circ\chi}$. Since there are uniform bounds for $(a\partial/\partial x_2)^j c_1(x_2)^{-1}$ for any j, it follows that

$$|a^j(\partial/\partial x_2)^j(\partial/\partial x_1)^i Q(0)|\leq C_{ij}M^{j(s+1)}\rho M^{i+1}. \tag{27.4.28}$$

With the notation

$$B_2=M^{s+1}/a=1/\rho A_2 \tag{27.4.29}$$

it follows that

$$|(\partial/\partial x_2)^j(\partial/\partial x_1)^i Q(0)| \leqq C_{ij}\rho M^{i+1}B_2^j. \tag{27.4.28$'$}$$

This is the main benefit derived from the change of variables. For later reference we also record that $\operatorname{mod}\partial/\partial\xi_1$

$$H_{Q^{(s)}}(0,x_2,0)=ac_1(x_2)\partial/\partial x_2, \quad C^{-1}\leqq|c_1|\leqq C, \tag{27.4.30}$$

which implies that $Q^{(s)}(0,x_2,0)=Q^{(s)}(0)$.

Next we must examine to what extent the basic estimates (27.4.1) have survived the change of variables. We claim that

$$|D_{\xi'}^{\alpha'}D_x^{\beta'}Q(x,\xi')|\leqq C'_{\alpha\beta}M^{\beta_1}a^{2-|\alpha'+\beta'|} \tag{27.4.31}$$

if $\alpha'+\beta'\neq 0$ and $|x_1 M|<1$, $|(x',\xi')|<ca$. For the proof we observe that when $D_{\xi'}^{\alpha'}D_{x'}^{\beta'}$ acts on $Q^{(\beta_1)}(x,\xi')=q^{(\beta_1)}(x,\chi(x',\xi'))$ the terms where $q^{(\beta_1)}$ is differentiated $i\geqq 2$ times can be estimated by

$$C\lambda^{i-2}a^{i-|\alpha'+\beta'|}\leqq C'a^{2-|\alpha'+\beta'|}$$

by (27.4.1) and (27.4.26). The terms with $i=1$ can be estimated by

$$C(|d_{x'\xi'}q^{(\beta_1)}(x_1,0)|+a)a^{1-|\alpha'+\beta'|}$$

since second order derivatives of $q^{(\beta_1)}(x,\xi')$ with respect to (x',ξ') are bounded. By (27.4.23) and (27.4.13) we have

$$|d_{x'\xi'}q^{(\beta_1)}(x_1,0)|\leqq aC_{\beta_1}M^{\beta_1-s}\leqq aC_{\beta_1}M^{\beta_1}, \quad \text{if } |x_1 M|<1,$$

which proves (27.4.31). A better bound is obtained for large β_1 if we note that a^2 and $ad_{x'\xi'}q^{(\beta_1)}$ can be estimated by $C/\lambda^2\leqq C_1 M^{k+1}\rho$, which gives

$$|D_\xi^\alpha D_x^\beta Q(x,\xi')|\leqq C'_{\alpha\beta}\rho M^{k+1}a^{-|\alpha'+\beta'|}, \tag{27.4.32}$$

$|x_1 M|<1$, $|(x',\xi')|<ca$; this is also true if $\alpha'+\beta'=0$. In particular (27.4.32) improves (27.4.28) when $j(s+1)+i>k$ and shows in view of (27.4.28)′ that $Q(x_1/M, x_2/B_2, 0)/\rho M$ is close to a polynomial with uniformly bounded coefficients and weight $\leqq k$ with the basic weights (27.2.14). We shall now prove that Lemma 27.4.3 gives an essential improvement of the preceding estimates. The notation $x''=(x_3,\dots,x_n)$ will then be used.

Lemma 27.4.6. *There are positive constants $C'_{\alpha\beta}$ such that for small λ*

$$\begin{aligned}|D_\xi^\alpha D_x^\beta(Q(x,\xi')-\xi_2\partial Q(x_1,x_2,0)/\partial\xi_2)|\\ \leqq C'_{\alpha\beta}\rho M^{\beta_1+1}b_2^{\beta_2}(M\rho)^{-|\alpha'+\beta''|/2}\end{aligned} \tag{27.4.33}$$

when $|x_1 M|<1$ and $|(x'',\xi')|<(\rho M)^{\frac12}$, $|x_2|<1/b_2=\max(1/B_2,(\rho M)^{\frac12})$.

Proof. Assume first that $1/b_2=(\rho M)^{\frac12}$ and set with $y=(x',\xi')$

$$F(t,y)=cQ(t/M, x'(\rho M)^{\frac12}, \xi'(\rho M)^{\frac12})/\rho M, \quad |t|<1, \ |y|<1.$$

If c is a sufficiently small constant it follows from (27.4.8), (27.4.9) that $|F(t,0)|\leqq 1$ and from (27.4.31) that $|F''_{yy}|\leqq 1$. Thus we can choose a unit vector ω in $\mathbb{R}^{2n-2}$ such that

$$|F'_y(t,0)-\omega\langle F'_y(t,0),\omega\rangle|\leqq 3.$$

By (27.4.13) we have a uniform bound for $\partial^j F'_y(t,0)/\partial t^j$ if $j\geqq k/2$, so $F'_y(t,0)$ differs from some polynomial of degree $k/2$ by a fixed amount. Since a bound for a polynomial implies a bound for its derivatives, we obtain

$$|\partial^{s+1}F(0,0)/\partial y\,\partial t^s-\omega\langle\partial^{s+1}F(0,0)/\partial y\,\partial t^s,\omega\rangle|\leqq C.$$

Here $|\partial^{s+1}F(0,0)/\partial y\,\partial t^s|=caM^{-s}(\rho M)^{-\frac12}$ by (27.4.22), and it follows from (27.4.23) and (27.4.13) that $|\partial F(t,0)/\partial y|\leqq CaM^{-s}(\rho M)^{-\frac12}$. Hence we conclude as in the remark after Lemma 27.4.5 that we may replace ω by the direction $\partial/\partial\xi_2$ of $\partial^{s+1}F(0,0)/\partial y\,\partial t^s$. Thus we have a uniform bound for $\partial G(t,y)/\partial y$ when $y=0$, if

$$G(t,y)=F(t,y)-\xi_2\partial F(t,0)/\partial\xi_2.$$

Since $\partial^2 G/\partial y^2=\partial^2 F/\partial y^2$ has norm $\leqq 1$ it follows that we have a fixed bound for $\partial G(t,y)/\partial y$ when $|t|<1$, $|y|<1$, hence also one for $G(t,y)$ since $|G(t,0)|=|F(t,0)|\leqq 1$. For t derivatives of G or $\partial G/\partial y$ of order $\geqq k$ we also have uniform bounds from (27.4.9) and (27.4.13), so all t derivatives of G and $\partial G/\partial y$ have uniform bounds. For all other derivatives of G a bound follows from (27.4.31) since $(M\rho)^{\frac12}\leqq a$. This completes the proof of (27.4.33) when $1/b_2=(\rho M)^{\frac12}$.

Next we assume that $b_2=B_2<(\rho M)^{-\frac12}$ and define

$$F(t,x_2,y)=cQ(t/M,x_2/B_2,x''(\rho M)^{\frac12},\xi'(\rho M)^{\frac12})/\rho M;$$

$|t|<1$, $|x_2|<1$, $|y|<1$; where $y=(x'',\xi')$. We regard x_2 as a parameter at first. F is well defined since

$$(\rho M)^{\frac12}<B_2^{-1}=a/M^{s+1}\ll a.$$

As already pointed out it follows from (27.4.28)′ and (27.4.32) that $F(t,x_2,0)$ is close to a polynomial of weight k and uniformly bounded coefficients. In particular $|F(t,x_2,0)|\leqq 1$ if $|t|<1$, $|x_2|<1$ when c is chosen small enough, and then we also have $|F''_{yy}|<1$ for $|t|<1$, $|x_2|<1$, $|y|<1$. Moreover, (27.4.23), (27.4.13) and (27.4.31) give

$$|F'_y(t,x_2,0)|\leqq Ca/(M^s(\rho M)^{\frac12})$$

since $1/B_2=a/M^{s+1}$. On the other hand, when $t=0$ we have

$$d_y\partial^s F(t,x_2,0)/\partial t^s=ac_1(x_2)/(M^s(\rho M)^{\frac12})\,d\xi_2$$

by (27.4.30). If we apply Lemmas 27.4.3 and 27.4.5 as in the first part of the proof it follows that there is a uniform bound for

$$G(t,x_2,y)=F(t,x_2,y)-\xi_2\partial F(t,x_2,0)/\partial\xi_2$$

when $|t|<1$, $|x_2|<1$, $|y|<1$. From (27.4.31) and (27.4.32) we obtain a uniform bound for all derivatives of sufficiently high order so all derivatives of G have uniform bounds. This completes the proof of (27.4.33).

For the coefficient of ξ_2 in (27.4.33) we have by (27.4.22), (27.4.23) and (27.4.13)

$$|D_{x_1}^j \partial Q(x_1,0)/\partial\xi_2| \leqq C_j \rho M^{j+1} A_2, \quad |x_1 M|<1; \tag{27.4.34}$$

the estimate is precise when $j=s$ and $x_1=0$ but can be improved when $j>k/2$. Thus $(\rho M A_2)^{-1}\partial Q(x_1/M,0)/\partial\xi_2$ is essentially a normalized polynomial in x_1 of degree $\leqq k/2$. From (27.4.31) we obtain

$$\begin{aligned}|D_x^\beta \partial(Q(x_1,0)-Q(x_1,x_2,0))/\partial\xi_2|\\ \leqq C_\beta \rho A_2 M^{\beta_1+1} b_2^{\beta_2}(R/(\rho M)^{\frac12})^{\beta_2}; \quad |x_1 M|<1,\ |x_2 b_2|<1.\end{aligned} \tag{27.4.35}$$

In fact, the estimate for $\beta_2=0$ follows from the estimate for $\beta_2=1$ with an additional factor $R/(\rho M)^{\frac12}$ bounded by a positive power of λ. By (23.4.21) we have

$$1/\rho A_2 M b_2 \leqq \max(1/M, R/(\rho M)^{\frac12}) = R/(\rho M)^{\frac12}$$

which proves (27.4.35) when $\beta_2=1$. Since $b_2 R/(\rho M)^{\frac12} \geqq 1/\rho M A_2 \geqq 1/a$ we obtain (27.4.35) from (27.4.31) also if $\beta_2>1$. Thus we conclude that $\partial Q(x_1/M, x_2/b_2, 0)/\partial\xi_2/(\rho M A_2)$ is almost independent of x_2 when $|x_2|<1$.

From (27.4.33), (27.4.34), (27.4.35) we obtain the estimate

$$|D_\xi^\alpha D_x^\beta Q(x,\xi')| \leqq C''_{\alpha\beta}\rho(1+|A_2\xi_2|) M^{\beta_1+1} b_2^{\beta_2} A_2^{\alpha_2}(\rho M)^{-|\alpha''+\beta''|/2} \tag{27.4.36}$$

when $|x_1 M|<1$, $|x_2 b_2|<1$, $|(x'',\xi')|<c(\rho M)^{\frac12}$. It is of the form (27.4.6) with $K=M(1+|A_2\xi_2|)$, B_2 replaced by b_2, and $A_j=B_j=(\rho M)^{-\frac12}$ when $j>2$. By the symplectic invariance of Poisson brackets the function $M_\rho^Q(x,\xi')$ defined by (27.4.4) with q replaced by Q is the composition $M_\rho(x_1,\chi(x',\xi'))$, and from Lemma 27.4.1 we obtain

$$M_\rho^Q(x,\xi') \leqq CM(1+|A_2\xi_2|), \tag{27.4.37}$$

for the same (x,ξ') as in (27.4.36). When $A_2\xi_2$ is large it is easy to get a similar lower bound. In fact, by (27.4.33) we have

$$\rho M_\rho^Q(x,\xi')^{j+1} \geqq |Q^{(j)}(x,\xi')| \geqq |\xi_2 \partial Q^{(j)}(x_1,x_2,0)/\partial\xi_2| - C\rho M^{j+1}$$

when $j\leqq k$. This implies that

$$\max_{j\leqq k} |(\xi_2/\rho)\partial Q^{(j)}(x_1,x_2,0)/\partial\xi_2|^{1/(j+1)} \leqq M_\rho^Q(x,\xi') + CM. \tag{27.4.38}$$

For some $j\leqq k/2$ we have

$$|\partial Q^{(j)}(x_1,x_2,0)/\partial\xi_2|/\rho \geqq c_1 M^{j+1} A_2$$

so the left-hand side of (27.4.38) is bounded from below by

$$2c_2 M|A_2\xi_2|^{2/(k+2)} \quad \text{if } |A_2\xi_2|>1.$$

Hence we obtain for some N

$$(27.4.39)\quad c_2 M|A_2\xi_2|^{2/(k+2)} \le \max_{j\le k} |(\xi_2/\rho)\partial Q^{(j)}(x_1,x_2,0)/\partial\xi_2|^{1/(j+1)} \le M_\rho^Q(x,\xi'), \quad \text{if } |A_2\xi_2|>N,$$

provided of course that $|x_1 M|<1$, $|x_2 b_2|<1$, $|(x'',\xi')|<c(\rho M)^{\frac12}$. When $M_\rho^Q(x,\xi')\le\frac12 M$ we obtain a bound for $|A_2\xi_2|$ and conclude that $(x_1,\chi(x',\xi'))$ belongs to case II if $|(x'',\xi'')|\ll(\rho M)^{\frac12}$.

From (27.4.39) it follows easily that $(x_1,\chi(x',\xi'))$ belongs to case I if $|\xi_2|\gg R$ and λ is small. In fact, by (27.4.33), (27.4.31) and (27.4.39) we have, since $RA_2>1$ by (27.4.21),

$$\begin{aligned}|d_{x',\xi'}Q^{(j)}(x,\xi')| &\le |\partial Q^{(j)}(x_1,x_2,0)/\partial\xi_2| \\ &\quad + |\xi_2\partial^2 Q^{(j)}(x_1,x_2,0)/\partial\xi_2\partial x_2 + CM^j(\rho M)^{\frac12} \\ &\le M_\rho^Q(x,\xi')^{j+1}\rho/|\xi_2| + C_1 M^j(\rho M)^{\frac12} \ll M_\rho^Q(x,\xi')^{j+1}/R\end{aligned}$$

if $|\xi_2|\gg R$. Since the first order derivatives of χ^{-1} are uniformly bounded the same conclusion holds for the differential of $q^{(j)}$. From Proposition 27.4.2 it follows, since $M\ll M_\rho^Q(x,\xi')$ and

$$|Q^{(j)}(x,\xi')| \le |\xi_2\partial Q^{(j)}(x_1,x_2,0)/\partial\xi_2| + C\rho M^{j+1},$$

that for another C

$$(27.4.40)\qquad M_\rho^Q(x,\xi') \le C\max_{j\le k}|(\xi_2/\rho)\partial Q^{(j)}(x_1,x_2,0)/\partial\xi_2|^{1/(j+1)}$$

if $|x_1 M|<1$, $|x_2 b_2|<1$, $|(x'',\xi')|<(\rho M)^{\frac12}$ and $|\xi_2|\gg R$.

To prove a lower bound for $M_\rho^Q(x,\xi')$ when $|A_2\xi_2|$ is not large requires the following lemma.

Lemma 27.4.7. *Let m_1 and m_2 be fixed positive integers, $m_2\le m_1$, and let $F_0(x_1,x_2)$, $G_0(x_1)$ be real valued polynomials of weight less than m_1, m_2 respectively if x_1 and x_2 are given the weights 1 and m_2. Set $L_1=\xi_1$, $L_2=F_0(x)+G_0(x_1)\xi_2$ and define $L_I=\{L_{i_1},\{L_{i_2},\dots\}\}$. Then $L_I=0$ when $|I|>m_1$. Assume that*

$$(27.4.41)\qquad \max|G_0^{(j)}(0)|\le C$$

for some fixed C, and that for some s we have

$$(27.4.42)\qquad |G_0^{(s)}(0)|\ge 1/C.$$

If $E_0=G_0\partial F_0/\partial x_1 - F_0\partial G_0/\partial x_1$ and $H_0(x_2)=(\partial^s F_0(0,x_2)/\partial x_1^s)/G_0^{(s)}(0)$, it follows, with the notation $\tilde E_0(x)=\sum|D^\alpha E_0(x)|$, that for some C_1,N

$$(27.4.43)\qquad \begin{aligned}C_1^{-1}(1+|x_1|)^{-N}(|\xi_2+H_0(x_2)|+\tilde E_0(x)) &\le \max_I|L_I(x,\xi)| \\ &\le C_1(1+|x_1|)^N(|\xi_2+H_0(x_2)|+\tilde E_0(x)) \quad \text{if } \xi_1=0,\end{aligned}$$

$$(27.4.43)'\qquad \begin{aligned}C_1^{-1}(1+|x_1|)^{-N}\tilde E_0(x) &\le \sum|D^\alpha(F_0(x)-G_0(x_1)H_0(x_2))| \\ &\le C_1(1+|x_1|)^N\tilde E_0(x).\end{aligned}$$

Proof. The Poisson brackets $\{L_1, \{L_1, \dots, \{L_1, L_2\}\}\}$ are of the form

$$F_0^{(j)} + G_0^{(j)}\xi_2 \tag{27.4.44}$$

where $F_0^{(j)} = \partial^j F_0/\partial x_1^j$. They vanish when $j \geqq m_1$. The Poisson bracket with L_2 is

$$G_0 \partial F_0^{(j)}/\partial x_2 - G_0^{(j)} \partial F_0/\partial x_2. \tag{27.4.45}$$

Taking Poisson brackets with L_2 is later on equivalent to applying $G_0\partial/\partial x_2$. The weight of (27.4.45) is $<m_1-1-j$. Every Poisson bracket with L_1 or L_2 decreases the weight so L_I is of weight at most $m_1-|I|$, which implies $L_I=0$ when $|I|>m_1$.

Taking $\xi_2+H_0(x_2)$ as a new variable instead of ξ_2 gives a symplectic transformation and replaces F_0 by $F_0(x)-G_0(x_1)H_0(x_2)$, of order $\leqq m_1+m_2$. Thus we may assume in the proof of (27.4.43), (27.4.43)' that $\partial^s F_0(0,x_2)/\partial x_1^s$ and $H_0(x_2)$ vanish identically.

Since

$$E_0 = G_0\{L_1, L_2\} - G_0' L_2$$

and the Hamilton field of $H_{L_1}^j L_2$ is $G_0^{(j)}\partial/\partial x_2 \mod \partial/\partial\xi$, we have

$$(G_0^{(j)})^{\beta_2}(\partial/\partial x)^\beta E_0 = ((\operatorname{ad} H_{L_1})^j H_{L_2})^{\beta_2} H_{L_1}^{\beta_1}(G_0 H_{L_1} L_2 - G_0' L_2).$$

The right-hand side is a linear combination of the Poisson brackets L_I with coefficients which are polynomials in x_1 with fixed bounds for the coefficients. By (27.4.42) and Taylor's formula we have (cf. (10.1.8))

$$\max_j |G_0^{(j)}(x_1)| \geqq C(1+|x_1|)^{-m_2},$$

so we obtain the estimate

$$\tilde{E}_0(x) \leqq C(1+|x_1|)^N \max_I |L_I(x,\xi)|.$$

We also have

$$\begin{aligned}|\xi_2| &\leqq C|F_0^{(s)}(0,x_2)+G_0^{(s)}(0)\xi_2| \\ &\leqq C_1(1+|x_1|)^{m_2}\sum |F_0^{(j)}(x)+G_0^{(j)}(x_1)\xi_2|\end{aligned}$$

which completes the proof of the first inequality (27.4.43).

To prove the second inequality it suffices to prove the second inequality (27.4.43)' for all the Poisson brackets are linear in F_0 and ξ_2. Let V be the finite dimensional vector space of polynomials f of degree $<m_1$ in one variable t, such that $f^{(s)}(0)=0$, and set $Tf=w$ where $w(t)=G_0(t)\,df(t)/dt - f(t)\,dG_0(t)/dt$. This is an injective linear transformation with values in the vector space W of polynomials of degree $<m_1+m_2-1$, for if $w=0$ it follows that $f=cG_0$ for some constant c, hence $0=f^{(s)}(0)=cG_0^{(s)}(0)\neq 0$. Thus $\|f\|/\|Tf\|$ is bounded. The maximum is a continuous function of G_0 so we have a bound independent of G_0 when (27.4.41), (27.4.42) hold with a fixed C. Taking $f=F_0(.,x_2)$ we obtain

$$\sum |D_{x_1}^j F_0(0,x_2)| \leqq C\sum |D_{x_1}^j E_0(0,x_2)|,$$

which implies the second part of (27.4.43)′ first when $x_1=0$ and then for general x_1 by Taylor's formula (cf. (10.1.8)). The first part of (27.4.3)′ is obvious which completes the proof.

Before summing up the local properties of M^Q_ρ we introduce another distance

$$R_2=\lambda^{-\kappa_2}, \quad \kappa<\kappa_2<1/(k+1). \tag{27.4.46}$$

Thus $R_1 \ll R \ll R_2 \ll (\rho M)^{\frac{1}{2}}$ where $\ll$ always means that the quotient is bounded by a positive power of λ. Set

$$1/B'_2=\max(1/B_2, R_2)\leqq \max(1/B_2, (\rho M)^{\frac{1}{2}})=1/b_2.$$

Proposition 27.4.8. *There exists a positive constant Υ such that either*

$$M^Q_\rho(x,\xi')\geqq \Upsilon M \quad \textit{when } |x_1 M|<1,\ |x_2 B'_2|<1,\ |(x'',\xi'')|<R_2, \tag{27.4.47}$$

or else, for $\Xi=Q^{(s)}(0)/(\partial Q^{(s)}(0)/\partial\xi_2)$,

$$M^Q_\rho(x,\xi')<\tfrac{1}{2}M \quad \textit{when } \xi_2=-\Xi, \quad |x_1 M|<1,\ |x_2 B'_2|<1,\ |(x'',\xi'')|<R_2. \tag{27.4.48}$$

Here $|\Xi|\leqq C/A_2\leqq CR\ll R_2$.

Proof. Choose N so that (27.4.39) gives $M^Q_\rho\geqq M$ when $|A_2\xi_2|\geqq N$. In what follows we assume $|A_2\xi_2|<N$. From (27.4.33) it follows by Taylor's formula that

$$Q(x_1/M, x_2/B_2, \xi_2 B_2, x'', \xi'')/M-F(x_1,x_2)-G(x_1,x_2)\xi_2$$

is $O(\lambda^\delta)$ for some $\delta>0$ when $|x_1|<1$, $|x_2|<1$, $|\xi_2|<\rho N$, $|(x'',\xi'')|<R_2$, where

$$F(x_1,x_2)=Q(x_1/M, x_2/B_2, 0)/M,$$
$$G(x_1,x_2)=B_2 M^{-1}\partial Q(x_1/M, x_2/B_2, 0)/\partial\xi_2.$$

The same is true for the derivatives and remains valid by (27.4.32), (27.4.35) if we replace F and G by the Taylor expansions F_0 and G_0 of $F(x_1,x_2)$ and $G(x_1,0)$ of order k and $k/2$ respectively. Thus F_0 and G_0 satisfy the hypotheses of Lemma 27.4.7 with $H_0=\Xi/B_2=\rho A_2\Xi$ constant by a comment after (27.4.30). For small λ we have

$$\max|L_I(0,0)|/\rho\geqq\tfrac{1}{2}$$

hence

$$|H_0|+\tilde{E}_0(0)\geqq C_2\rho$$

with the notation in (27.4.43). Two cases can now occur:

(i) If $\tilde{E}_0(0)<\varepsilon\rho$ it follows from (27.4.43) that

$$|L_I(x,\xi)|/\rho\leqq C_3\varepsilon \quad \text{when } \xi_2=-H_0 \text{ and } |x_1|<1,\ |x_2|<1.$$

If $C_3\varepsilon\leqq\frac{1}{4}$ this gives

$$|Q_I(x,\xi')/\rho|<(M/3)^{|I|} \quad \text{when } |I|\leqq k+1$$

provided that $|x_1 M|<1$, $|x_2 B_2|<1$, $|(x'', \xi'')|<R_2$ and $\xi_2=-\Xi$. If $B_2'<B_2$ we have $B_2'=1/R_2$, and since $R_2 \ll (\rho M)^{\frac{1}{2}}$ it follows from (27.4.36) and the proof of Lemma 27.4.1 that

$$|Q_I(x, \xi')/\rho|<(\tfrac{1}{2}M)^{|I|} \quad \text{when } |I|\leqq k+1,$$
$$|x_1 M|<1,\ |x_2 B_2'|<1,\ -\xi_2=\Xi \text{ and } |(x'', \xi'')|<R_2,$$

for $R_2 |\partial Q_I(x, \xi')/\partial \xi_2|/M^{|I|}=O(R_2 b_2) \ll 1$. This proves (27.4.48).

(ii) Assume now that $\tilde{E}_0(0) \geqq \rho/3C_3$. Then it follows from (27.4.43) that

$$\max_I |L_I(x, \xi)|/\rho \geqq C_4, \quad |x_1|<1,\ |x_2|<1,$$

for $\tilde{E}_0(0)/\tilde{E}_0(x)$ is bounded then. Going back to M_ρ^Q as in the proof of (27.4.48) we obtain (27.4.47). The proof is complete.

Points for which (27.4.48) occurs must be avoided in our localization so we introduce

Definition 27.4.9. If (27.4.47) is valid we shall say that

$$\Omega_0=\{(x_1, \chi(x', \xi')); |x_1 M|<1, |x_2 B_2'|<1, |\xi_2|<R_2, |(x'', \xi'')|<R_2\} \tag{27.4.49}$$

is an admissible neighborhood of type II with (0, 0) as center.

Every point belonging to case II is close to the center of an admissible neighborhood of type II:

Proposition 27.4.10. *There exists a constant C_a such that for every (x, ξ') in case II with $|x_1|<\frac{1}{2}$, $|(x', \xi')|<\frac{1}{2}\lambda^{-1}$ one can find (y', η') with $|(x'-y', \xi'-\eta')|\leqq C_a R$ such that (x_1, y', η') is the center of an admissible neighborhood.*

Proof. Let C_a be larger than the constant C in Proposition 27.4.8. If (x, ξ') is not itself the center of an admissible neighborhood, it follows from Proposition 27.4.8 that the minimum $\underline{M}$ of $M_\rho(x_1, y', \xi')$ when $|(x'-y', \xi'-\eta')|\leqq C_a R$ is $<\frac{1}{2}M_\rho(x, \xi')$. Choose (y', η') so that $M_\rho(x_1, y', \eta')=\underline{M}$. From the discussion after (27.4.39) we know that (x_1, y', η') is also in case II. To simplify notation we assume that (x_1, y', η') is the origin and change coordinates as above. Since $(0, x', \xi')$ belongs to case II the ξ_2 coordinate δ of $\chi^{-1}(x', \xi')$ is bounded by $C'R$, again by the discussion after (27.4.39). If (27.4.47) does not hold, with M replaced by $\underline{M}$, it follows from (27.4.48) and the definition of $\underline{M}$ that the distance $|\delta+\Xi|$ from $\chi^{-1}(x', \xi')$ to the plane $\xi_2=-\Xi$ is at least $C_a R/C''$ where C'' is a Lipschitz constant for χ. Since $|\Xi|\leqq CR$ we obtain a contradiction if $C_a>(C+C')C''$, which completes the proof.

For the admissible neighborhood Ω_0 in (27.4.49) we define $\Phi_0 \in C_0^\infty(\Omega_0)$ by

$$\Phi_0(x_1, \chi(x', \xi'))=\phi(Mx_1)\phi(B_2' x_2)\phi(\xi_2/R_2)\phi((|x''|^2+|\xi''|^2)/R_2^2), \tag{27.4.50}$$

where ϕ is the function used in (27.4.17) for case I. The estimates which we have proved in Ω_0 remain valid with some other constants in the "double" of Ω_0 or more generally the neighborhood obtained if $B_2'^{-1}$ and R_2 are multiplied by fixed constants in the definition. It follows that for every $(x, \xi') \in \Omega_0$ the minimum of $M_\rho(x_1, y', \eta')$ for $|y'-x'|^2+|\eta'-\xi'|^2 < CR_2^2$ is bounded from above and below by constants times $M_\rho(0)$ and is attained only at points mapped by χ^{-1} to the set where $|A_2\eta_2|<N$ for some fixed N. The admissible neighborhoods can be obtained essentially by taking through any such point an interval of parameter length ≈ 1 on the orbit of the Hamilton field of $q^{(s)}(x_1, x', \xi')/M^{s+1}$ for fixed x_1, with s chosen so that the differential is not accidentally small, and then taking the R_2 neighborhood in the x', ξ' variables and the $1/M$ neighborhood in the x_1 variable. It follows from (27.4.33) that (x'', ξ'') varies very little on such orbits. Hence it follows that if two admissible neighborhoods overlap, then there is a fixed bound for the values of M_ρ at their centers, and one is contained in the other enlarged by a fixed factor.

It is now a matter of routine to construct an appropriate covering of the set

$$\Omega=\{(x, \xi'); |x_1|<\tfrac{1}{2}, |(x', \xi')|<\tfrac{1}{2}\lambda^{-1}\}.$$

(Recall that we started from (27.4.1) in a set twice as large.) First we choose a (necessarily finite) maximal sequence of admissible neighborhoods $\Omega_1, \Omega_2, \dots$ of type II with centers in Ω such that the center of Ω_k is not in the inner half of Ω_j if $j<k$. For each of them we choose $\Phi_j \in C_0^\infty(\Omega_j)$ just as we defined Φ_0 by (27.4.50). Since overlapping neighborhoods have quite similar shape and become disjoint if they are shrunk by a fixed factor, it is clear that there is a fixed upper bound for the number of Ω_j which have a point in common (cf. the proof of Lemma 1.4.9). For later use we also define a function $\psi_j \in C_0^\infty$ with support in the set where $\Phi_j=1$ by writing for example

$$\psi_0(x_1, \chi(x', \xi'))=\Phi_0(4x_1/3, \chi(4x'/3, 4\xi'/3)) \tag{27.4.51}$$

when the center is at 0 and χ is the corresponding canonical transformation. Then $\psi_j=1$ at all points at distance $<cR_2$ in the $x'\xi'$ variables from the inner half of Ω_j, because $\frac{9}{16}>\frac{1}{2}$. It follows that the set

$$\omega=\{(x, \xi') \in \Omega;\ \psi_j(x, \xi')<1 \text{ for every } j\}$$

contains only points belonging to case I, for every point where case II occurs is at distance $O(R)$ in the $x'\xi'$ variables from the center of an admissible neighborhood of type II. We can therefore cover ω by standard neighborhoods ω_j of type I with corresponding $\phi_j \in C_0^\infty(\omega_j)$ chosen according to (27.4.17) so that only a fixed number of ω_j can overlap and

$$\omega \subset \{(x, \xi');\ \phi_j(x, \xi')=1 \text{ for some } j\}.$$

If we write

$$\Psi_j(x, \xi')=\psi_j(x, \xi')\prod_{i<j}(1-\psi_i(x, \xi')) \tag{27.4.52}$$

then $\operatorname{supp}\Psi_j$ is contained in the set where $\Phi_j=1$, and

(27.4.53) $$\Psi(x,\xi')=\sum \Psi_j(x,\xi')=1-\prod(1-\psi_j(x,\xi'))$$

is equal to 1 in $\Omega\setminus\omega$. The intersection of Ω with the support of $d\Psi$ is covered by the sets where $\Phi_j=1$ as well as by the sets where $\phi_k=1$, which will make it possible to use Ψ in Section 27.6 to separate the study of case I from the study of case II. We shall then need some information on the properties of Φ_j and Ψ_j as symbols with respect to the metric

(27.4.54) $$g_2=(|dx'|^2+|d\xi'|^2)/R_2^2.$$

We denote by $M_{\rho j}$ or M_j for short the value of M_ρ at the "center" of Ω_j.

Lemma 27.4.11. *Φ_j, Ψ_j, $M_j^{-1}(D_{x_1}+H_q)\Phi_j$ and $M_j^{-1}(D_{x_1}+H_q)\Psi_j$ are bounded in $S(1,g_2)$ uniformly with respect to j, x_1 and ρ for small λ.*

Proof. To simplify notation we assume that $j=0$ and that Φ_0, Ψ_0 are given by (27.4.50), (27.4.51). Since $a>M/R\gg R_2$ it follows from (27.4.26) and the similar estimate for χ^{-1} that it suffices to prove the statement for the composition with χ in the (x',ξ') variables. Since $B_2'\leq 1/R_2$ it is obvious that $\Phi_0(x_1,\chi(x',\xi'))$ and $\Psi_0(x_1,\chi(x',\xi'))$ are bounded in $S(1,g_2)$. The pullback of $M_0^{-1}(D_{x_1}+H_q)\Phi_0$ by χ is $M^{-1}(D_{x_1}+H_Q)\Phi_0(x_1,\chi(x',\xi'))$ where we have written $M=M_\rho(0,0)$ as before. It is obvious that we have a bound for $M^{-1}D_{x_1}\Phi_0(x_1,\chi(x',\xi'))$ in $S(1,g_2)$. For

$$M^{-1}\{Q(x,\xi')-\xi_2\partial Q(x_1,x_2,0)/\partial\xi_2,\ \Phi_0(x_1,\chi(x',\xi'))\}$$

it follows from (27.4.33) that we even have a bound in $S(\rho/R_2(M\rho)^{\frac{1}{k}},g_2)$. It remains to examine

$$-M^{-1}\partial Q(x_1,x_2,0)/\partial\xi_2\,\partial\Phi_0(x_1,\chi(x',\xi'))/\partial x_2$$
$$+M^{-1}\xi_2\{\partial Q(x_1,x_2,0)/\partial\xi_2,\ \Phi_0(x_1,\chi(x',\xi'))\}.$$

The first term is bounded in $S(\rho A_2B_2',g_2)$ and the second one is bounded in $S(R_2\rho A_2b_2/R_2,g_2)$ by (27.4.34) and (27.4.35). Since $b_2\leq B_2'\leq B_2$ and $\rho A_2B_2=1$, we obtain a bound in $S(1,g_2)$. The same is true for $\psi_0(x_1,\chi(x',\xi'))$.

To pass from ψ_j to Ψ_j we observe that in (27.4.52) all but a fixed number of factors $1-\psi_i$ are equal to 1 in a neighborhood of $\operatorname{supp}\psi_j$. Hence Ψ_j is bounded in $S(1,g_2)$ since all ψ_j are. We have

$$M_j^{-1}(D_{x_1}+H_q)\Psi_j=M_j^{-1}(D_{x_1}+H_q)\psi_j\prod_{i<j}(1-\psi_i(x,\xi'))$$
$$-\sum_{k<j}(M_k/M_j)\psi_j\prod_{k\neq i<j}(1-\psi_i(x,\xi'))M_k^{-1}(D_{x_1}+H_q)\psi_k.$$

There is a uniform bound for M_k/M_j when $\operatorname{supp}\psi_k\cap\operatorname{supp}\psi_j\neq\emptyset$, which completes the proof.

For ϕ_j we have analogous results with the metric

$$g_1 = (|dx'|^2 + |d\xi'|^2)/R_1^2. \tag{27.4.55}$$

We denote by m_j the value of M_ρ at the center of ω_j.

Lemma 27.4.12. *ϕ_j and $m_j^{-1}(D_{x_1} + H_q)\phi_j$ are bounded in $S(1, g_1)$ uniformly with respect to j, x_1 and ρ when λ is small.*

Proof. This is an immediate consequence of (27.4.14) since $\rho < R/R_1$ for small λ.

Lemmas 27.4.11 and 27.4.12 contain all the information we need in Section 27.6 to localize the estimates to be proved for $D_1 + iq(x, D')$. To study the localized operators in case II we must extract further information from (27.4.2) on the approximating differential operator

$$D_1 + i(Q(x_1, x_2, 0) + \partial Q(x_1, x_2, 0)/\partial \xi_2 D_2).$$

This requires a supplement to Lemma 27.4.3.

Lemma 27.4.13. *Let $F \in C^\infty(\Omega)$, $\Omega = \{(t, y) \in \mathbb{R}^2;\ |t| < 1,\ |y| < 1\}$, and assume that for some fixed positive integer k*

$$|\partial^2 F/\partial y^2| \leqq 1, \quad |\partial^3 F/\partial t\, \partial y^2| \leqq 1 \quad \text{in } \Omega, \tag{27.4.56}$$

$$|F_0(t)| \leqq 1, \quad |d^k F_1(t)/dt^k| \leqq 1, \quad -1 < t < 1, \tag{27.4.57}$$

where $F_j(t) = \partial^j F(t, 0)/\partial y^j$. In addition we assume that (27.4.2)′ holds. If $N = \sup_{|t|<1} F_1(t)$ is large positive, it follows with constants C_j depending only on k that

$$|F_0(s)| \leqq C_1 F_1(s)/N + C_1^2/2N^2, \quad |s| < \tfrac{1}{2}, \tag{27.4.58}$$

$$F_0(t)/F_1(t) - F_0(s)/F_1(s) \leqq C_2(1/F_1(s) + 1/F_1(t))/N^2, \tag{27.4.59}$$
$$\text{if } -\tfrac{1}{2} < t < s < \tfrac{1}{2} \text{ and } N \min(F_1(s), F_1(t)) > C_0,$$

$$F_1(s)\, d(F_0(s)/F_1(s))/ds \geqq -C_2(1 + |F_1'(s)/F_1(s)|)/N^2, \tag{27.4.60}$$
$$\text{if } |s| < \tfrac{1}{2} \text{ and } NF_1(s) > C_0.$$

If $G(s) = F_1(s) + C_1/N$ then $G(s) \geqq 0$ when $|s| < \frac{1}{2}$ and

$$F_0(t)/G(t) - F_0(s)/G(s) \leqq C_3(1/G(s) + 1/G(t))/N^2, \tag{27.4.59$'$}$$
$$\text{if } -\tfrac{1}{2} < t < s < \tfrac{1}{2} \text{ and } N \min(G(s), G(t)) > C_0 + 2C_1,$$

$$G(s)\, d(F_0(s)/G(s))/ds \geqq -C_3(1 + |G'(s)/G(s)|)/N^2 \tag{27.4.60$'$}$$
$$\text{if } |s| < \tfrac{1}{2} \text{ and } NG(s) > C_0 + 2C_1.$$

Proof. By Lemma 27.4.3 we know already that $F_1(t) \geqq -3$. For the Taylor polynomial T of order $k-1$ for F_1 at 0 we have $|T - F_1| \leqq 1/k!$, hence the maximum of T in $(-1,1)$ is at least $N/2$ if $N \geqq 1$, which implies with a positive constant $c(k)$ that

$$\max_{-1<t<\frac{1}{2}} |T(t)| \geqq 2c(k)N > 3$$

if $N > \frac{3}{2}c(k)^{-1}$ too. We can therefore choose $t \in (-1, -\frac{1}{2})$ so that $F_1(t) > c(k)N$. By Taylor's formula $|F(t,y) - F_0(t) - F_1(t)y| \leqq \frac{1}{2}$ so $F(t,y)$ has the sign of y if $c(k)N|y| > \frac{3}{2}$. We can find $t \in (\frac{1}{2}, 1)$ with the same property and conclude from (27.4.2)′ that $F(s,y)$ has the sign of y if $|s| < \frac{1}{2}$ and $|y| > 3/(2Nc(k))$. Hence we obtain (27.4.58) with $C_1 = 3/(2c(k))$. In particular it follows that $F_1(s) > -C_1/2N$ and that $|F_0(s)| \leqq 2C_1F_1(s)/N$ if $F_1(s) > C_1/N$.

If $|s| < \frac{1}{2}$ and $K = F_1(s)N > 2C_1$, the solution y of the linearized equation $F_0(s) + F_1(s)y = 0$ satisfies $|y| < 2C_1/N < \frac{1}{2}$ if $N > 4C_1$. When $|z| < |y|$ we have

$$F(s, y+z) = F_1(s)(z+R)$$

where

$$|R| \leqq (y+z)^2/2F_1(s) < 2y^2N/K < 4C_1|y|/K.$$

Thus $z+R$ has opposite signs for $z = \pm 4C_1 y/K$. If $K > 4C_1 = C_0$ we can therefore choose z with $|z| < 4C_1|y|/K < |y|$ so that $F(s, y+z) < 0$. Then it follows from (27.4.2)′ that $F(t, y+z) \leqq 0$ for $t \leqq s$, so

$$F_0(t) + F_1(t)(-F_0(s)/F_1(s) + z) \leqq (y+z)^2/2 \leqq 8C_1^2/N^2,$$

and $|z| < 4C_1|y|/K < 8C_1^2/NK$. This proves (27.4.59) with $C_2 = 8C_1^2$. In the preceding argument we could have chosen z with $|z| < 4C_1|y|/K$ so that $F(s, y+z) = 0$, and then it follows from (27.4.2)′ that

$$0 \leqq \partial F(s, y+z)/\partial s \leqq F_0'(s) + F_1'(s)(y+z) + 2y^2$$

which gives

$$\begin{aligned} F_0'(s) - F_1'(s)F_0(s)/F_1(s) &\geqq -|F_1'(s)|\,|z| - 2y^2 \\ &\geqq -4C_1|F_1'(s)/F_1(s)|\,|F_0(s)|/K - 8C_1^2/N^2. \end{aligned}$$

Thus we have proved (27.4.60).

That $G \geqq 0$ has already been observed. To derive (27.4.59)′ from (27.4.59) we first note that $NG(t) \geqq C_0 + 2C_1$ implies $NF_1(t) \geqq C_0 + C_1$ and

$$\begin{aligned} |F_0(t)(1/G(t) - 1/F_1(t))| &\leqq C_1|F_0(t)|/NF_1(t)G(t) \leqq 2C_1^2/G(t)N^2, \\ G(t)/F_1(t) &\leqq 1 + C_1/NF_1(t) \leqq 1 + C_1/(C_0 + C_1). \end{aligned}$$

This gives (27.4.59)′ with $C_3 = 2C_1^2 + 2(C_0 + 2C_1)C_2/(C_0 + C_1)$. Since

$$\begin{aligned} |F_0(t)G'(t)(1/F_1(t) - 1/G(t))| &\leqq C_1|G'(t)/G(t)|\,|F_0(t)/NF_1(t)| \\ &\leqq 2C_1^2|G'(t)/G(t)|/N^2 \end{aligned}$$

we obtain (27.4.60)′ in the same way, which completes the proof.

We can apply Lemma 27.4.13 to

$$F(t, y, x_2) = c_0 Q(2t/M, x_2, 0, y(\rho M)^{\frac{1}{2}}, 0)/\rho M$$

where x_2 is considered as a parameter, $|x_2 b_2|<1$. For a suitable positive constant c_0 it follows from (27.4.33) that (27.4.56) holds, and (27.4.57) is a consequence of (27.4.32), (27.4.33). The number N in Lemma 27.4.13 is equivalent to $A_2(\rho M)^{\frac{1}{2}}$ when $x_2=0$, by the definition of A_2 and Taylor's formula. Since $R \ll (\rho M)^{\frac{1}{2}}$ it follows from (27.4.35) that this remains true when $|x_2 b_2|<1$. Now we change the normalizations in Lemma 27.4.13 by introducing for $|x_1|<1$ and $|x_2 b_2|<B_2$

$$\begin{aligned} F_0(x_1, x_2) &= \rho F(x_1/2, 0, x_2/B_2)/c_0 = Q(x_1/M, x_2/B_2, 0)/M,\\ F_1(x_1, x_2) &= \partial F(x_1/2, 0, x_2/B_2)/\partial y (c_0 A_2(\rho M)^{\frac{1}{2}})^{-1}\\ &= (B_2/M)\partial Q(x_1/M, x_2/B_2, 0)/\partial \xi_2. \end{aligned}$$

This means that when $|x_1 M|<1$ and $|x_2 b_2|<1$ we have

$$\begin{aligned} &Q(x_1, x_2, 0) + \xi_2 \partial Q(x_1, x_2, 0)/\partial \xi_2\\ &\quad = M(F_0(Mx_1, B_2 x_2) + (\xi_2/B_2) F_1(Mx_1, B_2 x_2)). \end{aligned} \tag{27.4.61}$$

The symbol $\xi_1 + iM(F_0(Mx_1, B_2x_2) + (\xi_2/B_2) F_1(Mx_1, B_2x_2))$ is changed to $M(\xi_1 + i(F_0(x_1, x_2) + \xi_2 F_1(x_1, x_2)))$ by a symplectic dilation, which just means a change of scales for the corresponding first order differential operators. Writing

$$G(x_1, x_2) = F_1(x_1, x_2) + C/(A_2^2 \rho M) \tag{27.4.62}$$

for a suitable C we have $G \geqq 0$ and when $|x_1|<1$, $|x_2 b_2|<B_2$

$$|D^\alpha F_0| \leqq \rho C_\alpha, \tag{27.4.63}$$

$$|D^\alpha G| \leqq C_\alpha/\rho \quad \text{if } \alpha_1 > k \text{ or } \alpha_2 \neq 0, \tag{27.4.64}$$

$$\tfrac{1}{2} \leqq \max_{j \leqq k} |\partial^j G(x)/\partial x_1^j| \leqq 2, \tag{27.4.65}$$

$$\begin{aligned} &F_0(x_1, x_2)/G(x_1, x_2) - F_0(y_1, x_2)/G(y_1, x_2) > -1\\ &\quad \text{if } -1<y_1<x_1<1 \text{ and } \min(G(x_1, x_2), G(y_1, x_2)) > \rho^{-k-1}, \end{aligned} \tag{27.4.66}$$

$$\partial(F_0/G)/\partial x_1 > -1 \quad \text{if } G(x) > \rho^{-k-1}. \tag{27.4.67}$$

In fact, the bounds (27.4.59)′, (27.4.60)′ are applicable when $\min(G(x_1, x_2), G(y_1, x_2)) \geqq C/A_2^2 \rho M$ and they give a lower bound $O(1/GA_2^2 M)$ which is much more than required in (27.4.66), (27.4.67) since $1/A_2^2 M \leqq R^2/M \ll 1$ for small λ. The estimate (27.4.63) follows from (27.4.33), and (27.4.64), (27.4.65) follow if we use (27.4.35).

Assume now that $(0,0)$ is the center of an admissible neighborhood and write $L_1=\xi_1$, $L_2=F_0(x_1, x_2)+G(x_1, x_2)\xi_2$. Part (ii) of the proof of Proposition 27.4.8 is now applicable and gives in view of the symplectic invariance of Poisson brackets that

$$\max_{|I| \leqq k+1} |L_I(x, \xi)/\rho| \geqq c > 0 \quad \text{if } |x_1|<1,\ |x_2 B_2'|<B_2. \tag{27.4.68}$$

In fact, when $|\xi_2| \gg \rho$ we just have to use a Poisson bracket of the form $F_0^{(J)} + \xi_2 G^{(J)}$, and when $|\xi_2/\rho|$ is bounded the term $C/(A_2^2 \rho M)$ in G is irrelevant since it is $\ll 1$.

The preceding conditions are valid in a rectangle in $\mathbb{R}^2$ with sides at least equal to 1. When proving estimates for the inverse of $L_1(D)+iL_2(x,D)$ in Section 27.5 we shall first only consider a square, for the passage to the general rectangles just requires an additional partition of unity in x_2.

27.5. Local Subelliptic Estimates

In Section 27.4 we found that a subelliptic operator is microlocally well approximated by a differential operator in one or two variables (cases I and II respectively). For the ordinary differential operators which occur we can use Lemma 27.3.4. In this section we shall prove an analogue for the other case. By $\| \ \|$ we denote the L^2 norm.

Proposition 27.5.1. *Assume that $F, G \in C^\infty(\Omega)$, $\Omega = \{x \in \mathbb{R}^2; |x_1|<1, |x_2|<1\}$, that F is real valued and G non-negative, and that for some fixed C_α*

$$|D^\alpha F| \leqq C_\alpha \rho, \quad \forall \alpha, \tag{27.5.1}$$

$$|D^\alpha G| \leqq C_\alpha/\rho \quad \text{if } \alpha_1 > k \text{ or } \alpha_2 \neq 0, \tag{27.5.2}$$

$$\max_{j \leqq k} |\partial^j G(0)/\partial x_1^j| = 1, \tag{27.5.3}$$

where k is a fixed positive integer. Also assume that with $L_1 = \xi_1$ and $L_2 = F(x) + iG(x)\xi_2$

$$\max_{|I| \leqq k+1} |L_I(x,\xi)/\rho|^{1/|I|} \geqq 1 \quad \text{if } x \in \Omega,\ \xi \in \mathbb{R}, \tag{27.5.4}$$

$$\partial(F/G)/\partial x_1 > -1 \quad \text{when } G(x) > \rho^{-k-1}, \tag{27.5.5}$$

$$F(x_1,x_2)/G(x_1,x_2) - F(y_1,x_2)/G(y_1,x_2) > -1 \quad \text{if } -1<y_1<x_1<1 \text{ and } \min(G(x_1,x_2), G(y_1,x_2)) > \rho^{-k-1}. \tag{27.5.6}$$

If K is a compact subset of Ω it follows for large ρ that

$$\rho^{1/(k+1)} \|u\| + \|D_1 u\| \leqq C_K \|(D_1 + i(F(x)+G(x)D_2))u\|, \quad u \in C_0^\infty(K). \tag{27.5.7}$$

It is of course sufficient to assume (27.5.1), (27.5.2) for finitely many α but our proof gives no good bound for the required number. The conditions (27.5.5), (27.5.6) are of course inherited from the condition $(\bar{\Psi})$. The proof will follow from a series of lemmas. The first is essentially a case of Proposition 27.3.1 which will give control of the part of u which has high frequency with respect to x_2.

Lemma 27.5.2. *Let $0 \leqq G \in C^\infty(\Omega)$ and assume that for all α*

$$|D^\alpha G(x)| \leqq C_\alpha, \quad x \in \Omega, \tag{27.5.8}$$

$$\max_{j \leqq k} |D_1^j G(x)| \geqq 1, \quad x \in \Omega, \tag{27.5.9}$$

where k is also fixed. If K is a compact subset of Ω it follows that

$$\begin{aligned} &\|D_1 v\| + \|G D_2 v\| + \| \, |D_2|^{1/(k+1)} v\| \\ &\leqq C_K(\|(D_1 + iG(x) D_2) v\| + \|v\|), \quad v \in C_0^\infty(K). \end{aligned} \tag{27.5.10}$$

Proof. We can change G outside K so that the hypotheses are fulfilled in $\mathbb{R}^2$. Then we no longer need to restrict the support in the x_2 direction but we shall assume that $|x_1| < 1$ there. A repetition of the proof of (27.3.10) which we leave for the reader shows that

$$\begin{aligned} &\|D_1 v\| + \| \, |\xi_2|^{1/(k+1)} v\| \\ &\leqq C \|(D_1 + iG(x_1, x_2)\xi_2) v\|, \quad v \in C_0^\infty(-1, 1), \end{aligned} \tag{27.5.11}$$

provided that $\xi_2 \in \mathbb{R}$ and that $|\xi_2|$ is large. Next we observe that Lemma 7.7.2 gives

$$|\partial G / \partial x_2|^2 \leqq CG. \tag{27.5.12}$$

Now choose $\varepsilon \in (0, 1/(2k+2))$ and $\chi_j \in C_0^\infty(\mathbb{R})$ with $\sum \chi_j^2 = 1$ so that

$$\chi_j \in S(1, d\xi_2^2/(1+|\xi_2|)^{1+2\varepsilon}),$$

at most two supports overlap, and $|\xi_2 - t_j| < C t_j^{\frac{1}{2}+\varepsilon}$ for some t_j when $\xi_2 \in \operatorname{supp} \chi_j$. Then

$$\begin{aligned} (D_1 + iGt_j)\chi_j(D_2) v = \chi_j(D_2) f + i t_j [G, \chi_j(D_2)] \, v \\ + i(t_j - D_2)\chi_j(D_2) G v + \chi_j(D_2) \, \partial G / \partial x_2 \, v. \end{aligned}$$

Here $f = (D_1 + iG(x) D_2) v$. With $\|g\|_{(s)} = \|(1 + D_2^2)^{s/2} g\|$ we have

$$\sum \|\chi_j(D_2) f\|^2 = \|f\|^2, \qquad \sum \|\chi_j(D_2)(\partial_2 G) v\|^2 \leqq C\|v\|^2,$$
$$\sum \|(t_j - D_2)\chi_j(D_2) G v\|^2 \leqq C \|Gv\|^2_{(\frac{1}{2}+\varepsilon)}.$$

Moreover, $t_j[G, \chi_j(D_2)] = i t_j \chi_j'(D_2) \partial_2 G + \psi_j(x, D_2)$ where $\{\psi_j\}$ is a symbol in $S^0_{\frac{1}{2}+\varepsilon, \frac{1}{2}-\varepsilon}$ with values in l^2 which is uniformly bounded with respect to x_1. Hence

$$\sum \|t_j [G, \chi_j(D_2)] \, v\|^2 \leqq C(\|(\partial_2 G) v\|^2_{(\frac{1}{2}-\varepsilon)} + \|v\|^2).$$

The calculus of pseudodifferential operators of type 1, 0 gives

$$|((1 + D_2^2)^{\frac{1}{2}-\varepsilon}(\partial_2 G) v, (\partial_2 G) v) - ((\partial_2 G)^2 D_2 v, D_2(1 + D_2^2)^{-\frac{1}{2}-\varepsilon} v)| \leqq C\|v\|^2$$

so we obtain using (27.5.12)

$$\sum \|(D_1 + iGt_j)\chi_j(D_2) v\|^2 \leqq C(\|f\|^2 + \|Gv\|^2_{(\frac{1}{2}+\varepsilon)} + \|v\|^2 + \|G D_2 v\| \, \|v\|).$$

If (27.5.11) is applied to all terms with large j it follows that

$$\|D_1 v\|^2+\|v\|^2_{(1/(k+1))}+\|Gv\|^2_{(1)}\leq C(\|f\|^2+\|Gv\|^2_{(\frac{1}{2}+\varepsilon)}+\|v\|^2+\|Gv\|_{(1)}\|v\|)$$

since $iGD_2v=f-D_1v$ and $iD_2Gv=iGD_2v+(\partial_2G)v$. The right-hand side can be estimated by

$$C\|f\|^2+\tfrac{1}{2}\|Gv\|^2_{(1)}+C'\|v\|^2$$

which proves (27.5.10) after cancellation of $\frac{1}{2}\|Gv\|^2_{(1)}$.

Next we prove a lemma parallel to Lemma 27.3.2, restricting ourselves to polynomials.

Lemma 27.5.3. *Let m_1 and m_2 be fixed positive numbers, $m_2\leq m_1$, and let $F_0(x_1,x_2)$, $G_0(x_1)$ be real valued polynomials of weight less than m_1, m_2 respectively if x_1 and x_2 are given the weights 1 and m_2. If $L_1=\xi_1$ and $L_2=F_0(x)+G_0(x_1)\xi_2$, it follows that the Poisson brackets L_I vanish when $|I|>m_1$. If*

$$\max|G_0^{(j)}(0)|=1, \tag{27.5.13}$$

$$\min_\xi \max_I |L_I(0,\xi)|=\rho, \tag{27.5.14}$$

and ρ is large enough, it follows that with $\varepsilon=2^{1-m_1}$

$$\rho^\varepsilon\|u\|\leq C(\|D_1u\|+\|(F_0(x)+G_0(x_1)D_2)u\|),\quad u\in C_0^\infty(\Omega). \tag{27.5.15}$$

Proof. From (27.4.43) and (27.4.43)′ it follows that the coefficients of $F_0(x)-G_0(x_1)H_0(x_2)$ are $\leq C\rho$ for an appropriate polynomial H_0 such that the difference is of weight $\leq m_1+m_2$. Replacing $u(x)$ by $u(x)\cdot\exp(-i\int H_0(x_2)\,dx_2)$ in (27.5.15) will change $F_0(x)$ to $F_0(x)-G_0(x_1)H_0(x_2)$. For the symbols of L_I the effect is precisely that ξ_2+H_0 is introduced as a new symplectic coordinate instead of ξ_2, so the Poisson brackets are invariant. Thus we may assume in the proof that $\tilde{F}_0(0)\leq C\rho$. By (27.5.14) we have with the notation of Lemma 27.4.7

$$\rho\leq C\tilde{E}_0(0)\leq C_1\tilde{E}_0(x),\quad x\in 2\Omega. \tag{27.5.16}$$

Choose $\chi\in C_0^\infty(\mathbb{R})$ equal to 1 in $(-1,1)$ and set $P_1=D_1$, $P_2=\chi(x_2)(F_0(x)+G_0(x_1)D_2)$. The symbol of P_2 is uniformly bounded in $S(\rho+|\xi_2|,g)$ when $|x_1|<1$ if

$$g=|dx_2|^2+|d\xi_2|^2/(\rho+|\xi_2|)^2.$$

Regarding x_1 as a parameter and using the notation

$$\|v\|_{(s)}=\|(\rho^2+D_2^2)^{s/2}v\|,\quad v\in\mathscr{S},$$

we shall prove along the lines of Lemma 22.2.3 and Lemma 27.3.2 that for the commutators $P_I=[P_{i_1},[P_{i_2},\ldots]]\in \operatorname{Op}S(\rho+|\xi_2|,g)$ we have when $|x_1|<1$ in supp u

$$\|P_Iu\|_{(2^{1-|I|}-1)}\leq C(\|P_1u\|+\|P_2u\|+\|u\|). \tag{27.5.17}$$

This is trivial when $|I|=1$ and we shall prove (27.5.17) in general by induction with respect to $|I|$. Assuming (27.5.17) we shall thus estimate $[P_j, P_I]u$, $j=1,2$. To do so we set $\delta=2^{-|I|}$ and

$$A=(\rho^2+D_2^2)^{\delta-1}[P_j, P_I]\in \operatorname{Op}S((\rho+|\xi_2|)^{2\delta-1}, g).$$

Then

$$\begin{aligned}\|[P_j,P_I]u\|_{(\delta-1)}^2 &= ([P_j,P_I]u,Au)\\ &= (P_Iu,AP_j^*u)+(P_Iu,[P_j^*,A]u)\pm(P_j^*u,AP_Iu)\pm(P_j^*u,[P_I,A]u)\\ &\leqq C(\|P_Iu\|_{(2\delta-1)}(\|P_ju\|+\|u\|)+\|P_ju\|(\|P_Iu\|_{(2\delta-1)}+\|u\|_{(2\delta-1)}))\end{aligned}$$

which completes the inductive proof of (27.5.17).

From (27.5.16) it follows that

$$\rho\leqq C_2\sum(E_0^{(\alpha)}(x)/\rho)E_0^{(\alpha)}(x).$$

Here $E_0^{(\alpha)}/\rho$ is bounded in C^∞ and $E_0^{(\alpha)}$ is a linear combination with coefficients bounded in C^∞ of the commutators P_I, by the proof of Lemma 27.4.7. Hence we can write

$$\rho=\sum c_I(x)P_I, \quad x\in\Omega, \tag{27.5.18}$$

where c_I is in a bounded set in C_0^∞ and $|I|\leqq m_1$ in the sum. From (27.5.13) we obtain a bound for the first factor in

$$D_2=(\sum G_0^{(j)}(x_1)^2)^{-1}\sum G_0^{(j)}G_0^{(j)}D_2$$

and since $F_0^{(j)}+G_0^{(j)}D_2$ is among our P_I we obtain in view of (27.5.18) that also

$$D_2=\sum d_I(x)P_I, \quad x\in\Omega,$$

where d_I is bounded in C_0^∞. Hence it follows from (27.5.17) that

$$\|u\|_{(\varepsilon)}\leqq\|\rho u\|_{(\varepsilon-1)}+\|D_2u\|_{(\varepsilon-1)}\leqq C(\|D_1u\|+\|(F_0+G_0D_2)u\|+\|u\|)$$

when $u\in C_0^\infty(\Omega)$. Since $\rho^\varepsilon\|u\|\leqq\|u\|_{(\varepsilon)}$ the lemma is proved.

The proof of the following lemma is close to that of Lemma 27.3.3 but the result is only good for functions having a small high frequency component with respect to x_2.

Lemma 27.5.4. *Assume in addition to the hypotheses in Lemma 27.5.3 that G_0 is non-negative and F_0/G_0 almost increasing in the sense that*

$$\partial(F_0(x_1,x_2)/G_0(x_1))/\partial x_1>-1 \quad \textit{when } G_0(x_1)\geqq\rho^{-2},\quad |x_1|<1,\ |x_2|<1, \tag{27.5.19}$$

$$F_0(x_1,x_2)/G_0(x_1)-F_0(y_1,x_2)/G_0(y_1)>-1 \quad \textit{if } -1<y_1<x_1<1,\ |x_2|<1 \ \textit{and}\ \min(G_0(x_1), G_0(y_1))\geqq\rho^{-2}. \tag{27.5.20}$$

If ρ is large enough and $K=\{x\in\mathbb{R}^2;\ |x_1|\leqq\frac{8}{9}, |x_2|\leqq 1\}$, $\varepsilon=2^{1-m_1}$, then

$$\rho^{2\varepsilon}\int|v|^2\,dx+\int|D_1v|^2\,dx \leqq C\Big(\int|(L_1+iL_2)v|^2/\max(G_0,\rho^{-2})\,dx + \rho^{-2(1+1/m_2)}\int|D_1D_2v|^2\,dx\Big),\quad v\in C_0^\infty(K). \tag{27.5.21}$$

Proof. As in the proof of Lemma 27.5.3 we may assume a fixed bound for $\tilde F(0)/\rho$. We shall follow the proof of Lemma 27.3.3 closely. Thus we first observe that with $\delta=\rho^{-2}$ and $Q_0=F_0/G_0$ we have

$$\int_{G_0>\delta}|(|L_1v|^2+|L_2v|^2)/G_0\,dx=\int_{G_0>\delta}|(L_1+iL_2)v|^2/G_0\,dx+E_1+E_2 \tag{27.5.22}$$

where

$$E_1=2\operatorname{Re}\int_{G_0>\delta}\partial v/\partial x_1\overline{D_2v}\,dx=-2\operatorname{Re}\int_{G_0<\delta}\partial v/\partial x_1\overline{D_2v}\,dx, \tag{27.5.23}$$

$$E_2=2\operatorname{Re}\int_{G_0>\delta}\partial v/\partial x_1\overline{Q_0v}\,dx =-\int_{G_0>\delta}|v|^2\,\partial Q_0/\partial x_1\,dx-\int_{G_0=\delta}|v|^2Q_0\,dx_2. \tag{27.5.24}$$

Here we have considered the set $G_0=\delta$ as the boundary of the set where $G_0<\delta$. If a component of this set is defined by $a<x_1<b$ we have to estimate

$$\int|v(a,x_2)|^2Q_0(a,x_2)\,dx_2-\int|v(b,x_2)|^2Q_0(b,x_2)\,dx_2 \leqq\int|v(a,x_2)|^2\,dx_2+\int(|v(a,x_2)|^2-|v(b,x_2)|^2)\,Q_0(b,x_2)\,dx_2$$

where we have used (27.5.20). From (27.5.13) it follows that in each of the intervals $(\pm\frac{8}{9},\pm 1)$ there is some x_1 where G_0 is larger than a fixed constant. Hence $|F_0(x)/G_0(x_1)|<C\rho$ there, so the monotonicity (27.5.20) shows that

$$|Q_0(x)|<C\rho\quad\text{when } x\in K\ \text{ and }\ G_0(x_1)\geqq\delta. \tag{27.5.25}$$

Hence the second term above can be estimated by

$$2C\rho\int_{a<x_1<b}|vD_1v|\,dx.$$

Since $\{x_1;\ G_0(x_1)<\delta\}$ has at most m_2 components and the first term in E_2 can be estimated using (27.5.19), we obtain

$$\int_{G_0>\delta}(|L_1v|^2+|L_2v|^2)/G_0\,dx \leqq\int_{G_0>\delta}|(L_1+iL_2)v|^2/G_0\,dx+2\int_{G_0<\delta}|D_1v|\,|D_2v|\,dx +2C\rho\int_{G_0<\delta}|v\,D_1v|\,dx+\int_{G_0>\delta}|v|^2\,dx+m_2\max\int_{G_0=\delta}|v|^2\,dx_2. \tag{27.5.26}$$

In a maximal interval where $|G_0|<2\delta$ we have $|F_0|<2C\delta\rho$ when $\delta<G_0<2\delta$, in view of (27.5.25). Changing scales so that the interval becomes $(0,1)$, for example, one concludes immediately, since F_0 and G_0 are polynomials of fixed degrees, that $|F_0|<C_1\delta\rho$ when $|G_0|<2\delta$, that is,

$$|F_0(x)|\leq C_1\rho\max(G_0,\rho^{-2}) \quad \text{when } x\in K. \tag{27.5.25$'$}$$

Hence

$$\begin{aligned}\int_{G_0<\delta}(|L_1v|^2+|L_2v|^2)\,dx &\leq 2\int_{G_0<\delta}|(L_1+iL_2)v|^2\,dx+3\int_{G_0<\delta}|L_2v|^2\,dx\\ &\leq 2\int_{G_0<\delta}|(L_1+iL_2)v|^2\,dx+C\delta^2\int_{G_0<\delta}(|D_2v|^2+\rho^2|v|^2)\,dx.\end{aligned} \tag{27.5.27}$$

We divide by δ and add (27.5.26) where the integrals of $|vD_1v|$ and $|D_1v||D_2v|$ are estimated by means of the Cauchy-Schwarz inequality. This gives

$$\begin{aligned}\int(|L_1v|^2+|L_2v|^2)/\max(G_0,\delta)\,dx &\leq 2\int|(L_1+iL_2)v|^2/\max(G_0,\delta)\,dx\\ &\quad+\int_{G_0<\delta}(|D_1v|^2/2\delta+C\delta|D_2v|^2)\,dx\\ &\quad+C\int|v|^2\,dx+\max\int_{G_0=\delta}|v|^2\,dx_2.\end{aligned} \tag{27.5.28}$$

The integral of $|D_1v|^2/2\delta$ on the right-hand side can be cancelled against part of the left-hand side, so using Lemma 27.5.3 we obtain

$$\begin{aligned}\rho^{2\varepsilon}\int|v|^2\,dx+\int|D_1v|^2\,dx &\leq C\int|(L_1+iL_2)v|^2/\max(G_0,\delta)\,dx\\ &\quad+C\delta\int_{G_0<\delta}|D_2v|^2\,dx+C\int|v|^2\,dx+C\max\int_{G_0=\delta}|v|^2\,dx_2.\end{aligned} \tag{27.5.29}$$

By Cauchy-Schwarz again

$$|v(x)|^2\leq 2\int|v(t,x_2)D_1v(t,x_2)|\,dt\leq\int(\rho^\varepsilon|v(t,x_2)|^2+\rho^{-\varepsilon}|D_1v(t,x_2)|^2)\,dt.$$

If we integrate with respect to x_2 it follows that the last term in (27.5.29) is much smaller than the left-hand side for large ρ. The measure of the set $\{x_1\in(-1,1);\,G(x_1)<\delta\}$ can be estimated by $C\delta^{1/m_2}$ for it consists of at most m_2 intervals of length $<C\delta^{1/m_2}$ by (27.5.13) and any interpolation formula. Since

$$\int|D_2v|^2\,dx_2\leq\int|D_1D_2v|^2\,dx$$

for every x_1, the estimate (27.5.21) follows.

In the following complete analogue of Lemma 27.3.3 the hypotheses may look artificial but they are forced by the proof of Proposition 27.5.1 using scale changes. The sceptical reader might prefer to read that argument first.

Lemma 27.5.5. *Assume that F, $G \in C^\infty(\Omega)$, that F is real valued, $G \geqq 0$, and that for fixed positive integers $m_2 \leqq m_1$ and $L_1 = \xi_1$, $L_2 = F(x) + G(x)\xi_2$ we have with ρ and t large*

$$\max_{j<m_2} |D_1^j G(0)| = 1, \tag{27.5.13$'$}$$

$$\min_{\xi} \max_{|I| \leqq m_1} |L_I(0, \xi)| = t, \tag{27.5.14$'$}$$

$$|D^\alpha F| \leqq t^{-6} \quad \text{if } |\alpha| \leqq m_1 \text{ and } m_1 \leqq \alpha_1 + m_2\alpha_2, \tag{27.5.30}$$

$$|D^\alpha G| < \rho^{-1} < t^{-8} \quad \text{if } |\alpha| \leqq m_1 \text{ and } \alpha_1 \geqq m_2 \text{ or } \alpha_2 \neq 0, \tag{27.5.31}$$

$$|D^\alpha F| < \rho/t^6 \quad \text{if } \alpha_1 + m_2\alpha_2 < m_1, \tag{27.5.32}$$

$$\partial(F(x_1, x_2)/G(x_1, x_2))/\partial x_1 > -\tfrac{1}{2} \text{ when } G(x_1, x_2) > \tfrac{1}{2}t^2, \tag{27.5.19$'$}$$

$$F(x_1, x_2)/G(x_1, x_2) - F(y_1, x_2)/G(y_1, x_2) > -\tfrac{1}{2} \quad \text{if } -1 < y_1 < x_1 < 1,\ |x_2| < 1, \text{ and } \min(G(x_1, x_2), G(y_1, x_2)) > \tfrac{1}{2}t^{-2}. \tag{27.5.20$'$}$$

Also assume that G is in a fixed bounded subset of $C^\infty(\Omega)$, and set $\varepsilon = 2^{1-m_1}$. If K is a compact subset of $8\Omega/9$ it follows for large t that

$$\|D_1 v\|^2 + t^{2\varepsilon}\|v\|^2 \leqq C \int |(L_1 + iL_2)v|^2/\max(G, t^{-2})\, dx, \quad v \in C_0^\infty(K). \tag{27.5.33}$$

We also have

$$C^{-1} \leqq \min_{\xi} \max_{|I| \leqq m_1} |L_I(x, \xi)/t| \leqq C, \quad x \in \Omega. \tag{27.5.34}$$

Proof. By (27.5.31) we have $\rho > t^8$ and ρ may be very much larger than that so (27.5.32) is a very weak information which we must first improve. To do so we observe that if the minimum in (27.5.14)$'$ is attained for $\xi = \eta$, then

$$|F^{(j)}(0) + G^{(j)}(0)\eta_2| \leqq t, \quad j < m_2.$$

We can choose j so that $|G^{(j)}(0)| = 1$ and conclude using (27.5.32) that $|\eta_2| < 2\rho/t^6$. Now we make a preliminary symplectic change of variables just replacing ξ_2 by $\xi_2 + \eta_2$. This means that F is replaced by $F + G\eta_2$ which does not affect the hypotheses except that a constant factor is introduced in the right-hand side of (27.5.32). The corresponding change of (27.5.33) is obtained when v is replaced by $v \exp(ix_2\eta_2)$. Thus we assume from now on that the minimum in (27.5.14)$'$ is attained when $\xi = 0$.

Let $F_0(x_1, x_2)$ and $G_0(x_1)$ be the part of the Taylor expansions of $F(x)$ and of $G(x_1, 0)$ at 0 with weight $< m_1$ and $< m_2$ respectively. Then G_0 satisfies (27.5.13) and we shall prove that (27.5.14) is valid with ρ replaced by t and a slight change of the right-hand side for the Poisson brackets formed with $L_1^0 = L_1$, $L_2^0 = F_0(x) + G_0(x_1)\xi_2$. By induction it follows immediately that all L_I except L_1 are of the form

$$A(x_1, x_2) + B(x_1, x_2)\xi_2$$

where B is a polynomial in G and its derivatives while A is a linear combination of F and its derivatives with such coefficients. When $|I|\leq m_1$ only derivatives of order $<m_1$ occur so it follows from (27.5.30–32) and Taylor's formula that

$$|L_I(0,0)-L_I^0(0,0)|<C/t^6.$$

Hence $|L_I^0(0,0)|<2t$ so it follows from Lemma 27.4.7 that there is a polynomial $H_0(x_2)$ with coefficients $O(\rho/t^6)$ such that $F_0(x)-G_0(x_1)H_0(x_2)$ is of weight $<m_1$ and has coefficients $O(t)$. We replace F by $F-GH_0$ and F_0 by $F_0-G_0H_0$ in L_2 and L_2^0 if we take $\xi_2+H_0(x_2)$ as a new canonical variable instead of ξ_2 and replace v by $v\exp(\int -iH_0(x_2)\,dx_2)$ in (27.5.33). This does not affect the hypotheses or conclusions in the lemma except that (27.5.32) is now improved to

$$|D^\alpha F|\leq Ct, \quad \text{if } \alpha_1+\alpha_2 m_2<m_1. \tag{27.5.32'}$$

Lemma 27.5.4 is now applicable to F_0, G_0 with ρ replaced by t, for the bounds (27.5.30), (27.5.31) have been chosen so small that (27.5.19), (27.5.20) follow from (27.5.19)', (27.5.20)' when t is large. The verification is a simple exercise for the reader. Lemma 27.5.2 can also be used, and we shall prove (27.5.33) by splitting v into two parts corresponding to low and high frequencies with respect to x_2. The inequality (27.5.34) follows from (27.4.43) when L_I is replaced by L_I^0 which does not change the Poisson brackets by more than $O(t^{-6})$, so (27.5.34) is true.

Choose $\psi\in C_0^\infty(-\frac{8}{9},\frac{8}{9})$ such that $\psi(x_2)=1$ in a neighborhood of K, and let $\chi\in C_0^\infty(\mathbb{R})$ be equal to 1 in $(-1,1)$. With a fixed $\kappa\in(0,1/m_2)$ we set

$$\psi_t^{(1)}(x_2,\xi_2)=\psi(x_2)(1-\chi(\xi_2/t^{1+\kappa})),$$
$$\psi_t^{(2)}(x_2,\xi_2)=\psi(x_2)\chi(\xi_2/t^{1+\kappa}).$$

We shall apply Lemma 27.5.2 to $\psi_t^{(1)}(x_2,D_2)v$ and Lemma 27.5.4 to $\psi_t^{(2)}(x_2,D_2)v$ but first we must prove that

$$\begin{aligned}\int|(L_1+iL_2)\psi_t^{(2)}(x_2,D_2)v|^2/\max(G,t^{-2})\,dx\\ \leq 2\int|(L_1+iL_2)v|^2/\max(G,t^{-2})\,dx+C\|v\|^2, \quad v\in C_0^\infty(K).\end{aligned} \tag{27.5.35}$$

To do so we shall estimate the commutator $[L_2,\psi_t^{(2)}(x_2,D_2)]$ using the calculus of pseudo-differential operators with the metric

$$g=|dx_2|^2+|d\xi_2|^2/(t^{1+\kappa}+|\xi_2|)^2.$$

Since $L_2\in S(t^{1+\kappa}+|\xi_2|,g)$ and $\psi_t^{(2)}\in S(1,g)$, the symbol of the commutator is bounded in $S((t^{1+\kappa}+|\xi_2|)^{-1},g)\subset S(t^{-1-\kappa},g)$, so the norm is $O(t^{-1-\kappa})$, apart from the Poisson bracket term

$$-i\{L_2(x,\xi_2),\psi_t^{(2)}(x_2,\xi_2)\}.$$

The term where $\psi(x_2)$ is differentiated is also the sum of an operator in this class and an operator obtained when $\partial\psi/\partial x_2$ is moved next to v so that it annihilates v. Thus we only have to examine the symbol

$$i\psi(x_2)\,\partial L_2/\partial x_2 t^{-1-\kappa}\chi'(\xi_2/t^{1+\kappa}).$$

Since $|\partial G/\partial x_2|<1/t$ the norm of the operator

$$i\psi(x_2)\,\partial G/\partial x_2(D_2/t^{1+\kappa})\chi'(D_2/t^{1+\kappa})$$

is $O(1/t)$ so what remains is to study

$$it^{-1-\kappa}\psi(x_2)\,\partial F/\partial x_2\,\chi'(\xi_2/t^{1+\kappa}).$$

By (27.5.25)′ we have

$$|F_0(x)|\leqq Ct\max(G_0(x_1),t^{-2}),\qquad x\in\Omega,$$

and since F_0 is a polynomial in x_2 a similar estimate is valid for $\partial F_0/\partial x_2$. Thus

$$|\partial F/\partial x_2|\leqq Ct(G(x)+1/t^2)$$

which completes the proof of (27.5.35).

In the left-hand side of (27.5.35) we may replace L_j by L_j^0, for

$$\|(L_2-L_2^0)\psi_t^{(2)}(x_2,D_2)v\|\leqq C\|v\|/t^6.$$

When v is replaced by $\psi_t^{(2)}(x_2,D_2)v$ in (27.5.21) the last term (with ρ replaced by t) becomes the square of

$$t^{-1-1/m_2}\|D_1D_2\psi_t^{(2)}(x_2,D_2)v\|\leqq Ct^{\kappa-1/m_2}\|D_1v\|.$$

Since $\kappa<1/m_2$ and $\|D_1v\|$ occurs in the left-hand side of (27.5.33) this error term can be cancelled for large t.

When (27.5.10) is applied to $v_1=\psi_t^{(1)}(x_2,D_2)v$, we obtain

$$\begin{aligned}\text{(27.5.36)}\qquad &\|GD_2v_1\|+\|D_1v_1\|+\||D_2|^{1/m_2}v_1\|\\ &\leqq C(\|(L_1+iL_2)v_1\|+\|v\|+\|Fv_1\|).\end{aligned}$$

From the calculus of pseudo-differential operators it follows at once that

$$t^{1/m_2}\|v_1\|\leqq C(\||D_2|^{1/m_2}v_1\|+\|v\|).$$

To estimate $\|Fv_1\|$ we choose $h\in C^\infty(\mathbb{R})$ equal to 0 near 0 and equal to $1/\xi_2$ when $|\xi_2|>1$. If $h_t(\xi_2)=h(\xi_2/t^{1+\kappa})$ and $\Psi\in C_0^\infty(-1,1)$ is equal to 1 in $\operatorname{supp}\psi$, then the symbol of

$$\psi_t^{(1)}(x_2,D_2)-t^{-1-\kappa}\Psi(x_2)h_t(D_2)D_2\psi_t^{(1)}(x_2,D_2)$$

is in $S((t^{1+\kappa}+|\xi_2|)^{-1},g)$ so the norm is $O(t^{-1-\kappa})$. Since $|F|\leqq C(G_0t+1)$ and $h_t(D_2)D_2/t^{-1-\kappa}$ is uniformly bounded, we obtain by considering separately the sets where $G_0<1/t$ and $G_0\geqq 1/t$

$$\|Fv_1\|\leqq C(t^{-\kappa}\|GD_2v_1\|+\|v\|),$$

for in the latter set we can estimate G_0 by G. The first term on the right-hand side occurs also on the left-hand side of (27.5.36), with a larger coefficient, so it can be cancelled. Summing up, we have proved

$$\|D_1 v\|^2+t^{2\varepsilon}\|v\|^2 \leqq C(\int |(L_1+iL_2)v|^2/\max(G,t^{-2})\,dx+\|v\|^2)$$

which proves (27.5.33) when t is large. The proof is complete.

Proof of Proposition 27.5.1. We can now essentially copy the proof of Lemma 27.3.4. Let t be a number with $1 \ll t \ll \rho$ which will be fixed later on in the proof. We want to apply Lemma 27.5.5 with $m_1=k^2+k+1$ and $m_2=k+1$ so we introduce

$$M(x)=\min_{\xi}\max_{|I|\leqq m_1}|L_I(x,\xi)/t|^{1/|I|},$$
$$1/B(x)=\max_{j\leqq k}|\partial^j G(x)/\partial x_1^j|/M(x)^{j+1}.$$

For large ρ it follows from (27.5.1)–(27.5.4) in view of (27.3.5)′ that

$$(\rho/t)^{1/(k+1)}\leqq M(x)\leqq C\rho/t, \quad M(x)/3\leqq B(x)\leqq 3M(x)^{k+1}. \tag{27.5.37}$$

We shall now change scales at x in the ratios $M(x)$, $B(x)$. To simplify notation we take $x=0$ at first and write M and B instead of $M(0)$ and $B(0)$. After the scale change and division by M, the symbol of our operator becomes

$$\xi_1+i(F(x_1/M,x_2/B)/M+G(x_1/M,x_2/B)\xi_2 B/M). \tag{27.5.38}$$

The hypothesis (27.5.14)′ in Lemma 27.5.5 is valid for this operator by the definition of M, for symplectic dilations do not affect Poisson brackets, and (27.5.13)′ follows from the definition of B. Since

$$|D^\alpha F(x_1/M,x_2/B)/M|\leqq C_\alpha\rho M^{-\alpha_1-1}B^{-\alpha_2}\leqq C'_\alpha\rho M^{-|\alpha|-1}$$

we obtain an estimate of the form (27.5.30) for $|\alpha|>k$ if t is fixed and ρ is large. Since $m_1\leqq\alpha_1+m_2\alpha_2$ implies $k^2+k<(k+1)|\alpha|$, hence $|\alpha|>k$, this proves (27.5.30). The estimate (27.5.31) follows from (27.5.2) since $B/M^{k+1}<3$, and (27.5.32) is obvious. If $GB/M>t^{-2}/2$ then

$$G>M/2t^2B>1/(6t^2M^k)>\rho^{-k-1}$$

if ρ is large. Since $1/BM\ll 1/B\ll 1$ the conditions (27.5.19)′ and (27.5.20)′ follow from (27.5.5) and (27.5.6). Altogether this shows that (27.5.33) can be applied to the operator with symbol (27.5.38). As in the proof of Lemma 27.3.3 we modify (27.5.33) by substituting ψv for v, where $\psi\in C_0^\infty(K)$ is equal to one in $\omega=\{x;|x_1|<\frac{1}{2},|x_2|<\frac{1}{2}\}$ and G is uniformly bounded from below in $\operatorname{supp}\partial\psi/\partial x_1$. Then it follows that

$$\int_\omega(|D_1v|^2+t^{2\varepsilon}|v|^2)\,dx\leqq C\int_\Omega(t^2|(D_1+i(F(x_1/M,x_2/B)/M + G(x_1/M,x_2/B)(B/M)D_2))v|^2+|v|^2)\,dx. \tag{27.5.33$'$}$$

By (27.5.34) and (27.3.5)′ we know that M and B do not change by more than a fixed factor when $|x_1 M|<1$ and $|x_2 B|<1$. We can therefore return to the original variables in (27.5.33)′, multiply by $M^3 B$ evaluated at a variable point y where we make the preceding construction instead of at the origin, and integrate with respect to y. Then we obtain

$$\|D_1 u\|^2 + t^{2\varepsilon}\|Mu\|^2 \leqq C(t^2\|(D_1 + i(F(x)+G(x)D_2))u\|^2 + \|Mu\|^2).$$

Now we fix t so large that the last term can be cancelled because $2C<t^{2\varepsilon}$. In view of the first inequality (27.5.37) this completes the proof of (27.5.7) and of Proposition 27.5.1.

27.6. Global Subelliptic Estimates

As indicated at the beginning of Section 27.4 the sufficiency in Theorem 27.1.11 is an easy consequence of the following

Proposition 27.6.1. *Let $q\in C^\infty(\mathbb{R}^n \times \mathbb{R}^{n-1})$ be real valued, and assume that*

(27.6.1) $$|D_\xi^\alpha D_x^\beta q(x,\xi')| \leqq C_{\alpha\beta}\lambda^{|\alpha'+\beta'|-2};$$

(27.6.2) *$q(x,\xi')$ does not change sign from + to − for increasing x_1 when $|x_1|<1$ and $|(x',\xi')|<1/\lambda$;*

(27.6.3) $$\lambda^{-2} \leqq C' \sum_{|I|\leqq k+1} |q_I(x,\xi)| \quad \text{if } |x_1|<1,\ |(x',\xi')|<1/\lambda.$$

If $h\in C_0^\infty(\mathbb{R}^{2n-1})$ and $|x_1|\leqq\frac12$, $|(x',\xi')|<\frac12$ when $(x,\xi')\in\operatorname{supp} h$, it follows that for some C'' depending only on h, C' and $C_{\alpha\beta}$ we have when $\lambda C''<1$

(27.6.4) $$\lambda^{-2/(k+1)}\|h(x_1,\lambda x',\lambda D')u\| \leqq C''(\|D_1 u + iq(x,D')u\| + \|u\|), \quad u\in\mathscr{S}(\mathbb{R}^n).$$

Most of this section is devoted to the proof of Proposition 27.6.1. After the proof we shall show that it implies Theorem 27.1.11.

With $u\in\mathscr{S}(\mathbb{R}^n)$ we set

(27.6.5) $$U = h(x_1,\lambda x',\lambda D')u.$$

If $P = D_1 + iq(x,D')$ we have

(27.6.6) $$PU = h(x_1,\lambda x',\lambda D')Pu + [P, h(x_1,\lambda x',\lambda D')]u = f.$$

The calculus of pseudo-differential operators with the constant metric $\lambda^2(|dx'|^2+|d\xi'|^2)$ shows that the symbol of the commutator is uniformly bounded of weight 1 so the norm is uniformly bounded. Hence

(27.6.7) $$\|f\| \leqq C(\|D_1 u + iq(x,D')u\| + \|u\|).$$

Next we split U by means of the function Ψ constructed in (27.4.53). Thus we write $U=U_1+U_2$ where

$$U_1=(1-\Psi(x,D'))U, \quad U_2=\Psi(x,D')U. \tag{27.6.8}$$

With $g=[P,\Psi(x,D')]U$ we have

$$PU_1=(1-\Psi(x,D'))f-g, \quad PU_2=\Psi(x,D')f+g. \tag{27.6.9}$$

Lemma 27.6.2. *With the notation in Lemmas* 27.4.11 *and* 27.4.12 *we have*

$$\|g\|^2\leqq C(\sum\|m_j\phi_j(x,D')U_1\|^2+\sum\|M_j\Phi_j(x,D')U_2\|^2+\|u\|^2), \tag{27.6.10}$$

$$\|U_1\|^2\leqq C(\sum\|\phi_j(x,D')U_1\|^2+\lambda^2\|u\|^2), \tag{27.6.11}$$

$$\|U_2\|^2\leqq C(\sum\|\Phi_j(x,D')U_2\|^2+\lambda^2\|u\|^2). \tag{27.6.12}$$

Proof. First we shall prove that

$$\|g\|^2\leqq C(\sum\|M_j\Phi_j(x,D')U\|^2+\|U\|^2). \tag{27.6.10$'$}$$

To do so we observe that Ψ is bounded in $S(1,g_2)$ by Lemma 27.4.11, where g_2 is defined by (27.4.54), and that q is bounded in $S(\lambda^{-2},g_2)$ while the second order derivatives are even bounded in $S(1,g_2)$. It follows that the symbol of $[P,\Psi(x,D')]$ is bounded in $S(R_2^{-2},g_2)\subset S(1,g_2)$ apart from the term with symbol $(D_{x_1}+H_q)\Psi(x,\xi')$. We can write

$$(D_{x_1}+H_q)\Psi(x,\xi')=\sum r_j(x,\xi')M_j\Phi_j(x,\xi'),$$
$$r_j(x,\xi')=M_j^{-1}(D_{x_1}+H_q)\Psi_j(x,\xi'),$$

for $\Phi_j=1$ in $\operatorname{supp}\Psi_j$. By Lemma 27.4.11 $\{r_j\}$ is a symbol in $S(1,g_2)$ with values in $l^2=\mathscr{L}(l^2,\mathbb{C})$, so it follows that

$$\|\sum r_j(x,D')v_j\|^2\leqq C\sum\|v_j\|^2.$$

This proves (27.6.10)′ when we take $v_j=M_j\Phi_j(x,D')U$. (Recall that $\Phi_j=1$ in a neighborhood of $\operatorname{supp}\Psi_j$ so the composition series for $r_j(x,D')\Phi_j(x,D')$ has only one term.) If we set $U=U_1+U_2$ in (27.6.10)′ we obtain (27.6.10) if we prove that

$$\sum\|M_j\Phi_j(x,D')U_1\|^2\leqq C\sum\|m_k\phi_k(x,D')U_1\|^2+\|u\|^2). \tag{27.6.13}$$

Since $M_j\leqq C'm_k$ when $\operatorname{supp}\Phi_j\cap\operatorname{supp}\phi_k\neq\emptyset$ it suffices to show that

$$\Phi_j(x,D')U_1=\sum a_{jk}(x,D')\phi_k(x,D')U_1+S_j(x,D')u \tag{27.6.14}$$

where $\operatorname{supp}a_{jk}\subset\operatorname{supp}\Phi_j\cap\operatorname{supp}\phi_k$ and a_{jk} is bounded in $S(1,g_1)$ while the norm of S_j is bounded by as high a power of λ as we please. To prove (27.6.14) we first observe that

$$1=\sum\gamma_k(x,\xi')\phi_k(x,\xi') \quad \text{in } \omega=\Omega\cap\operatorname{supp}(1-\Psi), \tag{27.6.15}$$
$$\gamma_k=\phi_k\prod_{j<k}(1-\phi_j^2),$$

for $\prod(1-\phi_j^2)=0$ in ω. It is clear that γ_k is bounded in $S(1,g_1)$. A first approximation to (27.6.14) is

$$a_{jk}^0(x,\xi')=\Phi_j(x,\xi')\gamma_k(x,\xi'),$$

which gives (27.6.14) apart from an error $\Phi'_j(x,D')U_1$ where $\operatorname{supp}\Phi'_j\subset\operatorname{supp}\Phi_j$ and Φ'_j is bounded in $S(R_1^{-2},g_1)$. We iterate the construction with Φ_j replaced by Φ'_j and so on, which gives the desired remainder term in (27.6.14) after a finite number of steps.

Starting from (27.6.15) we can also find $\tilde{\gamma}_k$ bounded in $S(1,g_1)$ with $\operatorname{supp}\tilde{\gamma}_k\subset\operatorname{supp}\phi_k$ and

$$U_1=\sum\tilde{\gamma}_k(x,D')\phi_k(x,D')U_1+Su$$

where S is bounded in $S(\lambda,g_1)$. This gives the estimate (27.6.11). A similar argument, which we leave for the reader to carry out, gives (27.6.12) and completes the proof.

The estimate (27.6.10) shows that we have achieved a separation of the cases I and II just as in the proof of Lemma 27.5.5. To proceed we just decompose U_1 and U_2 using the functions ϕ_k and Φ_j respectively.

Lemma 27.6.3. *With the notation of Lemma* 27.6.2 *we have*

$$\sum\|P\phi_k(x,D')U_1\|^2\leq C(\|PU_1\|^2+\sum\|m_k\phi_k(x,D')U_1\|^2+\|u\|^2), \tag{27.6.16}$$

$$\sum\|P\Phi_j(x,D')U_2\|^2\leq C(\|PU_2\|^2+\sum\|M_j\Phi_j(x,D')U_2\|^2+\|u\|^2). \tag{27.6.17}$$

Proof. To prove (27.6.17) it suffices to estimate $f_j=[P,\Phi_j(x,D')]U_2$, for $P\Phi_j(x,D')U_2=f_j+\Phi_j(x,D')PU_2$ and

$$\sum\|\Phi_j(x,D')PU_2\|^2\leq C\|PU_2\|^2$$

by the L^2 continuity of (vector valued) operators with symbol in $S(1,g_2)$. We can write $[P,\Phi_j]$ as the sum of the operator with symbol $(D_{x_1}+H_q)\Phi_j(x,\xi')$ and an operator with l^2 valued symbol in $S(1,g_2)$, hence L^2 continuous. In $\operatorname{supp}\Psi$ we have

$$(D_{x_1}+H_q)\Phi_j=\sum a_{jk}^0M_k\Phi_k,$$

where

$$a_{jk}^0=((D_{x_1}+H_q)\Phi_j)\Gamma_k/M_k,\quad \Gamma_k=\Phi_k\prod_{j<k}(1-\Phi_j^2).$$

From Lemma 27.4.11 it follows that there is a uniform bound for a_{jk}^0 in $S(1,g_2)$, for M_j/M_k is bounded if $\operatorname{supp}\Phi_j\cap\operatorname{supp}\Phi_k\neq\emptyset$. By the iterative argument used to establish (27.6.14) we can find a_{jk} bounded in $S(1,g_2)$, with "leading term a_{jk}^0" and $\operatorname{supp}a_{jk}\subset\operatorname{supp}\Phi_j\cap\operatorname{supp}\Phi_k$ so that

$$((D_{x_1}+H_q)\Phi_j)(x,D')=\sum a_{jk}(x,D')M_k\Phi_k(x,D')U_2+S_j(x,D')u$$

where the norm of S_j is bounded by any desired power of λ. This proves (27.6.17). The similar and even somewhat simpler proof of (26.6.16) is left for the reader. This ends the proof.

Our final lemma gives estimates for the localizations:

Lemma 27.6.4. *For sufficiently large ρ we have*

(27.6.18) $\rho^{2/(k+1)}\sum \|m_j\phi_j(x,D')U_1\|^2 \leqq C(\sum \|P\phi_j(x,D')U_1\|^2+\|u\|^2);$

(27.6.19) $\rho^{2/(k+1)}\sum \|M_j\phi_j(x,D')U_2\|^2 \leqq C(\sum \|P\Phi_j(x,D')U_2\|^2+\|u\|^2).$

Proof. We shall begin by estimating $\Phi_j(x,D')U_2$. To simplify notation we assume at first that $j=0$ so that Φ_j is given by (27.4.50) and $Q(x,\xi')=q(x_1,\chi(x',\xi'))$. Set

$$L(x,\xi')=Q(x_1,x_2,0)+\xi_2\,\partial Q(x_1,x_2,0)/\partial\xi_2,$$
$$r_0(x,\xi')=q(x,\xi')-L\circ\chi^{-1}(x,\xi').$$

We have assumed here that χ and χ^{-1} are globally defined. This is legitimate since χ is of the form (27.4.24), possibly composed with a linear canonical transformation, and there is no difficulty in extending the functions $f_3,\dots,g_n$ so that (27.4.26) remains valid uniformly in $T^*(\mathbb{R}^{n-1})$. To estimate $r_0(x,D')\Phi_0(x,D')U_2$ we introduce another cutoff function

$$\hat\Phi_0(x_1,\chi(x',\xi'))=\Phi_0(x_1,\chi(x'/2,\xi'/2))$$

with support in the double of Ω_0 and equal to 1 in $\operatorname{supp}\Phi_0$. The crudest estimates show that $L\circ\chi^{-1}$ is bounded in $S(\lambda^{-2},g_2)$, which implies that the symbol of $(1-\hat\Phi_0(x,D'))r_0(x,D')\Phi_0(x,D')$ is bounded in $S(\lambda^N,g_2)$ for any N, since the composition series is equal to 0. If

$$R_0(x,\xi')=r_0(x_1,\chi(x',\xi'))=Q(x,\xi')-L(x,\xi')$$

we have by (27.4.33) and Taylor's formula

$$R_0(x,\xi')=R_0(x,\xi')-R_0(x_1,x_2,0)=\sum_3^n T_j(x,\xi')x_j+\sum_2^n S_j(x,\xi')\xi_j$$

where

$$\sum_j |D_\xi^\alpha D_x^\beta T_j|+\sum_j |D_\xi^\alpha D_x^\beta S_j|\leqq C_{\alpha\beta}M\rho(M\rho)^{-(1+|\alpha+\beta|)/2},\quad \beta_1=0.$$

In $\operatorname{supp}\hat\Phi_0(x_1,\chi(x',\xi'))$ we have $|x''|+|\xi'|<2R_2\ll(M/\rho)^{\frac12}$, so R_2T_j/M and R_2S_j/M are bounded in $S(1,g_2)$ there. It follows that the symbol of $\hat\Phi_0(x,D')r_0(x,D')$ is bounded in $S(M_0,g_2)$. Using the same result for every j we have proved that

(27.6.20) $\sum \|r_j(x,D')\Phi_j(x,D')U_2\|^2 \leqq C(\sum \|M_j\Phi_j(x,D')U_2\|^2+\|u\|^2).$

The symbol of $P-ir_0(x,D')$ composed with χ is ξ_1+iL. We shall prove in a moment that $P-ir_0(x,D')$ is essentially unitarily equivalent to $D_1+iL(x,D')$ but first we shall derive estimates for $D_1+iL(x,D')$. With F_0, F_1 and G defined by (27.4.61), (27.4.62) we have by Proposition 27.5.1

(27.6.21) $$\rho^{1/(k+1)}\|v\| \leqq C\|D_1v+i(F_0(x)+G(x)D_2)v\|$$

if $|x_1|<c$, $|x_2|<c$ in supp v for some fixed $c<1$. In fact, the hypotheses of Proposition 27.5.1 follow from (27.4.63)–(27.4.67). By a translation we find that the same estimate holds if $|x_1|<c$, $|x_2-y_2|<c$ in supp v for some y_2 with $|y_2|<B_2/b_2$. Decomposing v by a partition of unity in x_2 we conclude that (27.6.21) is valid if $|x_1|<c$, $|x_2|<B_2/b_2$ in supp v, provided that ρ is large enough. By (27.4.61), (26.4.62) this means that

$$\rho^{1/(k+1)}\|Mv\| \leqq C(\|D_1 v + iL(x,D')v\| + \|A_2^{-1}D_2 v\|) \tag{27.6.22}$$

when $|x_1 M|<c$, $|x_2 b_2|<1$ in supp v. Here v is a function in $\mathbb{R}^2$ and the norm is in $L^2(\mathbb{R}^2)$ but without change of constant we may of course let v depend on parameters $x_3, \dots, x_n$ and take the norm in $L^2(\mathbb{R}^n)$. We must now examine how one can pass to an estimate where L and D_2 are replaced by the composition with χ.

i) First assume that χ is defined by (27.4.24), (27.4.25). For the canonical transformation χ_1 in $T^*(\mathbb{R}^{n-1})$ induced by the measure preserving map $\mathbb{R}^{n-1} \ni x' \mapsto y' \in \mathbb{R}^{n-1}$ defined by

$$y_2 = x_2, \quad y_j = x_j - af_j(x_2/a), \quad j>2, \tag{27.6.23}$$

we have $\chi_1(x',\xi') = (y',\eta')$ if $\sum \eta_j dy_j = \sum \xi_j dx_j$, that is,

$$\eta_2 = \xi_2 + \sum_3^n \xi_j f_j'(x_2/a), \quad \eta_j = \xi_j \quad \text{for } j>2. \tag{27.6.24}$$

A simple calculation gives $\chi_1 \circ \chi = \chi_2$ where

$$\chi_2(x',\xi') = (x', \xi' + \partial H(x')/\partial x'),$$

$$H(x') = \sum_3^n a g_j(x_2/a) x_j + \int a g_2(x_2/a)\, dx_2 + \sum_3^n \int a^2 g_j(x_2/a)\, df_j(x_2/a).$$

To pass from (27.6.22) to the same estimate with the symbols on the right-hand side composed with $\chi^{-1} = \chi_2^{-1} \circ \chi_1$ we first replace v by ve^{-iH}, which gives the composition with χ_2^{-1}, and then change variables from y' to x' which gives the composition with χ_1. Note that in both cases we apply unitary operators to v. Thus we obtain

$$\begin{aligned}\rho^{1/(k+1)}\|M_0 \Phi_0(x,D')U_2\| \leqq C(&\|(P - ir_0(x,D'))\,\Phi_0(x,D')U_2\| \\ &+ \|A_2^{-1}\eta_2(x,D')\,\Phi_0(x,D')U_2\|.\end{aligned} \tag{27.6.25}$$

Here η_2 denotes the composition of ξ_2 with χ^{-1}. Since $R_2/A_2 < R_2/R \ll M_0$ the symbol of $M_0^{-1}A_2^{-1}\eta_2(x,D')\,\Phi_0(x,D')$ is bounded in $S(1,g_2)$.

ii) Now assume that $\chi = T_0 \circ \hat{\chi}$ where T_0 is the canonical transformation $(x',\xi') \mapsto (\xi', -x')$ and $\hat{\chi}$ has the form considered in (i). Apart from a change of labels of coordinate pairs this hypothesis will always be fulfilled if i) does not hold. Now we have

$$L \circ \chi^{-1} = \hat{L} \circ T_0^{-1}, \quad \hat{L} = L \circ \hat{\chi}^{-1}.$$

Thus we have proved in part i) that

$$(27.6.26)\quad \rho^{1/(k+1)}\|M_0 v\| \leqq C(\|D_1 v + i\hat{L}(x,D')v\| + \|A_2^{-1}\hat{\eta}_2(x,D')v\|)$$

where $\hat{\eta}_2$ is the composition of ξ_2 with $\hat{\chi}^{-1}$. It is preferable for reasons which will soon be apparent to replace L in (27.6.22) by the operator L_r obtained by putting the coefficient of D_2 to the right of D_2. This is possible since the difference is just a bounded function times M_0 which can be absorbed by the left-hand side for large ρ. Now we apply (27.6.26) to the Fourier transform of v in x'. This changes $\hat{L}_r(x,D')$ to $\hat{L}\circ T_0^{-1}(x,D') = L\circ\chi^{-1}(x,D')$ so (27.6.26) gives

$$(27.6.27)\quad \rho^{1/(k+1)}\|M_0 v\| \leqq C(\|(D_1 + i(L\circ\chi^{-1})(x,D'))v\| + \|A_2^{-1}\eta_2(x,D')v\|.$$

Unfortunately (27.6.22) is only valid when $|x_1 M| < c$ and $|x_2 b_2| < 1$ in $\operatorname{supp} v$, so (27.6.27) holds when $|x_1 M| < c$ and $|\xi_2 b_2| < 1$ in the support of the Fourier transform of $v(x_1, x')$ with respect to x'. This is not the case for $v = \Phi_0(x,D')U_2$. However, we can choose $h \in C_0^\infty(-1,1)$ so that $h(B_2' x_2) = 1$ in $\operatorname{supp}\Phi_0\circ\chi$. This means that $h(B_2'\xi_2) = 1$ in $\operatorname{supp}\Phi_0$, so the symbol of $(h(B_2' D_2) - 1)\Phi_0(x,D')$ is bounded in $S(\lambda^N, g_2)$ for any N. If we apply (27.6.27) to $v = h(B_2' D_2)\Phi_0(x,D')U_2$, which is legitimate, we therefore obtain

$$(27.6.25)'\quad \begin{aligned} &\rho^{1/(k+1)}\|M_0\Phi_0(x,D')U_2\| \\ &\leqq C(\|(P - i r_0(x,D'))\Phi_0(x,D')U_2\| \\ &\quad + \|A_2^{-1}\eta_2(x,D')\Phi_0(x,D')U_2\| + \lambda^N\|u\|). \end{aligned}$$

If we square the estimates corresponding to (27.6.25), (27.6.25)′ for a general j, sum and use (27.6.20), it follows that

$$(\rho^{2/(k+1)} - C)\sum \|M_j\Phi_j(x,D')U_2\|^2 \leqq C(\sum\|P\Phi_j(x,D')U_2\|^2 + \|u\|^2),$$

which proves (27.6.19) when ρ is large enough.

The proof of (27.6.18) is similar, with Proposition 27.5.1 replaced by Lemma 27.3.4 and without the complications caused by the canonical transformations. We shall therefore leave for the reader to complete the proof of the lemma.

Proof of Proposition 27.6.1. With the notation

$$N = \sum\|m_j\phi_j(x,D')U_1\|^2 + \sum\|M_j\Phi_j(x,D')U_2\|^2$$

we obtain from (27.6.16)–(27.6.19)

$$\rho^{2/(k+1)}N \leqq C_1(\|PU_1\|^2 + \|PU_2\|^2 + \|u\|^2 + N).$$

By (27.6.9) and (27.6.10) we have

$$\|PU_1\|^2 + \|PU_2\|^2 \leqq C_2(\|f\|^2 + N + \|u\|^2)$$

so it follows that

$$(\rho^{2/(k+1)} - C_3)N \leqq C_3(\|f\|^2 + \|u\|^2).$$

At last we now fix ρ so large that $\rho^{2/(k+1)} > 2C_3$ and obtain

$$N \leqq \|f\|^2 + \|u\|^2.$$

Since M_j and m_j can be estimated from below by $\lambda^{-2/(k+1)}$, by (27.6.3), it follows in view of (27.6.11), (27.6.12) that

$$\lambda^{-2/(k+1)} \|U\| \leqq C_4(\|f\| + \|u\|).$$

Using (27.6.7) we obtain (27.6.4) which completes the proof.

Proof of the Sufficiency in Theorem 27.1.11. We may assume that $X = \mathbb{R}^n$ and that $p(x,\xi) = \xi_1 + i p_2(x,\xi')$ when $|\xi'| > 1$ in the conic neighborhood V of $\gamma_0 = (0,\xi_0)$ in the theorem, $p_2 \in S^1(\mathbb{R}^n \times \mathbb{R}^{n-1})$. The coordinates can be chosen so that V contains $\{(x,\xi);\ |x| < 3,\ |\xi - \xi_0| < 2\}$. Choose $\phi_j \in C_0^\infty(\mathbb{R}^{2n-2})$, $j = 0, 1, 2$, so that

$$|x'| + |\xi' - \xi'_0| < 2^{1-j} \quad \text{if } (x',\xi') \in \operatorname{supp} \phi_j;$$
$$\phi_j(x',\xi') = 1 \quad \text{if } |x'| + |\xi' - \xi'_0| < 2^{-j}.$$

We can then apply Proposition 27.6.1 to

$$q(x,\xi') = \lambda^{-2}(\phi_0 p_2)(x_1, \lambda x', \lambda\xi' + \xi'_0).$$

(See the beginning of Section 27.4.) We take $h(x,\xi') = \phi_2(x',\xi' + \xi'_0)$ when $|x_1| < \frac{1}{3}$. If $u \in \mathscr{S}$ and $u(x) = 0$ when $|x_1| > \frac{1}{3}$ it follows that

$$\lambda^{-2/(k+1)} \|\phi_2(x_1, \lambda x', \lambda D' + \xi'_0) u\| \leqq C''(\|D_1 u + i q(x,D') u\| + \|u\|)$$

for $0 < \lambda \leqq \lambda_0$. By a change of scales and translation of $\hat{u}$ we obtain

$$\lambda^{-2/(k+1)} \|\phi_2(x, \lambda^2 D') u\| \leqq C''(\|D_1 u + i\lambda^{-2}(\phi_0 p_2)(x, \lambda^2 D') u\| + \|u\|).$$

Here we replace u by $\phi_1(x', \lambda^2 D') u$. To estimate

$$\lambda^{-2}(\phi_0 p_2)(x, \lambda^2 D') \phi_1(x', \lambda^2 D') u - \phi_1(x', \lambda^2 D') p_2(x, D') u \tag{27.6.28}$$

we can use pseudodifferential operators of type 1,0. Note that $\lambda^{-2N} \phi_j(x', \lambda^2 \xi')$ is a symbol in $S^N_{1,0}$ with values in $H = L^2((0,\lambda_0); d\lambda/\lambda)$ since $C_0 < |\lambda^2 \xi'| < C_1$ in the support (cf. Example 18.1.2). The leading terms in the compositions cancel since

$$\lambda^{-2}(\phi_0 p_2)(x, \lambda^2 \xi') \phi_1(x', \lambda^2 \xi') - \phi_1(x', \lambda^2 \xi') p_2(x,\xi') = 0$$

so it follows from the calculus that the symbol of (27.6.28) is in $S^0_{1,0}$ with values in H. Hence

$$\int_{\lambda < \lambda_0} |\phi_2(x', \lambda^2 D') u|^2 \lambda^{-4/(k+1)} d\lambda/\lambda \leqq C_3(\|D_1 u + i p_2(x, D') u\|^2 + \|u\|^2). \tag{27.6.29}$$

Again by the vector valued calculus the left-hand side is equal to $(\Phi(x', D') u, u)$ where $\Phi \in S^{2/(k+1)}_{1,0}$ has the principal symbol

$$\int_0^\infty |\phi_2(x', \lambda^2 \xi')|^2 \lambda^{-4/(k+1)} d\lambda/\lambda$$

which is positive at $(0,\xi_0')$. Choose $t_2 \in S^0_{1,0}(\mathbb{R}^n \times \mathbb{R}^n)$ non-characteristic at γ_0 so that $t_1 = t_2(\operatorname{Re}\Phi)^{\frac{1}{2}} \in S^{1/(k+1)}_{1,0}$, and apply (27.6.29) to $t_2(x,D)u$. (We assume $t_2=0$ when $|x_1|>\frac{1}{3}$.) Since $(D_1+ip_2(x,D'))t_2(x,D)-t_2(x,D)P$ and $t_2(x,D)^*\Phi(x',D')t_2(x,D)-t_1(x,D)^*t_1(x,D)$ are in Ψ^0, it follows that

$$\|t_1(x,D)u\|^2 \leqq C(\|Pu\|^2+\|u\|^2), \qquad u\in\mathscr{S}(\mathbb{R}^n). \tag{27.6.30}$$

Hence it follows from Lemma 27.1.5 that P is subelliptic with loss of $k/(k+1)$ derivatives at γ_0. The proof is complete.

Notes

The term "subelliptic operator" was introduced in Hörmander [17] for what is here called a subelliptic operator with loss of $\frac{1}{2}$ derivatives. Such operators were characterized by the condition $\{\operatorname{Re} p, \operatorname{Im} p\}>0$ at the zeros of the principal symbol, and it was shown that with current terminology subellipticity with loss of δ derivatives always requires that $\{\operatorname{Re} p, \operatorname{Im} p\}\geqq 0$ when $p=0$. The necessity of condition $(\overline{\Psi})$ was proved by Nirenberg-Trèves [2] and independently by Egorov [2] who announced a complete characterization of subelliptic operators. For operators satisfying condition (P) a complete proof was given by Trèves [8] but the first presentation of the general result was given by Egorov [3]. His proof of the sufficiency seems inadequate though. To supply the missing arguments a more detailed study of the local properties and the localization procedure was given in Hörmander [38]. The presentation here is only a slight modification in principle. Fefferman [1] has announced simplifications of the localization procedures but they have not yet appeared in print at the time when this chapter is being written.

Subellipticity with loss of $\frac{1}{2}$ derivative is well understood also for systems (see Hörmander [17, 44]). Subellipticity with loss of $\delta\in(\frac{1}{2},1)$ derivatives is much more complicated for systems as shown by the recent results of Catlin [1] concerning the $\bar{\partial}$ Neumann problem. In particular, any $\delta \in (\frac{2}{3},1)$ can then occur and not only reciprocals of integers as in the scalar case.

Chapter XXVIII. Uniqueness for the Cauchy Problem

Summary

In this chapter we resume the study of uniqueness theorems for differential operators with non-analytic coefficients started in Section 17.2. Section 28.1 is devoted to Calderón's uniqueness theorem which in its original form states that Theorem 17.2.1 remains valid when there are real characteristics, too. The proofs here start from scratch and rely on factorization in first order pseudo-differential operators. A careful study of these factors leads to more general forms of the Calderón uniqueness theorem.

As in Section 17.2 the basic tool is Carleman estimates. A systematic discussion of such estimates is given in Section 28.2 particularly for operators of real principal type or more generally for operators satisfying a stronger form of condition (P) (principally normal operators). The resulting uniqueness theorems which involve convexity conditions on the initial surface are presented in Section 28.3. More precise results in the second order case are discussed in Section 28.4.

28.1. Calderón's Uniqueness Theorem

Let X be an open set in $\mathbb{R}^n$ and let

$$p(x,D)=\sum_{|\alpha|=m} a_\alpha(x)\,D^\alpha$$

be a differential operator such that

(i) $a_\alpha \in C^\infty(X)$, $|\alpha|=m$,

(ii) $\{(x,\xi)\in T^*(X)\setminus 0;\ p(x,\xi)=0\}$ is a C^∞ hypersurface.

By Σ we denote the closed conic set (cf. (17.2.1))

(28.1.1) $\qquad \Sigma=\{(x,N)\in T^*(X)\setminus 0;\ p(x,\xi+\tau N)$

has a zero $\tau\in\mathbb{C}$ of multiplicity ≥ 2 with $\xi+\tau N\neq 0$ for some $\xi\in\mathbb{R}^n\}$.

That Σ is closed follows from the fact that $p^{-1}(0)\subset\Sigma$ and that it suffices to let ξ be a unit vector orthogonal to N if $p(x,N)\neq 0$.

L. Hörmander, *Classics in Mathematics*
The Analysis of Linear Partial Differential Operators IV,
DOI: 10.1007/978-3-642-00117-8_4,

Theorem 28.1.1. *Assume that p satisfies* (i) *and* (ii) *above. If* $u \in H^{loc}_{(m-1)}(X)$, $p(x,D)u \in L^2_{loc}(X)$, *and for every compact set* $K \subset X$ *we have a differential inequality*

$$(28.1.2) \qquad |p(x,D)u| \le C_K \sum_{|\alpha|<m} |D^\alpha u| \quad \text{in } K,$$

then $\bar{N}(\operatorname{supp} u) \subset \Sigma$ *where* Σ *is defined by* (28.1.1).

Thus it follows from Proposition 8.5.8 that there is unique continuation across any C^1 surface with normal outside Σ. The main difference from Theorem 17.2.1 is that we assume (ii) instead of ellipticity. As in Theorem 17.2.1 it would be sufficient to assume a_α Lipschitz continuous here, but we have assumed (i) above to allow the use of pseudo-differential operators. In the refinements later on in the section it is not at all clear what the best regularity assumptions should be.

As in Chapter XXIII we shall prove Theorem 28.1.1 using a factorization of $p(x,D)$ with factors which are first order differential operators in x_1 and pseudo-differential operators in $x' = (x_2, \dots, x_n)$. First we shall consider elliptic factors (cf. Proposition 17.2.3). Set

$$\phi(x) = x_1 + x_1^2/2, \qquad X_0 = \{x \in \mathbb{R}^n; |x_1| < \tfrac{1}{2}\}.$$

Proposition 28.1.2. *If* $a \in S^1(\mathbb{R}^n \times \mathbb{R}^{n-1})$ *and*

$$(28.1.3) \qquad |\xi'| \le C(|\operatorname{Im} a(0,x',\xi')| + 1), \quad (x',\xi') \in \mathbb{R}^{2n-2},$$

it follows for small positive ε *and* $1/\tau$ *that for* $u \in C_0^\infty(X_0)$

$$(28.1.4) \quad \sum_{|\alpha| \le 1} \tau^{2(1-|\alpha|)-1} \int |D^\alpha u|^2 e^{2\tau\phi}\, dx \le C \int |(D_1 - a(\varepsilon x, D'))u|^2 e^{2\tau\phi}\, dx.$$

Proof. Set $A = a(\varepsilon x, D')$. Since $D_1 u = (D_1 - A)u + Au$ and A is continuous from $H_{(1)}(\mathbb{R}^{n-1})$ to $L^2(\mathbb{R}^{n-1})$ for fixed x_1 and ε, with bound independent of x_1 and ε, we may assume $\alpha_1 = 0$ in the sum. With $v = ue^{\tau\phi}$ the estimate (28.1.4) then becomes

$$(28.1.4)' \quad \tau\|v\|^2 + \sum_2^n \|D_j v\|^2/\tau \le C\|D_1 v + i\tau\phi' v - Av\|^2, \quad v \in C_0^\infty(X_0).$$

To compare $P_\tau = D_1 + i\tau\phi' - A$ with its adjoint $P_\tau^* = D_1 - i\tau\phi' - A^*$ we note that

$$(28.1.5) \qquad [P_\tau^*, P_\tau] = 2\tau\phi'' + B, \quad B = [A^*, A] + [D_1, A^* - A].$$

Thus we have

$$(28.1.5)' \qquad 2\tau\|v\|^2 + (Bv, v) + \|P_\tau^* v\|^2 = \|P_\tau v\|^2.$$

Now the symbol of B/ε is uniformly bounded in S^1 since $a(\varepsilon x, \xi')$ is uniformly bounded in

$$S(1 + |\xi'|, |\varepsilon\, dx|^2 + |d\xi'|^2/(1 + |\xi'|^2))$$

so it follows that

$$2\tau \|v\|^2 \leqq \|P_\tau v\|^2 + C\varepsilon \|v\|_{(0,1)} \|v\| \tag{28.1.6}$$

where $\| \ \|_{(0,1)}$ is the norm in $H_{(0,1)}$ defined in (B.1.10). We also obtain

$$\|P_\tau^* v\|^2 \leqq \|P_\tau v\|^2 + C\varepsilon \|v\|_{(0,1)} \|v\|,$$

hence

$$\begin{aligned}\|(A^*-A)v\| &\leqq \|P_\tau^* v - P_\tau v\| + 2\tau \|\phi' v\| \\ &\leqq 2\|P_\tau v\| + 3\tau \|v\| + (C\varepsilon \|v\|_{(0,1)} \|v\|)^{\frac{1}{2}}.\end{aligned}$$

The symbol of A^*-A is $-2i \operatorname{Im} a(0, \varepsilon x', \xi') + \varepsilon R_\varepsilon(x, \xi')$ where R_ε is uniformly bounded in S^1 so it follows from Theorem 18.1.9 that

$$\|v\|_{(0,1)} \leqq C_1(\|(A^*-A)v\| + \|v\|) \leqq C_2(\|P_\tau v\| + \tau \|v\| + \varepsilon \|v\|_{(0,1)}/\tau).$$

When $C_2 \varepsilon/\tau < \frac{1}{2}$ it follows that

$$\|v\|_{(0,1)} \leqq 2C_2(\|P_\tau v\| + \tau \|v\|). \tag{28.1.7}$$

If the last term in (28.1.6) is estimated using (28.1.7) we obtain

$$2\tau \|v\|^2 \leqq C_3(\|P_\tau v\|^2 + \|v\|^2(1+\varepsilon\tau))$$

which gives

$$\tau \|v\|^2 \leqq C_3 \|P_\tau v\|^2$$

when $\varepsilon + 1/\tau < 1/C_3$. Another application of (28.1.7) yields

$$\|v\|_{(0,1)} \leqq 2C_2(1 + (C_3\tau)^{\frac{1}{2}}) \|P_\tau v\|$$

and completes the proof of the proposition.

To prove Theorem 28.1.1 we must also handle self-adjoint factors. To prepare for a later refinement we prove a somewhat more general result (cf. Lemma 23.1.1).

Proposition 28.1.3. *If* $a \in S^1(\mathbb{R}^n \times \mathbb{R}^{n-1})$ *and* $\operatorname{Im} a$ *is bounded from above it follows that for large* τ *and* $0 < \varepsilon < 1$ *we have*

$$\tau^2 \int |u|^2 e^{2\tau\phi} dx \leqq 16 \int |(D_1 - a(\varepsilon x, D'))u|^2 e^{2\tau\phi} dx, \quad u \in C_0^\infty(X_0). \tag{28.1.8}$$

Proof. With the notation in the proof of Proposition 28.1.2 we have

$$\operatorname{Im}(P_\tau v, v) = \tau(\phi' v, v) - \operatorname{Im}(Av, v).$$

Since $\phi' \geqq \frac{1}{2}$ in $\operatorname{supp} v$ it follows from the sharp Gårding inequality (Theorem 18.1.14) that

$$(\tau - C)\|v\|^2 \leqq 2\|P_\tau v\| \, \|v\|.$$

When $\tau > 2C$ the estimate (28.1.8) follows.

Proof of Theorem 28.1.1. To prove that $N(\operatorname{supp} u) \subset \Sigma$ means to show that if for some $x_0 \in X$ and $\psi \in C^\infty(X, \mathbb{R})$ with $\psi'(x_0) \neq 0$, $(x_0, \psi'(x_0)) \notin \Sigma$ we have

$u=0$ when $\psi(x)>\psi(x_0)$ then $u=0$ in a neighborhood of x_0. Since the hypotheses are invariant for changes of coordinates we may as well assume that $x_0=0$ and that $\psi(x)=x_1+|x'|^2/2$, that is,

$$u=0 \quad \text{when } x_1>-|x'|^2/2. \tag{28.1.9}$$

By hypothesis the equation $p(0,\tau,\xi')=0$ has simple roots $\tau_1,\dots,\tau_m$ when $0\neq\xi'\in\mathbb{R}^{n-1}$, and by condition (ii) the number of real roots is independent of ξ'. Assume at first that $n\neq 3$. The unit sphere in $\mathbb{R}^{n-1}$ is then simply connected so we can choose $\tau_1,\dots,\tau_m$ as continuous, hence C^∞ functions of ξ', such that $\tau_1,\dots,\tau_r$ are real and $\tau_{r+1},\dots,\tau_m$ are non-real for every ξ'. Assuming as we may that $p(x,1,0)=1$ in a neighborhood of 0 we conclude using the implicit function theorem that for some neighborhood U of 0

$$p(x,\xi)=\prod_1^m(\xi_1-\tau_j(x,\xi')), \quad x\in U,$$

where τ_j is a positively homogeneous function of ξ' of degree 1 belonging to $C^\infty(U\times(\mathbb{R}^{n-1}\setminus 0))$; $\tau_1,\dots,\tau_r$ are real but $\tau_{r+1},\dots,\tau_m$ are non-real everywhere. Choose $\chi\in C_0^\infty(\mathbb{R}^n)$ equal to 1 in a neighborhood of 0 and set

$$a_j^0(x,\xi')=\chi(x/\delta)\,\tau_j(x,\xi')+(1-\chi(x/\delta))\,\tau_j(0,\xi').$$

If δ is small enough then a_j^0 has the properties of τ_j just stated, and a_j^0 is globally defined. We choose $a_j\in S^1(\mathbb{R}^n\times\mathbb{R}^{n-1})$ equal to a_j^0 where $|\xi'|>1$ and obtain

$$p(x,\xi)=\prod_1^m(\xi_1-a_j(x,\xi'))$$

when $|\xi'|>1$ and x is in the neighborhood U_0 of 0 where $\chi(x/\delta)=1$. Let $0\in U_1\Subset U_0$. If we apply (28.1.4) when $j>r$ and (28.1.8) when $j\leq r$, it follows that for small ε and $1/\tau$, and $1\leq k\leq m$,

$$\tau\int|\prod_{j\neq k}(D_1-a_j(\varepsilon x,D'))u|^2 e^{2\tau\phi}dx\leq C\int|p(\varepsilon x,D)u|^2 e^{2\tau\phi}dx+M, \quad u\in C_0^\infty(U_1),$$
$$M=\sum_{|\alpha|<m}\int|D^\alpha u|^2 e^{2\tau\phi}dx.$$

In fact, if $\psi\in C_0^\infty(U_0)$ is equal to 1 in U_1 then $u=\psi u$ and

$$u\mapsto\left(\prod_1^m(D_1-a_j(\varepsilon x,D'))-p(\varepsilon x,D)\right)\psi u$$

is an operator of order $m-1$ for an arbitrary order of the factors in the product. Using an interpolation formula as in the proof of Lemma 23.2.1 we can if $|\alpha|=m-1$ write

$$D^\alpha=\sum g_k(x,D')\prod_{j\neq k}(D_1-a_j(\varepsilon x,D'))+R_\varepsilon(x,D)$$

where g_k is of order 0 and R_ε is of order $m-2$. (We choose $\sum g_k(x,D')=0$ or 1 exactly when $\alpha_1<m-1$ and $\alpha_1=m-1$ respectively.) Hence we obtain

$$\sum_{|\alpha|<m} \tau^{2(m-|\alpha|)-1} \int |D^\alpha u|^2 e^{2\tau\phi}\, dx$$
$$\leqq C\Big(\int |p(\varepsilon x, D)u|^2 e^{2\tau\phi}\, dx + \sum_{|\alpha|<m} \tau^{2(m-|\alpha|-1)} \int |D^\alpha u|^2 e^{2\tau\phi}\, dx\Big),$$

for it follows from (28.1.8) that the left-hand side can be estimated by a constant times the sum when $|\alpha| = m-1$ only. When τ is large enough the estimate

(28.1.10)
$$\sum_{|\alpha|<m} \tau^{2(m-|\alpha|)-1} \int |D^\alpha u|^2 e^{2\tau\phi}\, dx \leqq C \int |p(\varepsilon x, D)u|^2 e^{2\tau\phi}\, dx, \quad u \in C_0^\infty(U_1),$$

follows. When $n=3$ the estimate follows by localizing in the ξ' variables also, say for $\pm\xi_2 < |\xi_3|$. This small problem is left as an exercise for the reader.

By Friedrichs' lemma (Lemma 17.1.5) the estimate (28.1.10) remains valid if $u \in H_{(m-1)}^{\mathrm{comp}}$ and $p(\varepsilon x, D)u \in L^2$. The proof of Theorem 17.2.1 is then applicable with no change at all to show that (28.1.9) and (28.1.2) imply $u=0$ in a neighborhood of 0. The repetition is left for the reader.

Theorem 28.1.1 can be improved in several ways. It is not hard to relax the condition that the complex characteristics are simple. In fact, if a_1 and a_2 satisfy the hypotheses of Proposition 28.1.2 then we obtain for small ε and $1/\tau$

(28.1.11)
$$\sum_{|\alpha|\leqq 1} \tau^{2(1-|\alpha|)} \int |D^\alpha u|^2 e^{2\tau\phi}\, dx \leqq C\tau \int |(D_1 - a_2(\varepsilon x, D'))u|^2 e^{2\tau\phi}\, dx$$
$$\leqq C' \int |(D_1 - a_1(\varepsilon x, D'))(D_1 - a_2(\varepsilon x, D'))u|^2 e^{2\tau\phi}\, dx,$$

if $u \in C_0^\infty(X_0)$. Although this is a weaker estimate than (28.1.4) it can still be used to prove a uniqueness theorem where $p(x, \xi + \tau N)$ is allowed to have double non-real roots provided that they depend smoothly on ξ. However, this is a far too restrictive condition for we shall now prove that there is an estimate of the form (28.1.11) for second order elliptic operators having no smooth factorization.

Proposition 28.1.4. *Let* $a \in S^1(\mathbb{R}^n \times \mathbb{R}^{n-1})$, $q \in S^2(\mathbb{R}^n \times \mathbb{R}^{n-1})$, *and assume in addition to* (28.1.3) *the ellipticity condition*

(28.1.12)
$$(1+|\xi|)^2 \leqq C(|(\xi_1 - a(x,\xi'))^2 - q(x,\xi')| + 1, \quad (x,\xi) \in \mathbb{R}^{2n},$$

and that

(28.1.13)
$$|\partial q(x,\xi')/\partial\xi'|^2 + |\partial q(x,\xi')/\partial x'|^2/(1+|\xi'|^2) \leqq C(|q(x,\xi')| + |\xi'| + 1).$$

If ε *and* $1/\tau$ *are small it follows that for* $u \in C_0^\infty(X_0)$

(28.1.14)
$$\sum_{|\alpha|\leqq 1} \tau^{2(1-|\alpha|)} \int |D^\alpha u|^2 e^{2\tau\phi}\, dx \leqq C \int |((D_1 - a(\varepsilon x, D'))^2 - q(\varepsilon x, D'))u|^2 e^{2\tau\phi}\, dx.$$

Before the proof we observe that (28.1.3) is an ellipticity condition for $D_1 - a(x, D')$. If q is homogeneous then (28.1.13) means that the characteristic roots $\zeta_1 = a(x,\xi') \pm q(x,\xi')^{\frac{1}{2}}$ are uniformly Lipschitz continuous where $q \neq 0$ and $|\xi'|=1$. This is of course a much weaker condition than smoothness of the roots and does not even require that they can be chosen continuously. An example is

$$q(x,\xi') = (x_1 + i x_2)(x_1^2 + x_2^2)^{\frac{1}{2}} |\xi'|^2/(1+|x|^2).$$

It would suffice to assume (28.1.12) when $x_1 = 0$ and (28.1.13) for small $|x_1|$; (28.1.12) follows then for small x_1 and (28.1.12), (28.1.13) will be valid for all x_1 if we compose with a retraction in the x_1 variables chosen as in Lemma 19.2.11.

When $q=0$ the estimate (28.1.14) follows from (28.1.11). We shall essentially follow the same proof in general and show that the presence of q only causes errors which in some sense are small. As in the proof of Proposition 28.1.2 we write $A = a(\varepsilon x, D')$, and with $Q = q(\varepsilon x, D')$ we write

$$P_\tau = (D_1 + i\tau\phi' - A)^2 - Q.$$

Lemma 28.1.5. *From* (28.1.12) *it follows for* $v \in C_0^\infty(X_0)$ *that*

$$\sum_{|\alpha|\leq 2} \tau^{2-|\alpha|} \|D^\alpha v\|_{(0,-1)} \leq C(\|P_\tau v\|_{(0,-1)} + \tau\|v\|), \tag{28.1.15}$$

$$\|v\|_{(1)} \leq C(\|(|D|^2+\tau^2)^{-\frac{1}{2}} P_\tau v\| + \tau\|v\|). \tag{28.1.16}$$

Proof. By Proposition 20.1.11 we have

$$\|u\|_{(2)} \leq C(\|P_0 u\| + \|u\|_{(1)}), \quad u \in C_0^\infty.$$

Hence

$$\|u\|_{(2)} \leq C_1(\|P_\tau u\| + \tau^2\|u\| + \tau\|u\|_{(1)}),$$

and since $C_1 \tau \|u\|_{(1)} \leq \frac{1}{2}\|u\|_{(2)} + \frac{1}{2}(C_1\tau)^2\|u\|$, it follows that

$$\|(|D|^2+\tau^2)u\| \leq C_2(\|P_\tau u\| + \tau^2\|u\|), \quad u \in C_0^\infty(X_0). \tag{28.1.17}$$

This remains true when $u \in \mathscr{S}$ and $|x_1| < \frac{1}{2}$ in supp u, so we may take $u = (|D'|^2+\tau^2)^{-\frac{1}{2}} v$ if $v \in C_0^\infty(X_0)$. The commutator of P_τ and $(|D'|^2+\tau^2)^{-\frac{1}{2}}$ has L^2 norm $O(\tau)$, for the coefficient of τ^2 commutes and the commutators with A and Q are of order -1 and 0 respectively. Hence

$$\|(|D|^2+\tau^2)(|D'|^2+\tau^2)^{-\frac{1}{2}} v\| \leq C(\|(|D'|^2+\tau^2)^{-\frac{1}{2}} P_\tau v\| + \tau\|v\|). \tag{28.1.18}$$

This suffices for the proof of (28.1.15) in the frequency range where $|\xi'| > \tau$, but when $|\xi'| < \tau$ we must use an argument regarding P_τ as a perturbation not of P_0 but of $(D_1 + i\tau\phi')^2$. We have

$$\tau\|u\| \leq 2\|(D_1 + i\tau\phi')u\|, \quad u \in C_0^\infty(X_0),$$

by the proof of Proposition 28.1.3, hence

$$\|D_1 u\| + \tau\|u\| \leq \|(D_1 + i\tau\phi')u\| + 5\tau\|u\|/2 \leq 6\|(D_1 + i\tau\phi')u\|$$

which gives by iteration

$$\|D_1^2 u\| + 2\tau \|D_1 u\| + (\tau^2 - 6\tau)\|u\| \leqq 36 \|(D_1 + i\tau\phi')^2 u\|, \quad u \in C_0^\infty(X_0).$$

If we replace u by $(1+|D'|^2)^{-\frac{1}{2}} v$ it follows that for large τ

$$\sum_0^2 \tau^{2-j} \|D_1^j v\|_{(0,-1)} \leqq 40 \|(D_1 + i\tau\phi')^2 v\|_{(0,-1)}$$
$$\leqq 40 \|P_\tau v\|_{(0,-1)} + C \|(|D|^2 + \tau^2)^{\frac{1}{2}} v\|.$$

The last term can be estimated by means of (28.1.18) which gives

$$\|(D_1^2 + \tau^2)(1+|D'|^2)^{-\frac{1}{2}} v\| \leqq C(\|P_\tau v\|_{(0,-1)} + \tau \|v\|). \tag{28.1.19}$$

(28.1.15) follows from (28.1.18) and (28.1.19) since

$$(\tau^2 + |\xi|^2)(1+|\xi'|^2)^{-\frac{1}{2}} \leqq 2(\tau^2 + |\xi|^2)(\tau^2 + |\xi'|^2)^{-\frac{1}{2}}, \quad |\xi'| > \tau,$$
$$(\tau^2 + |\xi|^2)(1+|\xi'|^2)^{-\frac{1}{2}} \leqq 2(\xi_1^2 + \tau^2)(1+|\xi'|^2)^{-\frac{1}{2}}, \quad |\xi'| < \tau.$$

To prove (28.1.16) we choose $\chi \in C_0^\infty(\mathbb{R})$ with support in $(-1,1)$ and equal to 1 in $(-\frac{1}{2}, \frac{1}{2})$. The estimate (28.1.17) remains valid for some C_2 if $|x_1| < 1$ in $\operatorname{supp} u$, so we can take $u = \chi(x_1)(|D|^2 + \tau^2)^{-\frac{1}{2}} v$ where $v \in C_0^\infty(X_0)$. Then

$$(|D|^2 + \tau^2) u - (|D|^2 + \tau^2)^{\frac{1}{2}} v = (|D|^2 + \tau^2)[\chi, (|D|^2 + \tau^2)^{-\frac{1}{2}}] v,$$

and the calculus of pseudo-differential operators with the metric $|dx|^2 + |d\xi|^2/(|\xi|^2 + \tau^2)$ shows that the operator on the right is bounded. Furthermore, with $\chi_1(x_1) = \chi(x_1/2)$,

$$P_\tau u - \chi_1 (|D|^2 + \tau^2)^{-\frac{1}{2}} P_\tau v = \chi_1 [P_\tau \chi, (|D|^2 + \tau^2)^{-\frac{1}{2}}] v.$$

The norm of the commutator is uniformly bounded since

$$[D_1 + i\tau\phi', (|D|^2 + \tau^2)^{-\frac{1}{2}}] = \tau D_1 (|D|^2 + \tau^2)^{-\frac{1}{2}}$$

and the symbol of $[Q, (|D|^2 + \tau^2)^{-\frac{1}{2}}]$ is uniformly bounded in the symbol class $S(1, |dx|^2 + |d\xi|^2/(1+|\xi'|^2))$. Hence (28.1.16) follows from (28.1.17) and the proof is complete.

Proof of Proposition 28.1.4. To a large extent we shall follow the proof of Proposition 28.1.2. First we note as there that (28.1.14) is equivalent to

$$\sum_{|\alpha| \leqq 1} \tau^{2(1-|\alpha|)} \|D^\alpha v\|^2 \leqq C \|P_\tau v\|^2, \quad v \in C_0^\infty(X_0). \tag{28.1.14)'$$

Writing $T = D_1 + i\tau\phi' - A$, which has all the properties of P_τ in the proof of Proposition 28.1.2, we have

$$[P_\tau^*, P_\tau] = 2T^*[T^*, T]T + 2T[T^*, T]T^* + [T, [T^*, [T^*, T]]] \tag{28.1.20}$$
$$- [T^{*2}, Q] - [Q^*, T^2] + [Q^*, Q].$$

In fact, the sum of the first three terms is $[T^{*2}, T^2]$ since

$$\begin{aligned} T^* T^* TT &= T^*[T^*,T]T+[T^*,T]T^*T+TT^*T^*T \\ &= 2T^*[T^*,T]T+[[T^*,T],T^*]T+TT^*T^*T, \\ TTT^*T^* &= 2T[T,T^*]T^*+T[T^*,[T,T^*]]+TT^*T^*T. \end{aligned}$$

We shall at first use only the first two terms on the right-hand side of (28.1.20) and afterwards estimate the errors caused by the others.

By (28.1.5) we have $[T^*,T]=2\tau+B$ where B/ε is uniformly bounded in S^1. Hence

$$\begin{aligned} &2\tau(\|Tv\|^2+\|T^*v\|^2) \\ &\leqq 2((T^*[T^*,T]T+T[T^*,T]T^*)v,v) \\ &\quad + C\varepsilon(\|Tv\|_{(0,1)}\|Tv\|+\|T^*v\|_{(0,1)}\|T^*v\|). \end{aligned} \tag{28.1.21}$$

By (28.1.16) applied to Tv we have

$$\|Tv\|_{(0,1)} \leqq C(\|P_\tau v\|+\tau\|Tv\|),$$

for $(|D|^2+\tau^2)^{-\frac{1}{2}}T$ is uniformly bounded and

$$\|(|D|^2+\tau^2)^{-\frac{1}{2}}[P_\tau,T]v\| = \|(|D|^2_\varepsilon+\tau^2)^{-\frac{1}{2}}[-Q,D_1-A]v\| \leqq C_1\|v\|_{(1)} \leqq C_2\tau\|Tv\|$$

where the last inequality follows from Proposition 28.1.2. If we apply (28.1.16) to T^*v instead we obtain

$$\|T^*v\|_{(0,1)} \leqq C(\|P_\tau v\|+\tau\|T^*v\|+\tau^{\frac{1}{2}}\|Tv\|),$$

for

$$[P_\tau,T^*]=[-Q,T^*]+2[T,T^*]T+[T,[T,T^*]],$$

hence, again by (28.1.4),

$$\|(|D|^2+\tau^2)^{-\frac{1}{2}}[P_\tau,T^*]v\| \leqq C_3(\|v\|_{(1)}+\|Tv\|) \leqq C_4\tau^{\frac{1}{2}}\|Tv\|.$$

If we use the preceding estimates of $\|Tv\|_{(0,1)}$ and $\|T^*v\|_{(0,1)}$ in (28.1.21), it follows for small ε that

$$\begin{aligned} \tau(\|Tv\|^2+\|T^*v\|^2) &\leqq 2((T^*[T^*,T]T+T[T^*,T]T^*)v,v) \\ &\quad + C\tau^{-1}\|P_\tau v\|^2. \end{aligned} \tag{28.1.22}$$

By Proposition 28.1.2 we have as already observed

$$\tau^2\|v\|^2+\|v\|^2_{(1)} \leqq C\tau\|Tv\|^2, \tag{28.1.23}$$

so what remains is to show that the last four terms in (28.1.20) play no role.

First we shall prove that

$$(([Q^*,Q]-[T^{*2},Q]-[Q^*,T^2])v,v) \geqq -C\varepsilon(\|P_\tau v\|^2+\tau\|Tv\|^2), \tag{28.1.24}$$

when $v\in C_0^\infty(X_0)$. To prove this we shall use the Fefferman-Phong inequality (Theorem 18.6.8) with the metric

$$g=|dx|^2+h^2|d\xi|^2, \quad h^{-2}=1+|\xi'|^2.$$

If $0 \leqq c \in S((|\xi|+\tau)^4, g)$, it follows from Theorem 18.6.8 that

$$(c^w(x,D)v,v) \geqq -C((|D|^2+\tau^2)^2(|D'|^2+1)^{-1}v,v).$$

In fact, if $q(\xi)=(|\xi'|^2+1)^{\frac{1}{2}}(|\xi|^2+\tau^2)^{-1}$, this is equivalent to

$$(q(D)c^w(x,D)q(D)v,v) \geqq -C\|v\|^2,$$

and the Weyl symbol of $q(D)c^w(x,D)q(D)$ is equal to $q(\xi)^2 c(x,\xi)$ with an error in $S(1,g)$ since $\{q,c\}q+q\{c,q\}=0$. Let us therefore look for a lower bound for the Weyl symbol of

$$\varepsilon^{-1}([Q^*,Q]-[T^{*2},Q]-[Q^*,T^2])+P_\tau^* P_\tau. \tag{28.1.25}$$

If $t_{\tau,\varepsilon}$ is the Weyl symbol of T and q_ε is that of Q, then the Weyl symbol is

$$\varepsilon^{-1}(\{\bar{q}_\varepsilon,q_\varepsilon\}/i-4\operatorname{Im} t_{\tau,\varepsilon}\{t_{\tau,\varepsilon},\bar{q}_\varepsilon\})+|t_{\tau,\varepsilon}^2-q_\varepsilon|^2+\{\bar{t}_{\tau,\varepsilon}^2-\bar{q}_\varepsilon,t_{\tau,\varepsilon}^2-q_\varepsilon\}/2i \tag{28.1.26}$$

modulo terms of weight $(|\xi|^2+\tau^2)^2 h^2$. (Note that $[T^{*2},Q]=T^*[T^*,Q]+[T^*,Q]T^*$ where the symbol of $[T^*,Q]/\varepsilon$ is bounded in $S(1+|\xi'|^2,g)$). It is clear that (28.1.13) remains true for q_ε with $\partial q_\varepsilon/\partial x'$ divided by ε. Hence

$$|\{\bar{q}_\varepsilon,q_\varepsilon\}/\varepsilon| \leqq C(|q_\varepsilon|(1+|\xi'|)+(1+|\xi'|^2)),$$
$$|\{t_{\tau,\varepsilon},\bar{q}_\varepsilon\}/\varepsilon|=|\{t_{0,\varepsilon},\bar{q}_\varepsilon\}/\varepsilon| \leqq C(|q_\varepsilon|+1+|\xi'|)^{\frac{1}{2}}(1+|\xi'|),$$
$$|\{\bar{t}_{\tau,\varepsilon},t_{\tau,\varepsilon}\}| \leqq 2\tau+C\varepsilon(1+|\xi'|).$$

Hence we can estimate (28.1.26) from below by

$$\begin{aligned}&|t_{\tau,\varepsilon}^2-q_\varepsilon|^2-C(|q_\varepsilon|+1+|\xi'|+|t_{\tau,\varepsilon}|^2)(\tau+|\xi'|)\\ &\geqq |t_{\tau,\varepsilon}^2-q_\varepsilon|^2-C(|t_{\tau,\varepsilon}^2-q_\varepsilon|+1+|\xi'|+2|t_{\tau,\varepsilon}|^2)(\tau+|\xi'|)\\ &\geqq -C_1((\tau+|\xi'|)^2+|t_{\tau,\varepsilon}|^2(\tau+|\xi'|)).\end{aligned}$$

Since the principal symbol of $2T^*(\tau^2+|D'|^2)^{\frac{1}{2}}T$ is $\geqq |t_{\tau,\varepsilon}|^2(\tau+|\xi'|)$, it follows from the Fefferman-Phong inequality applied to the sum of (28.1.25) and

$$2C_1 T^*(\tau^2+|D'|^2)^{\frac{1}{2}}T+K(\tau^2+|D'|^2)$$

for some large K that the left-hand side of (28.1.24) is bounded from below by ε times

$$-\|P_\tau v\|^2-2C_1((|D'|^2+\tau^2)^{\frac{1}{2}}Tv,Tv)-C_2\|(|D|^2+\tau^2)(|D'|^2+1)^{-\frac{1}{2}}v\|^2.$$

Since $\|(|D'|^2+\tau^2)^{\frac{1}{4}}Tv\| \leqq C(\|P_\tau v\|+\tau\|Tv\|)$ as proved above, we obtain (28.1.24) if the last term is estimated by (28.1.15) and (28.1.23) is used.

Finally, the commutator $[T,[T^*,[T^*,T]]]$ is just an operator with symbol bounded in $S(\varepsilon^3(1+|\xi'|),g)$. Summing up (28.1.20), (28.1.22), (28.1.24) we obtain for small ε

$$2\|P_\tau v\|^2 \geqq \|P_\tau v\|^2+([P_\tau^*,P_\tau]v,v) \geqq \tfrac{1}{2}\tau(\|Tv\|^2+\|T^*v\|^2)-\varepsilon^3\|v\|_{(0,1)}^2 \geqq \tau\|Tv\|^2/3$$

where (28.1.23) has been used in the last estimate. This estimate combined with (28.1.23) completes the proof of Proposition 28.1.4.

Proposition 28.1.3 does not require that the principal symbol of a is real but by itself this does not improve Theorem 28.1.1. Indeed, if $\tau_j(x,\xi')$ is one of the roots then $-\tau_j(x,-\xi')$ is also among the roots, and if both have non-negative imaginary parts then τ_j is real. However, we get a useful result if $-\tau_j(x,\xi')$ satisfies the condition in the following proposition instead. (Note that it contains Proposition 28.1.2 but the proof is much less elementary so we preferred to give a separate proof of that result.)

Proposition 28.1.6. *Let $a\in S^1(\mathbb{R}^n\times\mathbb{R}^{n-1})$ and assume that for the Poisson bracket*

(28.1.27) $$b(x,\xi')=\{\xi_1-\bar{a},\xi_1-a\}/i=2\{\operatorname{Re}a,\operatorname{Im}a\}-2\partial\operatorname{Im}a/\partial x_1$$

we have either

(28.1.28) $$|b(x,\xi')|\leqq C_0|\operatorname{Im}a(x,\xi')|+C_1$$

or

(28.1.29) $$b(x,\xi')\geqq -C_0\operatorname{Im}a(x,\xi')-C_1$$

for some constants C_0, C_1. Then it follows that for small ε, $1/\tau$

(28.1.30) $$\tau\int|u|^2e^{2\tau\phi}dx\leqq 3\int|(D_1-a(\varepsilon x,D'))u|^2e^{2\tau\phi}dx,\quad u\in C_0^\infty(X_0).$$

Proof. We start as in the proof of Proposition 28.1.2, and keep the notation used there. To exploit (28.1.5)′ we must estimate (Bv,v) from below. A principal symbol of B is given by $\varepsilon b(\varepsilon x,\xi')$. If (28.1.29) holds then the principal symbol of $B+\varepsilon C_0(A-A^*)/2i$ is bounded below, so

(28.1.31) $$(Bv,v)+\varepsilon C_0\operatorname{Im}(Av,v)\geqq -C\|v\|^2$$

by the sharp Gårding inequality (Theorem 18.1.14). Set

$$M^2=\max(0,-(Bv,v)).$$

From (28.1.5)′ and (28.1.31) we obtain

(28.1.32) $$\|P_\tau^*v\|\leqq\|P_\tau v\|+M,$$

(28.1.33) $$M^2\leqq\tfrac12\varepsilon C_0\|(A^*-A)v\|\,\|v\|+C\|v\|^2.$$

Since

$$\|(A^*-A)v\|\leqq\|P_\tau^*v-P_\tau v\|+2\tau\|\phi'v\|\leqq 2\|P_\tau v\|+M+3\tau\|v\|$$

it follows for large τ that

$$M^2\leqq\varepsilon C_0(\|P_\tau v\|+2\tau\|v\|)\,\|v\|+\tfrac12 M^2.$$

Hence

$$M^2\leqq 2\varepsilon C_0(\|P_\tau v\|+2\tau\|v\|)\,\|v\|.$$

If this estimate is used in (28.1.5)′ we obtain

$$2\tau\|v\|^2 \leqq \|P_\tau v\|^2 + 2\varepsilon C_0(\|P_\tau v\| + 2\tau\|v\|)\|v\|.$$

When $2\varepsilon C_0 < \frac{1}{2}$ it follows that

$$\tau\|v\|^2 \leqq \|P_\tau v\|^2 + \tfrac{1}{2}\|P_\tau v\|\,\|v\| \leqq 2\|P_\tau v\|^2 + \|v\|^2$$

which proves (28.1.30).

Next assume that (28.1.28) holds. Then the real part of the symbol of

$$B^* B + \varepsilon^2 C_0^2 (A - A^*)^2/2 \tag{28.1.34}$$

is uniformly bounded above. In fact, the Weyl symbols of B and $(A-A^*)/2i$ are $\varepsilon b(\varepsilon x, \xi') + b_\varepsilon(x, \xi')$ and $\operatorname{Im} a(\varepsilon x, \xi') + a_\varepsilon(x, \xi')$ respectively, where a_ε and b_ε are bounded in S^0. Apart from an error which is uniformly bounded in S^0 the Weyl symbol of (28.1.34) is therefore

$$(\varepsilon b(x, \xi') + b_\varepsilon(x, \xi'))^2 - 2\varepsilon^2 C_0^2 (\operatorname{Im} a(\varepsilon x, \xi') + a_\varepsilon(x, \xi'))^2$$

which is bounded above by (28.1.28). Hence it follows from the Fefferman-Phong inequality (Theorem 18.6.8) that

$$\|Bv\|^2 - \varepsilon^2 C_0^2 \|(A - A^*)v\|^2/2 \leqq C\|v\|^2.$$

Thus

$$-(Bv, v) \leqq \varepsilon C_0 \|(A - A^*)v\|\,\|v\| + C^{\frac{1}{2}}\|v\|^2$$

which means that (28.1.33) is valid with ε replaced by 2ε. The rest of the proof in the first case is therefore applicable with no change. The proof is complete.

It is now fairly clear that Theorem 28.1.1 can be improved so that we get uniqueness whenever p can be factored into a product of operators to which Proposition 28.1.3, 28.1.4 or 28.1.6 is applicable. This leads us to introduce a smaller set than the set in (28.1.1). It is somewhat easier to define the complement as we shall do now.

Definition 28.1.7. Let X be a C^∞ manifold and let $p \in C^\infty(T^*(X))$ be a homogeneous polynomial of degree m in the fibers. Then $\Gamma(p)$ is the set of all $(x, N) \in T^*(X) \setminus 0$ such that

(i) $p(x, N) \neq 0$;

(ii) for every $\xi \in T_x^* \setminus \mathbb{R}N$ the real zeros τ of $p(x, \xi + \tau N)$ are simple and the complex zeros are at most double;

(iii) for every $\xi \in T_x^* \setminus 0$ with $p(x, \xi) = 0$ there is a neighborhood U in $T^*(X)$ such that either

$$\{\bar{p}, p\}/i \geqq -C|p| \quad \text{in } U \tag{28.1.35}$$

for some constant C, or else p is microhyperbolic with respect to N in U;

(iv) for every $\zeta=\xi+\tau N$ such that $\xi\in T_x^*$, $\operatorname{Im}\tau\neq 0$ and $P(x,\zeta)=0$ there is a neighborhood U in the complexified cotangent bundle such that with respect to some local coordinates

$$|p'_{y,\eta}(y,\eta)|\leq C|\langle p'_\eta(y,\eta),N\rangle| \quad \text{if } (y,\eta)\in U, p(y,\eta)=0. \tag{28.1.36}$$

Some comments on conditions (iii) and (iv) may be useful. If we pass from ξ to $-\xi$ in (iii) we conclude that

$$|\{\bar{p},p\}|\leq C|p| \quad \text{in } U \tag{28.1.35$'$}$$

unless p is microhyperbolic with respect to $-N$ in U. (28.1.35)$'$ is a very strong version of condition (P), for on a bicharacteristic for $\operatorname{Re}p$ it shows that $|H_{\operatorname{Re}p}\operatorname{Im}p|\leq C|\operatorname{Im}p|$ so $\operatorname{Im}p$ cannot have a zero without vanishing identically. In the microhyperbolic case the condition (P) is of course also valid; this is no surprise since Carleman estimates imply local solvability. However, it would not suffice to assume condition (P) in Proposition 28.1.6. Note that condition (iii) is *not* invariant under a change of sign for N.

Condition (iv) is automatically fulfilled if τ is a simple zero. Otherwise it means that the projection in the complexified tangent space of X of the normalized Hamilton vector at a simple zero close to $\xi+\tau N$ cannot be nearly orthogonal to N. This is a natural substitute for the condition that the zeros are simple. Note that it is always true if the zero remains double when x, ξ varies, or more generally if the complex zeros of $p(y,\eta+\tau N)$ are C^1 functions of (y,η). It is obvious that $\Gamma(p)$ is open, for the zeros of $p(x,\xi+\tau N)$ depend continuously on x,ξ,N, and conditions (i), (iii), (iv) obviously hold in open sets.

In the following uniqueness theorem we consider solutions of a differential equation with lower order coefficients in L^∞, which is clearly equivalent to a differential inequality (28.1.2).

Theorem 28.1.8. *If P is a differential operator of order m with C^∞ principal symbol p and all coefficients in L^∞_{loc}, then $u\in H^{\text{loc}}_{(m-1)}$ and $Pu=0$ implies $N_e(\operatorname{supp}u)\cap\Gamma(p)=\emptyset$.*

Proof. As in the proof of Theorem 28.1.1 we may assume that P is defined in a neighborhood of $0\in\mathbb{R}^n$, and that $(0,N)\in\Gamma(p)$ if $N=(1,0,\dots,0)$; we must show that $0\notin\operatorname{supp}u$ if $Pu=0$ and u satisfies (28.1.9). To do so we first discuss the condition $(0,N)\in\Gamma(p)$.

Let $0\neq\xi'_0\in\mathbb{R}^{n-1}$, and let $p(0,\tau_0,\xi'_0)=0$. The following cases can occur:

(a) τ_0 is real. Then τ_0 is a simple zero by Definition 28.1.7 (ii). Hence there is a unique solution $\tau(x,\xi')$ of the equation $p(x,\tau,\xi')=0$ which is a C^∞ function of (x,ξ') in a neighborhood U of $(0,\xi'_0)$ and satisfies $\tau(0,\xi'_0)=\tau_0$. Thus

$$p(x,\xi)=(\xi_1-\tau(x,\xi'))\,p_1(x,\xi), \quad (x,\xi')\in U,$$

where p_1 is a polynomial in ξ_1 of degree $m-1$.

(a′) If p is microhyperbolic with respect to N then $\operatorname{Im}\tau \leqq 0$ in some neighborhood U_1 of $(0,\xi_0')$, which is a microlocal form of the condition in Proposition 28.1.3.

(a″) If p is not microhyperbolic with respect to N, we obtain from (28.1.35) since $p_1(0,\tau_0,\xi_0') \neq 0$ that

$$\{\xi_1-\bar\tau, \xi_1-\tau\}/i \geqq -C|\xi_1-\tau|$$

in a neighborhood of $(0,\tau_0,\xi_0')$. The left-hand side is independent of ξ_1. Taking $\xi_1 = \operatorname{Re}\tau(x,\xi')$ we conclude that

$$\{\xi_1-\bar\tau, \xi_1-\tau\}/i \geqq -C|\operatorname{Im}\tau|$$

in some neighborhood of $(0,\xi_0')$. If p is microhyperbolic with respect to $-N$ then $\operatorname{Im}\tau \geqq 0$ in some neighborhood U_2, so

$$\{\xi_1-\bar\tau, \xi_1-\tau\}/i \geqq -C\operatorname{Im}\tau(x,\xi'), \quad (x,\xi') \in U_2,$$

which is a microlocal form of condition (28.1.29) in Proposition 28.1.6. On the other hand, if p is not microhyperbolic with respect to $-N$ we can apply (28.1.35) to the root $-\tau(x,-\xi')$ too and obtain for some neighborhood U_3 of $(0,\xi_0')$

$$|\{\xi_1-\bar\tau, \xi_1-\tau\}| \leqq C|\operatorname{Im}\tau(x,\xi')|, \quad (x,\xi') \in U_3,$$

which is a microlocal form of condition (28.1.28) in Proposition 28.1.6.

(b) τ_0 is a non-real simple root. Then p has an elliptic factor $\xi_1-\tau(x,\xi')$ where $\tau \in C^\infty$ in a conic neighborhood of $(0,\xi_0')$ and $\tau(0,\xi_0')=\tau_0$. This is the microlocalized hypothesis of Proposition 28.1.2.

(c) τ_0 is a double root, hence non-real. Then p has a second order polynomial factor,

$$p(x,\xi) = (\xi_1^2 - 2a(x,\xi')\xi_1 + b(x,\xi'))\,p_1(x,\xi),$$

where $a, b \in C^\infty$ in a neighborhood of $(0,\xi_0')$ and $a(0,\xi_0')=\tau_0$, $b(0,\xi_0')=\tau_0^2$, and p_1 is a polynomial in ξ_1 of degree $m-2$. This follows from the Weierstrass preparation theorem at $(0,\tau_0,\xi_0')$ or rather from its proof, for p is analytic in ξ_1. With the notation $q=a^2-b$ the second order factor can be written

$$(\xi_1 - a(x,\xi'))^2 - q(x,\xi').$$

Since $p_1 \neq 0$ at $(0,\tau_0,\xi_0')$ the condition (28.1.36) means that

$$|\partial q/\partial x'| + |\partial q/\partial \xi'| \leqq C|\xi_1 - a| \quad \text{if } (\xi_1-a)^2 = q,$$

in a neighborhood of $(0,\tau_0,\xi_0')$ in $\mathbb{R}^n \times \mathbb{C}^n$ now. Thus

$$|\partial q/\partial x'| + |\partial q/\partial \xi'| \leqq C|q|^{\frac{1}{2}}.$$

In view of the homogeneity we obtain (28.1.13) microlocally; the ellipticity of $(\xi_1-a)^2-q$ and ξ_1-a is also obvious then.

Summing up, for (x, ξ') in a conic neighborhood of $(0, \xi_0')$ we can write

$$p(x,\xi)=\prod_1^k(\xi_1-\tau_j(x,\xi'))\prod_1^l((\xi_1-a_j(x,\xi'))^2-q_j(x,\xi'))$$

where τ_j satisfies some of the conditions in Proposition 28.1.3 or 28.1.6, the second order factors satisfy the hypotheses of Proposition 28.1.4, and different factors have no common zeros.

Let us first assume that such a factorization holds in $\mathbb{R}^{2n}$ when $|x_1|$ is small enough, and that $p(x,\xi)$ is independent of x when $|x|$ is large. Then it follows that

$$\sum_{|\alpha|<m}\tau^{2(m-1-|\alpha|)}\int|D^\alpha u|^2 e^{2\tau\phi}\,dx\leq C\int|p(\varepsilon x,D)u|^2 e^{2\tau\phi}\,dx, \tag{28.1.37}$$

if $u\in C_0^\infty(X_0)$, $X_0=\{x\in\mathbb{R}^n, |x_1|<1\}$ and ε is small enough. The proof differs so little from that of (28.1.10) that we shall only sketch it briefly. After modifying $\tau_j(x,\xi')$, $a_j(x,\xi')$ and $q_j(x,\xi')$ to smooth functions for $|\xi'|<1$ we let $p_j(\varepsilon x, D)$, $j=1,\dots,k+l$, denote the operators $D_1-\tau_j(\varepsilon x, D')$ and the operators $(D_1-a_j(\varepsilon x,D'))^2-q_j(\varepsilon x,D')$ in some arbitrary order. Then

$$\int|p(\varepsilon x,D)u-\prod p_j(\varepsilon x,D)u|^2 e^{2\tau\phi}\,dx\leq C\varepsilon^2 N$$

where N denotes the left-hand side of (28.1.37). Placing p_k first in the product we have by Propositions 28.1.3, 28.1.4, 28.1.6

$$\sum_{r+|\alpha|<m_k}\tau^{2r}\int|D^\alpha\prod_{j\neq k}p_j(\varepsilon x,D)u|^2e^{2\tau\phi}\,dx\leq C\int|\prod p_j(\varepsilon x,D)u|^2e^{2\tau\phi}\,dx$$

where m_j is the degree of p_j. By partial fraction decomposition we have for $|\xi'|>1$ and $|\alpha|=m-1$

$$\xi^\alpha=\sum q_k(\varepsilon x,\xi)\prod_{j\neq k}p_j(\varepsilon x,\xi)$$

where q_k is homogeneous of degree m_k-1 in ξ and a polynomial in ξ_1 of degree $< m_k$. Using (28.1.8) also when $|\alpha|<m-1$ we conclude as in the proof of (28.1.10) that

$$N(1-C\varepsilon^2)\leq C\int|p(\varepsilon x,D)u|^2 e^{2\tau\phi}\,dx$$

which gives (28.1.37) when ε is small enough.

Our local factorization gives a similar result. If $\psi(x',\xi')\in S^0$ has support in a sufficiently small conic neighborhood of $(0,\xi_0')$ then

$$\begin{aligned}\sum_{|\alpha|<m}\tau^{2(m-1-|\alpha|)}\int|D^\alpha\psi(\varepsilon x',D')u|^2 e^{2\tau\phi}\,dx \\ \leq C\Big(\int|\psi(\varepsilon x',D')p(\varepsilon x,D)u|^2 e^{2\tau\phi}\,dx \\ +\varepsilon^2\sum_{|\alpha|<m}\tau^{2(m-1-|\alpha|)}\int|D^\alpha u|^2 e^{2\tau\phi}\,dx.\end{aligned} \tag{28.1.38}$$

(We may assume that the coefficients of p are in $C_0^\infty(\mathbb{R}^n)$.) For the proof one just has to establish microlocal versions of Propositions 28.1.3, 28.1.4, 28.1.6.

In the first two cases this can be done by extending the symbols globally so that the full hypotheses of these results hold; in Proposition 28.1.4 one just has to cut off q by multiplication with the square of a cutoff function and then piece together a with $\pm i|\xi'|$. In the case of Proposition 28.1.6 we obtain a microlocal version by inspecting the proof.

If we now choose $\psi_1, \dots, \psi_N$ such that (28.1.38) holds for each of them and $\psi_1 + \dots + \psi_N = 1$ in $U \times \mathbb{R}^{n-1}$ for some neighborhood U of 0, it follows for small ε that (28.1.37) is valid for $u \in C_0^\infty(X_1)$ if $X_1 \subset X_0$ is a bounded neighborhood of 0. By Friedrichs' lemma (28.1.37) remains valid if $u \in H_{(m-1)}^{\mathrm{comp}}(X_1)$ and $p(\varepsilon x, D) u \in L^2$.

Let us now return to the function u at the beginning of the proof. It vanishes when $x_1 > -|x'|^2/2$ and satisfies the equation

$$p(x,D)u + \sum_{|\alpha|<m} a_\alpha(x) D^\alpha u = 0$$

in a neighborhood X_1 of 0, with all a_α bounded. Then $u_\varepsilon(x) = u(\varepsilon x)$ satisfies the equation

$$p(\varepsilon x, D) u_\varepsilon + \sum_{|\alpha|<m} a_\alpha(\varepsilon x) \varepsilon^{m-|\alpha|} D^\alpha u_\varepsilon = 0$$

in X_1/ε, and $u_\varepsilon = 0$ when $x_1 > -\varepsilon|x'|^2/2$. Choose $\chi \in C_0^\infty(X_1)$ equal to 1 in a neighborhood of 0 and apply (28.1.37) to $v_\varepsilon = \chi u_\varepsilon$ noting that $\phi(x_1) < -c$ for some $c > 0$ in $\operatorname{supp} u_\varepsilon \cap \operatorname{supp} d\chi$. Then it follows that

$$\sum_{|\alpha|<m} \tau^{2(m-1-|\alpha|)} \int |D^\alpha v_\varepsilon|^2 e^{2\tau\phi} dx (1 - C_1 \varepsilon^2)$$
$$\leqq C_2 \sum_{|\alpha|<m} \int_{\phi<-\varepsilon c} |D^\alpha u_\varepsilon|^2 e^{2\tau\phi} dx.$$

Taking $\varepsilon < C_1^{-\frac{1}{2}}$ we conclude when $\tau \to \infty$ that $v_\varepsilon = 0$ when $\phi > -c\varepsilon$. Thus $u = 0$ in a neighborhood of 0 which completes the proof.

28.2. General Carleman Estimates

In Section 28.1 we confined ourselves from the start to Carleman estimates for operators which can be decomposed into first or at least second order factors with respect to a distinguished variable. However, this is not appropriate for the uniqueness theorems with convexity conditions in Section 28.3. We shall therefore take an entirely new look at Carleman estimates now, starting with necessary conditions for estimates of the form (28.1.10).

Theorem 28.2.1. *Let X be an open set in $\mathbb{R}^n$, $\phi \in C^\infty(X, \mathbb{R})$, and let P be a differential operator of order m in X with C^∞ principal symbol p and all coefficients in $L^\infty_{\mathrm{loc}}(X)$. Assume that*

$$(28.2.1) \qquad \sum_{|\alpha|<m} \tau^{2(m-|\alpha|)-1} \int |D^\alpha u|^2 e^{2\tau\phi} dx \leqq K \int |Pu|^2 e^{2\tau\phi} dx, \qquad u \in C_0^\infty(X),$$

when $\tau>\tau_0$. Then it follows that

(28.2.2) $$\sum_{|\alpha|<m} \tau^{2(m-|\alpha|)-1} |(\xi+i\tau\phi'(x))^\alpha|^2 \leqq K\{\bar{p}(x,\xi-i\tau\phi'(x)), p(x,\xi+i\tau\phi'(x))\}/i,$$

if $(x,\xi)\in T^*(X)$, $\tau>0$ and $p(x,\xi+i\tau\phi'(x))=0$. Moreover, p does not vanish of second order at any point in $T^*(X)\smallsetminus 0$.

Before the proof we observe that with the notation $\zeta=\xi+i\tau\phi'(x)$ we have

(28.2.3) $$\{\bar{p}(x,\xi-i\tau\phi'(x)), p(x,\xi+i\tau\phi'(x))\}/i\tau = 2\Big(\sum_{j,k=1}^{n} \partial^2\phi/\partial x_j\partial x_k\, p^{(j)}(x,\zeta)\overline{p^{(k)}(x,\zeta)} + \operatorname{Im}\sum_1^n p_{(j)}(x,\zeta)\overline{p^{(j)}(x,\zeta)}/\tau\Big).$$

When $\phi'(x)$ is fixed the condition (28.2.2) is therefore a convexity condition on ϕ. For the proof of Theorem 28.2.1 we need a lemma.

Lemma 28.2.2. *If $L(x,\xi)=\sum A_j x_j+\sum B_j\xi_j$, with coefficients in $\mathbb{C}$ and $(x,\xi)\in\mathbb{R}^{2n}$, then*

(28.2.4) $$\kappa\|\psi\|^2 \leqq \|L(x,D)\psi\|^2, \quad \psi\in C_0^\infty(\mathbb{R}^n)$$

is equivalent to

(28.2.5) $$\kappa \leqq \max(\{\bar{L},L\}/i, 0).$$

Proof. The symplectic invariance proved in Theorem 18.5.9 shows that both (28.2.4) and (28.2.5) are unchanged if L is replaced by $L\circ\chi$ where χ is a real linear symplectic map. Write $L=L_1+iL_2$ where L_1 and L_2 are real, thus $\{\bar{L},L\}/i=2\{L_1,L_2\}$. If $\{L_1,L_2\}=0$ it follows from Proposition 21.1.3 and Theorem 18.5.9 that we may assume $L=A_1x_1+A_2x_2$. When the support of ψ is taken close to 0 we see at once that (28.2.4) implies $\kappa\leqq 0$, which proves the lemma in this case. If $\{L_1,L_2\}\neq 0$ we may assume $L_1=\xi_1$ and $L_2=ax_1$ where $a=\{L_1,L_2\}$. Then (28.2.4) means that

$$\kappa\|\psi\|^2 \leqq \|(D_1+iax_1)\psi\|^2, \quad \psi\in C_0^\infty.$$

If $a<0$ we can let ψ approach $e^{ax_1^2/2}$ in $\mathscr{S}$ and conclude that $\kappa\leqq 0$ if (28.2.4) holds, which agrees with (28.2.5). If $a>0$ the inequality (28.2.4) is equivalent to

$$(\kappa-2a)\|\psi\|^2 \leqq \|(D_1-iax_1)\psi\|^2, \quad \psi\in C_0^\infty$$

which follows by a partial integration. This is true if and only if $\kappa-2a\leqq 0$, which completes the proof.

Proof of Theorem 28.2.1. We may assume in the proof that $x=0$ and also that $\tau=1$ in (28.2.2), for the inequality is homogeneous. It is no restriction to assume that $\phi(0)=0$, thus

$$\phi(x)=\langle x, N\rangle + A(x) + O(|x|^3)$$

where A is a quadratic form. Let $\xi\in\mathbf{R}^n$ and $p(0,\xi+iN)=0$, which implies $p(0,\tau\xi+i\tau N)=0$ for every τ, and set

$$u(x)=e^{i\langle x,\tau\xi+i\tau N\rangle}\psi(x\tau^{\frac{1}{2}}),\quad \psi\in C_0^\infty(\mathbf{R}^n).$$

Then $u\in C_0^\infty(X)$ for large τ, and with $\zeta=\xi+iN$ we have

$$\begin{aligned}
&e^{-i\langle x,\tau\zeta\rangle}P(x,D)u(x)\\
&=p(x,\tau\zeta)\psi(x\tau^{\frac{1}{2}})+\sum p^{(j)}(x,\tau\zeta)D_j\psi(x\tau^{\frac{1}{2}})+O(\tau^{m-1})\\
&=\tau^{m-\frac{1}{2}}(\sum p_{(j)}(0,\zeta)x_j\tau^{\frac{1}{2}}\psi(x\tau^{\frac{1}{2}})+\sum p^{(j)}(0,\zeta)(D_j\psi)(x\tau^{\frac{1}{2}}))+O(\tau^{m-1}).
\end{aligned}$$

If we divide (28.2.1) by $\tau^{2m-1-n/2}$ and introduce $x\tau^{\frac{1}{2}}$ as a new variable, it follows when $\tau\to\infty$ that

$$\sum_{|\alpha|<m}|\zeta^\alpha|^2\|e^A\psi\|^2\leqq K\|e^A(\sum p_{(j)}(0,\zeta)x_j+\sum p^{(j)}(0,\zeta)D_j)\psi\|^2.$$

If we replace ψ by $e^{-A}\psi$ it follows that

$$\sum_{|\alpha|<m}|\zeta^\alpha|^2\|\psi\|^2\leqq K\|(\sum p_{(j)}(0,\zeta)x_j+\sum p^{(j)}(0,\zeta)(D_j-(D_jA))\psi\|^2.$$

In view of (28.2.3) we obtain (28.2.2) by applying Lemma 28.2.2.

Now assume that $\xi\in\mathbf{R}^n\setminus 0$ and that $p^{(\alpha)}_{(\beta)}(0,\xi)=0$ when $|\alpha+\beta|\leqq 1$. We shall apply (28.2.1) directly to

$$u(x)=e^{i\tau^2\langle x,\xi\rangle}\psi(\tau x),\quad \psi\in C_0^\infty(\mathbf{R}^n).$$

In the support we have $|\phi(x)-\phi(0)|=O(|x|)=O(1/\tau)$, and

$$|p(x,D)u(x)|=\tau^{2m}|p(y/\tau,\xi+D/\tau)\psi(y)|=O(\tau^{2m-2}),\qquad y=\tau x.$$

The right-hand side of (28.2.1) is therefore $O(\tau^{4m-4-n})$. The sum in the left-hand side with $|\alpha|=m-1$ grows as $\tau^{1+4(m-1)-n}$ which gives a contradiction completing the proof.

Remark. The proof of Theorem 28.2.1 also gives information on weaker kinds of Carleman estimates. In fact, suppose that the right-hand side of (28.2.2) is negative for some ξ and τ with $p(0,\xi+i\tau N)=0$ when $x=0$, and choose for ψ the function in $\mathscr{S}$ provided by the proof of Lemma 28.2.2 multiplied by $\chi(x/\tau^{\frac{1}{2}})$ for some fixed $\chi\in C_0^\infty(X)$ equal to 1 in a neighborhood of 0. Then it follows that (28.2.1) is not even valid with a factor $o(\tau)$ in the right-hand side. Thus even very weak Carleman estimates require that the right-hand side of (28.2.2) is non-negative when $p(x,\xi+i\tau\phi'(x))=0$.

We shall now prove a converse of Theorem 28.2.1 where the hypothesis is of the form (28.2.2) but not restricted to the characteristic set.

Theorem 28.2.3. *Let X be an open set in $\mathbb{R}^n$, $\phi \in C^\infty(X,\mathbb{R})$, and let $p \in C^\infty(T^*(X))$ be a polynomial of degree m in the fibers. If for $(x,\xi)\in T^*(X)$ and $\tau>\tau_0$*

$$(28.2.6)\quad \sum_{|\alpha|<m} \tau^{2(m-|\alpha|)-1} |(\xi+i\tau\phi'(x))^\alpha|^2 \leqq K(|p(x,\xi+i\tau\phi'(x))|^2+\{\bar{p}(x,\xi-i\tau\phi'(x)), p(x,\xi+i\tau\phi'(x))\}/2i),$$

it follows that for every $Y \Subset X$ there is a constant C such that

$$(28.2.7)\quad \sum_{|\alpha|<m} \tau^{2(m-|\alpha|)-1} \int |D^\alpha u|^2 e^{2\tau\phi}\,dx \leqq K(1+C/\tau^{\frac12})\int |p(x,D)u|^2 e^{2\tau\phi}\,dx, \quad u\in C_0^\infty(Y),\ \tau>1.$$

Proof. With $v=ue^{\tau\phi}$ we rewrite (28.2.7) in the form

$$(28.2.7)'\quad \sum_{|\alpha|<m} \tau^{2(m-|\alpha|)-1} \|(D+i\tau\phi')^\alpha v\|^2 \leqq K(1+C/\tau^{\frac12})\|p(x,D+i\tau\phi')v\|^2.$$

Choose $\chi\in C_0^\infty(X)$ equal to 1 in Y. We can then replace v by χv when $v\in C_0^\infty(Y)$, which gives coefficients of compact support everywhere. The Weyl symbol of $D_j+i\tau\partial\phi/\partial x_j$ is of course $\xi_j+i\tau\partial\phi/\partial x_j$, so $p(x,D+i\tau\phi')\chi$ differs from the operator P_τ with Weyl symbol $p(x,\xi+i\tau\phi'(x))\chi$ only by terms of order $m-1$ in D and in τ. The Weyl symbol of

$$T=KP_\tau^*P_\tau-\sum_{|\alpha|<m}\tau^{2(m-|\alpha|)-1}\chi(D-i\tau\phi')^\alpha(D+i\tau\phi')^\alpha\chi$$

is equal to χ^2 times the difference between the right-hand side and the left-hand side in (28.2.6) with an error in

$$S((|\xi|+\tau)^{2m-2},g);\qquad g=|dx|^2+|d\xi|^2/(\tau^2+|\xi|^2).$$

Hence we obtain

$$(Tv,v)\geqq -C((\tau^2+|D|^2)^{m-1}v,v)$$

if we apply the Fefferman-Phong inequality (Theorem 18.6.8) to

$$(\tau^2+|D|^2)^{(1-m)/2}T(\tau^2+|D|^2)^{(1-m)/2}.$$

If M is the left-hand side of (28.2.7)′, it follows that

$$(M(1-C_1/\tau))^{\frac12}\leqq K^{\frac12}\|Pv\|\leqq K^{\frac12}\|p(x,D+i\tau\phi')v\|+C_2\tau^{-\frac12}M^{\frac12},$$

which proves (28.2.7).

The proof of Theorem 28.2.1 could also yield an estimate similar to (28.2.6), but it is more important here to discuss the possibility of passing from (28.2.2) to (28.2.6). In doing so we shall assume a strong version of condition (P), the two sided form of (28.1.35).

Definition 28.2.4. A differential operator P of order m in the open set $X \subset \mathbb{R}^n$ with C^1 principal symbol p will be called principally normal if for every compact set $K \subset X$

$$|\{\bar{p}, p\}(x,\xi)| \leq C_K |p(x,\xi)|\,|\xi|^{m-1}, \quad x \in K,\ \xi \in \mathbb{R}^n. \tag{28.2.8}$$

The definition is of course invariant so it makes sense also for a differential operator on a manifold X. The importance of this condition is that it allows us to control the second sum in (28.2.3) when $\tau \to 0$.

Lemma 28.2.5. *If p is principally normal in X we have for every compact set $K \subset X$ if $x \in K$ and $\tau > 0$*

$$\begin{aligned} &|\operatorname{Im} \sum p_{(j)}(x, \xi + i\tau N)\overline{p^{(j)}(x, \xi + i\tau N)}/\tau - \operatorname{Re} \sum N_k \{\bar{p}(x,\xi), p^{(k)}(x,\xi)\}| \\ &\quad \leq C_K(\tau |N|^2 |\xi + i\tau N|^{2m-3} + |p(x, \xi + i\tau N)|\,|\xi + i\tau N|^{m-1}/\tau \\ &\qquad + |\sum p^{(j)}(x, \xi + i\tau N) N_j|\,|\xi + i\tau N|^{m-1}). \end{aligned} \tag{28.2.9}$$

Proof. By Taylor's formula we have

$$\begin{aligned} &|2 \operatorname{Im} \sum p_{(j)}(x, \xi + i\tau N)\overline{p^{(j)}(x, \xi + i\tau N)} - \{\bar{p}(x,\xi), p(x,\xi)\}/i \\ &\quad - 2\tau \operatorname{Re} \sum (p^{(k)}_{(j)}(x,\xi)\overline{p^{(j)}(x,\xi)} - p_{(j)}(x,\xi)\overline{p^{(jk)}(x,\xi)}) N_k \\ &\quad \leq C|\xi + i\tau N|^{2m-3} |\tau N|^2. \end{aligned}$$

(28.2.9) follows since by hypothesis and Taylor's formula again

$$\begin{aligned} |\{\bar{p}(x,\xi), p(x,\xi)\}| &\leq C|p(x,\xi)|\,|\xi|^{m-1} \\ &\leq C(|p(x, \xi + i\tau N)| + |\sum p^{(j)}(x, \xi + i\tau N)\tau N_j| \\ &\quad + C_1 |\xi + i\tau N|^{m-2} |\tau N|^2)|\xi|^{m-1}. \end{aligned}$$

The proof is complete.

Lemma 28.2.5 allows us to give a supplement to Theorem 28.2.1 concerning real characteristics.

Theorem 28.2.1'. *If P is principally normal and* (28.2.1) *holds, $\phi'(x) \neq 0$, then*

$$\sum_{|\alpha| = m-1} |\xi^\alpha|^2 \leq 2K \operatorname{Re} \{\bar{p}, \{p, \phi\}\}(x,\xi), \tag{28.2.2)'$$

if $(x,\xi) \in T^*(X)$, $p(x,\xi) = 0$ and $\{p, \phi\}(x,\xi) = 0$.

Proof. First note that $\{p, \phi\} = \sum p^{(j)} \partial\phi/\partial x_j$ and that

$$\{\bar{p}, \{p, \phi\}\} = \sum_{j,k=1}^{n} \partial^2 \phi/\partial x_j \partial x_k\, p^{(j)} \bar{p}^{(k)} + \sum \{\bar{p}, p^{(j)}\}\, \partial\phi/\partial x_j. \tag{28.2.10}$$

If $m = 1$ then $p(x, \xi + i\tau\phi'(x)) = 0$ for all τ. Hence (28.2.2)' follows from (28.2.2), (28.2.3) and (28.2.9). For $m > 1$ it suffices to prove (28.2.2)' when

$\xi \neq 0$. If we rewrite (28.2.2) using (28.2.3), it follows from Lemma 28.2.5 that (28.2.2)′ is valid if (x, ξ) is in the closure of

$$\{(y, \eta + i\tau\phi'(y));\ \tau \neq 0,\ p(y, \eta + i\tau\phi'(y)) = 0\}.$$

We may assume that ϕ is linear, that is, $\phi' = N$ is constant. If (x, ξ) is not in the closure then $p(y, \eta)$ is microhyperbolic with respect to $(0, N)$ at (x, ξ). Since $p'_{x,\xi}(x, \xi) \neq 0$ by Theorem 28.2.1 it follows from Lemma 8.7.2 that $\langle p'_\xi(x, \xi), N\rangle \neq 0$. (Real analyticity was assumed in Lemma 8.7.2 but it suffices to have analyticity in the direction θ as we have here.) This is a contradiction completing the proof.

28.3. Uniqueness Under Convexity Conditions

The results of Section 28.2 shall now be used to prove unique continuation theorems across a surface defined by

$$\psi(x) = \psi(x_0) \tag{28.3.1}$$

where ψ is a real valued C^2 function in a neighborhood of x_0 and $\psi'(x_0) \neq 0$. Then (28.3.1) defines in a neighborhood of x_0 a C^2 oriented hypersurface. Our purpose is to show that solutions of a differential equation $Pu = 0$ vanishing on the positive side $\{x;\ \psi(x) > \psi(x_0)\}$ must vanish in a full neighborhood of x_0 when suitable convexity conditions are fulfilled. These must only depend on the surface (28.3.1) and not on the function ψ used to represent it. If ϕ is another such function then $\phi'(x_0) = a\psi'(x_0)$ for some $a > 0$, and

$$\sum \partial^2 \phi(x_0)/\partial x_j \partial x_k t_j \bar{t}_k = a \sum \partial^2 \psi(x_0)/\partial x_j \partial x_k t_j \bar{t}_k$$

if $\langle t, \psi'(x_0)\rangle = 0$, but not for all $t \in \mathbb{C}^n$. This is the reason why the following definition contains only a part of the necessary conditions for Carleman estimates established in Theorems 28.2.1 and 28.2.1′.

Definition 28.3.1. Let P be principally normal in $X \subset \mathbb{R}^n$, with principal symbol p, and let $\psi \in C^2(X)$, $\psi'(x_0) \neq 0$. Then the oriented hypersurface defined by (28.3.1) is called strongly pseudo-convex at x_0 with respect to P if

$$\operatorname{Re}\{\bar{p}, \{p, \psi\}\}(x_0, \xi) > 0; \tag{28.3.2}$$
$$\xi \in \mathbb{R}^n \setminus 0,\quad p(x_0, \xi) = \langle p'_\xi(x_0, \xi), \psi'(x_0)\rangle = 0;$$

$$\{\bar{p}(x, \xi - i\tau\psi'(x)), p(x, \xi + i\tau\psi'(x))\}/2i\tau > 0 \ \text{ if } x = x_0,\ \xi \in \mathbb{R}^n,\ \tau > 0 \tag{28.3.3}$$
$$\text{and } p(x, \xi + i\tau\psi'(x)) = \langle p'_\xi(x, \xi + i\tau\psi'(x)), \psi'(x)\rangle = 0.$$

Let M be the set of all (Ψ, x, ξ, τ, N) such that Ψ is a real and symmetric $n \times n$ matrix, $x \in X$, $\xi \in \mathbb{R}^n$, $\tau \geq 0$, $N \in \mathbb{R}^n \setminus 0$, $p(x, \xi + i\tau N) = 0$,

$\langle p'_\xi(x,\xi+i\tau N),N\rangle=0$, and $|\xi|^2+\tau^2=1$. By Lemma 28.2.5 the function F defined in M by

$$F(\Psi,x,\xi,\tau,N)=\sum \Psi_{jk}p^{(j)}(x,\xi+i\tau N)\bar{p}^{(k)}(x,\xi-i\tau N) + \operatorname{Im}\sum p_{(j)}(x,\xi+i\tau N)\bar{p}^{(j)}(x,\xi-i\tau N)/\tau, \quad \tau>0;$$

$$F(\Psi,x,\xi,0,N)=\sum \Psi_{jk}p^{(j)}(x,\xi)\bar{p}^{(k)}(x,\xi)+\operatorname{Re}\sum N_k\{\bar{p}(x,\xi),p^{(k)}(x,\xi)\};$$

is continuous. Definition 28.3.1 means in view of (28.2.3) and (28.2.10) that $F>0$ in the compact subset of M where $\Psi=\psi''(x_0)$, $N=\psi'(x_0)$, $x=x_0$. Hence $F>0$ in a neighborhood, which proves the stability of strong pseudo-convexity:

Proposition 28.3.2. *Suppose that the level surface of ψ is strongly pseudo-convex at x_0 with respect to P. Then there is a neighborhood U of x_0 and some $\varepsilon>0$ such that every $\phi\in C^2(U)$ with $|D^\alpha(\phi-\psi)|<\varepsilon$ in U, $|\alpha|\leqq 2$, has strongly pseudo-convex level surfaces at every point in U.*

We shall now prove that strong pseudo-convexity is not far from the hypothesis of Theorem 28.2.3.

Proposition 28.3.3. *Let P be principally normal in the open set $X\subset\mathbf{R}^n$, and assume that the level surfaces of $\psi\in C^\infty(X,\mathbf{R})$ are strongly pseudo-convex at every point in X. If $Y\Subset X$ and λ, K are large positive constants, it follows that (28.2.6) is valid for $\phi=e^{\lambda\psi}$ when $x\in Y$.*

Proof. The Poisson bracket in (28.2.6) is given explicitly by (28.2.3) where

$$\sum \partial^2\phi/\partial x_j\partial x_k p^{(j)}\bar{p}^{(k)}=\lambda\phi\left(\sum \partial^2\psi/\partial x_j\partial x_k p^{(j)}\bar{p}^{(k)}+\lambda|\langle p'_\xi,\psi'\rangle|^2\right).$$

Since $\phi'=\lambda\phi\psi'$ it follows that (28.2.6) is equivalent to

$$\tau(|\xi|^2+\tau^2)^{m-1}\leqq K\Big(|p(x,\xi+i\tau\psi'(x))|^2+\tau\big(\lambda|\langle p'_\xi(x,\xi+i\tau\psi'),\psi'\rangle|^2 + \sum \partial^2\psi/\partial x_j\partial x_k p^{(j)}(x,\xi+i\tau\psi')\bar{p}^{(k)}(x,\xi-i\tau\psi') + \operatorname{Im}\sum p_{(j)}(x,\xi+i\tau\psi')\bar{p}^{(j)}(x,\xi-i\tau\psi')\big)\Big), \tag{28.3.4}$$

apart from the size of the constant.

From condition (28.3.2) we obtain that for some $c>0$ and C

$$2c|\xi|^{2m-2}\leqq\sum \partial^2\psi/\partial x_j\partial x_k p^{(j)}(x,\xi)\bar{p}^{(k)}(x,\xi) +\operatorname{Re}\sum\{\bar{p}(x,\xi),p^{(k)}(x,\xi)\}\,\partial\psi/\partial x_k+C|p(x,\xi)|\,|\xi|^{m-2} +C|\langle p'_\xi(x,\xi),\psi'\rangle|\,|\xi|^{m-1};\quad x\in Y,\ \xi\in\mathbf{R}^n.$$

In fact, we can choose $c>0$ so that the estimate is valid when $|\xi|=1$ and the last two terms vanish. Then the estimate remains valid in a neighborhood for a smaller c, and C can then be chosen to make it hold everywhere. When $\tau\ll|\xi|$ we may replace ξ by $\xi+i\tau\psi'(x)$ in any one of the terms, for the error then committed is $O(\tau|\xi|^{2m-3})$ and can be absorbed by a part of the left-hand side. Using (28.2.9) we therefore obtain when $x\in Y$ and $\tau/|\xi|$ is small

$$\begin{aligned}c(|\xi|^2+\tau^2)^{m-1}\leqq &\sum \partial^2\psi/\partial x_j\partial x_k p^{(j)}(x,\xi+i\tau\psi')\bar{p}^{(k)}(x,\xi-i\tau\psi')\\ &+\operatorname{Im}\sum p_{(j)}(x,\xi+i\tau\psi')\bar{p}^{(j)}(x,\xi-i\tau\psi')/\tau+C(|p(x,\xi+i\tau\psi')|/\tau\\ &+|\langle p'_\xi(x,\xi+i\tau\psi'),\psi'\rangle|)(|\xi|^2+\tau^2)^{(m-1)/2}.\end{aligned}$$

We can estimate the last two terms by

$$\begin{aligned}&|p(x,\xi+i\tau\psi')|^2/\tau+\lambda|\langle p'_\xi(x,\xi+i\tau\psi'),\psi'\rangle|^2\\ &+C^2(1/\tau+1/\lambda)(|\xi|^2+\tau^2)^{m-1}.\end{aligned}$$

This gives the estimate (28.3.4) if λ and τ are sufficiently large and $\tau/|\xi|<\delta$, say.

Next assume that $|\xi|\leqq\delta\tau$ for some fixed $\delta>0$. Then the estimate (28.3.4) is valid even with $\lambda=0$ and without the first term on the right in a conic neighborhood of (x,ξ,τ) if

$$p(x,\xi+i\tau\psi'(x))=\langle p'_\xi(x,\xi+i\tau\psi'(x)),\psi'(x)\rangle=0.$$

For sufficiently large λ it is obviously valid at infinity in a conic neighborhood of any other point with $\tau\neq 0$. The proof is complete.

We are now prepared for the proof of the main theorem of this section.

Theorem 28.3.4. *Let P be a principally normal differential operator of order m in the open set $X\subset\mathbb{R}^n$, with C^∞ principal symbol p and all coefficients in $L^\infty_{\text{loc}}(X)$. Let $\psi\in C^2(X,\mathbb{R})$, $\psi'(x_0)\neq 0$, and assume that the level surface of ψ is strongly pseudo-convex at $x_0\in Y$. Then there is an open neighborhood $Y\subset X$ of x_0 such that if $u\in H^{\text{loc}}_{(m-1)}(Y)$ and $Pu=0$ in Y, $u=0$ in $\{x\in Y;\ \psi(x)>\psi(x_0)\}$ it follows that $u=0$ in Y.*

Proof. We may assume that $x_0=0$ and that $\psi(0)=0$. If $\varepsilon>0$ it follows from Taylor's formula that

$$\psi_\varepsilon(x)=\sum_{|\alpha|\leqq 2} x^\alpha\partial^\alpha\psi(0)/\alpha!-\varepsilon|x|^2<\psi(x),\quad 0\neq x\in X_\varepsilon,$$

where $X_\varepsilon\Subset X$ is a neighborhood of 0. By Proposition 28.3.2 we can choose ε and X_ε so small that the level surfaces of ψ_ε are strongly pseudo-convex in a neighborhood of $\bar{X}_\varepsilon$. With $\delta>0$ so small that $\psi_\varepsilon(x)\leqq\psi(x)-\delta$ if $x\in\partial X_\varepsilon$, we set

$$Y=\{x\in X_\varepsilon;\ \psi_\varepsilon(x)>-\delta\}.$$

In $Y\cap\operatorname{supp}u$ we have $\psi_\varepsilon(x)\leqq\psi(x)\leqq 0$, and

$$\operatorname{supp}u\cap\{x\in Y;\ \psi_\varepsilon(x)\geqq -t\}\tag{28.3.5}$$

is compact if $t<\delta$, for in this set we have

$$0\geqq\psi(x)\geqq\psi_\varepsilon(x)\geqq -t,$$

hence $\psi(x)-\psi_\varepsilon(x)\leq t<\delta$, so the closure cannot contain any point in ∂X_ε by the definition of δ.

By Proposition 28.3.3 and Theorem 28.2.3 we can find a large positive λ such that for $\phi=e^{\lambda\psi_\varepsilon}$ there is a Carleman estimate

$$\sum_{|\alpha|<m} \tau^{2(m-|\alpha|)-1}\int|D^\alpha v|^2 e^{2\tau\phi}\,dx\leq C\int|Pv|^2 e^{2\tau\phi}\,dx, \quad v\in C_0^\infty(Y). \tag{28.3.6}$$

By Friedrichs' lemma the estimate (28.3.6) remains valid when $v\in H^{\text{comp}}_{(m-1)}(Y)$ and $Pv\in L^2$. If u satisfies the hypotheses in the theorem we can apply (28.3.6) to $v=\chi u$ where $\chi\in C_0^\infty(Y)$ is equal to 1 in (28.3.5). As in the proof of Theorem 17.2.1 it follows at once when $\tau\to\infty$ that $u=0$ in Y when $\psi_\varepsilon(x)>-t$. The proof is complete.

The strong pseudo-convexity condition is fulfilled for all ψ with $\psi'(x_0)=N$ if and only if $p(x_0,N)\neq 0$ and the equation $p(x_0,\xi+\tau N)=0$ has only simple roots τ when $\xi\in T^*_{x_0}\setminus\mathbb{R}N$. Thus Theorem 28.3.4 contains Theorem 28.1.1 but not the refinement in Theorem 28.1.8 which is based on weaker Carleman estimates.

28.4. Second Order Operators of Real Principal Type

In this special case the strong pseudo-convexity introduced in Definition 28.3.1 has a very simple geometrical interpretation. In fact, if the level surface of ψ is non-characteristic at x_0 then all non-real zeros τ of $p(x,\xi+i\tau\psi'(x))$ are simple since they occur in complex conjugate pairs, so the condition (28.3.3) is void. Condition (28.3.2) becomes

$$H_p^2\psi(x_0,\xi)>0 \quad \text{if } p(x_0,\xi)=H_p\psi(x_0,\xi)=0.$$

This means that ψ restricted to a bicharacteristic curve is strictly convex at x_0 if the curve is a tangent to the level surface of ψ there. Corollary 13.6.7 shows that the convexity condition is quite natural and that uniqueness may fail if there is concavity on some bicharacteristic curve.

The convexity condition implies in particular that $\psi(x)>\psi(x_0)$ for $x\neq x_0$ in a neighborhood of x_0 on the bicharacteristic curve. For the characteristic analytic Cauchy problem Theorem 8.5.9 gave a uniqueness theorem when only this weaker condition is valid for the bicharacteristic defined by the characteristic conormal vector. The purpose of this section is to prove a similar although weaker result in the non-characteristic C^∞ case.

Let P be a second order differential operator defined in a neighborhood of $0\in\mathbb{R}^n$ with real C^∞ principal symbol p and of principal type in the strong sense that

$$p'_\xi(0,\xi)\neq 0 \quad \text{if } \xi\in\mathbb{R}^n\setminus 0. \tag{28.4.1}$$

By Corollary C.5.3 we can choose local coordinates such that

$$p(x, \xi)=\xi_1^2-r(x, \xi'), \quad \xi'=(\xi_2, \dots, \xi_n), \tag{28.4.2}$$

and the plane $x_1=0$ is any given non-characteristic surface. Then the condition (28.4.1) means that

$$\partial r(0, \xi')/\partial \xi' \neq 0 \quad \text{if } \xi' \in \mathbb{R}^{n-1} \setminus 0. \tag{28.4.1)'$$

The main analytical task is to prove the following Carleman estimate where we use the notation

$$X=\{x \in \mathbb{R}^n, -\tfrac{1}{2}<x_1<0, |x'|<1\}$$

and write $\phi(x)=x_1+x_1^2/2$ as in Section 28.1.

Proposition 28.4.1. *Assume that $r(x, \xi')$ is a real valued quadratic form in $\xi' \in \mathbb{R}^{n-1}$ with C^∞ coefficients satisfying (28.4.1)' and*

$$\partial r(x, \xi')/\partial x_1 \geqq 0 \quad \textit{if } x \in \varepsilon X \textit{ and } r(x, \xi')=0, \tag{28.4.3}$$

for some $\varepsilon>0$. Then it follows for sufficiently small ε that

$$\int \Big(\tau^3 |u|^2+\tau |D_1 u|^2+\sum_2^n |D_j u|^2\Big) e^{2\tau\phi} dx \tag{28.4.4}$$
$$\leqq C \int |(D_1^2-r(\varepsilon x, D'))u|^2 e^{2\tau\phi} dx, \quad u \in C_0^\infty(X).$$

Proof. Writing $v=ue^{\tau\phi}$ we find that (28.4.4) follows from

$$\tau^3 \|v\|^2+\tau^2 \|D_1 v\|^2+\sum_2^n \|D_j v\|^2 \leqq C \|P_\tau v\|^2, \quad v \in C_0^\infty(X), \tag{28.4.5}$$

where $P=(D_1+i\tau\phi')^2-r(\varepsilon x, D')$. To shorten notation we shall write $T=D_1+i\tau\phi'$ and $R=r(\varepsilon x, D')$. We have the commutation relations

$$[T^*, T]=2\tau, \quad [T^{*2}, T^2]=4\tau(T^* T+TT^*)$$

(see the proof of Proposition 28.1.4). Hence

$$[P_\tau^*, P_\tau]=[T^{*2}-R^*, T^2-R] \tag{28.4.6}$$
$$=4\tau(T^* T+TT^*)+[R^*, R]-[T^{*2}, R]-[R^*, T^2].$$

Here $[R^*, R]=[R^*-R, R]$ is of second order with coefficients $O(\varepsilon)$, and

$$[T^2, R^*]-[T^{*2}, R]=[T^2, R^*-R]+[4i\tau\phi' D_1, R]$$

since

$$T^2-T^{*2}=(T-T^*)(T+T^*)+[T^*, T]=4i\tau\phi' D_1+2\tau.$$

We can write

$$[T^2, R^*-R]=T[D_1, R^*-R]+[D_1, R^*-R]T,$$

for $[T, R^*-R]=[D_1, R^*-R]$ which is of first order with coefficients $O(\varepsilon^2)$. Hence

$$\|P_\tau^* v\|^2+4\operatorname{Re}\tau(\phi'\varepsilon r_{(1)}(\varepsilon x, D')v, v)+4\tau(\|Tv\|^2+\|T^*v\|^2) \le \|P_\tau v\|^2+C(\varepsilon\|v\|^2_{(0,1)}+\varepsilon^3\|Tv\|^2+\varepsilon^3\|T^*v\|^2). \tag{28.4.7}$$

The main problem is to estimate the scalar product from below using (28.4.3), but before doing so we shall examine which quantities can be estimated in terms of $P_\tau v$, $P_\tau^* v$, Tv and T^*v. First we note that since $(T-T^*)v=2i\tau\phi' v$ and $\phi'\ge\frac{1}{2}$ in X, we have

$$\tau\|v\|\le\|Tv\|+\|T^*v\|. \tag{28.4.8}$$

The computation of T^2-T^{*2} above gives

$$P_\tau v-P_\tau^* v=(4i\tau\phi' D_1+2\tau)v+(R^*-R)v,$$

and since R^*-R is a first order operator with coefficients $O(\varepsilon)$ it follows that

$$2\tau\|D_1 v\|\le\|P_\tau v\|+\|P_\tau^* v\|+2\tau\|v\|+C\varepsilon\|v\|_{(0,1)}. \tag{28.4.9}$$

On the other hand,

$$P_\tau+P_\tau^*=2D_1^2-2\tau^2\phi'^2-(R+R^*)$$

which gives

$$2\|D_1^2 v-Rv\|\le\|P_\tau v\|+\|P_\tau^* v\|+2\tau^2\|v\|+C\varepsilon\|v\|_{(0,1)}. \tag{28.4.10}$$

If (28.4.3) is valid in $2\delta X$, we can find $a\in C^\infty(\mathbb{R}^n\times(\mathbb{R}^{n-1}\setminus 0))$ homogeneous of degree 0 in the second variable, such that

$$\partial r(x,\xi')/\partial x_1-a(x,\xi')r(x,\xi')\ge 0, \quad x\in\delta X. \tag{28.4.11}$$

In fact, in a neighborhood of a zero (x_0,ξ_0') of r where $\partial r/\partial\xi_n\ne 0$, say, the equation $r=0$ defines ξ_n as a function $f(x,\xi'')$ where $\xi''=(\xi_2,\dots,\xi_{n-1})$. In this neighborhood

$$a(x,\xi')=(r_{(1)}(x,\xi')-r_{(1)}(x,\xi'',f(x,\xi'))/r(x,\xi')$$

is a C^∞ function and

$$r_{(1)}(x,\xi')-a(x,\xi')r(x,\xi')=r_{(1)}(x,\xi'',f(x,\xi''))\ge 0.$$

Where $r\ne 0$ we may take $a=r_{(1)}/r$. Piecing together such local constructions by a partition of unity in $\mathbb{R}^n\times S^{n-2}$, we obtain a with the desired properties and vanishing for large $|x|$. Define $A\in C^\infty$ equal to a when $|\xi'|\ge 1$, and choose $\chi\in C_0^\infty(\mathbb{R}^{n-1})$ equal to 1 when $|x'|<1$ and equal to 0 when $|x'|>2$. If $2\varepsilon<\delta$, then

$$\chi(x')r_{(1)}(\varepsilon x,\xi')\ge\chi(x')A(\varepsilon x,\xi')r(\varepsilon x,\xi') \quad \text{if } x_1\in(-\tfrac{1}{2},0),\ |\xi'|>1.$$

The real part of the Weyl symbol of $r_{(1)}(\varepsilon x, D')\chi$ (resp. of $A(\varepsilon x, D')r(\varepsilon x, D')\chi$) is equal to the left-hand side (resp. right-hand side) apart from an operator of order 0 with coefficients $O(\varepsilon)$. Hence it follows from the sharp Gårding inequality (Theorem 18.1.14) that

$$\operatorname{Re}(r_{(1)}(\varepsilon x, D')v, v) \geqq \operatorname{Re}(A(\varepsilon x, D')Rv, v) - C\|v\|\,\|v\|_{(0,1)}.$$

(Note that $\|v\|^2_{(0,\frac12)} \leqq \|v\|\,\|v\|_{(0,1)}$.) Since A is of order 0 we have

$$\begin{aligned}|(A(\varepsilon x, D')D_1^2 v, v)| &\leqq |(A(\varepsilon x, D')D_1 v, D_1 v)| + |([A(\varepsilon x, D'), D_1]D_1 v, v)|\\ &\leqq C(\|D_1 v\|^2 + \|v\|^2).\end{aligned}$$

Using (28.4.10) and (28.4.7) we now obtain for large τ

$$\begin{aligned}&\|P_\tau^* v\|^2 + 2\tau(\|Tv\|^2 + \|T^* v\|^2)\\ &\quad\leqq \|P_\tau v\|^2 + C\varepsilon\big(\|v\|^2_{(0,1)} + \tau\|D_1 v\|^2 + \tau\|v\|\,\|v\|_{(0,1)}\\ &\qquad + (\|P_\tau v\| + \|P_\tau^* v\| + 2\tau^2\|v\| + C\varepsilon\|v\|_{(0,1)})\tau\|v\|\big).\end{aligned}$$

Since $\frac12\tau^3\|v\|^2 \leqq \tau(\|Tv\|^2 + \|T^* v\|^2)$ by (28.4.8) and $\|D_1 v\|$ can be estimated by means of (28.4.9), we obtain for small ε

$$\|P_\tau^* v\|^2 + 2\tau(\|Tv\|^2 + \|T^* v\|^2) \leqq 2\|P_\tau v\|^2 + C\varepsilon\|v\|^2_{(0,1)}. \tag{28.4.12}$$

We shall finally estimate $\|v\|_{(0,1)}$ by standard energy integral arguments. To do so we set $Q = -x_j r^{(j)}(\varepsilon x, D')$ for some $j = 2, \dots, n$, and form

$$((P_\tau v, Q^* v) - (Qv, P_\tau^* v))/i = ([Q, P_\tau]v, v)/i.$$

Since

$$[Q, T^2] = [Q, T]T + T[Q, T] = [Q, D_1]T + T[Q, D_1],$$

we have

$$|([Q, T^2]v, v)| \leqq (\|Tv\| + \|T^* v\|)\|v\|_{(0,1)}.$$

Finally $[Q, -R]/i$ has the principal symbol

$$r^{(j)}(\varepsilon x, \xi')^2 - x_j\{r^{(j)}(\varepsilon x, \xi'), r(\varepsilon x, \xi')\}.$$

The second term is a quadratic form in ξ' with coefficients $O(\varepsilon)$, but it follows from (28.4.1)$'$ that for small ε

$$|\xi'|^2 \leqq C\sum r^{(j)}(\varepsilon x, \xi')^2, \quad x \in X.$$

Adding the result now obtained for $j = 2, \dots, n$ we have

$$\|v\|^2_{(0,1)} \leqq C((\|P_\tau v\| + \|P_\tau^* v\| + \|Tv\| + \|T^* v\|)\|v\|_{(0,1)} + \varepsilon\|v\|^2_{(0,1)}).$$

When ε is small we conclude that

$$\|v\|_{(0,1)} \leqq 2C(\|P_\tau v\| + \|P_\tau^* v\| + \|Tv\| + \|T^* v\|). \tag{28.4.13}$$

If this estimate is entered in the right-hand side of (28.4.12) we have proved that for small ε

$$\|P_\tau^* v\|^2 + 2\tau(\|Tv\|^2 + \|T^* v\|^2) \leqq 3\|P_\tau v\|^2.$$

Hence (28.4.13) gives the estimate (28.4.5) for $\|v\|_{(0,1)}$; the estimate for $\|v\|$ follows by (24.4.8) and that for $\|D_1 v\|$ is a consequence of (28.4.9) now. The proof is complete.

The end of the proof of Theorem 28.1.8 can be combined with (28.4.4) to prove a uniqueness theorem: If $Pu=0$ in a neighborhood of 0, the principal symbol of P is p, and all coefficients of P are bounded, then $u=0$ in a neighborhood of 0 if $\{x\in\varepsilon\bar{X}\cap\operatorname{supp} u;\ x_1\geqq 0\}$ is a compact subset of $S=\{x; x_1=0\}$. To put the result in an invariant form we must study the condition (28.4.3) further. Since

$$\{p, x_1\}=2\xi_1, \quad \{p,\{p,x_1\}\}=2\partial r/\partial x_1$$

the condition is that in εX we have

$$H_p^2 x_1\geqq 0 \quad \text{if } p=H_p x_1=0. \tag{28.4.14}$$

The flaw of this condition is that it does not only depend on the oriented surface S but also on the function x_1 used to represent it. In fact, $\phi(x)=\psi(x)x_1$ defines the same oriented surface if $\psi>0$, and

$$H_p\phi=\psi H_p x_1+x_1H_p\psi, \quad H_p^2\phi=\psi H_p^2x_1+2(H_p\psi)H_px_1+x_1H_p^2\psi,$$

so (28.4.14) implies

$$H_p^2\phi\geqq 0 \quad \text{if } p=\phi=H_p\phi=0. \tag{28.4.15}$$

Conversely, (28.4.15) implies (28.4.14) when $x_1=0$. The lack of invariance of (28.4.14) outside S can be exploited:

Lemma 28.4.2. *Let p satisfy* (28.4.1), (28.4.2), *and assume that for some $\varepsilon>0$*

$$\partial r(x,\xi')/\partial x_1\geqq 0 \quad \text{if } x\in\varepsilon\bar{X},\ x_1=0, \text{ and } r(x,\xi')=0. \tag{28.4.16}$$

Set $\phi(x)=x_1(1-A|x'|^2)$. If A is sufficiently large it follows that for small $\delta>0$

$$H_p^2\phi(x,\xi)\geqq 0 \quad \text{if } x\in\delta X,\ \phi(x)\leqq 0 \text{ and } p(x,\xi)=H_p\phi(x,\xi)=0. \tag{28.4.17}$$

Proof. Set $\psi(x')=1-A|x'|^2$. Then

$$\{p,\phi\}=\{\xi_1^2-r,\psi x_1\}=2\xi_1\psi-x_1\{r,\psi\}=0,$$

means that

$$\xi_1=x_1\sum_2^n r^{(j)}(x,\xi')Ax_j/\psi(x'). \tag{28.4.18}$$

A simple computation gives

$$\begin{aligned}\{p,\{p,\phi\}\}&=\{\xi_1^2-r, 2\xi_1\psi-x_1\{r,\psi\}\}\\&=2r_{(1)}\psi-4\xi_1\{r,\psi\}+x_1(\{r,\{r,\psi\}\}-2\xi_1\{r_{(1)},\psi\}).\end{aligned}$$

The hypothesis (28.4.16) and Taylor's formula give in view of (28.4.1)′

$$r_{(1)}(x,\xi')\geqq C(x_1|\xi'|^2-|r(x,\xi')|) \quad \text{if } x\in\delta X,$$

if δ is small enough. When $p=\{p,\phi\}=0$ we obtain using (28.4.18)

$$r(x,\xi')=\xi_1^2\leqq(C_1x_1|\xi'|A|x'|)^2<-x_1|\xi'|^2,$$
$$r_{(1)}(x,\xi')\geqq 2Cx_1|\xi'|^2, \quad \text{if } x\in\delta X.$$

Moreover,

$$4\xi_1\{r,\psi\}=2x_1\{r,\psi\}^2/\psi\leqq 0,$$

which proves that

$$\{p,\{p,\phi\}\}\geqq x_1(4C|\xi'|^2+\{r,\{r,\psi\}\}-2\xi_1\{r_{(1)},\psi\}).$$

Again for $\delta<\delta_A$ we have by (28.4.1)'

$$\{r,\{r,\psi\}\}=-2A(\sum r^{(j)}(x,\xi')^2+\sum x_j\{r,r^{(j)}\}(x,\xi'))\leqq -AC_2|\xi'|^2.$$

It follows from (28.4.18) that for small δ

$$-2\xi_1\{r_{(1)},\psi\}\leqq AC_2|\xi'|^2/2.$$

Hence

$$\{p,\{p,\phi\}\}(x,\xi)\geqq -x_1(AC_2/2-4C)|\xi'|^2$$

if $p(x,\xi)=\{p,\phi\}(x,\xi)=0$ and $x\in\delta X$. When $AC_2>8C$ the proof of (28.4.17) is complete.

One can introduce new local coordinates y such that $\phi(x)=y_1$ and $p(x,\xi)$ is transformed to $c(\eta_1^2-\tilde{r}(y,\eta'))$ in the new variables. To do so we just have to integrate the vector field t defined by

$$p(x;\xi,\phi'(x))=\langle t(x),\xi\rangle$$

where $p(x;.,.)$ is the polarized form of $p(x,.)$. If we choose $y_2,\dots,y_n$ constant on the orbits and $\tilde{p}$ is the new principal symbol then

$$\tilde{p}(y;\eta,e_1)=0 \quad \text{if } \eta_1=0,$$

that is, $\tilde{p}$ has the desired form. (This argument is of course an alternative proof of Lemma C.5.3.) From Propositions 28.4.1 and Lemma 28.4.2 we now obtain:

Theorem 28.4.3. *Let P be a second order differential operator in the open set $X\subset\mathbb{R}^n$ with C^∞ real principal symbol p and all coefficients in $L^\infty_{\text{loc}}(X)$. Let $x_0\in X$, $\psi\in C^\infty(X,\mathbb{R})$, $p(x_0,\psi'(x_0))\neq 0$, and assume that*

$$H_p^2\psi(x,\xi)\geqq 0 \quad \text{if } x\in X,\ \psi(x)=\psi(x_0),\ p(x,\xi)=H_p\psi(x,\xi)=0. \tag{28.4.19}$$

If Y is a sufficiently small neighborhood of x_0 and $u\in H^{\text{loc}}_{(1)}(Y)$ satisfies the equation $Pu=0$, it follows that $u=0$ in a neighborhood of x_0 if $\psi(x)\leqq\psi(x_0)$ in supp u *and*

$$\{x\in Y\cap \operatorname{supp} u;\ \psi(x)=\psi(x_0)\}$$

is compact.

For the geometrical meaning of the condition (28.4.19) the reader should recall the definition of diffractive and gliding points given in Section 24.3.

Notes

In the notes for Chapter XVII we mentioned the earlier work on the uniqueness of the Cauchy problem including the fundamental paper by Calderón [1] where Theorem 28.1.1 was established. That complex characteristic roots may be allowed to be double was first recognized by Pederson [1] and Mizohata [1] (see also Hörmander [8]) who allowed products of elliptic factors with simple complex roots. Calderón [2] (as earlier Pederson [1]) restricted himself, some years later, to double complex roots of constant multiplicity, and this hypothesis was kept by Nirenberg [5] who proved the part of Proposition 28.1.6 which corresponds to condition (28.1.29). The conditions on the double complex roots have been successively relaxed in the literature. Pederson [2] proved uniqueness for elliptic operators when the complex double roots are locally C^1 functions. The Fefferman-Phong estimate has been used here to prove Proposition 28.1.4 which contains a geometrically more natural condition (see (28.1.36)) and does not require that the roots can be distinguished even locally. For nearly real characteristics Proposition 28.1.6 gives another improvement which depends on the Fefferman-Phong inequality. This tool is also used in Section 28.2 to improve the results of Chapter VIII in the predecessor of this book.

There an operator with principal symbol p was said to be principally normal if $\{p, \bar{p}\}$ is in the ideal generated by p and $\bar{p}$ over polynomials in the ξ variable. Lerner [1] has shown that one can use the ring of C^∞ symbols instead. The original reason for the condition was that 20 years ago such a strong form of condition (P) was required to prove local solvability. Already in Hörmander [17] the proof of the sharp Gårding inequality made it possible to prove local solvability when $|\{p, \bar{p}\}/p|$ is locally bounded. Here we use this condition to extend the class of principally normal operators. Apart from this improvement the results of Sections 28.2 and 28.3 were all proved in the predecessor of this book, with weaker smoothness assumptions since the same techniques were used as in Section 17.2 here. It is not clear how smooth the coefficients must be for the theorems proved in Section 28.3 to hold. Nor is it known precisely how much the condition (P) has to be strengthened in these results; this seems related to the open problems on the sufficiency of condition (Ψ).

The more precise uniqueness theorems for second order operators of real principal type given in Section 28.4 are essentially due to Lerner and Robbiano [1]. Fur further references and non-uniqueness theorems which show the importance of the conditions imposed the reader should consult Alinhac [3], Zuily [1], as well as the references given in this context in Chapter XIII.

Chapter XXIX. Spectral Asymptotics

Summary

This chapter is devoted to the asymptotic properties of the eigenvalues and the spectral function for self-adjoint elliptic operators. In Sections 29.1 and 29.2 we study operators on a compact manifold without boundary. If P is positive and of order m then $P^{1/m}$ is a pseudo-differential operator with the eigenfunctions for the eigenvalue λ equal to those of P for λ^m. We shall therefore generalize the problem by studying first order pseudo-differential operators. This is actually a simplification from the geometrical as well as from the analytical point of view. In fact, as emphasized in Chapter XXVI, the Hamilton flow defined by the principal symbol is a flow on the cosphere bundle when the order is one. Moreover, the corresponding unitary group $t\mapsto e^{-itP}$ is the solution operator for the hyperbolic equation $(D_t+P)u=0$ when the order of P is one. This operator was studied by means of energy estimates in Section 23.1. Here we identify the solution operator as a Fourier integral operator, which we actually know from Chapter XXVI also. This gives a perfect substitute for the Hadamard construction used in Section 17.5 and allows us to determine the singularities of the kernel of e^{-itP} for any t. As a result we obtain not only an analogue of the asymptotic formulas proved there in the second order case but also an improved formula with a second term if the set of closed orbits of the Hamilton field is of measure 0. In Section 29.3 we give a similar improvement for the Dirichlet problem for a second order elliptic operator in a manifold with boundary.

Section 29.2 is devoted to the opposite case where all bicharacteristics are closed with the same period Π. Then the eigenvalues tend to cluster around an arithmetic sequence with difference $2\pi/\Pi$ determined by a Maslov class and spread out in a manner determined by the subprincipal symbol. These results illuminate the example given in Section 17.5 where we used spherical harmonics to study the Laplacean on the unit sphere.

29.1. The Spectral Measure and its Fourier Transform

As mentioned in the summary we shall begin by studying a first order pseudo-differential operator P on a compact connected C^∞ manifold X without boundary. More precisely we assume that $P\in\Psi^1(X;\Omega^{\frac12},\Omega^{\frac12})$, where

L. Hörmander, *Classics in Mathematics*
The Analysis of Linear Partial Differential Operators IV,
DOI: 10.1007/978-3-642-00117-8_5, © Springer-Verlag Berlin Heidelberg 2009

$\Omega^{\frac{1}{2}}$ is the half density bundle, and that P is symmetric:

$$(Pu, v)=(u, Pv); \quad u, v\in C^\infty(X;\Omega^{\frac{1}{2}}). \tag{29.1.1}$$

It will be convenient to assume also that P has polyhomogeneous symbol. Note that L^2 scalar products are intrinsically defined on half densities since their products are densities. If as in Section 17.5 we had been given an operator $Q\in\Psi^1(X)$ between scalar functions and a positive density ρ such that Q is symmetric with respect to the corresponding scalar product, then

$$Pu=\rho^{\frac{1}{2}}Q(u\rho^{-\frac{1}{2}}), \quad u\in C^\infty(X,\Omega^{\frac{1}{2}})$$

defines $P\in\Psi^1(X;\Omega^{\frac{1}{2}},\Omega^{\frac{1}{2}})$ satisfying (29.1.1), with the same eigenvalues as Q and the eigenfunctions just multiplied by $\rho^{\frac{1}{2}}$. Thus the hypotheses made in Section 17.5 fit into our present framework.

Now assume that P is elliptic and let $\mathscr{P}$ be the operator in $L^2(X,\Omega^{\frac{1}{2}})$ defined by restricting P to all $u\in L^2(X,\Omega^{\frac{1}{2}})$ such that $Pu\in L^2(X,\Omega^{\frac{1}{2}})$. This is a self-adjoint operator. In fact, every u in the domain belongs to $H_{(1)}$ by Theorem 18.1.29, and C^∞ is dense in $H_{(1)}$, so (29.1.1) extends by continuity to all u, v in the domain of $\mathscr{P}$. On the other hand it follows from (29.1.1) which remains valid for $v\in\mathscr{D}'$ that the adjoint of $\mathscr{P}$ is contained in $\mathscr{P}$, which proves the statement.

We shall always assume that the dimension n is at least equal to two, so the ellipticity implies that the principal symbol $p(x,\xi)$ has a fixed sign when $(x,\xi)\in T^*(X)\smallsetminus 0$, say positive. Let $\|\ \|_{(s)}$ be a Hilbert space norm defining the topology in $H_{(s)}(X,\Omega^{\frac{1}{2}})$. Then

$$\|u\|^2_{(\frac{1}{2})}\leqq C_1(Pu,u)+C_2\|u\|^2), \quad u\in\mathscr{D}_{\mathscr{P}}, \tag{29.1.2}$$

for if $Q\in\Psi^{\frac{1}{2}}(X;\Omega^{\frac{1}{2}},\Omega^{\frac{1}{2}})$ has principal symbol $p^{\frac{1}{2}}$ then Q^*Q-P is of order 0 and

$$\|u\|^2_{(\frac{1}{2})}\leqq C_1(\|Qu\|^2+\|u\|^2)\leqq C_1(Pu,u)+C_2\|u\|^2.$$

Hence the spectrum of $\mathscr{P}$ is discrete (cf. Section 17.5). Let $\lambda_1\leqq\lambda_2\leqq\dots$ be the eigenvalues and $\phi_1,\phi_2,\dots$ corresponding orthonormal eigenfunctions, and define the spectral function as the kernel of the spectral projection E_λ

$$e(x,y,\lambda)=\sum_{\lambda_j\leqq\lambda}\phi_j(x)\otimes\overline{\phi_j(y)}\in C^\infty(X\times X,\Omega^{\frac{1}{2}}).$$

The proof of Theorem 17.5.3 shows that for any differential operator Q of order μ in $C^\infty(X\times X,\Omega^{\frac{1}{2}})$ we have

$$|Qe(.,.,\lambda)|\leqq C_Q\lambda^{n+\mu}. \tag{29.1.3}$$

We just have to observe that if $u=E_\lambda u$ then

$$\|u\|_{(k)}\leqq C_k\|\mathscr{P}^k u\|\leqq C_k\lambda^k\|u\|$$

and apply Lemma 17.5.2 in the simple case of a manifold without boundary. If

$$N(\lambda)=\operatorname{Tr}E_\lambda=\int_X e(x,x,\lambda)$$

is the counting function for the eigenvalues, it follows that $N(\lambda)=O(\lambda^n)$. Our purpose is to improve this result.

In Section 17.5 we studied the cosine transform of the spectral function with λ replaced by λ^2 using that it satisfies the wave equation. Here we shall just consider the Fourier transform of the spectral measure,

$$\hat{E}(t)=\int e^{-it\lambda}\,dE_\lambda=e^{-it\mathscr{P}},$$

that is, the unitary group generated by $\mathscr{P}$, and use that it satisfies the equation $(D_t+P)\hat{E}=0$. Note that $\hat{E}(t)$ defines a strongly continuous family of bounded operators in $H_{(s)}$ for every s, for $\mathscr{P}^{-s}$ is an isomorphism of L^2 on $H_{(s)}$ and $\mathscr{P}^s\hat{E}(t)\mathscr{P}^{-s}=\hat{E}(t)$ if $s\geqq 0$. By (29.1.3) and the positivity of dE_λ the kernel $e(x,y,\lambda)$ of E_λ is a uniformly temperate measure, and its Fourier transform is a continuous function of (x,y) with values in $\mathscr{D}'(\mathbb{R})\otimes\Omega^{\frac12}_{x,y}$. The following description of the singularities of the corresponding distribution kernel $\in\mathscr{D}'(\mathbb{R}\times X\times X;\,\Omega^{\frac12}(X\times X))$ of the map

$$C^\infty(X;\Omega^{\frac12}(X))\ni f\mapsto \hat{E}(t)f\in C^\infty(\mathbb{R}\times X;\Omega^{\frac12}(X))$$

is a global analogue of Theorem 17.5.5 (see also Theorem 26.1.14).

Theorem 29.1.1. *The kernel of $\hat{E}$ is in $I^{-\frac14}(\mathbb{R}\times X\times X, C';\,\Omega(X\times X)^{\frac12})$ where C is the canonical relation*

$$C=\{(t,x,y;\tau,\xi,\eta);\ \tau+p(x,\xi)=0 \text{ and } (x,\xi)=\exp(tH_p)(y,\eta)\}. \tag{29.1.4}$$

The last condition means that the orbit of H_p which is at (y,η) when the parameter is 0 arrives at (x,ξ) when the parameter is t.

Proof. C is the bicharacteristic relation for the symbol $\tau+p(x,\xi)$ defined in $T^*(\mathbb{R}\times X)$, when the second component in $\mathbb{R}$ is fixed at 0, so it follows from Proposition 26.1.13 that C is a canonical relation. It is the flow-out of $\{(0,x,x;\,-p(x,\xi),\xi,\xi)\}$ by the Hamilton field of $\tau+p(x,\xi)$ lifted to $T^*(\mathbb{R}\times X\times X)$ by the projection omitting the second X factor. We can also argue as follows. In view of the homogeneity $\chi_t(x,\xi)=\exp(tH_p)(x,\xi)$ is a homogeneous symplectic map defined in $T^*(X)\smallsetminus 0$ for small t (see (6.4.12)). We can extend the definition of χ_t to all t by the group property which gives $\chi_t=\chi_{t/N}\circ\ldots\circ\chi_{t/N}$ with a large number N of factors. When $(x,\xi)=\chi_t(y,\eta)$ we have $|\xi|\leqq e^{C|t|}|\eta|$ for some C, with norms taken with respect to some Riemannian metric, which proves that C is closed in $T^*(\mathbb{R}\times X\times X)\smallsetminus 0$; that C is Lagrangian follows at once from the fact that χ_t is canonical for each t and that $\partial/\partial t+H_p$ is orthogonal to the tangent plane since $\tau+p(x,\xi)=0$ in C.

If $U\in I^{-\frac14}(\mathbb{R}\times X\times X, C';\,\Omega(X\times X)^{\frac12})$ and i_t is the injection $X\times X\ni(x,y)\mapsto(t,x,y)\in\mathbb{R}\times X\times X$, then

$$i_t^*U\in I^0(X\times X, C_t';\ \Omega(X\times X)^{\frac12})$$

where C_t is the graph of χ_t. This follows from Theorem 25.2.2 and the fact that the kernel $\delta(s-t)\,\delta(x-x')\,\delta(y-y')$ where $(x, y, s, x', y')\in X\times X\times \mathbb{R}\times X\times X$ is a conormal distribution in $I^{\frac{1}{4}}$ with respect to the surface of codimension $2n+1$ in the $4n+1$ dimensional space $X\times X\times\mathbb{R}\times X\times X$ defined by $s=t$, $x=x'$, $y=y'$. (Note that $4n+1-2(2n+1)=-1$.) In particular, $i_0^* U$ is a pseudo-differential operator, and its principal symbol determines that of U when $t=0$.

Now it follows from Theorem 25.2.4 that the principal symbol of $i(D_t+P_x)\,U(t, x, y)$ is equal to $\mathscr{L}_{H_{\tau+p}}+ip^s$ applied to the principal symbol of U. This would be quite clear if P_x were a pseudo-differential operator in the product $\mathbb{R}\times X\times X$. For the general proof we observe that by Theorems 18.1.35 and 18.1.36 we can for any $\delta>0$ write $P=P_1+P_2$ where P_1 is a pseudo-differential operator in $\mathbb{R}\times X$ and

$$(t, x, \tau, \xi;\, t', x', \tau', \xi')\in WF'(P_2)\Rightarrow|\xi|+|\xi'|<\delta\min(|\tau|, |\tau'|).$$

By Theorem 8.2.14 it follows that $P_2\circ C\in C^\infty$ provided that δ is small enough, and $P_1\circ C$ can be calculated using Theorem 25.2.4.

By solving the transport equation $(\mathscr{L}_{H_{\tau+p}}+ip^s)a=0$ in C_t we can determine the principal symbol a of $U_0\in I^{-\frac{1}{4}}(\mathbb{R}\times X\times X, C')$ so that the principal symbol of $i_0^* U$ is equal to 1 and

$$(D_t+P_x)\,U_0=F_1\in I^{-\frac{5}{4}}(\mathbb{R}\times X\times X, C').$$

Next we determine $U_1\in I^{-\frac{5}{4}}(\mathbb{R}\times X\times X, C')$ so that the principal symbol of $(D_t+P_x)\,U_1$ is equal to that of F_1 and the principal symbol of $i_0^* U_1$ is equal to the principal symbol of the identity minus $i_0^* U_0$. Continuing in this way and taking $U\sim\sum U_j$ we obtain $U\in I^{-\frac{1}{4}}(\mathbb{R}\times X\times X, C';\, \Omega(X\times X)^{\frac{1}{2}})$ such that

$$(D_t+P_x)\,U=R\in C^\infty(\mathbb{R}\times X\times X;\, \Omega(X\times X)^{\frac{1}{2}}). \tag{29.1.5}$$

We may assume that $i_0^* U$ is the identity for the difference is in C^∞ by the construction and can be extended to a C^∞ function.

$\hat{E}(t)$ is the exact solution of the Cauchy problem

$$(D_t+P)\,\hat{E}(t)=0,\qquad \hat{E}(0)=Id.$$

If $f\in C^\infty(X, \Omega^{\frac{1}{2}})$ then (29.1.5) means that

$$(D_s+P)\,U(s)f=R(s)f$$

if $U(s)$ is the operator with kernel $i_s^* U$ and similarly for R. Moreover $U(0)f=f$. If we multiply by $\hat{E}(t-s)$ and integrate from 0 to s using that

$$i\int_0^t \hat{E}(t-s)\,D_s U(s)f\,ds=i\int_0^t D_t\hat{E}(t-s)\,U(s)f\,ds+U(t)f-\hat{E}(t)f,$$

we obtain since $P\hat{E}(t-s)=\hat{E}(t-s)P$

$$(U(t)-\hat{E}(t))f=i\int_0^t \hat{E}(t-s)R(s)f\,ds.$$

The right-hand side is a continuous map from $\mathscr{D}'$ to C^∞ since $\hat{E}$ is continuous from C^∞ to C^∞. Hence $U(.)-\hat{E}(.)$ has a C^∞ kernel (Theorem 5.2.6), which completes the proof.

The proof just given shows that every point in C' is non-characteristic for U. Hence we have proved in particular that

$$WF'(\hat{E})=C. \tag{29.1.6}$$

(The inclusion $WF'(\hat{E})\subset C$ is essentially contained in Theorem 23.1.4.) While this is qualitatively very satisfactory and will be important later on, we must also make some more explicit calculations which will yield the first two terms corresponding to those in Theorem 17.5.5. In doing so we shall in fact give another proof of Theorem 29.1.1 for small t; by the group property $\hat{E}(t+s)=\hat{E}(t)\hat{E}(s)$ and Theorem 25.2.3, the global result in t is an immediate consequence.

Since the variable τ dual to t never vanishes in C' it follows from (29.1.6) that $\hat{E}$ has a restriction $\in\mathscr{D}'(\mathbb{R},\Omega(X\times X)^{\frac12}_{x,y})$ to any fiber of $\mathbb{R}\times X\times X$ over $X\times X$. The restriction is in C^∞ at t if no orbit of H_p passes from $T_x^*(X)\smallsetminus 0$ to $T_y^*(X)\smallsetminus 0$ in time t. In particular, the restriction is smooth in a deleted neighborhood of $0\in\mathbb{R}$ if $x=y$. Set $\Delta=\{(x,x);\, x\in X\}\subset X\times X$.

Proposition 29.1.2. *Let $B\in\Psi^0_{\rm phg}(X;\Omega^{\frac12},\Omega^{\frac12})$ have principal symbol b and subprincipal symbol b^s while the principal and subprincipal symbols of P are p and p^s. Then the restriction K of the kernel of $\hat{E}(.)B$ to $\mathbb{R}\times\Delta$ is conormal with respect to $\{0\}\times\Delta$ in a neighborhood of this submanifold; for small $|t|$ we have*

$$K(t,y)=\int \partial A(y,\lambda)/\partial\lambda\, e^{-i\lambda t}\,d\lambda$$

where $A\in S^n(\Delta\times\mathbb{R};\Omega(\Delta))$ and $A(y,0)=0$,

$$A(y,\lambda)-(2\pi)^{-n}\Big(\int_{p(y,\eta)<\lambda}(b+b^s)\,d\eta-\partial/\partial\lambda\int_{p(y,\eta)<\lambda}(p^s b+\tfrac12 i\{b,p\})\,d\eta\Big)\in S^{n-2}.$$

The integrals on the right are C^∞ densities, homogeneous of degree n with respect to λ, defined so that the scalar product with $\phi\in C(X)$ is the integral of the product by ϕ with respect to the symplectic measure $dy\,d\eta$.

Proof. We may assume that the kernel of B has compact support in $Y\times Y$ where Y is a coordinate patch which we identify with an open set in $\mathbb{R}^n$. The operators P and B can then be written $P(x,D)$ and $B(x,D)$ where $P(x,\xi)\sim p(x,\xi)+p_0(x,\xi)+p_{-1}(x,\xi)+\dots$ and $B(x,\xi)\sim b(x,\xi)+b_{-1}(x,\xi)+\dots$,

$$p^s=p_0+\tfrac12 i\sum p^{(j)}_{(j)},\qquad b^s=b_{-1}+\tfrac12 i\sum b^{(j)}_{(j)}.$$

It will be convenient to replace B by B^* in the proof; since the Weyl symbol of B is $b+b^s \bmod S^{-2}$ this means a complex conjugation also of b and of b^s.

The projection

$$C' \ni (t, x, y, \tau, \xi, \eta) \mapsto (t, x, \eta)$$

is a diffeomorphism in a neighborhood of $\{0\} \times Y \times Y$. Hence it follows from Theorem 21.2.18 that C' is defined by a phase function of the form $\langle y, \eta\rangle - S(t, x, \eta)$,

$$\begin{aligned} C' &= \{(t, x, \partial S/\partial\eta, -\partial S/\partial t, -\partial S/\partial x, \eta)\} \\ &= \{(t, x, \partial\psi/\partial\eta, \partial\psi/\partial t, \partial\psi/\partial x, -\eta)\} \end{aligned}$$

if $\psi(t, x, \eta) = -S(t, x, -\eta)$. Thus we have the defining phase function $\psi(t, x, \eta) - \langle y, \eta\rangle$ for the canonical relation. Since $\tau + p(x, \xi) = 0$ in C, we have

$$\partial\psi/\partial t + p(x, \partial\psi/\partial x) = 0. \tag{29.1.7}$$

The restriction of C to $t=0$ is the diagonal so we have $d\psi(0, x, \eta) = d\langle x, \eta\rangle$, hence $\psi(0, x, \eta) = \langle x, \eta\rangle$ since ψ is homogeneous of degree 1. Thus ψ is just the solution of the Cauchy problem for (29.1.7) with initial value $\langle x, \eta\rangle$, so we could have started with solving that problem instead. When $t=0$ we have $\partial\psi/\partial t = -p(x, \eta)$ by (29.1.7), hence

$$\partial^2\psi/\partial t^2 = -\sum p^{(k)}(x, \eta)\, \partial^2\psi/\partial x_k \partial t = \sum p^{(k)}(x, \eta)\, p_{(k)}(x, \eta).$$

Hence there is a C^∞ function μ with

$$\psi(t, y, \eta) - \langle y, \eta\rangle = -t\mu(t, y, \eta), \tag{29.1.8}$$

$$\mu(0, y, \eta) = p(y, \eta); \quad \partial_t\mu(0, y, \eta) = -\tfrac{1}{2}\langle \partial p(y, \eta)/\partial\eta, \partial p(y, \eta)/\partial y\rangle.$$

μ is homogeneous in η of degree 1.

For small t we know that on any compact subset of $Y \times Y$ we have for the kernel U of $\hat{E}(t)$

$$U(t, x, y) = (2\pi)^{-n} \int e^{i(\psi(t, x, \eta) - \langle y, \eta\rangle)} a(t, x, \eta)\, d\eta$$

where $a \in S^0_{\text{phg}}$. (That a can be taken independent of y is seen just as in the discussion following Proposition 25.3.3 which actually proves this when t is fixed.) Putting $t=0$ we obtain $a(0, x, \eta) = 1$. The equations $(D_t^{k+1} + PD_t^k)U = 0$ when $t=0$ give recursion formulas for determining the t derivatives of a when $t=0$; for $k=0$ we obtain

$$-i\partial_t a(0, x, \eta) - p(x, \eta) + P(x, \eta) = 0,$$

that is, modulo terms of order $\leqq -1$

$$-i\partial a(0, y, \eta)/\partial t \equiv -p^s(y, \eta) + \tfrac{1}{2} i \sum p^{(j)}_{(j)}(y, \eta). \tag{29.1.9}$$

For $v \in C_0^\infty$ we have

$$U(t)v(x) = (2\pi)^{-n} \int e^{i\psi(t, x, \eta)} a(t, x, \eta)\, \hat{v}(\eta)\, d\eta$$

and shall replace v by $B^* v$. Thinking of $B(x, D)$ as the composition $\beta\mathscr{F}$ of an integral operator β and the Fourier transformation we have $\mathscr{F}B^*$

$=(2\pi)^n \beta^*$ by the Fourier inversion formula, hence

$$U(t)B^* v(x)=(2\pi)^{-n}\iint e^{i(\psi(t,x,\eta)-\langle y,\eta\rangle)} a(t,x,\eta)\bar{B}(y,\eta)v(y)\,d\eta\,dy.$$

Even without integration with respect to y the oscillatory integral exists and defines a distribution on $\mathbb{R}$, for $\partial\psi/\partial t=-p(\partial\psi/\partial x)\neq 0$ (see Section 7.8). In particular the restriction $K(t,y)$ in the proposition is for small t given by

$$K(t,y)=(2\pi)^{-n}\int e^{-it\mu(t,y,\eta)} a(t,y,\eta)\bar{B}(y,\eta)\,d\eta. \tag{29.1.10}$$

To proceed we need a lemma.

Lemma 29.1.3. *If $f(t,y,\eta)\in S^0_{\mathrm{phg}}(\mathbb{R}^{n+1}\times\mathbb{R}^n)$ vanishes for $|\eta|<1$ and for y outside a compact subset of Y, then*

$$F(t,y,\lambda)=\int_{\mu<\lambda} f(t,y,\eta)\,d\eta\in S^n_{\mathrm{phg}}(\mathbb{R}^{n+1}\times\mathbb{R}) \tag{29.1.11}$$

for small t, and when $t=0$ we have

$$F=\int_{p<\lambda} f\,d\eta,\quad \partial_t F=\int_{p<\lambda}\left(\partial_t f+\tfrac{1}{2}\sum\partial(f\partial p/\partial y_j)/\partial\eta_j\right)d\eta. \tag{29.1.12}$$

Proof. From (29.1.8) it follows that $F(t,y,\lambda)=0$ for small $|\lambda|$ and that $|F(t,y,\lambda)|\leqq C\lambda^n$. Writing

$$F(t,y,\lambda)=\int f(t,y,\eta)H(\lambda-\mu(t,y,\eta))\,d\eta$$

where H is the Heaviside function, we obtain

$$\partial F/\partial t=\langle\partial f/\partial t, H(\lambda-\mu)\rangle-\langle f\partial\mu/\partial t,\delta(\lambda-\mu)\rangle.$$

For $t=0$ the second term becomes, by (29.1.8) and Theorem 6.1.5,

$$\tfrac{1}{2}\int_{p=\lambda} f\langle\partial p/\partial y,\partial p/\partial\eta/|\partial p/\partial\eta|\rangle\,dS=\tfrac{1}{2}\int_{p<\lambda}\sum\partial(f\partial p/\partial y_j)/\partial\eta_j\,d\eta$$

by the Gauss-Green formula, which proves (29.1.12). For arbitrary small t we obtain using polar coordinates and writing

$$\delta(\lambda-\mu)r^{n-1}\,dr\,d\omega=(\lambda/\mu_1)^{n-1}\mu_1^{-1}\delta(r-\lambda/\mu_1)\,dr\,d\omega,\quad \mu_1=\mu(t,y,\omega),$$

that the second term is equal to

$$\lambda^{n-1}\int (f\partial_t\mu)(t,y,\lambda\omega/\mu(t,y,\omega))\,d\omega/\mu(t,y,\omega)^n$$

which is clearly in S^n_{phg} since $f\partial_t\mu$ is in S^1_{phg}. Any derivative $D^\alpha_{t,y}F$ is a sum of one term of this type and the term

$$\langle D^\alpha f, H(\lambda-\mu)\rangle$$

which is obviously bounded by $C_\alpha\lambda^n$. We have

$$\partial F(t,y,\lambda)/\partial\lambda=\langle f,\delta(\lambda-\mu)\rangle$$

which is in S^{n-1}_{phg} by the argument just given. This completes the proof that $F\in S^n_{\mathrm{phg}}$.

End of Proof of Proposition 29.1.2. Choose $\chi \in C_0^\infty(\mathbb{R})$ equal to 1 in a neighborhood of 0 so that (29.1.10) holds in supp χ. Assuming as we may that $\bar{B}(y, \eta) = 0$ when $|\eta| < 1$ and writing

$$f(t, y, \eta) = \chi(t)\, a(t, y, \eta)\, \bar{B}(y, \eta)/(2\pi)^n$$

we obtain from (29.1.10) and (29.1.11) that

$$\chi(t) K(t, y) = \int e^{-it\lambda}\, \partial F(t, y, \lambda)/\partial \lambda\, d\lambda.$$

By Lemma 18.2.1 this implies that

$$\chi(t) K(t, y) = \int e^{-it\lambda}\, \partial A(y, \lambda)/\partial \lambda\, d\lambda,$$
$$A(y, \lambda) = e^{iD_t D_\lambda} F(t, y, \lambda)|_{t=0} \sim \sum i^{-j} (\partial_t \partial_\lambda)^j F(t, y, \lambda)/j!|_{t=0}.$$

Thus
$$A(y, \lambda) - F(0, y, \lambda) + i\partial_t \partial_\lambda F(0, y, \lambda) \in S^{n-2}$$

and by Lemma 29.1.3

$$(2\pi)^n (F(0, y, \eta) - i\partial_t \partial_\lambda F(0, y, \lambda))$$
$$= \int_{p<\lambda} \bar{B}\, d\eta - i\partial_\lambda \int_{p<\lambda} (\bar{B}\partial_t a + \tfrac{1}{2} \sum \partial(\bar{B}\partial p/\partial y_j)/\partial \eta_j)\, d\eta.$$

Here
$$\bar{B} = \bar{b} + \bar{b}^s + \tfrac{1}{2} i \sum \bar{b}^{(j)}_{(j)} \bmod S^{-2}$$

and $-i\partial_t a$ is given by (29.1.9). Gauss-Green's formula gives

$$\tfrac{1}{2} i \sum_{p<\lambda} \int \bar{b}^{(j)}_{(j)}\, d\eta = \tfrac{1}{2} i \langle \sum \bar{b}_{(j)} p^{(j)}, \delta(p-\lambda)\rangle = \tfrac{1}{2} i \partial_\lambda \int_{p<\lambda} \sum \bar{b}_{(j)} p^{(j)}\, d\eta.$$

This proves that

$$(2\pi)^n A(y, \lambda) - \int_{p<\lambda} (\bar{b} + \bar{b}^s)\, d\eta + \partial_\lambda \int_{p<\lambda} (\bar{b} p^s + \tfrac{1}{2} i \{\bar{b}, p\})\, d\eta \in S^{n-2}.$$

Replacing $A(y, \lambda)$ by $A(y, \lambda) - A(y, 0)$ we complete the proof.

If we identify Δ with X then the restriction K to Δ of the kernel of $\hat{E}B$ is the pullback by the map $\mathbb{R} \times X \ni (t, x) \mapsto (t, x, x)$. From (29.1.6), Theorem 25.2.3 and Theorem 8.2.4 we therefore obtain

$$WF(K) \subset \{(t, x, \tau, \xi - \eta);\ (t, x, x, \tau, \xi, \eta) \in C, (x, \eta) \in WF(B)\}. \tag{29.1.13}$$

That $(t, x, x, \tau, \xi, \eta) \in C$ means that the orbit of H_p starting at (x, η) at time 0 arrives at (x, ξ) at time t, and that $\tau = -p(x, \xi) = -p(x, \eta)$. For the integral k of the density K over X it follows from Theorem 8.2.12 that

$$WF(k) \subset \{(t, \tau);\ (t, x, x, \tau, \xi, \xi) \in C, (x, \xi) \in WF(B)\}. \tag{29.1.13$'$}$$

Now introduce

$$\Pi(x) = \inf\{t > 0;\ (t, x, x, \tau, \xi, \eta) \in C, (x, \eta) \in WF(B)\}, \quad x \in X,$$
(29.1.14)
$$\Pi^*(x, \xi) = \inf\{t > 0;\ (t, x, x, -p(x, \xi), \xi, \xi) \in C\}, \quad (x, \xi) \in T^*(X) \setminus 0,$$

where Π and Π^* are defined as $+\infty$ if no such t exists. Π^* is homogeneous of degree 0, and in the definition of Π we may restrict (x,η) to the unit sphere bundle. Thus it is clear that these functions are lower semicontinuous.

Theorem 29.1.4. *If $P\in\Psi^1_{\text{phg}}(X;\Omega^{\frac12},\Omega^{\frac12})$ is a positive elliptic self-adjoint operator and e is the spectral function then*

$$\varlimsup_{\lambda\to\infty}\lambda^{1-n}|(e(x,x,\lambda)-(2\pi)^{-n}(\int_{p<\lambda}d\eta-\partial_\lambda\int_{p<\lambda}p^s\,d\eta))/\int_{p<1}d\eta| \le C\Pi(x)^{-1} \tag{29.1.15}$$

where C only depends on the dimension n of X. (The estimated quantity is a quotient of two densities, hence a scalar.)

Proof. If $x_0\in X$ and $\Pi<\Pi(x_0)$ then $\Pi(x)>\Pi$ for all x in a neighborhood Y of x_0. Choose $\chi\in C_0^\infty(Y)$ equal to 1 in a neighborhood of x_0 and apply Proposition 29.1.2 with $B=\chi(x)$. Then it follows that the Fourier transform of $\chi(x)\,de(x,x,\lambda)$ with respect to λ is equal to the Fourier transform of $\partial A(x,\lambda)/\partial\lambda$ in $\{t;\,|t|<\delta\}$ for some positive constant δ, and both are in C^∞ when $0<|t|<\Pi$. We shall now apply Lemma 17.5.6 where we recall that

$$0\le\phi\in\mathscr{S},\quad \hat\phi\in C_0^\infty(-1,1),\quad \int\phi(\lambda)\,d\lambda=1.$$

Let $\mu(\lambda)=e(x,x,\lambda)$ and $v(\lambda)=A(x,\lambda)$. If in terms of local coordinates we write

$$M_0=(2\pi)^{-n}\int_{p<1}d\eta,$$

and $\chi(x)=1$, then

$$|v'(\lambda)|=nM_0\lambda^{n-1}+O(\lambda^{n-2})\le nM_0(\lambda+a_0)^{n-1}$$

for some a_0, which gives the first condition in (17.5.13). With $a=1/\Pi$ we have $(d\mu-dv)*\phi_a\in\mathscr{S}$ which gives the second condition (17.5.13) with $\kappa=0$. Hence (17.5.14) gives

$$\varlimsup_{\lambda\to\infty}\lambda^{1-n}|\mu(\lambda)-v(\lambda)|\le CnM_0/\Pi,$$

which proves (29.1.15) with the constant equal to that in Lemma 17.5.6 multiplied by $n/(2\pi)^n$. The proof is complete.

(29.1.15) is uniform in x in the sense that if $\underline{\Pi}$ is a continuous function with $\underline{\Pi}(x)<\Pi(x)$ for every x (recall that Π is lower semi-continuous, hence the supremum of such functions) then

$$\lambda^{1-n}|(e(x,x,\lambda)-(2\pi)^{-n}(\int_{p<\lambda}d\eta-\partial_\lambda\int_{p<\lambda}p^s\,d\eta))/\int_{p<1}d\eta|\le C\underline{\Pi}(x)^{-1}$$

for λ larger than some number independent of x. We may therefore integrate (29.1.15) over X to obtain an estimate for the number of eigenvalues. However, we can do better by a microlocal argument involving the microlocal period function Π^*.

Theorem 29.1.5. *If $P \in \Psi^1_{\text{phg}}(X; \Omega^{\frac12}, \Omega^{\frac12})$ is a positive elliptic self-adjoint operator and $N(\lambda)$ is the number of eigenvalues $<\lambda$, then*

$$(29.1.15)' \quad \overline{\lim_{\lambda\to\infty}} |N(\lambda)-(2\pi)^{-n}(\iint_{p<\lambda} dx\,d\xi - \partial_\lambda \iint_{p<\lambda} p^s\,dx\,d\xi)|/\lambda^{n-1} \leqq C \iint_{p<1} \Pi^*(x,\xi)^{-1}\,dx\,d\xi,$$

where $dx\,d\xi$ is the symplectic volume element and C only depends on the dimension n of X.

Proof. Let $\Gamma_1, \dots, \Gamma_N$ be a covering of $T^*(X)\setminus 0$ by small open cones, and let $\Pi_j < \Pi^*$ in Γ_j. We can choose $B_j \in \Psi^0_{\text{phg}}(X; \Omega^{\frac12}, \Omega^{\frac12})$ with $WF(B_j) \subset \Gamma_j$ and

$$\sum_1^N B_j B_j^* = I + R$$

where $R \in \Psi^{-\infty}$. To do so we first choose B_j with principal symbol b_j satisfying $\sum |b_j|^2 = 1$ and $\operatorname{supp} b_j \subset \Gamma_j$. We can then choose an operator Q with principal symbol 1 such that Q^*Q is a parametrix for $\sum B_j B_j^*$ (see the proof of Theorem 18.1.11 or Proposition 29.1.9). Then

$$\sum (QB_j)(QB_j)^* - I \in \Psi^{-\infty}$$

so we just have to replace B_j by QB_j. Now

$$N(\lambda) = \operatorname{Tr} E_\lambda = \sum_1^N \operatorname{Tr} E_\lambda B_j B_j^* - \operatorname{Tr} E_\lambda R.$$

Here $E_\lambda R$ is for every s uniformly bounded as an operator from $H_{(-s)}$ to $H_{(s)}$ so the kernel and the trace are uniformly bounded. If $u \in C^\infty(X)$ then

$$E_\lambda B_j B_j^* u = \sum_{\lambda_\nu<\lambda} \phi_\nu (B_j B_j^* u, \phi_\nu) = \sum_{\lambda_\nu<\lambda} \phi_\nu (u, B_j B_j^* \phi_\nu),$$

hence

$$\operatorname{Tr}(E_\lambda B_j B_j^*) = \sum_{\lambda_\nu<\lambda} (\phi_\nu, B_j B_j^* \phi_\nu) = \sum_{\lambda_\nu<\lambda} \|B_j^* \phi_\nu\|^2$$

is increasing. The kernel of the Fourier-Stieltjes transform is of the form discussed in Proposition 29.1.2 with $B = B_j B_j^*$, hence smooth for $0<|t|\leqq \Pi_j$. If we apply Lemma 17.5.6 with $\mu(\lambda) = \operatorname{Tr}(E_\lambda B_j B_j^*)$, $\nu(\lambda) = \int A(x,\lambda)\,dx$ where A is given by Proposition 29.1.2, and $a = 1/\Pi_j$, we obtain

$$\overline{\lim_{\lambda\to\infty}} \lambda^{1-n} |\operatorname{Tr}(E_\lambda B_j B_j^*) - (2\pi)^{-n}(\iint_{p<\lambda} (b_j + b_j^s)\,dx\,d\xi - \partial_\lambda \iint_{p<\lambda} (p^s b_j + \tfrac12 i\{b_j, p\})\,dx\,d\xi)| \leqq C\Pi_j^{-1} \iint_{p<1} b_j\,dx\,d\xi$$

where b_j and b_j^s are the principal and subprincipal symbols of $B_j B_j^*$. These add up to 1 and 0 for $\sum B_j B_j^* - I \in \Psi^{-\infty}$, so we obtain (29.1.15)′ with the integral in the right-hand side replaced by

$$\iint_{p<1} \sum b_j \Pi_j^{-1}\, dx\, d\xi.$$

Here $0 \leq b_j$ and $\sum b_j = 1$ so we have an upper Riemann sum for the integral of an upper semi-continuous function which occurs in the right-hand side of (29.1.15)'. When the covering is refined the estimate (29.1.15)' follows.

Corollary 29.1.6. *If the closed orbits of H_p form a set of measure 0 in $T^*(X) \smallsetminus 0$ then*

$$N(\lambda) - (2\pi)^{-n}\Big(\iint_{p<\lambda} dx\, d\xi - \partial_\lambda \iint_{p<\lambda} p^s\, dx\, d\xi\Big) = o(\lambda^{n-1}), \quad \lambda \to \infty. \tag{29.1.15}''$$

Remark. We have assumed throughout that P has a positive lower bound. However, since the conclusions in Theorems 29.1.4, 5 and Corollary 29.1.6 remain invariant when P is replaced by $P+c$ for some positive constant c, this is also true when we take c negative. Thus one only has to assume that P has a lower bound, which follows from the ellipticity and positivity of the principal symbol.

The preceding results suggest that in the first approximation P acts as multiplication by p on step functions in the cotangent bundle partitioned in sets of volume $(2\pi)^n$. This heuristic argument leads to the following "Szegö type" theorem on the contraction of other pseudo-differential operators to the eigenspaces.

Theorem 29.1.7. *Let $P \in \Psi^1_{\mathrm{phg}}(X; \Omega^{\frac12}, \Omega^{\frac12})$ be a positive elliptic self-adjoint operator with principal symbol p, and denote the spectral measure of $(-\infty, \lambda)$ by E_λ. Let $B \in \Psi^0_{\mathrm{phg}}(X; \Omega^{\frac12}, \Omega^{\frac12})$ be self-adjoint with principal symbol b, and denote by ρ_λ the counting measure of the contraction $E_\lambda B E_\lambda$ to $E_\lambda L^2(X, \Omega^{\frac12})$, that is, the sum of the Dirac measures at the eigenvalues. Then ρ_λ/λ^n converges weakly as $\lambda \to +\infty$ to the pushforward ρ of the measure $(2\pi)^{-n} dx\, d\xi$ in $\{(x,\xi) \in T^*(X); p(x,\xi) < 1\}$ by the map b to $\mathbb{R}$, that is,*

$$\rho(f) = (2\pi)^{-n} \iint_{p(x,\xi)<1} f(b(x,\xi))\, dx\, d\xi, \quad f \in C(\mathbb{R}). \tag{29.1.16}$$

Proof. Since the measures ρ_λ are positive and supported by the finite interval $(-\|B\|, \|B\|)$, it suffices to prove that $\rho_\lambda(f)/\lambda^n \to \rho(f)$ for every polynomial f, that is,

$$\mathrm{Tr}(E_\lambda B E_\lambda)^j/\lambda^n \to (2\pi)^{-n} \iint_{p(x,\xi)<1} b(x,\xi)^j\, dx\, d\xi, \quad j = 0, 1, \ldots. \tag{29.1.17}$$

When $j=0$ this is a weak version of Theorem 29.1.5. Next we shall prove (29.1.17) when $j=1$, that is,

$$\mathrm{Tr}(E_\lambda B E_\lambda)/\lambda^n \to (2\pi)^{-n} \iint_{p(x,\xi)<1} b(x,\xi)\, dx\, d\xi. \tag{29.1.17}'$$

The trace of the finite rank operator $E_\lambda BE_\lambda$ is given by

$$\mu(\lambda)=\operatorname{Tr}(E_\lambda BE_\lambda)=\sum_{\lambda_k<\lambda}(B\phi_k,\phi_k).$$

By Proposition 19.1.13 the operator $E_\lambda B$ has the same trace for it induces the same operator in $E_\lambda L^2$ and in $L^2/E_\lambda L^2$. By the discussion following the statement of Theorem 19.3.1 it follows that $\mu(\lambda)$ is the integral of the kernel of $E_\lambda B$ over the diagonal. Hence the Fourier transform of $d\mu(\lambda)$ is equal to $\int K(t,x)\,dx$ for small t with the notation in Proposition 29.1.2. If A is defined as there and

$$v(\lambda)=\int A(x,\lambda)\,dx,$$

it follows that the Fourier transform of $d\mu-dv$ is in C^∞ when $|t|<\min\Pi(x)$, where Π is defined by (29.1.14). μ is increasing if B is positive. We can then apply Lemma 17.5.5 with $1/a<\min\Pi(x)$, which gives

$$\mu(\lambda)-v(\lambda)=O(\lambda^{n-1}).$$

This is a stronger result than (29.1.17)′. If B is not positive we replace B by $B+\|B\|$ and obtain (29.1.17)′ anyway, since (29.1.17) is valid when $j=0$.

We can replace B by B^j and b by b^j in (29.1.17)′. This shows that (29.1.17) is equivalent to

(29.1.18) $$\operatorname{Tr}((E_\lambda BE_\lambda)^j-E_\lambda B^jE_\lambda)/\lambda^n\to 0 \quad \text{as } \lambda\to\infty.$$

Let $E'_\lambda=I-E_\lambda$ be the projection orthogonal to E_λ. Then

$$E_\lambda B^jE_\lambda=E_\lambda B(E_\lambda+E'_\lambda)B\dots(E_\lambda+E'_\lambda)BE_\lambda.$$

If we expand this we get apart from the term $(E_\lambda BE_\lambda)^j$ only terms containing at least one factor $E'_\lambda BE_\lambda$. Thus the proof is completed by the following lemma:

Lemma 29.1.8. *Under the hypotheses of Theorem* 29.1.7 *the trace class norm of* $E'_\lambda BE_\lambda$ *is* $O(\lambda^{n-\frac12})$.

Proof. With $\Lambda>\lambda$ we can write $E'_\lambda=E'_\Lambda+E_\Lambda-E_\lambda$. If $\|\ \|_1$ denotes the trace class norm (see Definition 19.1.12) then

$$\begin{aligned}\|(E_\Lambda-E_\lambda)BE_\lambda\|_1&\le\|E_\Lambda-E_\lambda\|_1\,\|B\|\,\|E_\lambda\|=\|B\|(\operatorname{Tr}E_\Lambda-\operatorname{Tr}E_\lambda)\\&\le C\Lambda^{n-1}(1+\Lambda-\lambda)\end{aligned}$$

where the last estimate follows from Theorem 29.1.5. We have

$$\|E'_\Lambda BE_\lambda\|_1\le\|E'_\Lambda BE_\lambda\|_2\,\|E_\lambda\|_2$$

where $\|\ \|_2$ is the Hilbert-Schmidt norm. Thus $\|E_\lambda\|_2=(\operatorname{Tr}E_\lambda)^{\frac12}$ and

$$\|E'_\Lambda BE_\lambda\|_2^2=\sum_{\lambda_\mu\ge\Lambda}\sum_{\lambda_\nu<\lambda}|(\phi_\mu,B\phi_\nu)|^2.$$

Since

$$(\lambda_\mu-\lambda_\nu)(\phi_\mu,B\phi_\nu)=(P\phi_\mu,B\phi_\nu)-(\phi_\mu,BP\phi_\nu)=(\phi_\mu,[P,B]\phi_\nu)$$

and $[P,B]\in\Psi^0$ is a bounded operator in L^2, we obtain

$$\|E'_\Lambda B E_\lambda\|_2^2 \leqq (\Lambda-\lambda)^{-2}\sum_{\lambda_\nu<\lambda}\|[P,B]\phi_\nu\|^2 \leqq C^2(\Lambda-\lambda)^{-2}\operatorname{Tr}E_\lambda.$$

Hence

$$\|E'_\lambda B E_\lambda\|_1 \leqq C(\Lambda^{n-1}(1+\Lambda-\lambda)+(\Lambda-\lambda)^{-1}\lambda^n).$$

With $\Lambda=\lambda+\lambda^{\frac{1}{2}}$ the right-hand side is $O(\lambda^{n-\frac{1}{2}})$ which completes the proof of Lemma 29.1.8 and of Theorem 29.1.7.

So far we have assumed that P is of order one. The following proposition will show that the step to an arbitrary positive order is short.

Proposition 29.1.9. *Let $P\in\Psi^m_{\text{phg}}(X;\Omega^{\frac{1}{2}},\Omega^{\frac{1}{2}})$ be a positive elliptic symmetric operator. Then P defines a positive self-adjoint operator $\mathscr{P}$ in $L^2(X,\Omega^{\frac{1}{2}})$. If $m>0$ and $a\in\mathbb{R}$ then $\mathscr{P}^a$ is also defined by a pseudo-differential operator in $\Psi^{am}_{\text{phg}}(X;\Omega^{\frac{1}{2}},\Omega^{\frac{1}{2}})$, with principal and subprincipal symbols p^a and $a\,p^{a-1}p^s$ if p and p^s are those of P.*

Proof. $\mathscr{P}$ is bounded if $m\leqq 0$, hence self-adjoint. If $m>0$ then $\mathscr{P}$ is the restriction of P to all $u\in L^2$ with $Pu\in L^2$. Since this implies $u\in H_{(m)}$ by Theorem 18.1.29 and C^∞ is dense in $H_{(m)}$, it follows that $\mathscr{P}$ is self-adjoint. The resolvent $R(z)=(\mathscr{P}-z)^{-1}$ is defined and analytic in z except at the eigenvalues of $\mathscr{P}$ on $\mathbb{R}_+$, and the L^2 operator norm can be estimated by the reciprocal of the distance to the eigenvalues. If $a<0$ it follows from the spectral theorem that with absolute convergence in L^2

$$\mathscr{P}^a u = -(2\pi i)^{-1}\int_{-i\infty}^{i\infty} z^a R(z)u\,dz, \quad u\in L^2,$$

where z^a is analytic in the right half plane and equal to 1 when $z=1$. Since $\mathscr{P}^{a+1}=\mathscr{P}^a\mathscr{P}u$ when u is in the domain of $\mathscr{P}$ and $\operatorname{Re}a\leqq 0$ it follows that the distribution kernel of $\mathscr{P}^a$ is an entire analytic function of a. By approximating R with a parametrix for $P-z$ we shall determine the singularities for $a<0$ and then by analytic continuation for $a\geqq 0$ as well.

Let $Y\subset X$ be a local coordinate patch identified with an open set in $\mathbb{R}^n$. In the local coordinates the symbol of $P-z$ is uniformly bounded with respect to z in

$$S((1+|z|+|\xi|^m),g); \quad g=|dx|^2+|d\xi|^2/(1+|\xi|^2).$$

The reciprocal is uniformly bounded in $S((1+|z|+|\xi|^m)^{-1},g)$ if $\operatorname{Re}z=0$ as we assume from now on. (We may assume without restriction that $\operatorname{Re}P(x,\xi)>0$ everywhere, for $p>0$.) Hence

$$(P(x,D)-z)(P-z)^{-1}(x,D)=I-Q_z(x,D)$$

where Q_z is uniformly bounded in $S((1+|\xi|)^{m-1}(1+|\xi|^m+|z|)^{-1},g)$. (Note that multiplication by z does not require the calculus of pseudo-differential operators and introduces no error term!) We have

$$Q_z(x,\xi)\sim \sum_{\alpha\neq 0} P^{(\alpha)}(x,\xi)\, D_x^\alpha (P(x,\xi)-z)^{-1}/\alpha!.$$

Let E_z be the asymptotic sum of the symbols of

$$(P-z)^{-1}(x,D)(Q_z(x,D))^N, \qquad N=0,1,\dots.$$

Then we have

$$(P(x,D)-z)\,E_z(x,D)=I-W_z(x,D) \tag{29.1.19}$$

where W_z is uniformly bounded in $S((1+|z|)^{-1}(1+|\xi|)^{-N},g)$ for any N.

To be quite precise we should observe that so far we have only worked in the coordinate patch Y. Choose $\psi, \Psi \in C_0^\infty(Y)$ with $\Psi=1$ in $\operatorname{supp}\psi$. Then $\Psi E_z(x,D)\psi u$ is defined in X for every u in $\mathscr{D}'(X)$, and

$$(P-z)\,\Psi E_z(x,D)\,\psi u=\psi u - W_z(x,D)\,u$$

for another W_z such that $(1+|z|)\,W_z$ is uniformly bounded in $\Psi^{-\infty}$. Taking a covering of X by coordinate patches Y_j and corresponding $\psi_j, \Psi_j \in C_0^\infty(Y_j)$ with $\Psi_j=1$ in $\operatorname{supp}\psi_j$ and $\sum \psi_j=1$, we obtain when summing over j a parametrix E_z satisfying (29.1.19) globally in X.

Multiplication of (29.1.19) with $R(z)$ gives

$$R(z)=E_z+R(z)\,W_z.$$

Here $(1+|z|)\,W_z$ is for arbitrary s, t uniformly bounded as an operator from $H_{(s)}$ to $H_{(t)}$, and R_z has norm $\leqq C/(1+|z|)$ as an operator in $H_{(t)}$ since $\|u\|_{(t)}$ is equivalent to $\|\mathscr{P}^{t/m}u\|_{(0)}$ at least if t is a positive multiple of m. Thus we have for $a<0$

$$\mathscr{P}^a=-(2\pi i)^{-1}\int_{-i\infty}^{i\infty} z^a E_z\,dz+T(a)\,u$$

where $T(a)$ is an operator with kernel analytic in a when $\operatorname{Re} a<1$, with values in $C^\infty(X\times X;\Omega^{\frac12}(X\times X))$. Every term in the symbol of E_z is of the form $-(z-P)^{-k-1}q$ where $q\in S^{mk-\kappa}$ for some $\kappa\geqq 0$. We have

$$-(2\pi i)^{-1}\int_{-i\infty}^{i\infty} z^a(z-P)^{-k-1}q\,dz=a(a-1)\dots(a-k+1)\,P^{a-k}q/k!$$

which is a symbol in $S^{am-\kappa}$ and depends analytically on a. There are just a finite number of terms for fixed κ. If we break off the series defining E_z after sufficiently many terms the error replacing $T(a)$ will have as many derivatives as we wish when $\operatorname{Re} a<1$. Hence $\mathscr{P}^a$ is a pseudo-differential operator when $a<1$. Calculating the terms with $\kappa=0$ or $\kappa=1$ above we find that mod S^{am-2} the symbol is equal to

$$P^a+a(a-1)/2i\sum p^{(j)}\,p_{(j)}\,p^{a-2}.$$

Since P is congruent to $p+p^s-\frac{1}{2}i\sum p^{(j)}_{(j)}$ it follows that P^a is congruent to

$$p^a+a\,p^{a-1}(p^s-\tfrac{1}{2}i\sum p^{(j)}_{(j)}).$$

To obtain the sum of the principal and subprincipal symbols of $\mathscr{P}^a$ we must add $\frac{1}{2}i\sum \partial^2 p^a/\partial x_j\partial\xi_j$ to the symbol of $\mathscr{P}^a$, which proves the statement on the subprincipal symbol when $a<1$. Now it follows immediately from the Weyl calculus that if $\mathscr{P}^a$ is defined by a pseudo-differential operator then this is true for $\mathscr{P}^{2a}$ also, and that the principal and subprincipal symbols of this operator are obtained by multiplying those of $\mathscr{P}^a$ by (twice) the principal symbol of $\mathscr{P}^a$. This extends our results from all $a<1$ to all $a\in\mathbb{R}$, and completes the proof.

If $e(x,y,\lambda)$ is the spectral function of $\mathscr{P}$ then $e(x,y,\lambda^m)$ is the spectral function of $\mathscr{P}^{1/m}$. To rewrite Theorems 29.1.4, 5 and Corollary 29.1.6 for elliptic operators of arbitrary positive order m, we first observe that in the first order case

$$\partial_\lambda \int_{p<\lambda} p^s\,d\eta=\partial_\lambda\int H(\lambda-p)\,p^s\,d\eta=\int\delta(\lambda-p)\,p^s\,d\eta.$$

If $\mathscr{P}$ is of order $m>0$ and (29.1.15) is applied to $\mathscr{P}^{1/m}$, we obtain if e is the spectral function of $\mathscr{P}$

$$(29.1.15)_m\qquad \overline{\lim_{\lambda\to\infty}}\,\lambda^{1-n}|(e(x,x,\lambda^m)-$$
$$(2\pi)^{-n}\int(H(\lambda^m-p)-\delta(\lambda^m-p)p^s)\,d\eta)/\int_{p<1}d\eta|\leqq C\Pi(x)^{-1},$$

where Π is the period function for $p^{1/m}$. In fact,

$$\int\delta(\lambda-p^{1/m})\,p^{1/m-1}\,p^s\,d\eta/m=\int p^s\,|dS|/|p'|=\int\delta(\lambda^m-p)\,p^s\,d\eta$$

by Theorem 6.1.5, if dS is the area element on the surface $p=\lambda^m$ for some Euclidean metric in T^*_x. The asymptotic formulas (29.1.15)′ and (29.1.15)″ can be written in the same way, and Theorem 29.1.7 remains valid for the counting measure of $E_{\lambda^m}BE_{\lambda^m}$.

In the case of differential operators m has to be even so p^s is odd and therefore drops out in the integrations which occur in $(29.1.15)_m$ and the analogues of (29.1.15)′, (29.1.15)″.

29.2. The Case of a Periodic Hamilton Flow

As in most of Section 29.1 we denote by P a positive elliptic operator in $\Psi^1_{phg}(X;\Omega^{\frac{1}{2}},\Omega^{\frac{1}{2}})$ with principal and subprincipal symbol p and p^s. However, we shall now assume that all orbits of H_p are closed with the same period Π, that is, that the function $\Pi^*(x,\xi)$ in (29.1.14) is identically equal to Π. An example is $(-\Delta+c)^{\frac{1}{2}}$ on the sphere S^n, $c>0$, with the standard metric. From Section 17.5 we know that the eigenvalues are then $(k^2+k(n-1)+c)^{\frac{1}{2}}=k+(n-1)/2$ if $c=(n-1)^2/4$ and $k=0,1,\dots$. The multiplicities are given by a polynomial in k. For other values of c there is just a shift $O(1/k)$ of the eigenvalue.

With the notation $\hat{E}(t)=e^{-it\mathscr{P}}$ used in Section 29.1 it follows from Theorem 29.1.1 that the kernel of $\hat{E}(.+\Pi)$ is in $I^{-\frac{1}{4}}(\mathbb{R}\times X\times X, C'; \Omega(X\times X)^{\frac{1}{2}})$, for C has period Π in t. (This conclusion requires only that Π is a period of all the orbits of H_p, not that it is the minimal period.) Before discussing the principal symbol of $\hat{E}(\Pi)$ we must make some remarks on the Maslov bundle M_C of C defined in Section 25.2.

M_C is trivial as a $\mathbb{Z}_4$ bundle. In fact, the restriction to $t=0$ is the diagonal Δ^* in $(T^*(X)\setminus 0)\times(T^*(X)\setminus 0)$ with the trivial bundle. Moreover, the maps

$$C\ni(t,\exp tH_p(y,\eta),y,\eta)\mapsto(st,\exp stH_p(y,\eta),y,\eta)\in C$$

define for $0\leqq s\leqq 1$ a retraction of C to Δ^* so the assertion follows from the homotopy lifting theorem. However, the Maslov bundle over the canonical relation on the cotangent bundle of $(\mathbb{R}/\mathbb{Z}\Pi)\times X\times X$ defined by C is not trivial in general. Let μ be the locally constant section of the Maslov bundle on C which is equal to 1 when $t=0$. The pullback of μ by the translation $t\mapsto t+\Pi$ in C can be considered as a section of M_C and must be equal to $e^{\pi i\alpha/2}\mu$ where the Maslov index $\alpha\in\mathbb{Z}$ is the value on an orbit period of the Maslov class defined by (21.6.10).

We shall now return to the symbol of $\hat{E}(\Pi)$ assuming that the subprincipal symbol p^s is identically 0. From the proof of Theorem 29.1.1 we know that the principal symbol is then locally constant along the orbits of H_p. Hence the symbol of $\hat{E}(\Pi)$ as a pseudo-differential operator is equal to $e^{\pi i\alpha/2}$. If we change our assumption on p^s to

$$p^s=\pi\alpha/2\Pi. \tag{29.2.1}$$

then the principal symbol of $\hat{E}(\Pi)$ is the identity, and $\hat{E}(\Pi)$ is a unitary operator. Since $\hat{E}(\Pi)-I$ is compact, the eigenvalues $e^{i\Pi\lambda_k}$ converge to 1 as $k\to\infty$.

Lemma 29.2.1. *If* $\exp \Pi H_p$ *is the identity and* (29.2.1) *holds, then one can find a self-adjoint* $Q\in\Psi^{-1}_{\text{phg}}(X;\Omega^{\frac{1}{2}},\Omega^{\frac{1}{2}})$ *commuting with* P *such that* $e^{-i\Pi\mathscr{P}}=e^{i\Pi\mathscr{Q}}$.

Proof. The operator $A=e^{-i\Pi\mathscr{P}}-I$ commutes with $\mathscr{P}$ and $A\in\Psi^{-1}_{\text{phg}}(X;\Omega^{\frac{1}{2}},\Omega^{\frac{1}{2}})$ since the principal symbol is 0. We must define $i\Pi Q$ as a logarithm of $I+A$. To do so we choose a disc $\Gamma\subset\mathbb{C}$ with center at 0 and radius <1 such that none of the eigenvalues $e^{-i\Pi\lambda_k}-1$ of A is on the boundary, and we set

$$i\Pi Q_1=-(2\pi i)^{-1}\int_{\partial\Gamma}R(z)\log(1+z)\,dz \tag{29.2.2}$$

where $R(z)=(A-zI)^{-1}$ is the resolvent and $\log(1+z)$ is defined by the power series expansion. By Cauchy's integral formula

$$i\Pi Q_1\phi_k=\log e^{-i\Pi\lambda_k}\phi_k \quad \text{if } e^{-i\Pi\lambda_k}-1\in\Gamma,$$

and $Q_1\phi_k=0$ otherwise. Here $\mathscr{P}\phi_k=\lambda_k\phi_k$. If we define

$$Q_2u=-\sum\lambda_k\phi_k(u,\phi_k)$$

with summation for the finitely many k such that $e^{-i\Pi\lambda_k}\notin\Gamma$, then it is clear that $Q=Q_1+Q_2$ is self-adjoint and that $e^{i\Pi Q}=e^{-i\Pi P}$. The operator Q_2 has a C^∞ kernel, so it only remains to show that $Q_1\in\Psi^{-1}_{\text{phg}}$. To do so we choose a parametrix E_z of $A-zI$ which is bounded in $\Psi^0(X;\Omega^{\frac12},\Omega^{\frac12})$ when $z\in\partial\Gamma$; we just have to take $E_z\sim\sum_0^\infty A^jz^{-j-1}$. Then

$$(A-zI)E_z=I+W_z$$

where W_z is bounded in $\Psi^{-\infty}(X;\Omega^{\frac12},\Omega^{\frac12})$. Multiplication by $R(z)$ gives

$$E_z-R(z)=R(z)W_z$$

where the kernel of $R(z)W_z$ is uniformly bounded in $C^\infty(X\times X;\Omega^{\frac12})$ since $R(z)$ is uniformly bounded in the $H_{(s)}$ topology for every s. Hence it follows that

$$i\Pi Q_1+(2\pi i)^{-1}\int_{\partial\Gamma}E_z\log(1+z)\,dz$$

is an operator with C^∞ kernel. The integral here is a pseudo-differential operator with symbol asymptotically given by the symbols in the series $\sum-(-A)^n/n$, which completes the proof.

The commuting operators $\mathscr{P}$ and $\mathscr{P}+\mathscr{Q}$ have the same eigenfunctions, and since $e^{i\Pi(\mathscr{P}+\mathscr{Q})}=I$, the eigenvalues of $\mathscr{P}+\mathscr{Q}$ are integer multiples of $2\pi/\Pi$. Since $Q\in\Psi^{-1}$ the operator $\mathscr{P}^{\frac12}\mathscr{Q}\mathscr{P}^{\frac12}$ is bounded. If the norm is M then

$$-M\mathscr{P}^{-1}\leq\mathscr{Q}\leq M\mathscr{P}^{-1}.$$

If λ is an eigenvalue of $\mathscr{P}$ it follows that $2k\pi/\Pi\in(\lambda-M/\lambda,\lambda+M/\lambda)$ for some integer k. When λ is large then k is large, and we have

$$|\lambda-2k\pi/\Pi|\leq C/k. \tag{29.2.3}$$

Thus the eigenvalues are close to the arithmetic sequence $(2\pi/\Pi)\mathbf{Z}$, and for large k the number of eigenvalues of $\mathscr{P}$ in the interval (29.2.3) is equal to the multiplicity of $2k\pi/\Pi$ as an eigenvalue of $\mathscr{P}+\mathscr{Q}$.

Theorem 29.2.2. *Let $P\in\Psi^1_{\text{phg}}(X;\Omega^{\frac12},\Omega^{\frac12})$ be a positive elliptic self-adjoint operator with principal symbol p and subprincipal symbol p^s. Assume that all orbits of H_p have the same minimal period Π and that the subprincipal symbol p^s is given by (29.2.1) where α is the Maslov index of C for these closed orbits.*

Then there are constants C, k_0 and a polynomial μ of degree $n-1$ such that $\mathscr{P}$ has $\mu(k)$ eigenvalues satisfying (29.2.3) *for every integer $k>k_0$ and only finitely many other eigenvalues. We have*

$$\mu(k)=n(k-\Pi p^s/2\pi)^{n-1}\Pi^{-n}\iint_{p<1} dx\,d\xi+O(k^{n-3}). \tag{29.2.4}$$

Proof. If we replace P by $P+Q$ where Q is defined by Lemma 29.2.1 then the principal and subprincipal symbols do not change. By the discussion preceding the statement it is therefore sufficient to prove that the multiplicity of $2k\pi/\Pi$ as an eigenvalue of $\mathscr{P}$ is a polynomial in k for large positive k if $e^{-i\Pi\mathscr{P}}=I$. Since the Fourier transform $e^{-it\mathscr{P}}$ of the spectral measure is then periodic with period Π we know that the inverse Fourier transform dE_λ is a measure concentrated at $2\pi\mathbb{Z}/\Pi$ where the spectral mass is the Fourier coefficient

$$E(\{2\pi k/\Pi\})=\int e^{2\pi ikt/\Pi}\hat{\phi}(t)\hat{E}(t)\,dt/\Pi.$$

Here $\hat{\phi}\in C_0^\infty(-\Pi,\Pi)$ is chosen so that $\sum_j \hat{\phi}(t+j\Pi)=1$ (see Section 7.2). Taking the trace we obtain for the multiplicity $\mu(k)$ of the eigenvalue $2\pi k/\Pi$

$$\mu(k)=\int e^{2\pi ikt/\Pi}\hat{\phi}(t)\operatorname{Tr}\hat{E}(t)\,dt/\Pi.$$

Now the restriction of the kernel of $\hat{E}(t)$ to the diagonal is the kernel K in Proposition 29.1.2 with $B=1$. It is equal to the Fourier transform of $\partial A/\partial\lambda$ for small t and is in C^∞ for $0<|t|<\Pi$. Hence Fourier's inversion formula gives that $\mu(k)-\nu(k)$ is rapidly decreasing as $k\to\infty$ if

$$\nu(k)=\iint \partial A(x,\lambda)/\partial\lambda\,\phi(2\pi k/\Pi-\lambda)(2\pi/\Pi)\,d\lambda\,dx.$$

Since $\partial A/\partial\lambda\in S^{n-1}_{\mathrm{phg}}$ we have $\nu(k)=\nu_0(k)+O(1/k)$ where ν_0 is a polynomial. The following lemma will show that $\mu(k)=\nu_0(k)$ for large k. This will complete the proof, for the first two terms in ν_0 are obtained from the first two terms in A given in Proposition 29.1.2.

Lemma 29.2.3. *Let $g(k)$ be a polynomial in k such that the fractional part of $g(k)$ tends to 0 as the integer $k\to+\infty$, that is, $e^{2\pi ig(k)}\to 1$. Then $g(k)$ is an integer for every integer k.*

Proof. If g is a constant then the constant is an integer by the hypothesis. Assume now that g is of order N and that the lemma is already proved for polynomials of order $N-1$. Then $h(k)=g(k)-g(k-1)$ is a polynomial taking only integer values for integer k, and $g(k)=h(k)+\ldots+h(1)+g(0)$ so $e^{2\pi ig(0)}=1$. Thus $g(0)$ is an integer which completes the proof.

So far we have just studied the number of eigenvalues in the intervals (29.2.3) and not their distribution. The following more precise analogue of the "Szegö type" theorem 29.1.7 will easily give information on how the $\mu(k)$ eigenvalues are distributed.

Theorem 29.2.4. *Let P satisfy the hypotheses in Theorem 29.2.2 and assume in addition that $e^{-i\Pi\mathscr{P}}=I$. Let $B\in\Psi^0_{phg}(X;\Omega^{\frac12},\Omega^{\frac12})$ commute with P and let ρ_k be the counting measure for the eigenvalues of B restricted to the eigenspace $E(\{2\pi k/\Pi\})L^2$ of $\mathscr{P}$. Then $\rho_k/\mu(k)$ converges weakly to the measure ρ defined by*

$$\rho(f)=\iint_{p<1} f(b)\,dx\,d\xi/\iint_{p<1} dx\,d\xi,\quad f\in C(\mathbb{R}), \tag{29.2.5}$$

where b is the principal symbol of B.

Proof. Since the measures ρ_k are positive and supported by the finite interval $(-\|B\|,\|B\|)$ it suffices to prove that $\rho_k(f)/\mu(k)\to\rho(f)$ for every polynomial f. Since $\rho_k(1)=\mu(k)$ and $\rho(1)=1$ the statement is thus equivalent to

$$\operatorname{Tr}(B^j|E(\{2\pi k/\Pi\})L^2)/\mu(k)\to\iint_{p<1} b^j\,dx\,d\xi/\iint_{p<1} dx\,d\xi \tag{29.2.6}$$

for $j=1,2,\dots$. If this is proved for $j=1$ we just have to replace B by B^j to obtain the general result so the proof will be completed when (29.2.6) is proved for $j=1$.

The Fourier transform of the operator valued measure $B\,dE_\lambda$ is the period function $e^{-itP}B$, so the Fourier transform of the measure on $\mathbb{R}$

$$\sum\operatorname{Tr}(B|E(\{2\pi k/\Pi\})L^2)\,\delta_{2\pi k/\Pi}$$

is the periodic distribution $\operatorname{Tr}e^{-itP}B$. For small $|t|$ the restriction of the kernel K of $e^{-itP}B$ to the diagonal is described in Proposition 29.1.2, and it is in C^∞ when $0<|t|<\Pi$. With $\hat\phi\in C_0^\infty(-\Pi,\Pi)$ chosen as in the proof of Theorem 29.2.2 we have by the usual formula for the Fourier coefficients

$$\operatorname{Tr}(B|E(\{2\pi k/\Pi\})L^2)=\iint e^{2\pi ikt/\Pi}\hat\phi(t)K(x,t)\,dx\,dt/\Pi,$$

which is equal to

$$2\pi/\Pi\int\partial A(x,\lambda)/\partial\lambda\,dx,\quad \lambda=2\pi k/\Pi,$$

with a rapidly decreasing error. This proves that

$$\operatorname{Tr}(B|E(\{2\pi k/\Pi\})L^2)=(2\pi/\Pi)(2\pi k/\Pi)^{n-1}n(2\pi)^{-n}\iint_{p<1} b\,dx\,d\xi+O(k^{n-2}).$$

(Proposition 29.1.2 gives of course a full asymptotic expansion with an explicit value for the coefficient of k^{n-2}.) Division by (29.2.4) gives (29.2.6) with $j=1$ and completes the proof.

Theorem 29.2.5. *Let P satisfy the hypotheses in Theorem 29.2.2, and let $\Lambda(k)$ be the eigenvalues of P in the interval (29.2.3), $\mu(k)$ the cardinality of $\Lambda(k)$. Choose Q according to Lemma 29.2.1 and let b be the principal symbol of*

$-PQ$. *Then*

$$\sum_{\lambda\in\Lambda(k)} f(2\pi k\Pi^{-1}(\lambda-2k\pi/\Pi))/\mu(k) \to \iint_{p<1} f(b)\,dx\,d\xi / \iint_{p<1} dx\,d\xi, \quad k\to\infty, \tag{29.2.7}$$

when $f\in C(\mathbb{R})$.

Proof. Set $A=P+Q$. Then A satisfies the hypotheses of Theorem 29.2.4, $P=A-Q$ and Q commutes with A. The eigenvalues of P in $\Lambda(k)$ are $\lambda=2\pi k/\Pi+\varepsilon$ where ε is an eigenvalue of $-Q$ restricted to the eigenspace V_k of A corresponding to the eigenvalue $2\pi k/\Pi$, and $2\pi k\Pi^{-1}\varepsilon$ are the eigenvalues of $B=-AQ$ in this space. Thus the left-hand side of (29.2.7) is the normalized counting measure of the eigenvalues of B restricted to V_k, so (29.2.7) follows from (29.2.5). The proof is complete.

(29.2.7) does not suffice display the fine structure of the eigenvalue clusters $\Lambda(k)$ if the measure in the right-hand side is a Dirac measure, that is, if b is a constant. That means that $Q+c_1A^{-1}\in\Psi^{-2}_{\text{phg}}$ for some constant c_1. Let us assume more generally that for some integer $N>0$

$$P=A+\sum_1^N c_jA^{-j}+Q \tag{29.2.8}$$

where $Q\in\Psi^{-N-1}_{\text{phg}}$ commutes with A. Set $B=A^{N+1}Q$. Then the proof of Theorem 29.2.5 shows that

$$\sum_{\lambda\in\Lambda(k)} f\Big((2\pi k/\Pi)^{N+1}\Big(\lambda-2\pi k/\Pi-\sum_1^N c_j(2\pi k/\Pi)^{-j}\Big)\Big)\Big/\mu(k) \to \iint_{p<1} f(b)\,dx\,d\xi/\iint_{p<1} dx\,d\xi, \quad f\in C(\mathbb{R}) \tag{29.2.9}$$

where b is the principal symbol of B. If the right-hand side is still a Dirac measure we have a decomposition (29.2.8) with N replaced by $N+1$. Thus there are two alternatives: Either one can find N so that the limit in (29.2.9) is a measure with support equal to an interval of positive length, or else there is a sequence c_j such that for every N

$$\sup_{\lambda\in\Lambda(k)} |\lambda-2\pi k/\Pi-\sum_1^N c_j(2\pi k/\Pi)^j|\,\lambda^N\to 0 \quad \text{as } k\to\infty.$$

So far we have always assumed that the subprincipal symbol p^s of p is given by (29.2.1). Another constant value does not affect the results much apart from the fact that the arithmetic sequence $2\pi\mathbb{Z}/\Pi$ where the eigenvalues cluster is shifted to $p^s-\pi\alpha/2\Pi+2\pi\mathbb{Z}/\Pi$. However, new problems occur when p^s is variable.

It is convenient to write the operator to be studied in the form $P+Q$ where P satisfies the full hypothesis of Theorem 29.2.4 and

$Q\in\Psi^0_{\text{phg}}(X;\Omega^{\frac12},\Omega^{\frac12})$. We wish to find a unitarily equivalent operator of the form $P+B$ where $B\in\Psi^0_{\text{phg}}$ and B commutes at least approximately with P. To do so we first observe that by Theorem 29.2.1 and Theorem 25.2.3

$$Q_t=e^{it\mathscr{P}}Q\,e^{-it\mathscr{P}}$$

is a self-adjoint pseudo-differential operator of order 0, which is a C^∞ function of t. In fact, the kernel of $e^{-is\mathscr{P}}Q\,e^{-it\mathscr{P}}$ is in $I^{-\frac12}(\mathbb{R}\times X\times\mathbb{R}\times X,\tilde{C}')$ where $\tilde{C}$ is the full bicharacteristic relation, and Q_t is obtained by freezing the two variables in $\mathbb{R}$. Hence the average

$$B=\int_0^{\Pi}Q_t\,dt/\Pi \tag{29.2.10}$$

is in $\Psi^0_{\text{phg}}(X;\Omega^{\frac12},\Omega^{\frac12})$. We have $[B,e^{-is\mathscr{P}}]=0$ for every s, hence $[B,P]=0$. Thus Theorem 29.2.4 contains very precise information on the eigenvalues of $P+B$.

By Taylor's formula we have

$$B-Q=\int_0^{\Pi}dt\int_0^t dQ_s/ds\,ds/\Pi.$$

and $dQ_s/ds=[iP,Q_s]$. Hence

$$B-Q=[iS,P]$$

where

$$S=-\int_0^{\Pi}dt\int_0^t Q_s\,ds\in\Psi^0(X;\Omega^{\frac12},\Omega^{\frac12}) \tag{29.2.11}$$

is a self-adjoint operator. Thus S is continuous in all Sobolev spaces $H_{(s)}$, so e^{itS} is continuous in $H_{(s)}$ for all s and t. It is a unitary operator in $L^2(X,\Omega^{\frac12})$.

If $A\in\Psi^\mu$ then $e^{iS}A\,e^{-iS}$ is a pseudo-differential operator and the symbol is asymptotic to the sum of the symbols of the terms in the formal series

$$\sum_0^\infty(\operatorname{ad} iS)^j A/j!$$

where $(\operatorname{ad} iS)A=[iS,A]$. In fact, if $A(t)=e^{itS}A\,e^{-itS}$ then

$$A'(t)=e^{itS}[iS,A]\,e^{-itS},\qquad A^{(j)}(t)=e^{itS}((\operatorname{ad} iS)^j A)\,e^{-itS}.$$

This proves that $A^{(j)}(0)=(\operatorname{ad} iS)^j A\in\Psi^{\mu-j}$ and that $A^{(j)}(t)$ is continuous from $H_{(s_1)}$ to $H_{(s_2)}$ if $s_2\leqq s_1+j-\mu$. Thus

$$A(1)-\sum_0^N(\operatorname{ad} iS)^j A/j!$$

has for every ν a kernel in $C^\nu(X\times X)$ if $N>N_\nu$, which proves the claim.

Now $e^{iS}(P+Q)\,e^{-iS}$ is unitarily equivalent to $P+Q$, and since

$$Q+[iS,P]=B,$$

we have

$$e^{iS}(P+Q)e^{-iS}-(P+B)\in\Psi_{\mathrm{phg}}^{-1}. \tag{29.2.12}$$

If the principal symbol of B is a constant then the eigenvalues of $P+B$ are described in great detail in Theorem 29.2.5 and in (29.2.9). Otherwise we shall just use that

$$P+B-CP^{-1}\leqq e^{iS}(P+Q)e^{-iS}\leqq P+B+CP^{-1} \tag{29.2.13}$$

for some constant C. (The operators should here be interpreted as the self-adjoint operators defined in L^2.) We can apply Theorem 29.2.4 to estimate the eigenvalues of the operators to the left and to the right.

Let θ_k and θ be the increasing integrals of the measures $\rho_k/\mu(k)$ and ρ in Theorem 29.2.4 which vanish at $-\infty$, thus $\rho_k=\mu(k)\,d\theta_k$ and $\rho=d\theta$. The counting function of the eigenvalues of $P+B+CP^{-1}$ restricted to $E(\{2\pi k/\Pi\})L^2$ is $\theta_k(\lambda-2\pi k/\Pi-C\Pi/2\pi k)\mu(k)$ for $k>k_0$. If $N(\lambda)$ is the number of eigenvalues $\leqq\lambda$ of $P+Q$, it follows that

$$\begin{aligned} N_0+\sum_{k>k_0}\theta_k(\lambda-2\pi k/\Pi-C\Pi/2\pi k)\mu(k)&\\ \leqq N(\lambda)\leqq N_0+\sum_{k>k_0}\theta_k(\lambda-2\pi k/\Pi+C\Pi/2\pi k)\mu(k)& \end{aligned} \tag{29.2.14}$$

where N_0 is the dimension of $E(\{2\pi k_0/\Pi\})L^2$. To exploit this we need a simple lemma.

Lemma 29.2.6. *For every $\varepsilon>0$ one can find $k_\varepsilon>0$ such that*

$$\theta(\lambda-\varepsilon)-\varepsilon\leqq\theta_k(\lambda)\leqq\theta(\lambda+\varepsilon)+\varepsilon, \quad \text{if } k>k_\varepsilon,\ \lambda\in\mathbb{R}. \tag{29.2.15}$$

Proof. The supports of the measures ρ_k are contained in the interval $(-\|B\|,\|B\|)$ so $\theta_k(\lambda)=\theta(\lambda)=0$ for $\lambda<-\|B\|$ and $\theta_k(\lambda)=\theta(\lambda)=1$ for $\lambda>\|B\|$. The statement is obvious then so we may assume that $|\lambda|\leqq\|B\|$. Choose $\psi\in C^\infty$ so that $0\leqq\psi\leqq1$, $\psi=0$ in $(-\infty,-\varepsilon)$, and $\psi=1$ in $(0,\infty)$. Then

$$\theta_k(\lambda)\leqq\int\psi(\lambda-t)\,d\rho_k(t)/\mu(k)\leqq\int\psi(\lambda-t)\,d\rho(t)+\varepsilon, \quad k>k_\varepsilon,$$

for the convolution converges in C^∞, hence locally uniformly. The right-hand side is at most equal to $\theta(\lambda+\varepsilon)+\varepsilon$ which proves the second inequality (29.2.15). The first inequality follows in the same way if we replace ψ by $\psi(\cdot-\varepsilon)$, which completes the proof.

If we combine (29.2.14) with (29.2.15) it follows that for every $\varepsilon>0$

$$\overline{\lim_{\lambda\to\infty}}\,\lambda^{1-n}(N(\lambda)-N_a(\lambda+\varepsilon))\leqq0\leqq\underline{\lim}_{\lambda\to\infty}\,\lambda^{1-n}(N(\lambda)-N_a(\lambda-\varepsilon)) \tag{29.2.16}$$

where

$$N_a(\lambda)=\sum_{k>k_0}\theta(\lambda-2\pi k/\Pi)\,n(k-\Pi p^s/2\pi)^{n-1}\,\Pi^{-n}\iint_{p<1}dx\,d\xi. \tag{29.2.17}$$

Here we have also used (29.2.4). Note that

$$N_a(\lambda+\varepsilon)-N_a(\lambda-\varepsilon)\leq C\lambda^{n-1}\sum(\theta(\lambda+\varepsilon-2\pi k/\Pi)-\theta(\lambda-\varepsilon-2\pi k/\Pi)),$$

where the sum is actually finite since θ is constant outside $(-\|B\|, \|B\|)$. If θ is continuous then $\lambda^{1-n}(N_a(\lambda+\varepsilon)-N_a(\lambda-\varepsilon))\to 0$ as $\varepsilon\to 0$, uniformly for large λ, and (29.2.16) can then be replaced by

$$\lim_{\lambda\to\infty}\lambda^{1-n}(N(\lambda)-N_a(\lambda))=0. \tag{29.2.18}$$

Summing up, we have proved:

Theorem 29.2.7. *Let P satisfy the hypotheses of Theorem 29.2.4 and let $Q\in\Psi^0_{phg}(X;\Omega^{\frac12},\Omega^{\frac12})$ be self-adjoint with principal symbol q. Set*

$$b(x,\xi)=\int_0^{\Pi} q((\exp tH_p)(x,\xi))\,dt/\Pi,$$

and let ρ be the positive measure defined by (29.2.5). Set $\rho=d\theta$ where θ is increasing and equal to 0 at $-\infty$. Then the counting function $N(\lambda)$ for the eigenvalues of $P+Q$ satisfies (29.2.16) with N_a defined by (29.2.17). When ρ is a continuous measure then (29.2.18) holds.

Thus we have obtained a result of the same precision as Corollary 29.1.6, but the asymptotic approximation $N_a(\lambda)$ to $N(\lambda)$ is now considerably more complicated.

29.3. The Weyl Formula for the Dirichlet Problem

Let P be a positive, self-adjoint, elliptic second order differential operator acting on half densities in a C^∞ manifold X with boundary ∂X. Denote the principal symbol by p, and let $N(\lambda)$ be the number of eigenvalues $\leq\lambda$ of the Dirichlet problem. Our purpose is to prove an analogue of Theorem 29.1.5 and Corollary 29.1.6. The proof will essentially be a repetition combined with Corollary 17.5.11 and the results of Chapter XXIV on the propagation of singularities for D_t^2-P. We start with reviewing the basic constructions made in Section 29.1 in the present context.

In analogy with (29.1.4) we introduce

$$\begin{aligned}C=\{(t,x,y,\tau,\xi,\eta)\in T^*(\mathbb{R}\times X\times X)\smallsetminus 0;\ &\tau=\pm p(x,\xi)^{\frac12}\\ &\text{and } (t,x,\tau,\xi),\ (0,y,\tau,\eta) \text{ lie on a}\\ &\text{generalized bicharacteristic of } \tau^2-p(x,\xi)\}.\end{aligned}\tag{29.3.1}$$

Lemma 29.3.1. *C is closed in $T^*(\mathbb{R}\times X\times X)\smallsetminus 0$ and so is*

$$C_\Delta=\{(t,x,x,\tau,\xi,\xi)\in C,\ t\neq 0\}.$$

Proof. That C is closed follows from Proposition 24.3.12 and the discussion before it was stated. To prove the second statement assume that there is a sequence of closed generalized bicharacteristics from $(0, y_\nu, p(y_\nu, \eta_\nu)^{\frac{1}{2}}, \eta_\nu)$ to $(t_\nu, y_\nu, p(y_\nu, \eta_\nu)^{\frac{1}{2}}, \eta_\nu)$ such that $t_\nu \to 0$. If we change scales as in Proposition 17.4.4 in the ratio t_ν to 1 it follows when $\nu \to \infty$ through some subsequence that for the wave equation in a half space in $\mathbb{R}^n$ there is a generalized bicharacteristic from $(0, y, |\eta|, \eta)$ to $(1, y, |\eta|, \eta)$ for some $\eta \neq 0$. However, this is a contradiction for a bicharacteristic returning to the same point must have been reflected orthogonally, and then it returns with the opposite frequency η. Thus C_Δ is closed and the lemma is proved.

From Lemma 29.3.1 it follows that

$$\Pi^*(x, \xi) = \inf\{t > 0; (t, x, \tau, \xi) \in C_\Delta\} \tag{29.3.2}$$

is a strictly positive, lower semi-continuous function homogeneous of degree 0 on $T^*(X) \setminus 0$. Thus Π^* has a positive lower bound. If we define *generalized geodesics* as projections in $T^*(X) \setminus 0$ of generalized bicharacteristics of $D_t^2 - P$, then Π^* is the minimal length of a closed generalized geodesic through (x, ξ).

We shall now prove a substitute for (29.1.6):

Proposition 29.3.2. *If $F(t, x, y)$ is the restriction to the interior of $\mathbb{R} \times X \times X$ of the kernel of the cosine transform* $\cos(t\sqrt{\mathscr{P}})$, then

$$WF'(F) \subset C. \tag{29.3.3}$$

Proof. Assume that $(t_0, x_0, y_0, \tau_0, \xi_0, \eta_0) \notin C$ and that x_0, y_0 are in $X^\circ = X \setminus \partial X$. By Lemma 29.3.1 we can then find an open conic neighborhood Γ_1 of (y_0, η_0) and another Γ_2 of $(t_0, x_0, \tau_0, \xi_0)$ such that

$$(t, x, y, \tau, \xi, \eta) \notin C \quad \text{if } (t, x, \tau, \xi) \in \Gamma_2 \text{ and } (y, \eta) \in \Gamma_1. \tag{29.3.4}$$

(We may assume that τ_0, ξ_0, η_0 are all different from 0, for the equations $(D_t^2 - P_x) F = 0$, $(D_t^2 - P_y) \bar{F} = 0$ imply that $\tau^2 = p(x, \xi) = p(y, \eta)$ in the wave front set of F.) Choose $A \in \Psi^0(X^\circ; \Omega^{\frac{1}{2}}, \Omega^{\frac{1}{2}})$ and $Y \Subset X$ so that the kernel of A has compact support in $Y \times Y$, $WF(A) \subset \Gamma_1$ and A is non-characteristic at (y_0, η_0). Also choose $B \in \Psi^0(X^\circ \times \mathbb{R}; \Omega^{\frac{1}{2}}, \Omega^{\frac{1}{2}})$ with $WF(B) \subset \Gamma_2$ so that B is non-characteristic at $(t_0, x_0, \tau_0, \xi_0)$ and the kernel has compact support in $X^\circ \times \mathbb{R} \times X^\circ \times \mathbb{R}$. If $f \in C_0^\infty(Y)$ and $\|f\|_{(-2N)} \leqq 1$ for a fixed integer $N > 0$ then $\cos(t\sqrt{\mathscr{P}}) f$ is uniformly bounded in $\bar{H}^{\text{loc}}_{(-2N)}(X^\circ)$. In fact, we can write $f = P^N g + h$ where g and h have compact support in X° and are uniformly bounded in L^2 by just taking $g = Ef$ where E is a properly supported parametrix of P^N in X°. Then

$$\cos(t\sqrt{\mathscr{P}}) f = \mathscr{P}^N \cos(t\sqrt{\mathscr{P}}) g + \cos(t\sqrt{\mathscr{P}}) h$$

where $\cos(t\sqrt{\mathscr{P}})g$ and $\cos(t\sqrt{\mathscr{P}})h$ are uniformly bounded in $L^2(X)$ which proves the contention. Hence the map

$$T\colon f \mapsto \cos(t\sqrt{\mathscr{P}})Af$$

can be extended from $C^\infty(X,\Omega^{\frac12})$ to a continuous map from $H^{\mathrm{loc}}_{(-2N)}(Y,\Omega^{\frac12})$ to $C(\mathbb{R}, \bar{H}^{\mathrm{loc}}_{(-2N)}(X^\circ,\Omega^{\frac12}))$. We have $(D_t^2-P)Tf=0$ in $\mathbb{R}\times X^\circ$, $Tf=0$ on $\mathbb{R}\times\partial X$, $D_t Tf=0$ and $Tf=Af$ when $t=0$. Hence it follows from (29.3.4) and Theorems 23.2.9, 24.5.3 that $WF(Tf)\cap\Gamma_2=\emptyset$. Thus the composed operator

$$BT\colon H^{\mathrm{loc}}_{(-2N)}(Y)\ni f\mapsto BTf\in C^\infty(X\times\mathbb{R})$$

is everywhere defined and obviously closed, which proves that it is continuous. It follows that the kernel

$$B(t,x,D_t,D_x)\,{}^tA(y,D_y)\,F(t,x,y)$$

is in C^∞. If we multiply by a pseudo-differential operator in (t,x,y) with symbol equal to 0 in a conic neighborhood of the subspaces where $\eta=0$ or $(\tau,\xi)=0$, it follows by Theorem 18.1.35 that $(t_0,x_0,y_0,\tau_0,\xi_0,-\eta_0)\notin WF(F)$, which proves the proposition.

Remark. The inclusion $WF'(\hat{E})\subset C$ contained in (29.1.6) can similarly be deduced from Theorem 23.1.4.

We can now state and prove an analogue of Theorem 29.1.5.

Theorem 29.3.3. *If $N(\lambda)$ is the number of eigenvalues $<\lambda$ of the Dirichlet problem in X, then*

$$\varlimsup_{\lambda\to\infty} \lambda^{(1-n)/2}\,|N(\lambda)-(2\pi)^{-n}C_n\operatorname{Vol}(X)\lambda^{n/2} + (2\pi)^{1-n}C_{n-1}\operatorname{Vol}(\partial X)\lambda^{(n-1)/2}/4| \leq C\iint_{p(x,\xi)<1}\Pi^*(x,\xi)^{-1}\,dx\,d\xi. \tag{29.3.5}$$

Here $dx\,d\xi$ is the symplectic volume element and C only depends on the dimension n of X; C_ν is the volume of the Euclidean unit ball in $\mathbb{R}^\nu$ and $\operatorname{Vol}(X)$, $\operatorname{Vol}(\partial X)$ *are the Riemannian volumes.*

Proof. Let $\psi\in C^\infty(X)$ be equal to 1 in a neighborhood of ∂X, and let $\Gamma_1,\dots,\Gamma_N$ be a covering of $T^*(X)\setminus 0$ restricted to $\operatorname{supp}(1-\psi)$ by small open cones. Let Π_j be positive and $\Pi_j<\Pi^*$ in Γ_j. We can choose $B_j\in\Psi^0_{\mathrm{phg}}(X^\circ;\Omega^{\frac12},\Omega^{\frac12})$ with kernel of compact support in $X^\circ\times X^\circ$, $WF(B_j)\subset\Gamma_j$, and

$$\sum_1^N B_jB_j^*+\psi=I+R$$

where $R\in\Psi^{-\infty}$. If we choose ψ so that $(1-\psi)^{\frac12}\in C^\infty$ this follows from exactly the same argument that we used in the proof of Theorem 29.1.5. As there it is clear that $\operatorname{Tr} E_\lambda B_j B_j^*$ is increasing, and it follows from Proposition 29.3.2 that the cosine transform $\operatorname{Tr}\cos(t\sqrt{\mathscr{P}})B_j B_j^*$ is smooth for $0<|t|<\Pi_j$. The singularity when $t=0$ is the same as the even part of the Fourier transform discussed in the proof of Proposition 29.1.2. Hence Lemma 17.5.6 gives after summation over j

$$\overline{\lim_{\lambda\to\infty}}\, \lambda^{(1-n)/2}\,|\operatorname{Tr}(E_\lambda \textstyle\sum B_j B_j^*)-(2\pi)^{-n}\iint\limits_{p<\lambda}(1-\psi)\,dx\,d\xi|$$
$$\leqq C\iint\limits_{p<1}\textstyle\sum \Pi_j^{-1} b_j\,dx\,d\xi$$

where $\sum b_j = 1-\psi$ and $\operatorname{supp} b_j \subset \Gamma_j$. In fact, the symbol of $\sum B_j B_j^*$ is $1-\psi \bmod S^{-\infty}$. If we add (17.5.21) it follows that (29.3.5) is valid with the right-hand side replaced by

$$C'\int\limits_X |\psi|\,\Upsilon dx + C\iint\limits_{p<1}\textstyle\sum b_j \Pi_j^{-1}\,dx\,d\xi$$

where C' is the constant in Corollary 17.5.11. When we refine the covering the last integral converges to the integral of $(1-\psi(x))/\Pi^*(x,\xi)$, and (29.3.5) follows when the support of ψ shrinks to ∂X. The proof is complete.

Corollary 29.3.4. *If the closed generalized geodesics form a set of measure* 0 *in* $T^*(X)\setminus 0$ *then*

$$N(\lambda)-(2\pi)^{-n}C_n \operatorname{Vol}(X)\lambda^{n/2}+(2\pi)^{1-n}C_{n-1}\operatorname{Vol}(\partial X)\lambda^{(n-1)/2}/4$$
$$=o(\lambda^{(n-1)/2})\quad \text{as } \lambda\to\infty.$$

Notes

In the notes to Chapter XVII we have already discussed some of the work devoted to the asymptotic properties of eigenvalues and eigenfunctions. Fourier integral operators were introduced by Hörmander [22] to prove the estimate (29.1.15)″ with $O(\lambda^{n-1})$ in the right-hand side; such an estimate had been proved previously by Avakumovič and Lewitan in the second order case. Only a local version of Fourier integral operators was required in that context. It was Chazarain [2] and Duistermaat-Guillemin [1] who first exploited the fact that the global theory of Fourier integral operators developed in Hörmander [26] gives control of all the singularities of the Fourier transform of the spectral measure. Corollary 29.1.6 was proved in Duistermaat-Guillemin [1] but the presentation here is closer to the arguments of Ivrii [3]. In fact, Theorem 29.1.5 and the analogous Theorem 29.3.3 are implicit in Ivrii's proof of Corollary 29.3.4. Theorem 29.1.7 is analogous to a classical theorem of Szegö [1] on the contraction of a multiplication operator to the space of trigonometrical polynomials of fixed high order. It was proved by Guillemin [2] using Lemma 29.1.8 which comes from Widom [1]. Seeley [5] has proved far more general results than Proposition 29.1.9 on the relation between standard functional calculus and the calculus of pseudo-differential operators. (See also Taylor [4].)

Both Chazarin [2] and Duistermaat-Guillemin [1] studied the contributions to the spectral asymptotics given by closed orbits of H_p which are smoothly arranged in various ways. Weak versions of Theorem 29.2.2 were given in the latter paper. As it is stated here Theorem 29.2.2 is due to Colin de Verdière [1] who also determined the contributions from closed orbits of smaller period. Formulas for the polynomial μ in Theorem 29.2.2 related to the Atiyah-Singer index formula have been given by Boutet de Monvel and Guillemin [1]. Theorem 29.2.5 is basically due to Weinstein [2] who studied the perturbation of the Laplacean on the sphere by a potential. (See also Guillemin [3].) Here we have followed the arguments of Colin de Verdière [1]. However, the ideas of Weinstein [2] are used instead in the proof of Theorem 29.2.7. We refer to Guillemin [1] for a construction of Riemannian metrics on the sphere such that all geodesics are closed with the same period.

Chapter XXX. Long Range Scattering Theory

Summary

In Chapter XIV we studied the spectral properties of short range perturbations of fairly general differential operators $P_0(D)$ with constant coefficients in $\mathbb{R}^n$. The short range condition imposed on the perturbing differential operator $V(x,D)$ was designed to allow a study of the resolvent of the perturbed operator by compactness arguments. Roughly speaking it required that the coefficients of V decrease as fast as an integrable function of $|x|$ and that for x frozen at x_0 the operator $V(x_0,D)$ is compact with respect to $P_0(D)$ in the sense that $\tilde{V}(x_0,\xi)/\tilde{P}_0(\xi) \to 0$ as $\xi \to \infty$. In this section we shall relax the hypotheses on $V(x,\xi)$ both when x is large and when ξ is large. However, to bring out the main points as simply as possible we shall assume that P_0 is *elliptic*, of order m. Our first condition on V is then that V is of order m, that the coefficients of the terms of order m are continuous and that $P_0(D)+V(x,D)$ is also elliptic. Furthermore, we assume that $V = V^S(x,\xi)+V^L(x,\xi)$ where as in Chapter XIV the coefficients of the short range term V^S decrease as fast as an integrable function of $|x|$, while the long range term V^L has the bound

$$|D_x^\alpha V^L(x,\xi)| \le C(1+|x|)^{-|\alpha|-\varepsilon}(1+|\xi|)^m, \quad |\alpha| \le 2, \tag{30.1}$$

for some $\varepsilon>0$. A more precise discussion of these conditions is given in Section 30.1. A sufficient condition for (30.1) is of course that V^L is homogeneous in x of degree $-\varepsilon$ for $|x|>1$; the Coulomb potential for example verifies (30.1) with $\varepsilon=1$. However, (30.1) also allows oscillatory behavior of V^L as a function of x if it is compensated for by a faster decrease.

In Section 30.2 we just assume that (30.1) is valid when $|\alpha| \le 1$. This is only slightly more than is required in order that the ξ variable have a limit at infinity on the orbits of the Hamilton field of $P_0(\xi)+V^L(x,\xi)$. This hypothesis suffices to prove that $P_0(D)+V(x,D)$ with domain C_0^∞ has a self-adjoint closure H and that the resolvent $R(z)=(H-z)^{-1}$ has a limit when $z \to \lambda \pm i0$ except when λ is in a countable set which is discrete except at the set $Z(P_0)$ of critical values of P_0. The spectrum of P is absolutely continuous except at this set. The essential point in the proof is an a priori estimate proved by applying the same methods as in Section 26.7 to the operator $P_0(D)+V^L(x,D)-z$. This is possible because the imaginary part of

L. Hörmander, *Classics in Mathematics*
The Analysis of Linear Partial Differential Operators IV,
DOI: 10.1007/978-3-642-00117-8_6, © Springer-Verlag Berlin Heidelberg 2009

$P_0(D)+V^L(x,D)-z$ has a definite sign when $\operatorname{Im} z \neq 0$. (In this argument we split V so that $V^L(x,D)$ is self-adjoint. Below we shall use a splitting such that V^L is real.)

To examine the spectral properties of H we need some detailed information on the Hamilton flow of $P_0(\xi)+V^L(x,\xi)$ under the full condition (30.1) which enables us to find a suitable Lagrangian Λ with

$$\Lambda \subset \{(x,\xi);\, P_0(\xi)+V^L(x,\xi)=\lambda\} \tag{30.2}$$

and to solve the Hamilton-Jacobi equation

$$\partial W(\xi,t)/\partial t = P_0(\xi)+V^L(\partial W/\partial \xi, \xi). \tag{30.3}$$

In Section 30.1 we prove that V^L can be chosen so that all derivatives have estimates similar to (30.1), and in Section 30.3 we examine how these estimates are inherited by the solutions Λ and W of (30.2) and (30.3). In Section 30.4 we use the solution of (30.3) to prove the existence of the modified wave operators

$$\tilde{W}_{\pm} u = \lim_{t\to\pm\infty} e^{itH} e^{-iW(D,t)} u, \quad u \in L^2(\mathbb{R}^n). \tag{30.4}$$

When $t\to\infty$ we have $W(\xi,t)=t P_0(\xi)+O(t^{1-\varepsilon})$ with ε as in (30.1), but in general it is not possible to replace $W(D,t)$ by $t P_0(D)$ as in the definition of the wave operator in Section 14.4. Already for the Coulomb potential this gives rise to a logarithmic divergence. Nevertheless $\tilde{W}_{\pm}$ is always an isometric operator interwining H and $H_0=P_0(D)$,

$$e^{isH}\tilde{W}_{\pm} = \tilde{W}_{\pm} e^{isH_0}, \quad s\in\mathbb{R}. \tag{30.5}$$

Section 30.5 contains the proof of asymptotic completeness, that is, that the range of $\tilde{W}_{\pm}$ is equal to the orthogonal space of the eigenvectors. As in Section 14.6 the key to the proof is the construction of distorted Fourier transforms $F_{\pm}$. There we defined $F_{\pm}f(\xi)=\hat{f}_{\lambda\pm i0}(\xi)$ on $M_\lambda=\{\xi;\, P_0(\xi)=\lambda\}$ where $f\in B$ and $f_{\lambda\pm i0}$ were chosen so that

$$(H-\lambda\mp i0)^{-1} f = (H_0-\lambda\mp i0)^{-1} f_{\lambda\pm i0}.$$

By Theorem 14.2.2 this implies that

$$\lim_{R\to\infty} \int |\chi(D)R(\lambda\pm i0)f|^2 \phi(x/R)\,dx/R = (2\pi)^{1-n}\int |\chi F_{\pm}f|^2 \Big(\int_{t\gtrless 0} \phi(tP_0')\,dt\Big)\,dS/|P_0'|. \tag{30.6}$$

Also for long range perturbations a study of the asymptotic properties of $R(\lambda\pm i0)f$ will yield functions $\tilde{F}_{\pm}f\in L^2(M_\lambda)$ such that (30.6) is valid. In this context it is essential to relate the asymptotic behavior not to the conormal bundle of M_λ as in the short range case but to the Lagrangian Λ in (30.2).

30.1. Admissible Perturbations

As mentioned in the introduction we shall consider perturbations

$$V(x,D)=V^S(x,D)+V^L(x,D)$$

of an elliptic constant coefficient operator $P_0(D)$ of order m such that V^S is of short range and V^L is of long range. The precise definition of the short range condition will be given below (Definition 30.1.3) but we observe already now that an operator of order m is always of short range if the coefficients are $O(|x|^{-1-\varepsilon})$ for some $\varepsilon>0$ as $x\to\infty$. We shall assume that for some integer $\kappa>0$ the coefficients f of the long range part are in C^κ, and that

$$|D^\alpha f(x)|\leqq C(1+|x|)^{-|\alpha|-\varepsilon},\quad x\in\mathbb{R}^n,\ |\alpha|\leqq\kappa, \tag{30.1.1}$$

for some $\varepsilon\in(0,1)$. If (30.1.1) were only assumed when $|\alpha|=\kappa$ then f would be the sum of a polynomial and a function satisfying (30.1.1), so (30.1.1) follows for $|\alpha|\leqq\kappa$ if $f(x)\to 0$ as $x\to\infty$. By a regularization we can also obtain bounds on the higher order derivatives of f:

Lemma 30.1.1. *Let $f\in C^\kappa$ satisfy (30.1.1), and let $0<\delta<\varepsilon$. Then we can write $f=f^S+f^L$ where*

$$|f^S(x)|\leqq C(1+|x|)^{-1+\delta-\varepsilon},\quad x\in\mathbb{R}^n, \tag{30.1.2}$$

thus f^S is of short range, and for all α

$$|D^\alpha f^L(x)|\leqq C_\alpha(1+|x|)^{-m(|\alpha|)},\quad x\in\mathbb{R}^n, \tag{30.1.3}$$

where $m(j)$ is the concave sequence defined by

$$\begin{aligned} m(j)&=\delta+j &&\text{when } j\leqq\kappa,\\ m(j)&=1+(\delta+\kappa-1)j/\kappa &&\text{when } j\geqq\kappa. \end{aligned} \tag{30.1.4}$$

Proof. Choose $\psi\in C_0^\infty(\{x;|x|<2\})$ with $\psi(x)=1$, $|x|<1$, and set

$$f_0(x)=\psi(x)f(x),\quad f_\nu(x)=(\psi(2^{-\nu}x)-\psi(2^{1-\nu}x))f(x),\quad \text{if } \nu\neq 0.$$

Then $f=\sum f_\nu$, we have $2^{\nu-1}\leqq|x|\leqq 2^{\nu+1}$ in $\operatorname{supp} f_\nu$, if $\nu\neq 0$, and

$$|D^\alpha f_\nu(x)|\leqq C' 2^{-\nu(|\alpha|+\varepsilon)},\quad |\alpha|\leqq\kappa.$$

Let $\chi\in C_0^\infty$, $\int\chi\,dy=1$ and $\int y^\beta\chi(y)\,dy=0$ for $0<|\beta|<\kappa$, and set $\chi_\nu(y)=\chi(2^{-\nu\rho}y)/2^{n\nu\rho}$ where $\rho=(\kappa-1+\delta)/\kappa<1$. Then

$$2^{\nu-2}<|x|<2^{\nu+2}\quad\text{in } \operatorname{supp}\chi_\nu * f_\nu$$

if ν is large enough, and

$$\begin{aligned} |D^\alpha\chi_\nu * f_\nu(x)|&\leqq C'' 2^{-\nu(|\alpha|+\varepsilon)}, && |\alpha|\leqq\kappa,\\ |D^\alpha\chi_\nu * f_\nu(x)|&\leqq C_\alpha 2^{-\nu(\kappa+\varepsilon+\rho(|\alpha|-\kappa))}, && |\alpha|>\kappa. \end{aligned}$$

In fact, we can let κ derivatives fall on f_ν and let the remaining ones act on χ_ν. Hence (30.1.3) is valid for $f^L=\sum\chi_\nu * f_\nu$ since x can only be in the support of 4 terms when $|x|$ is large. If we expand $f_\nu(x-2^{\nu\rho}y)$ by Taylor's formula with remainder term after terms of degree $\kappa-1$, we obtain

$$|f_\nu(x)-\chi_\nu * f_\nu(x)| = |\int (f_\nu(x)-f_\nu(x-2^{\nu\rho}y))\chi(y)\,dy| \leq C2^{-\nu(\kappa+\varepsilon)+\kappa\nu\rho}.$$

Now $\kappa+\varepsilon-\kappa\rho=\kappa+\varepsilon-\kappa+1-\delta=1+\varepsilon-\delta$ so (30.1.2) is valid for $f^S=f-f^L$. The proof is complete.

Remark. If $\varepsilon>1/(\kappa+1)$ then $m(\kappa+1)>\kappa+1$ if δ is close to ε, for $(1/(\kappa+1)+\kappa-1)(\kappa+1)/\kappa=\kappa$. Hence f^L satisfies (30.1.1) with κ replaced by $\kappa+1$. We shall therefore usually assume that $\varepsilon\leq 1/(\kappa+1)$, which means that in Lemma 30.1.1 $\delta<1/(\kappa+1)$. This implies $m(j)>j$ when $j\leq\kappa$ and $m(j)<j$ when $j>\kappa$.

The precise short range condition we shall use is suggested by an example discussed after Lemma 14.4.3, where we actually proved the following

Lemma 30.1.2. *If $0\leq|\alpha|<m$ and V is a measurable function, then*

$$\|VD^\alpha u\|_{(0)}\leq C\|V\|_{L^p}\|u\|_{(m)}, \quad u\in C_0^\infty(\{x;|x-y|<1\}) \tag{30.1.5}$$

if $p=n/(m-|\alpha|)>2$ or $p=2$ and $n<2(m-|\alpha|)$ or $p>2$ and $n=2(m-|\alpha|)$.

It is of course sufficient to take the L^p norm over the unit ball with center at y. To shorten some proofs we shall make the definition of short range perturbation somewhat more restrictive at infinity than in Chapter XIV.

Definition 30.1.3. A differential operator

$$V(x,D)=\sum_{|\alpha|\leq m} V_\alpha(x)D^\alpha$$

with measurable coefficients, continuous when $|\alpha|=m$, will be said to be of short range if for some $\delta>0$

$$|V_\alpha(x)|\leq C(1+|x|)^{-1-\delta}, \quad |\alpha|=m; \tag{30.1.6}$$

$$\Big(\int_{|y|<1}|V_\alpha(x+y)|^p\,dy\Big)^{1/p}\leq C(1+|x|)^{-1-\delta}, \quad |\alpha|<m,$$

for all $x\in\mathbb{R}^n$. Here p depends on α; we have $p=n/(m-|\alpha|)$ if this is >2, while $p=2$ if $n<2(m-|\alpha|)$ and p is any number >2 if $n=2(m-|\alpha|)$. We shall say that $V(x,D)$ is a κ-admissible perturbation if V is symmetric, that is,

$$(V(x,D)u,v)=(u,V(x,D)v); \quad u,\ v\in C_0^\infty;$$

$P_0(D)+V(x,D)$ is elliptic, and $V=V^S+V^L$ where V^S is of short range, the coefficients V_α^L of V^L are real, and for some $\varepsilon>0$

$$|D^\beta V_\alpha^L(x)|\leq C(1+|x|)^{-|\beta|-\varepsilon}, \quad |\beta|\leq\kappa. \tag{30.1.7}$$

Using Lemma 30.1.1 we can change the splitting in such a way that $V_\alpha^L \in C^\infty$ and for all β

$$|D^\beta V_\alpha^L(x)| \leq C_\beta(1+|x|)^{-m(|\beta|)} \tag{30.1.7$'$}$$

where $m(j)$ is defined by (30.1.4) with some $\delta \in (0,1)$. In particular this implies that $V^L(x,D)-V^L(x,D)^*$ is of short range. Hence

$$V(x,D)=(V^L(x,D)+V^L(x,D)^*)/2+((V^L(x,D)-V^L(x,D)^*)/2+V^S(x,D))$$

gives a splitting of $V(x,D)$ with symmetric terms satisfying (30.1.6) and (30.1.7)′ respectively but with complex valued coefficients in the long range part. This splitting will be used in Section 30.2. (Alternatively one might use the Weyl operator $V^{Lw}(x,D)$ as the long range part.)

In Section 30.3 we shall work a great deal with functions satisfying estimates such as (30.1.3). The following calculus lemmas will be useful in keeping track of such conditions.

Lemma 30.1.4. *Let f and g be C^∞ functions in $\mathbb{R}^n$ and assume that for some $x \in \mathbb{R}^n$ and some $t>1$ we have for all multi-indices α and β*

$$|D^\alpha f(x)| \leq C_\alpha t^{a(|\alpha|)}, \quad |D^\beta g(x)| \leq C'_\beta t^{b(|\beta|)}.$$

If a and b are convex sequences it follows that

$$|D^\gamma(fg)(x)| \leq C''_\gamma t^{c(|\gamma|)}$$

where $C''_\gamma = 2^{|\gamma|} \max_{|\alpha|+|\beta|=|\gamma|} C_\alpha C'_\beta$ and

$$\begin{aligned} c(k) &= \max(a(k)+b(0), a(0)+b(k)) \\ &= a(0)+b(0)+\max(a(k)-a(0), b(k)-b(0)). \end{aligned}$$

Proof. $D^\gamma(fg)(x)$ is a sum of $2^{|\gamma|}$ terms, each of which is the product of a derivative of f of order j and one of g of order $|\gamma|-j$. In view of the convexity the exponent $a(j)+b(|\gamma|-j)$ of t in the estimate of such a term must take its maximum for $0 \leq j \leq |\gamma|$ at an end point. Hence the maximum is equal to $c(|\gamma|)$.

Next we shall consider composite functions.

Lemma 30.1.5. *Let $\psi\colon \mathbb{R}^{n_1} \to \mathbb{R}^{n_2}$ be a C^∞ map defined in a neighborhood of $x \in \mathbb{R}^{n_1}$ such that*

$$|D^\alpha \psi(x)| \leq C_\alpha t^{a(|\alpha|)} \tag{30.1.8}$$

where a is a convex sequence and $t>1$. If f is a C^∞ function at $\psi(x)$ then

$$D^\gamma(f \circ \psi)(x) = \sum_{0<|\beta|\leq|\gamma|} (D^\beta f)(\psi(x))\, \Psi_{\gamma,\beta}(x), \quad |\gamma|>0,$$

where

$$|\Psi_{\gamma,\beta}(x)| \leq C'_\gamma t^{|\beta| a(1)+a(|\gamma|-|\beta|+1)-a(1)} \tag{30.1.9}$$

with C'_γ determined by C_α, $|\alpha|\leqq|\gamma|$. If for another convex sequence $b(k)$

$$|(D^\beta f)(\psi(x))|\leqq C''_\beta t^{b(|\beta|)}$$

then

$$|D^\gamma(f\circ\psi)(x)|\leqq C'''_\gamma t^{c(|\gamma|)} \tag{30.1.10}$$

where

$$c(k)=\max(b(k)+k\,a(1),\, b(1)+a(k)), \qquad k>0.$$

Proof. Every term in $\Psi_{\gamma,\beta}$ is a product of $j=|\beta|$ derivatives of components of ψ of orders $k_1,\dots,k_j$, each at least equal to 1 and with sum $k=|\gamma|$. We can estimate it by a constant times t to the power

$$a(k_1)+\dots+a(k_j).$$

In the convex set of all $(k_1,\dots,k_j)\in\mathbb{R}^j$ with $k_1\geqq 1,\dots,k_j\geqq 1$ and $k_1+\dots+k_j=k$ the extreme points are those where $j-1$ coordinates are equal to 1 and the remaining one is $k-j+1$. In fact, if two coordinates are >1 we can increase one and decrease the other by the same small amount. Since we can interpolate the sequence a to a piecewise linear convex function it follows that the maximum in question is taken at an extreme point, so it is equal to $(j-1)a(1)+a(k-j+1)$. This proves (30.1.9). Since the convex sequence $b(j)+(j-1)a(1)+a(|\gamma|-j+1)$ takes its maximum for $1\leqq j\leqq|\gamma|$ at an end point, the maximum is equal to $c(|\gamma|)$ which proves (30.1.10).

A special case worth noticing is that

$$c(k)\leqq b(k) \quad \text{if } a(k)\leqq b(k)-b(1),\ a(1)\leqq 0. \tag{30.1.11}$$

Taking $a=b$ we see in particular that estimates of the form (30.1.8) are preserved under composition if $a(1)=0$. If ψ satisfies (30.1.8) and $\psi'(x)$ is invertible we conclude from this that the inverse $\phi=\psi^{-1}$ satisfies an estimate of the form (30.1.8) at $y=\psi(x)$ with constants only depending on C_β and a bound for $\psi'(x)^{-1}$. In fact, if we differentiate the identity $\psi\circ\phi(y)=y$ using (30.1.9) to estimate the terms where ψ is differentiated at least twice, we obtain

$$|\psi'(x)D^\gamma\phi(y)|\leqq C' t^{a(|\gamma|)}$$

if estimates for lower order derivatives of ϕ have already been proved. The assertion therefore follows by induction.

Another case of (30.1.11) which will occur frequently is that where $b(k)$ is convex and increasing, $b(1)\leqq 0$, and $a(k)=\max(0,b(k))$. Since $b(k)\leqq b(k)-b(1)$ and $0\leqq b(k)-b(1)$ when $k\geqq 1$, the validity of (30.1.11) is obvious. Also note that (30.1.11) remains valid if a constant is added to the sequence b.

30.2 The Boundary Value of the Resolvent, and the Point Spectrum

Let $P_0(D)$ be a formally self-adjoint differential operator with constant coefficients in $\mathbb{R}^n$ which is *elliptic* of order m. The purpose of this chapter is to show that the results on short range scattering in Chapter XIV can be extended with some modifications to $P=P_0(D)+V(x,D)$ if V is a 2-admissible perturbation. However, in this section we only need that V is 1-admissible, and even less is assumed in the first result:

Theorem 30.2.1. *Let* $V(x,D)=\sum_{|\alpha|\leq m} V_\alpha(x)D^\alpha$ *be a symmetric differential operator of order m such that* V_α *is continuous when* $|\alpha|=m$, *measurable for all* α, *and with p as in Definition* 30.1.3

$$\lim_{x\to\infty} V_\alpha(x)=0 \quad \text{if } |\alpha|=m; \tag{30.2.1}$$

$$\lim_{x\to\infty}\int_{|y|<1} |V_\alpha(x+y)|^p\,dy=0 \quad \text{if } |\alpha|<m.$$

If $P_0(D)+V(x,D)$ *is elliptic it follows that* $P_0(D)+V(x,D)$ *with domain* $H_{(m)}$ *is a self-adjoint operator H in* $L^2(\mathbb{R}^n)$.

Proof. Choose $\chi\in C_0^\infty(\{x;|x|<1\})$ real valued with $\chi(x)=1$ when $|x|<\frac{1}{2}$. If we apply (30.1.5) to $\chi(.-y)u$ it follows in view of (30.2.1) that

$$\int_{|x-y|<\frac{1}{2}} |V(x,D)u|^2\,dx \leq C(y)\|\chi(.-y)u\|^2_{(m)}, \quad u\in\mathscr{S},$$

where $C(y)\to 0$ when $y\to\infty$. Integration with respect to y gives

$$\|V(x,D)u\|\leq C\|u\|_{(m)}, \quad u\in\mathscr{S}, \tag{30.2.2}$$

and since $\mathscr{S}$ is dense in $H_{(m)}$ it follows that P is continuous from $H_{(m)}$ to L^2. Thus H is well defined and symmetric. If we prove that $H+it$ is surjective when $t\in\mathbb{R}$ and $|t|$ is large, we can conclude that H is self-adjoint.

For any $\varepsilon>0$ we can choose a splitting

$$V(x,D)=V_0(x,D)+V_1(x,D), \quad V_j(x,D)=\sum_{|\alpha|\leq m} V_{j\alpha}(x)D^\alpha$$

such that $V_0(x,D)$ is symmetric and has coefficients in C_0^∞ while

$$|V_{1\alpha}(x)|\leq\varepsilon, \quad |\alpha|=m; \tag{30.2.1'}$$

$$\Big(\int_{|y|<1} |V_{1\alpha}(x+y)|^p\,dy\Big)^{1/p}\leq\varepsilon, \quad |\alpha|<m,$$

with the usual choice of p. In fact, if $\chi_\delta(x)=\chi(\delta x)$ then

$$V(x,D)-\chi_\delta(x)V(x,D)\chi_\delta(x)=(1-\chi_\delta(x))V(x,D)+\chi_\delta(x)V(x,D)(1-\chi_\delta(x))$$

satisfies (30.2.1)′ with ε replaced by $\varepsilon/2$ if δ is small enough. The operator $\chi_\delta V\chi_\delta$ is symmetric and so are its translations. The coefficient of D^α is in L^p_{comp} with the usual choice of p. Hence we obtain (30.2.1)′ if we take for V_0 the operator obtained by regularizing the coefficients of $\chi_\delta V(x,D)\chi_\delta$ as in Theorem 4.1.4. In what follows we assume ε so small that $P_0(D)+V_0(x,D)$ is elliptic.

The operator H_0 with domain $H_{(m)}$ defined by $P_0(D)+V_0(x,D)$ is self-adjoint. In fact, if $u\in L^2$ and $H_0^*u=f\in L^2$, then

$$(P_0(D)+V_0(x,D))u=f$$

in the sense of distribution theory. Hence $u\in H_{(m)}^{\text{loc}}$ since the operator P_0+V_0 is elliptic, so $P_0(D)u\in L^2$ since the coefficients of V_0 are in C_0^∞. But this implies that $u\in H_{(m)}$.

Since

$$(H+it)(H_0+it)^{-1}=I+V_1(x,D)(H_0+it)^{-1}$$

it will follow that $H+it$ is surjective if we show that

$$\|V_1(x,D)(H_0+it)^{-1}\|<1 \tag{30.2.3}$$

for small ε and large $|t|$. To do so we let $g\in L^2$ and set $(H_0+it)^{-1}g=v\in H_{(m)}$. Then

$$\|V_1(x,D)(H_0+it)^{-1}g\|=\|V_1(x,D)v\|\leqq C\varepsilon\|v\|_{(m)}$$

by (30.2.1)′ and the argument used to prove (30.2.2). We have $\|v\|\leqq\|g\|/|t|$. To examine how $\|v\|_{(m)}$ depends on ε we observe that

$$(P_0(D)+V_0(x,D)+it)v=g.$$

Set

$$E_t(x,\xi)=(1+|\xi|^2)^{m/2}(P_0(\xi)+V_0(x,\xi)+it)^{-1}.$$

E_t is bounded in S^0 when $t\to\infty$, with bounds depending on ε though, and multiplication with $E_t(x,D)$ gives

$$(1+|D|^2)^{m/2}v=E_t(x,D)g+R_t(x,D)v \tag{30.2.4}$$

where R_t is bounded in S^{m-1} when $t\to\infty$. For large t we have

$$\|E_t(x,D)\|\leqq 2\sup|\xi|^m/|p_0(\xi)+v_0(x,\xi)|$$

where p_0 and v_0 are the principal symbols of P_0 and of V_0. This follows if we apply Theorem 18.1.15 to $E_t(x,D)(1-\chi(\delta D))$ and observe that the norm of $E_t(x,D)\chi(\delta D)$ obviously tends to 0 when $|t|\to\infty$ for fixed δ. Hence (30.2.4) shows that for large $|t|$

$$\|v\|_{(m)}\leqq C\|g\|+C_\varepsilon\|v\|_{(m-1)}$$

where C is independent of ε and t and C_ε is independent of t. Since

$$\|v\|_{(m-1)}\leqq\|v\|_{(m)}^{(m-1)/m}\|v\|^{1/m}\leqq(2C_\varepsilon)^{-1}\|v\|_{(m)}+C'_\varepsilon\|v\|,$$

we obtain after cancellation of a term $\|v\|_{(m)}/2$ that

$$\|v\|_{(m)}\leqq 2C\|g\|+2C'_\varepsilon\|v\|\leqq 2(C+C''_\varepsilon/t)\|g\|<3C\|g\|$$

for large $|t|$. Hence

$$\|V_1(x,D)(H_0+it)^{-1}\|\leqq C'\varepsilon$$

for large $|t|$, so (30.2.3) follows if $\varepsilon C'<1$. The proof is complete.

Let $R(z)=(H-z)^{-1}$, $\operatorname{Im} z \neq 0$, be the resolvent of H, and let

$$Z(P_0)=\{P_0(\xi);\ \xi \in \mathbb{R}^n,\ P_0'(\xi)=0\}$$

be the finite set of critical values of P_0. We shall now study the resolvent when z approaches a point $\lambda \in \mathbb{R} \setminus Z(P_0)$ from one of the half planes and prove that an analogue of Theorems 14.5.2, 14.5.4 and 14.5.5 is valid. In doing so we must strengthen the hypothesis in Theorem 30.2.1 by assuming that $V(x, D)$ is 1-admissible in the sense of Definition 30.1.3. We recall that this means that $V=V^L+V^S$ where V^L has real coefficients satisfying

$$|D^\beta V_\alpha^L(x)| \leq C_\beta(1+|x|)^{-m(|\beta|)} \tag{30.2.5}$$

where $m(0)=\delta$, $m(k)=1+k\delta$, $k>0$. P_0+V^L is elliptic if V^L is chosen equal to 0 in a sufficiently large set. Replacing V^L by $(V^L+V^{L*})/2$ we may assume that $V^L(x, D)$ is symmetric, with coefficients not necessarily real.

(30.2.5) means that

$$V^L(x, \xi)=\sum V_\alpha^L(x)\xi^\alpha$$

is a symbol of weight $(1+|x|)^{-\delta}(1+|\xi|)^m$ with respect to the σ-temperate metric

$$G_\delta=|dx|^2(1+|x|^2)^{-\delta}+|d\xi|^2(1+|\xi|^2)^{-1} \tag{30.2.6}$$

and that $D_x^\beta D_\xi^\alpha V$ for $\beta \neq 0$ has the same bound as if V^L were of weight $(1+|x|)^{-1}(1+|\xi|)^m$. (Here we are of course using the terminology of the advanced calculus of pseudo-differential operators in Section 18.5.) Before proceeding we must make some remarks on spaces related to the weights $(1+|x|^2)^{t/2}(1+|\xi|^2)^{s/2}$ which for $t=0$ reduce to the spaces $H_{(s)}$ and for $s=0$ to the spaces L_t^2 of Section 7.9.

Definition 30.2.2. By $H_{(s),t}(\mathbb{R}^n)$ we shall denote the space of $u \in \mathscr{S}'(\mathbb{R}^n)$ such that $(1+|x|^2)^{t/2}(1+|D|^2)^{s/2} u \in L^2$, with the norm

$$\|u\|_{(s),t}=\|(1+|x|^2)^{t/2}(1+|D|^2)^{s/2} u\|_{L^2}.$$

It is clear that $(1+|x|^2)^{t/2}$ and $(1+|\xi|^2)^{s/2}$ are symbols with the same weights with respect to the metric G_1 even. Hence the operator

$$(1+|D|^2)^{s/2}(1+|x|^2)^{t/2}(1+|D|^2)^{-s/2}(1+|x|^2)^{-t/2}$$

is of weight 1. It is therefore L^2 continuous which shows that we obtain a weaker norm if the factors $(1+|x|^2)^{t/2}$ and $(1+|D|^2)^{s/2}$ in Definition 30.2.2 are interchanged. In fact, it is equivalent since the order of the factors in the preceding argument is irrelevant. As in the proof of Theorem 18.1.13 it follows at once from the calculus and the L^2 continuity of operators with symbol in $S(1, G_\delta)$ that operators with symbol in $S((1+|x|)^\tau(1+|\xi|)^\mu, G_\delta)$ map $H_{(s),t}$ continuously to $H_{(s-\mu),t-\tau}$.

Already from the hypotheses of Theorem 30.2.1 it follows that

$$\|u\|_{(m),t} \leq C_t(\|Pu\|_{(0),t}+\|u\|_{(0),t}) \quad \text{if } u \in \mathscr{D}_H. \tag{30.2.7}$$

For $t=0$ this is part of Theorem 30.2.1. To prove (30.2.7) for $t\neq 0$ we set

$$v=F_\varepsilon(x)^t u, \quad F_\varepsilon(x)=(1+|x|^2)^{\frac{1}{2}}(1+\varepsilon|x|^2)^{-\frac{1}{2}}.$$

It is clear that $v\in H_{(m)}$, and we have

$$Pv=P(F_\varepsilon^t u)=F_\varepsilon^t Pu+[P,F_\varepsilon^t]F_\varepsilon^{-t}v.$$

Here $[P,F_\varepsilon^t]F_\varepsilon^{-t}$ is an operator of order $m-1$, and since F_ε is bounded in $S(F_\varepsilon, G_1)$ the coefficients can be estimated by higher order coefficients in P. Hence it follows from (30.1.5) that

$$\|Pv\|\leqq\|F_\varepsilon^t Pu\|+C\|v\|_{(m-1)}.$$

From (30.2.7) with $t=0$ and u replaced by v we obtain

$$\|v\|_{(m)}\leqq C(\|F_\varepsilon^t Pu\|+\|v\|_{(m-1)})\leqq C\|F_\varepsilon^t Pu\|+\|v\|_{(m)}/2+C'\|v\|_{(0)}.$$

Hence

$$\|v\|_{(m)}\leqq 2C\|F_\varepsilon^t Pu\|+2C'\|v\|_{(0)}$$

which gives (30.2.7) when $\varepsilon\to 0$.

We shall now estimate $u=R(z)f$ when $f\in L^2$ and $\operatorname{Im} z\neq 0$. This is the unique solution $u\in H_{(m)}$ of the equation

$$(P_0(D)+V(x,D)-z)u=f. \tag{30.2.8}$$

To be able to use the calculus of pseudo-differential operators we shall first consider the equation

$$(P_0(D)+V^L(x,D)-z)u=f \tag{30.2.8$'$}$$

instead. Fix $\lambda\notin Z(P_0)$ and choose a compact neighborhood U of

$$M_\lambda=\{\xi;\, P_0(\xi)=\lambda\}.$$

The symbol $P_0(\xi)+V^L(x,\xi)-z$ is different from 0 for large $|x|$ if $\xi\notin U$ and $|z-\lambda|<r_U$. The absolute value is also bounded from below by $c(1+|\xi|)^m$ for large $|\xi|$ and arbitrary x since $P_0(\xi)+V^L(x,\xi)$ is elliptic. If $\chi_0\in C_0^\infty(\mathbb{R}^n)$ is equal to 1 in U, it follows that

$$e_z(x,\xi)=(1-\chi_0(\xi))(P_0(\xi)+V^L(x,\xi)-z)^{-1} \tag{30.2.9}$$

is well defined for large $|x|+|\xi|$. By multiplication with a suitable cutoff function we make e_z defined in $\mathbb{R}^{2n}$. Since $P_0(\xi)+V^L(x,\xi)-z$ is in $S((1+|\xi|)^m, G_\delta)$ we obtain by Lemma 18.4.3

$$e_z(x,\xi)\in S((1+|\xi|)^{-m}, G_\delta),$$

uniformly with respect to z. Thus

$$e_z(x,D)(P_0(D)+V^L(x,D)-z)=I-\chi_0(D)-r_z(x,D)$$

where r_z is bounded in $S((1+|x|)^{-\delta}(1+|\xi|)^{-1}, G_\delta)$. If $\chi\in C_0^\infty$ is equal to 1 in $\operatorname{supp}\chi_0$ and E_z is the asymptotic sum of the symbols of $(1-\chi(D))r_z(x,D)^j e_z(x,D)$, which belong to $S((1+|x|)^{-j\delta}(1+|\xi|)^{-j}, G_\delta)$, we obtain

$$E_z(x,D)(P_0(D)+V^L(x,D)-z)u=(I-\chi(D))u+R_z(x,D)u$$

where for every N

$$|D_x^\beta D_\xi^\alpha R_z(x,\xi)|\leqq C_{\alpha\beta N}(1+|\xi|)^{-N}(1+|x|)^{-N},$$

which means that the kernel of $R_z(x,D)$ is in $\mathscr{S}(\mathbb{R}^{2n})$. (To define the asymptotic sum we have here used an obvious extension of Proposition 18.1.3 where cutoffs are introduced in the x variables as well.) We have now proved

Proposition 30.2.3. *Let $\lambda\in\mathbb{R}$ and let $\chi\in C_0^\infty(\mathbb{R}^n)$ be equal to 1 in a neighborhood of $M_\lambda=\{\xi;\, P_0(\xi)=\lambda\}$. If $u\in\mathscr{S}'$ is a solution of* (30.2.8)$'$ *with $f\in H_{(s),t}$ it follows then that $u-\chi(D)u\in H_{(s+m),t}$ if $|z-\lambda|<r_\chi$. For arbitrary s',t' there is a constant C such that*

$$\|u-\chi(D)u\|_{(s+m),t}\leqq C(\|f\|_{(s),t}+\|u\|_{(s'),t'}). \tag{30.2.10}$$

The preceding arguments are of course perfectly parallel to the standard analysis of elliptic operators, and they do not at all use the full strength of (30.2.5). We shall now proceed to estimate $\chi(D)u$ with the methods used in Section 26.6 to study the propagation of singularities for operators of principal type when the imaginary part of the principal symbol has a fixed sign. Here the imaginary part of the operator $P_0(D)+V^L(x,D)-z$ is of course $-i\operatorname{Im} z$ so we are precisely in such a situation. This will give the following substitute for Proposition 30.2.3.

Proposition 30.2.4. *If $u\in H_{(m)}$ is a solution of* (30.2.8)$'$ *with $f\in B$ and $\operatorname{Im} z\neq 0$, and if $\chi\in C_0^\infty(\{\xi;\, P_0'(\xi)\neq 0\})$ then*

$$\|\chi(D)u\|_{B^*}\leqq C_\chi(\|\chi(D)f\|_B+\|u\|_{(0),-(1+\delta)/2}). \tag{30.2.11}$$

Proof. With P denoting the operator $P_0(D)+V^L(x,D)$, which is self-adjoint by hypothesis, we start from the identity (cf. (26.6.3))

$$\operatorname{Im}(Q(P-z)u,Qu)=\operatorname{Im}([Q,P]u,Qu)-\operatorname{Im} z\,\|Qu\|^2 \tag{30.2.12}$$

where $Q\in\operatorname{Op}S(1,G_1)$ will be chosen in a moment. We may assume that $\operatorname{Im} z>0$ for otherwise we can just change the signs of P and of z. Then it follows that

$$\operatorname{Im}([P,Q]u,Qu)\leqq -\operatorname{Im}(Q(P-z)u,Qu). \tag{30.2.12$'$}$$

For large $|x|$ the leading term in the symbol of $Q^*[P,Q]/i$ is

$$-q(x,\xi)\langle\partial P_0(\xi)/\partial\xi,\partial q(x,\xi)/\partial x\rangle \tag{30.2.13}$$

if $q(x,\xi)$ is the symbol of Q and q is real valued. We shall choose q so that this is a non-negative quantity.

Let $\rho\in C_0^\infty(-\frac{1}{2},\frac{1}{2})$ be a decreasing function of t^2 with $\rho(0)=1$, and set $\tilde\rho(x,y)=\rho(1-\langle x,y\rangle/|x||y|)$; $x,y\in\mathbb{R}^n\setminus 0$. Then $\langle x,y\rangle/|x||y|\ge\frac{1}{2}$ in $\operatorname{supp}\tilde\rho$, that is, the angle between x and y is at most $\pi/3$, $\tilde\rho(x+t\,y,y)$ is an increasing function of t and

$$|D_x^\alpha D_y^\beta\tilde\rho(x,y)|\le C_{\alpha\beta}|x|^{-|\alpha|}|y|^{-|\beta|}$$

in view of the homogeneity. Choose $\psi\in C_0^\infty(\{x\in\mathbb{R}^n;\ \frac{1}{2}<|x|<\frac{5}{2}\})$ with $0\le\psi\le 1$ everywhere and $\psi(x)=1$ when $1\le|x|\le 2$, and set

$$\Psi(x,y)=\psi*\tilde\rho(x,y)=\int\psi(x-z)\,\tilde\rho(z,y)\,dz=\int\psi(z)\,\tilde\rho(x-z,y)\,dz.$$

Then $\Psi(x,y)$ is homogeneous in y of degree 0 and

$$\Psi(x,y)\ge 0,\qquad \Psi'(x,y)=\langle y,\partial\Psi(x,y)/\partial x\rangle\ge 0$$

with strict inequalities when $\psi(x)>0$; the equality is a definition. We can therefore choose $\phi\in C_0^\infty$ so that $\phi(x)$ is a constant $\neq 0$ when $1\le|x|\le 2$ and

$$|\phi(x)|^2\le\Psi'(x,y)\,\Psi(x,y)/|y|. \tag{30.2.14}$$

Also note that

$$|D_x^\alpha D_y^\beta\Psi(x,y)|\le C_{\alpha\beta}(1+|x|)^{-|\alpha|}|y|^{-|\beta|} \tag{30.2.15}$$

since we can let x derivatives act on ψ when $|x|<3$.

We shall apply (30.2.12)′ with $Q=q_R(x,D)$ where

$$q_R(x,\xi)=\Psi(x/R,\,-P_0'(\xi))\,\chi(\xi).$$

Here $R\ge 1$ and $\chi\in C_0^\infty(\{\xi;P_0'(\xi)\neq 0\})$ is real valued. q_R is bounded in $S(1,G_1)$ when $R\ge 1$ (see (30.2.6)), and

$$-q_R(x,\xi)\langle P_0'(\xi),\partial q_R(x,\xi)/\partial x\rangle=R^{-1}\Psi(x/R,-P_0'(\xi))\,\Psi'(x/R,-P_0'(\xi))\,\chi(\xi)^2$$

so it follows from (30.2.14) if $|P_0'(\xi)|\ge 1$ in $\operatorname{supp}\chi$, as we may assume, that

$$S_R(x,\xi)=-q_R(x,\xi)\langle P_0'(\xi),\partial q_R(x,\xi)/\partial x\rangle-R^{-1}|\phi(x/R)\,\chi(\xi)|^2\ge 0. \tag{30.2.16}$$

Since $S_R(x,\xi)$ is uniformly bounded in $S((1+|x|)^{-1},G_1)$ it follows from the sharp Gårding inequality (Theorem 18.1.14) that

$$\operatorname{Re}(S_R(x,D)u,u)\ge -C\|u\|^2_{(0),-1}. \tag{30.2.17}$$

Strictly speaking, Theorem 18.1.14 is applied to the adjoint of the conjugate of $S_R(x,D)$ by the Fourier transformation. Alternatively we could appeal to Theorem 18.6.7.

The symbol of $Q^*[P_0,Q]/i$ is bounded in $S((1+|x|)^{-2},G_1)$ apart from the leading term $-q_R\langle P_0',\partial q_R/\partial x\rangle$ just discussed. The symbol of $Q^*[V^L,Q]/i$ is bounded in $S((1+|x|)^{-1-\delta},G_\delta)$, for V^L is of weight $(1+|\xi|)^m(1+|x|)^{-\delta}$ while the x derivatives are of weight $(1+|\xi|)^m(1+|x|)^{-1-\delta}$. Thus the symbol of

$$Q^*[P,Q]/i-S_R(x,D)-R^{-1}\chi(D)\,\phi(x/R)^2\,\chi(D)$$

is bounded in $S((1+|x|)^{-1-\delta}, G_\delta)$ so it follows from (30.2.12)′ and (30.2.17) that

$$(30.2.18)\quad R^{-1}\|\phi(x/R)\chi(D)u\|^2 \leqq \|Qf\|_B\|Qu\|_{B^*} + C\|u\|^2_{(0),-(1+\delta)/2}.$$

Here $Q=Q'\chi(D)$ where Q' is defined as Q but with χ replaced by another $\chi' \in C_0^\infty(\{\xi;\, P_0'(\xi)\neq 0\})$ which is equal to 1 in $\operatorname{supp}\chi$. It follows from Theorem 14.1.4 that Q' is continuous in B and in B^*. Taking the supremum with respect to $R>1$ in (30.2.18) we therefore obtain

$$\|\chi(D)u\|^2_{B^*} \leqq C\|\chi(D)f\|_B\|\chi(D)u\|_{B^*} + C\|u\|^2_{(0),-(1+\delta)/2},$$

which proves (30.2.11).

When piecing together the estimates in Propositions 30.2.3 and 30.2.4 we shall also take the short range perturbation V^S into account. The proof of (30.2.2) gives in view of (30.1.6)

$$(30.2.19)\qquad \|V^S(x,D)u\|_{(0),t+1+\delta} \leqq C\|u\|_{(m),t}, \qquad u\in H_{(m),t},\ t\in\mathbb{R}.$$

Since V^S is symmetric we obtain if $u, v\in\mathscr{S}$

$$|(V^S(x,D)u, v)| = |(u, V^S(x,D)v)| \leqq \|u\|_{(0),t}\|V^S(x,D)v\|_{(0),-t}$$
$$\leqq C\|u\|_{(0),t}\|v\|_{(m),-t-1-\delta}.$$

Hence

$$(30.2.20)\qquad \|V^S(x,D)u\|_{(-m),t+1+\delta} \leqq C\|u\|_{(0),t}, \qquad u\in\mathscr{S}.$$

Since $\mathscr{S}$ is dense in $H_{(0),t}$ it follows that there is a unique continuous extension of $V^S(x,D)$ to an operator $H_{(0),t}\to H_{(-m),t+1+\delta}$, and (30.2.20) remains valid for the extension.

Theorem 30.2.5. *Let $V(x,D)$ be a 1-admissible perturbation of the self-adjoint elliptic operator $P_0(D)$ and denote by H the self-adjoint operator in L^2 defined by $P_0(D)+V(x,D)$. For every $\lambda\in\mathbb{R}\setminus Z(P_0)$ one can then find $r>0$ such that*

$$(30.2.21)\qquad \sum_{|\alpha|\leqq m}\|D^\alpha u\|_{B^*} \leqq C(\|f\|_B + \|u\|_{(0),-1})$$

if $u=(H-z)^{-1}f$ where $f\in B$ and $|z-\lambda|<r$, $\operatorname{Im} z\neq 0$.

Proof. Choose $\chi\in C_0^\infty(\{\xi;\, P_0'(\xi)\neq 0\})$ equal to 1 in a neighborhood of M_λ. Application of (30.2.10) to the equation $(P_0(D)+V^L(x,D)-z)u = f-V^S(x,D)u$ gives

$$\|(1-\chi(D))u\|_{(m),0} \leqq C(\|f-V^S(x,D)u\|_{(0),0} + \|u\|_{(0),-1})$$
$$\leqq C(\|f\| + C\|u\|_{(m),-1-\delta} + \|u\|_{(0),-1})$$
$$\leqq C'(\|f\| + \|u\|_{(0),-1}).$$

Here we have used first (30.2.19) and then (30.2.7). Hence

$$(30.2.22)\qquad \sum_{|\alpha|\leqq m}\|D^\alpha(1-\chi(D))u\|_{B^*} \leqq C'(\|f\|_B + \|u\|_{(0),-1}).$$

Next we note that

$$\|D^\alpha \chi(D)u\|_{B^*} \leq C\|\chi(D)u\|_{B^*}$$

for if $\chi' \in C_0^\infty$, $\chi'=1$ in $\operatorname{supp}\chi$, then $D^\alpha\chi(D)=D^\alpha\chi'(D)\chi(D)$, and $D^\alpha\chi'(D)$ is continuous in L_s^2 for every s, hence also in B^*. From (30.2.11) we therefore obtain

$$\|D^\alpha \chi(D)u\|_{B^*} \leq C(\|\chi(D)(f-V^S(x,D)u)\|_B + \|u\|_{(0),-(1+\delta)/2}). \tag{30.2.23}$$

If $\mu>\frac{1}{2}$ we have with constants depending on μ

$$\begin{aligned}\|\chi(D)V^S(x,D)u\|_B &\leq C\|\chi(D)V^S(x,D)u\|_{(0),\mu}\\ &\leq C'\|V^S(x,D)u\|_{(-m),\mu} \leq C''\|u\|_{(0),\mu-1-\delta}.\end{aligned} \tag{30.2.24}$$

Choosing $\mu=(1+\delta)/2$ and summing up (30.2.22)–(30.2.24) we obtain

$$\sum_{|\alpha|\leq m}\|D^\alpha u\|_{B^*} \leq C(\|f\|_B + \|u\|_{(0),-(1+\delta)/2}).$$

Now the estimate (30.2.21) follows since

$$2c\|u\|_{(0),-\frac{1}{2}-\delta/4} \leq \|u\|_{B^*} \leq \sum_{|\alpha|\leq m}\|D^\alpha u\|_{B^*}$$

for some $c>0$ and we have

$$C\|u\|_{(0),-(1+\delta)/2} \leq c\|u\|_{(0),-\frac{1}{2}-\delta/4} + C'\|u\|_{(0),-1}.$$

To pass from Theorem 30.2.5 to existence and estimates of boundary values of the resolvent we need an analogue of Theorem 14.3.4 for the limit of the graph of $H-z$ as $z\to\lambda\pm i0$. Recall that the space $\mathring{B}^*$ there is the space of all $u\in L^2_{\text{loc}}$ with

$$R^{-1}\int_{|x|<R}|u|^2\,dx\to 0, \quad R\to\infty.$$

Since pseudo-differential operators of weight 1 are continuous in B^* and in $\mathscr{S}$, they are also continuous in the closure $\mathring{B}^*$ of $\mathscr{S}$ in B^*.

Theorem 30.2.6. *Let* $\operatorname{Im} z_j>0$, $z_j\to\lambda\in\mathbb{R}\setminus Z(P_0)$, *and* $(H-z_j)u_j=f_j\in B$. *Assume that* $f_j\to f$ *in B norm and that* $D^\alpha u_j\to D^\alpha u$ *in the weak* topology of* B^* *when* $j\to\infty$ *and* $|\alpha|\leq m$. *Let* $a\in S((1+|\xi|)^m, G_1)$. *Then* $a(x,D)u\in\mathring{B}^*$ *if* $a(x,\xi)=0$ *on the positive normal bundle* $N_+(M_\lambda)$ *of* M_λ, *defined by*

$$N_+(M_\lambda)=\{(tP_0'(\xi),\xi);\ \xi\in M_\lambda,\ t>0\}.$$

Proof. Since u_j is bounded in $H_{(m),-\mu}$ if $\mu>\frac{1}{2}$ it follows from (30.2.19) that $V^S(x,D)u_j$ is bounded in $H_{(0),1+\delta-\mu}$ then. Moreover, f_j is bounded in $H_{(0),\frac{1}{2}}$ so it follows from Proposition 30.2.3 with the notation used there that $(1-\chi(D)^2)u_j$ is bounded in $H_{(m),\frac{1}{2}}$. Hence $(1-\chi(D)^2)u\in H_{(m),\frac{1}{2}}$ so

$$a(x,D)(1-\chi(D)^2)u\in H_{(0),\frac{1}{2}}\subset\mathring{B}^*.$$

To study $a(x,D)\chi(D)^2 u$ we must modify the proof of Proposition 30.2.4 so that the symbol q_R vanishes when $|x|/R$ is small.

In addition to the function $\rho\in C_0^\infty(\mathbb{R})$ in the proof of Proposition 30.2.4 we choose $\rho_1,\rho_2\in C_0^\infty(\mathbb{R})$ even, with values in $[0,1]$, such that $\rho_1=1$ in a neighborhood of $\operatorname{supp}\rho$, $\rho_2=1$ in a neighborhood of $\operatorname{supp}\rho_1$, and ρ_2 has its support close to zero. With $\tilde\rho_j$ and ψ defined as in the proof of Proposition 30.2.4 we set

$$\psi_1(x,y)=(1-\tilde\rho_1(x,-y))\psi(x),\qquad \Psi(x,y)=\int\psi_1(x-z,y)\tilde\rho(z,y)\,dz.$$

Since $\psi_1(x,y)=0$ when x is in a neighborhood of the cone $-\operatorname{supp}\tilde\rho(.,y)$ it is clear that $\Psi(x,y)=0$ when $|x|<c$, where c is a positive constant. Assume the function ϕ in the proof of Proposition 30.2.4 chosen so that $\psi>0$ in $\operatorname{supp}\phi$, and set

$$\phi_1(x,y)=k(1-\tilde\rho_2(x,-y))\phi(x).$$

Then $\psi_1(x,y)>0$ if $x\in\operatorname{supp}\phi_1(.,y)$ so if the constant k is small we have

$$|\phi_1(x,y)|^2\leqq\Psi'(x,y)\Psi(x,y)/\langle y\rangle. \tag{30.2.14$'$}$$

It is clear that (30.2.15) remains valid. If we define $q_R(x,\xi)$ as before and set

$$\Phi_R(x,\xi)=\phi_1(x/R,-P_0'(\xi))\chi(\xi),$$

it follows if $|P_0'(\xi)|\geqq 1$ in $\operatorname{supp}\chi$, as we may assume, that

$$S_R(x,\xi)=-q_R(x,\xi)\langle P_0'(\xi),\partial q_R(x,\xi)/\partial x\rangle-R^{-1}|\Phi_R(x,\xi)|^2\geqq 0. \tag{30.2.16$'$}$$

As before S_R is uniformly bounded in $S((1+|x|)^{-1},G_1)$. However, since $S_R(x,\xi)=0$ when $|x|<cR$ it follows that $R^{2\gamma}S_R$ is also bounded in $S((1+|x|)^{2\gamma-1},G_1)$ when $R\to\infty$, if $\gamma\geqq 0$. Thus

$$R^{2\gamma}\operatorname{Re}(S_R(x,D)u,u)\geqq -C\|u\|^2_{(0),\gamma-1}. \tag{30.2.17$'$}$$

Similarly we can trade a factor $R^{2\gamma}$ for an increase in the degree in the later estimates leading to (30.2.18), so we have

$$R^{-1}\|\Phi_R(x,D)u\|^2\leqq \operatorname{Im}-(q_R(x,D)f_0,q_R(x,D)u) + CR^{-2\gamma}\|u\|^2_{(0),\gamma-(1+\delta)/2} \tag{30.2.18$'$}$$

if $u\in H_{(m)}$ and $f_0=(P_0(D)+V^L(x,D)-z)u$, $\operatorname{Im} z>0$. Fix $\gamma<\delta/2$, take $u=u_j$, $z=z_j$, and let $j\to\infty$. Then $u_j\to u$ in the norm of $H_{(m-1),\mu}$ if $\mu<-\frac12$, and $q_R(x,D)V^S(x,D)u_j\to q_R(x,D)V^S(x,D)u$ in the norm of $H_{(0),\mu+1+\delta}$ by (30.2.19). When $\mu+1+\delta>\frac12$ we have convergence in B, and it follows that (30.2.18)$'$ remains valid for the limit u in the theorem, with $f_0=f-V^S(x,D)u\in B$. When $R\to\infty$ we have

$$\|q_R(x,D)f_0\|_B\to 0$$

for $q_R(x,D)f_0\to 0$ in $\mathscr{S}$ if $f_0\in\mathscr{S}$, which is a dense subspace of B, and q_R has uniformly bounded norm in B and in B^*. Hence

$$R^{-1}\|\Phi_R(x,D)u\|^2\to 0,\qquad R\to\infty. \tag{30.2.25}$$

Let c be the constant value of $\phi(x,y)$ when $1<|x|<2$ and $\tilde{\rho}_2(x,-y)=0$. If $a(x,\xi)=0$ in a conic neighborhood of $N_+(M_\lambda)$ and we set $b=a\chi/c$, then $b\in S(1,G_1)$ and

$$a(x,\xi)\,\chi(\xi)^2=b(x,\xi)\,\Phi_R(x,\xi) \quad \text{when } R<|x|<2R,$$

provided that supp ρ_2 is sufficiently close to 0 and supp χ is sufficiently close to M_λ. Regarding $R\Phi_R$ as a symbol which is bounded in $S((1+|x|),G_1)$, we obtain

$$a(x,D)\,\chi(D)^2=b(x,D)\,\Phi_R(x,D)+\Gamma_R(x,D), \qquad R<|x|<2R,$$

where $R\Gamma_R(x,\xi)$ is bounded in $S(1,G_1)$. From (30.2.25), the boundedness of $b(x,D)$ in L^2 and the uniform bound for $R\Gamma_R(x,D)u$ in B^*, it follows that

$$R^{-1}\int_{R<|x|<2R}|a(x,D)\,\chi(D)^2u|^2\,dx\to 0, \qquad R\to\infty.$$

If a is just equal to 0 on $N_+(M_\lambda)$ we can write $a=a_0+a_1$ where a_0 has the preceding properties and $\sup|a_1(x,\xi)\,\chi(\xi)^2|$ is as small as we please. The upper limit as $R\to\infty$ of the norm of $\psi(x/R)\,a_1(x,D)\,\chi(D)^2$ in L^2_s is $\leqq\sup|\psi|\sup|a_1\chi^2|$ by Theorem 18.1.15. The norm in B^* has a similar bound by Theorem 14.1.4 so we have

$$\overline{\lim}\, R^{-1}\int_{R<|x|<2R}|a_1(x,D)\,\chi(D)^2u|^2\,dx\leqq C\sup|a_1(x,\xi)\,\chi(\xi)^2|^2$$

which can be made as small as we please. Hence $a(x,D)\,\chi(D)^2\in\mathring{B}^*$ which completes the proof.

Theorem 30.2.6 means that u is in $\mathring{B}^*$ except at $N_+(M_\lambda)$ if $\mathring{B}^*$ is microlocalized in the same way as $H_{(s)}$ was in Section 18.1 (with the roles of the x and ξ variables reversed). However, we shall not develop this point of view further in spite of its appeal to intuition, for we would have few occasions to exploit it here.

Next we prove a substitute for Corollary 14.3.10.

Theorem 30.2.7. *Let $D^\alpha u\in B^*$, $|\alpha|\leqq m$, and assume that*

$$(P_0(D)+V(x,D)-\lambda)u=f\in B$$

where $\lambda\in\mathbb{R}\smallsetminus Z(P_0)$. We assume that $a(x,D)u\in\mathring{B}^*$ when $a\in S((1+|\xi|)^m,G_1)$ and $a=0$ on $N_+(M_\lambda)$. Then it follows that

$$\lim_{R\to\infty} R^{-1}\int_{|x|<R}|a(x,D)u|^2\,dx=2\,\mathrm{Im}\,(u,f) \tag{30.2.26}$$

if a is in the same class but

$$|a(x,\xi)|^2=\langle x/|x|,\,P_0'(\xi)\rangle \quad \textit{when } (x,\xi)\in N_+(M_\lambda) \textit{ and } |x| \textit{ is large.} \tag{30.2.27}$$

Proof. Since $(u, V^S u)=(V^S u, u)$ (cf. (14.4.7)′) the right-hand side of (30.2.26) is equal to $2\,\mathrm{Im}(u,(P_0(D)+V^L(x,D)-\lambda)u)$. Let $\psi\in C_0^\infty(\mathbb{R})$ be a real valued function which is 1 in a neighborhood of 0 and set $\psi_R(x)=\psi(|x|/R)$. Then we obtain

$$\begin{aligned}&2i\,\mathrm{Im}(u,f)\\&=\lim_{R\to\infty}(\psi_R u,(P_0(D)+V^L(x,D)-\lambda)u)-((P_0(D)+V^L(x,D)-\lambda)u,\psi_R u)\\&=\lim_{R\to\infty}([P_0(D)+V^L(x,D),\psi_R(x)]u,u).\end{aligned}$$

The symbol of

$$[P_0(D)+V^L(x,D),\psi_R(x)]/i+\sum\psi'(|x|/R)(x_j/R|x|)P_0^{(j)}(D)$$

is bounded in $S((1+|\xi|)^{m-1}(1+|x|)^{c-\delta-1}R^{-c},G_\delta)$ if $c>0$, for ψ_R may be replaced by ψ_R-1 which is bounded in $S((1+|x|)^c R^{-c},G_1)$. Choose $c>0$ and $\mu>\frac12$ so that $c+\mu<\delta+\frac12$. Since $u\in H_{(m-1),-\mu}$ it follows that

$$R^c([P_0(D)+V^L(x,D),\psi_R(x)]/i+\psi'(|x|/R)\sum(x_j/R|x|)P_0^{(j)}(D))u$$

is bounded in $H_{(0),\delta+1-c-\mu}$ and therefore in B since $\delta+1-c-\mu>\frac12$. Hence

$$2\,\mathrm{Im}(u,f)=\lim_{R\to\infty}-\sum(\psi'(|x|/R)(x_j/R|x|)P_0^{(j)}(D)u,u).$$

From (30.2.27) and the hypothesis it follows if $a\in S(1,G_1)$, as we may well assume, that

$$\sum x_j/|x|\,P_0^{(j)}(D)u-a(x,D)^*a(x,D)u\in\mathring{B}^*.$$

By Cauchy-Schwarz' inequality we therefore have

$$\begin{aligned}2\,\mathrm{Im}(u,f)&=\lim_{R\to\infty}-(\psi'(|x|/R)a(x,D)^*a(x,D)u,u)/R\\&=\lim_{R\to\infty}-\int\psi'(|x|/R)|a(x,D)u|^2\,dx/R.\end{aligned}$$

The last equality follows from the fact that the symbol of the commutator of $\psi'(|x|/R)/R$ and $a(x,D)^*$ is bounded in

$$S((1+|\xi|)^{m-1}(1+|x|)^{-1-c}R^{c-1},G_1)$$

for any c; we choose $c\in(0,1)$. For any $\varepsilon>0$ we can choose ψ so that

$$0\leqq-\psi'\leqq(1+\varepsilon)\phi\quad\text{or}\quad\phi/(1+\varepsilon)\leqq-\psi',$$

where ϕ is the characteristic function of $(1,2)$. This proves (30.2.26) with integration over the set $R<|x|<2R$ instead. Splitting the integration when $|x|<R$ into integration over annuli with radii $R, R/2, R/4,\dots$ we immediately obtain (30.2.26).

There exist functions satisfying (30.2.27), for $\langle x/|x|, P_0'(\xi)\rangle=|P_0'(\xi)|>0$ on $N_+(M_\lambda)$. This gives the following

Corollary 30.2.8. *If the hypotheses of Theorem* 30.2.7 *are fulfilled and in addition* $\mathrm{Im}(u,f)=0$, *then* $D^\alpha u\in\mathring{B}^*$ *when* $|\alpha|\leqq m$.

Proof. If $a(x,D)u\in \mathring{B}^*$ then $b(x,D)a(x,D)u\in \mathring{B}^*$, if $b\in S(1,G_1)$. Using the symbols in the hypothesis and in the ideal generated by a symbol satisfying (30.2.27) we can therefore conclude that $a(x,D)u\in \mathring{B}^*$ for every a in $S((1+|\xi|)^m, G_1)$, since this is true for all operators of lower order with respect to x and ξ.

If the limit of the graph of $R(z_j)$ discussed in Theorem 30.2.6 is not a graph, that is, contains a pair (u,f) with $u\neq 0$ and $f=0$, it follows that $D^\alpha u\in \mathring{B}^*$ for $|\alpha|\leqq m$. Our next goal is to extend Theorem 14.5.2 by proving that $u\in H_{(m),t}$ for every t. The result is applicable in particular to the eigenfunctions of H with eigenvalue $\lambda\notin Z(P_0)$. It is locally uniform so it allows us to conclude that the eigenvalues are discrete and of finite multiplicity. The estimate we need is essentially contained in the proof of Theorem 30.2.6 but we have to reconsider several points in it.

Theorem 30.2.9. *Let K be a compact subset of $\mathbb{R}\setminus Z(P_0)$, let $\lambda\in K$, $D^\alpha u\in \mathring{B}^*$, $|\alpha|\leqq m$, and*

$$(P_0(D)+V(x,D)-\lambda)u=f.$$

Set $X(x)=(1+|x|^2)^{\frac12}$. If $X^\gamma f\in B$ for some $\gamma\geqq 0$ it follows that $X^\gamma D^\alpha u\in B^*$, $|\alpha|\leqq m$, and that

$$\sum_{|\alpha|\leqq m}\|X^\gamma D^\alpha u\|_{B^*}\leqq C_\gamma(\|X^\gamma f\|_B+\sum_{|\alpha|\leqq m}\|D^\alpha u\|_{B^*}). \tag{30.2.28}$$

Proof. It suffices to prove the result for some compact neighborhood K of an arbitrary point $\lambda_0\in\mathbb{R}\setminus Z(P_0)$. We can then choose χ according to Proposition 30.2.3 and obtain

$$\begin{aligned}\sum_{|\alpha|\leqq m}\|X^\gamma D^\alpha(1-\chi(D))u\|_{B^*}&\leqq C\|(1-\chi(D))u\|_{(m),\gamma}\\ &\leqq C'(\|f-V^S(x,D)u\|_{(0),\gamma}+\|u\|_{(0),-1}).\end{aligned}$$

By (30.2.19)

$$\|V^S(x,D)u\|_{(0),\gamma}\leqq C\|u\|_{(m),\gamma-1-\delta}\leqq C'\sum_{|\alpha|\leqq m}\|X^{\gamma-\frac12}D^\alpha u\|_{B^*}.$$

If the statement is already proved with γ replaced by $\gamma-\frac12$ it follows that $(1-\chi(D))u$ has the stated properties.

To estimate $\chi(D)u$ we shall return to (30.2.18)'. Actually we have proved that for λ close to λ_0, $\gamma\geqq 0$ and $u\in H_{(m),0}$ we have

$$\begin{aligned}R^{-1}\|\Phi_R(x,D)u\|^2\leqq&|(q_R(x,D)(f-V^S(x,D)u),q_R(x,D)u)|\\ &+CR^{-2\gamma}\|u\|^2_{(0),\gamma-(1+\delta)/2}.\end{aligned}$$

By (30.2.19)

$$\|q_R(x,D)V^S(x,D)u\|_{(0),\gamma+(1+\delta)/2}\leqq C\|u\|_{(m),\gamma-(1+\delta)/2},$$

and since $R^{2\gamma}q_R\in S((1+|x|)^{2\gamma},G_1)$ we have

$$R^{2\gamma}\|q_R(x,D)u\|_{(0),-\gamma-(1+\delta)/2}\leqq C\|u\|_{(m),\gamma-(1+\delta)/2}.$$

Hence

$$(30.2.18)'' \qquad R^{-1}\|\Phi_R(x,D)u\|^2 \le \|q_R(x,D)f\|_B\|q_R(x,D)u\|_{B^*} + CR^{-2\gamma}\|u\|^2_{(m),\gamma-(1+\delta)/2}.$$

The estimate (30.2.18)″ remains valid under the hypothesis $D^\alpha u \in \mathring{B}^*$, $|\alpha|\le m$, made in the theorem. In fact, let $\zeta\in C_0^\infty(\{x;|x|<2\})$ be equal to 1 in the unit ball and set $\zeta_t(x)=\zeta(x/t)$, apply (30.2.18)″ to $\zeta_t u$ and let $t\to\infty$. The symbol of $[P_0(D)+V(x,D),\zeta_t]$ is

$$\sum_{\alpha\neq 0}(D_x^\alpha\zeta_t)(P_0^{(\alpha)}(\xi)+V^{(\alpha)}(x,\xi))/\alpha!.$$

Here $|D_x^\alpha\zeta_t|\le C/t$ and $t<|x|<2t$ in the support, so the proof of (30.2.2) gives

$$\|[P_0(D)+V(x,D),\zeta_t]u\|_B \le C(t^{-1}\sum\int_{|x|<2t}|D^\alpha u|^2\,dx)^{\frac12}\to 0, \quad t\to\infty,$$

for the function u in the theorem, since $D^\alpha u\in\mathring{B}^*$. If we note that $\zeta_t u\to u$ in B^* and that $\zeta_t f\to f$ in B as $t\to\infty$ it follows that (30.2.18)″ is valid for u.

Let $\varepsilon>0$ and multiply (30.2.18)″ by the increasing function $R^{2\gamma}(1+\varepsilon R)^{-2\gamma}$. (Later on ε will tend to 0.) We have

$$\|R^\gamma q_R(x,D)f\|_B\le C\|X^\gamma f\|_B$$

for $R^\gamma q_R(x,D)X^{-\gamma}$ is bounded in $\operatorname{Op}S(1,G_1)$ since $R^\gamma q_R(x,\xi)$ is bounded in $S(X^\gamma,G_1)$. Next we prove that

$$\|R^\gamma(1+\varepsilon R)^{-\gamma}q_R(x,D)u\|_{B^*}\le C\|X^\gamma(1+\varepsilon X)^{-\gamma}\chi(D)u\|_{B^*}.$$

To do so we define q'_R as q_R with χ replaced by a function $\chi_1\in C_0^\infty$ which is 1 in $\operatorname{supp}\chi$ but does not have much larger support. Then

$$\begin{aligned}&R^\gamma(1+\varepsilon R)^{-\gamma}q_R(x,D)u\\&\quad=R^\gamma(1+\varepsilon R)^{-\gamma}q'_R(x,D)X^{-\gamma}(1+\varepsilon X)^\gamma(X^\gamma(1+\varepsilon X)^{-\gamma}\chi(D)u).\end{aligned}$$

Here $R^\gamma(1+\varepsilon R)^{-\gamma}q'_R(x,\xi)$ is bounded in $S(X^\gamma(1+\varepsilon X)^{-\gamma},G_1)$ because $t^\gamma(1+\varepsilon t)^{-\gamma}$ is an increasing function of t, so $R^\gamma(1+\varepsilon R)^{-\gamma}q'_R(x,D)X^{-\gamma}(1+\varepsilon X)^\gamma$ is bounded in $\operatorname{Op}S(1,G_1)$. Hence we obtain

$$\begin{aligned}(30.2.29)\quad &R^{-1}\|R^\gamma(1+\varepsilon R)^{-\gamma}\Phi_R(x,D)u\|^2\\&\le C(\|X^\gamma f\|_B\|X^\gamma(1+\varepsilon X)^{-\gamma}\chi(D)u\|_{B^*}+\|u\|^2_{(m),\gamma-(1+\delta)/2}).\end{aligned}$$

If we apply (30.2.29) with P and λ replaced by $-P$ and $-\lambda$ we obtain (30.2.29) with Φ_R replaced by $\phi_1(x/R,P_0'(\xi))\chi(\xi)$. Since $0\le\rho_2\le1$ and $\operatorname{supp}\rho_2\subset(-\frac12,\frac12)$, we have $\rho_2(1+t)+\rho_2(1-t)\le1$, hence

$$\phi(x)\le\phi_1(x,y)+\phi_1(x,-y).$$

Since $\phi(x)\neq0$ when $1\le|x|\le2$ it follows that

$$\begin{aligned}(30.2.19)'\quad &R^{-1}\|R^\gamma(1+\varepsilon R)^{-\gamma}\phi(x/R)\chi(D)u\|^2\\&\le C(\|X^\gamma f\|_B\|X^\gamma(1+\varepsilon X)^{-\gamma}\chi(D)u\|_B+\|u\|^2_{(m),\gamma-(1+\delta)/2}).\end{aligned}$$

Here R is equivalent to X in $\operatorname{supp}\phi(x/R)$ so taking the supremum for $R>1$ we obtain if $N_\varepsilon=\|X^\gamma(1+\varepsilon X)^{-\gamma}\chi(D)u\|_{B^*}$

$$N_\varepsilon^2 \leqq C(\|X^\gamma f\|_B N_\varepsilon+\|u\|^2_{(m),\gamma-(1+\delta)/2}).$$

N_ε is finite so estimating $C\|X^\gamma f\|_B N_\varepsilon$ by $C^2\|X^\gamma f\|_B^2/2+N_\varepsilon^2/2$ gives

$$N_\varepsilon^2 \leqq C^2\|X^\gamma f\|_B^2+2C\|u\|^2_{(m),\gamma-(1+\delta)/2}.$$

ε has now fulfilled its mission; letting $\varepsilon\to 0$ we conclude that $X^\gamma\chi(D)u\in B^*$ and that (with a new C of course)

$$\|X^\gamma\chi(D)u\|_{B^*}\leqq C(\|X^\gamma f\|_B+\|u\|_{(m),\gamma-(1+\delta)/2})$$

if the right-hand side is finite. We have similar estimates for $D^\alpha\chi(D)u$ for if $\chi_1\in C_0^\infty$ is equal to 1 in $\operatorname{supp}\chi$ then

$$X^\gamma D^\alpha\chi(D)u=X^\gamma D^\alpha\chi_1(D)X^{-\gamma}X^\gamma\chi(D)u$$

and $X^\gamma D^\alpha\chi_1(D)X^{-\gamma}\in \operatorname{Op}S(1,G_1)$. Summing up, we have

$$\sum_{|\alpha|\leqq m}\|X^\gamma D^\alpha u\|_{B^*}\leqq C_{\gamma,\gamma'}(\|X^\gamma f\|_B+\sum_{|\alpha|\leqq m}\|X^{\gamma'}D^\alpha u\|_{B^*})$$

if $0\leqq\gamma'<\gamma<\gamma'+\delta/2$; thus $X^\gamma D^\alpha u\in B^*$ when $|\alpha|\leqq m$ if $X^{\gamma'}D^\alpha u\in B^*$ when $|\alpha|\leqq m$. Increasing γ in steps $<\delta/2$ from 0 we obtain the theorem.

Summing up the preceding results we now obtain a version of the limiting absorption principle extending the results of Section 14.5:

Theorem 30.2.10. *Let $V(x,D)$ be a 1-admissible perturbation of the elliptic self-adjoint constant coefficient operator $P_0(D)$ of order m. Let $Z(P_0)$ be the set of critical values of P_0 and let H be the self-adjoint operator with domain $H_{(m)}$ defined by $P_0(D)+V(x,D)$. Then the eigenvalues $\lambda\in\mathbb{R}\setminus Z(P_0)$ are of finite multiplicity, and form a set Λ which is discrete in $\mathbb{R}\setminus Z(P_0)$. The eigenfunctions are in $H_{(m),t}$ for every t. Every compact subset K of $\mathbb{R}\setminus(\Lambda\cup Z(P_0))$ has a neighborhood $K'\subset\mathbb{C}$ such that*

$$\sum_{|\alpha|\leqq m}\|D^\alpha u\|_{B^*}\leqq C\|f\|_B \tag{30.2.30}$$

if $f\in B$, $z\in K'$, $\operatorname{Im}z\neq 0$ and $u=(H-z)^{-1}f=R(z)f$. When $z\to\lambda\in K$ while $\operatorname{Im}z \gtrless 0$ then $R(z)f$ has a weak limit $R(\lambda\pm i0)f$ in B^*; moreover $D^\alpha R(z)f\to D^\alpha R(\lambda\pm i0)f$ in the weak* topology of B^* if $|\alpha|\leqq m$. The solution $u=R(\lambda\pm i0)f$ of the equation $(P_0(D)+V(x,D)-\lambda)u=f$ is characterized by the fact that $D^\alpha u\in B^*$, $|\alpha|\leqq m$, and that $a(x,D)u\in \mathring{B}^*$ for every $a\in S((1+|\xi|)^m,G_1)$ vanishing on*

$$N_\pm(M_\lambda)=\{(tP_0'(\xi),\xi);\ P_0(\xi)=\lambda,\ t\gtrless 0\}.$$

u is a weakly continuous function of λ. If $\chi\in C_0(\mathbb{R}\setminus(Z(P_0)\cup\Lambda))$ and dE_λ is the spectral family of H, then

$$\int\chi(\lambda)\,d(E_\lambda f,f)=\pm 1/\pi\int\chi(\lambda)\operatorname{Im}(R(\lambda\pm i0)f,f)\,d\lambda,\quad f\in B. \tag{30.2.31}$$

Proof. (a) If $u_1, u_2, \ldots$ are orthonormal eigenfunctions of H with eigenvalues $\lambda_1, \lambda_2, \ldots \in K$, then u_j and Hu_j are bounded in L^2 so u_j is bounded in $H_{(m)}$. Hence it follows from Theorem 30.2.9 with $u=u_j$, $\lambda=\lambda_j$, $f=0$ that the sequence u_j is precompact in L^2. This is not possible for an infinite orthonormal sequence which proves the statements on the eigenvalues and eigenfunctions.

(b) If $\lambda \notin \Lambda \cup Z(P_0)$ and (30.2.30) is not valid in any neighborhood of λ, we can find a sequence $z_j \to \lambda$ with $\operatorname{Im} z_j \neq 0$ and $f_j \in B$ with $\|f_j\|_B < 1/j$ so that

$$\sum_{|\alpha| \leqq m} \|D^\alpha u_j\|_{B^*} = 1 \quad \text{if } u_j = R(z_j) f_j.$$

Passing to a subsequence if necessary we may assume that $D^\alpha u_j$ has a weak* limit $D^\alpha u$ in B^* when $|\alpha| \leqq m$. Then $u_j \to u$ in the norm of $H_{(0),-1}$ so $u \neq 0$ by (30.2.21). If $\operatorname{Im} z_j > 0$ for every j, as we may assume, the hypotheses of Theorem 30.2.6 are then fulfilled with $f=0$ so Theorem 30.2.6 and Corollary 30.2.8 show that $D^\alpha u \in \mathring{B}^*$, $|\alpha| \leqq m$. Hence $u \in H_{(m),t}$ for every t by Theorem 30.2.9 so $\lambda \in \Lambda$ against our hypothesis. This proves that every $\lambda \notin \Lambda \cup Z(P_0)$ has a neighborhood where (30.2.30) is valid. If $f \in B$ the sequence $u_j = R(z_j) f$ is then relatively weakly compact in B^* for every sequence $z_j \to \lambda + i0$ (or $\lambda - i0$), and every limit u satisfies the equation

$$(P_0(D) + V(x, D) - \lambda) u = f,$$

$D^\alpha u \in B^*$ when $|\alpha| \leqq m$, and $a(x, D) u \in \mathring{B}^*$ for every $a \in S((1+|\xi|)^m, G_1)$ vanishing on $N_\pm(M_\lambda)$. By Corollary 30.2.8 and Theorem 30.2.9 this characterizes u since $\lambda \notin \Lambda$, so the weak limit of $z \to D^\alpha R(z) f$ exists in B^* for $|\alpha| \leqq m$ when $z \to \lambda \pm i0$. In particular this proves that the boundary values $R(\lambda \pm i0) f$ are weakly continuous functions of λ.

c) The formula (30.2.31) is an immediate consequence of Lemma 14.6.1 and the continuity of $(R(z) f, f)$ in $\mathbb{C}^\pm \setminus (\Lambda \cup Z(P_0))$. This completes the proof.

Remark. When λ is an eigenvalue the proof above shows that $R(z) f$ has a limit when $z \to \lambda \pm i0$ if $f \in B$ is orthogonal to the corresponding space Z_λ of eigenfunctions. The limit is also orthogonal to Z_λ. The limit of the graph discussed in Theorem 30.2.6 is therefore the sum of $Z_\lambda \times \{0\}$ and the graph of this limit defined in the orthogonal space of Z_λ.

30.3. The Hamilton Flow

In this section we assume that the long range part V^L of the perturbation satisfies the conditions in Definition 30.1.3 for some $\kappa \geqq 2$. Only the case $\kappa = 2$ will actually be used here but the greater generality does not lengthen the proofs significantly. As pointed out in the introduction we shall have to construct a Lagrangian Λ satisfying (30.2) and a solution of the Hamilton-

Jacobi equation (30.3). In both cases this requires integration of the Hamilton equations

(30.3.1) $\quad dx/dt=\partial(P_0(\xi)+V^L(x,\xi))/\partial\xi, \quad d\xi/dt=-\partial V^L(x,\xi)/\partial x.$

We shall therefore begin by discussing the Cauchy problem for (30.3.1). Recall that for ξ in a compact subset of $\mathbb{R}^n$ we have by (30.1.7)'

(30.3.2) $$|D_\xi^\alpha D_x^\beta V^L(x,\xi)|\leqq C_{\alpha\beta}(1+|x|)^{-m(|\beta|)}$$

where m is the concave sequence

(30.3.3) $\quad m(j)=\delta+j,\ j\leqq\kappa; \quad m(j)=1+(\delta+\kappa-1)j/\kappa \quad \text{when } j\geqq\kappa.$

As remarked after the proof of Lemma 30.1.1 we may assume $\delta<1/(\kappa+1)$, and then we have $m(j)<j$ when $j>\kappa$. To avoid discussion of separate cases where logarithms enter the estimates, this condition will always be assumed.

When $V^L=0$ the solutions of (30.3.1) are $\xi=$constant and $x-tP_0'(\xi)$ $=$constant. To compare with the unperturbed situation we introduce a new variable z,

$$x=t(P_0'(\xi)+z), \quad U(t,z,\xi)=V^L(t(P_0'(\xi)+z),\xi)/t.$$

Then

$$\begin{aligned} t\,dz/dt&=dx/dt-P_0'(\xi)-z-tP_0''(\xi)\,d\xi/dt=\partial V^L/\partial\xi+tP_0''(\xi)\partial V^L/\partial x-z\\ &=t\,\partial U(t,z,\xi)/\partial\xi-z \end{aligned}$$

so the equations (30.3.1) assume the form

(30.3.1)' $\quad dz/dt=-z/t+\partial U(t,z,\xi)/\partial\xi, \quad d\xi/dt=-\partial U(t,z,\xi)/\partial z.$

If ξ is in a compact set K where $|P_0'(\xi)|>c$ and $|z|<c/2$ then

(30.3.4) $$|D_{\xi,z}^\alpha U(t,z,\xi)|\leqq C_\alpha t^{|\alpha|-m(|\alpha|)-1}$$

when t is large. In fact, $D_{\xi,z}^\alpha U$ is a linear combination with bounded coefficients of terms of the form

$$t^{|\alpha'|}D_x^{\alpha'}D_\xi^{\alpha''}V^L(x,\xi)/t, \quad x=t(P_0'(\xi)+z),\ |\alpha'|+|\alpha''|\leqq|\alpha|.$$

Since $|\alpha'|-m(|\alpha'|)\leqq|\alpha|-m(|\alpha|)$ the estimate (30.3.4) follows.

Lemma 30.3.1. *Let ω be an open subset of $\mathbb{R}^{2n}$ such that $(z,\xi)\in\omega\Rightarrow(\lambda z,\xi)\in\omega$, $0\leqq\lambda\leqq1$, and (30.3.4) is valid for large t. If $\omega'\Subset\omega$ it follows that the equations (30.3.1)' with Cauchy data*

(30.3.5) $$(z,\xi)=(w,\eta)\in\omega' \quad \text{when } t=T$$

for sufficiently large T have a unique solution $(z(t,w,\eta),\xi(t,w,\eta))\in\omega$ when $t\geqq T$. With $\mu(k)=k+1-m(k+1)$ and constants independent of T we have

(30.3.6) $\quad |D_{w,\eta}^\alpha(z(t,w,\eta),\xi(t,w,\eta))|\leqq C_\alpha t^{\mu(|\alpha|)}, \quad |\alpha|\geqq\kappa,$

(30.3.7) $\quad |D_{w,\eta}^\alpha(\xi(t,w,\eta)-\eta)|\leqq CT^{\mu(|\alpha|)}, \quad |\alpha|<\kappa,$

(30.3.8) $\quad |D_{w,\eta}^\alpha z(t,w,\eta)|\leqq|D_{w,\eta}^\alpha w|\,T/t+Ct^{\mu(|\alpha|)}, \quad |\alpha|<\kappa.$

Proof. As long as the solution of (30.3.1)′ with initial data (30.3.5) exists and lies in ω, we have

$$|t\,dz/dt+z|\leqq Ct^{1-m(1)}, \quad |d\xi/dt|\leqq Ct^{-m(1)}.$$

Thus

$$|tz-Tw|\leqq C(t^{2-m(1)}-T^{2-m(1)})/(2-m(1)), \quad |\xi-\eta|\leqq CT^{1-m(1)}/(m(1)-1),$$

for $m(1)=1+\delta<2$. The first estimate implies (with another C)

$$|z-(T/t)w|\leqq Ct^{1-m(1)}\leqq CT^{1-m(1)}, \quad t\geqq T.$$

If T is large enough it follows that (z,ξ) will remain in a compact subset of ω for $t>T$, so the solution always exists. The preceding estimates also prove (30.3.7), (30.3.8) when $\alpha=0$.

Now we differentiate (30.3.1)′ with respect to w or η, using the notation $(z^\alpha,\xi^\alpha)=D^\alpha_{w,\eta}(z,\xi)$. When $|\alpha|=1$ we obtain

$$\begin{aligned} &(30.3.9) \quad & dz^\alpha/dt&=-z^\alpha/t+(\partial^2 U/\partial\xi\,\partial\xi)\xi^\alpha+(\partial^2 U/\partial\xi\,\partial z)z^\alpha,\\ && d\xi^\alpha/dt&=-(\partial^2 U/\partial z\,\partial\xi)\xi^\alpha-(\partial^2 U/\partial z\,\partial z)z^\alpha. \end{aligned}$$

If we take the scalar product with z^α and ξ^α and sum, we obtain

$$N\,dN/dt\leqq CN^2t^{1-m(2)}, \quad N^2=\sum_{|\alpha|=1}(|z^\alpha|^2+|\xi^\alpha|^2).$$

Since $m(2)>2$ we conclude that

$$N(t)/N(T)\leqq\exp\left(\int_T^\infty Cs^{1-m(2)}\,ds\right)<\infty, \quad t\geqq T; \quad N(T)=(2n)^{\frac12};$$

so $N(t)$ is bounded. Hence (30.3.9) gives

$$|d(tz^\alpha)/dt|\leqq Ct^{2-m(2)}, \quad |d\xi^\alpha/dt|\leqq Ct^{1-m(2)}.$$

Since $2<m(2)<3$, an integration of these inequalities gives

$$|z^\alpha(t)|\,t\leqq|z^\alpha(T)|\,T+C't^{3-m(2)}, \quad |\xi^\alpha(t)-\xi^\alpha(T)|\leqq C'T^{2-m(2)},$$

which proves (30.3.7), (30.3.8) when $|\alpha|=1$.

Assume now that k is an integer $\geqq 2$ and that (30.3.6)–(30.3.8) are already proved for $|\alpha|<k$. Differentiation of (30.3.1)′ gives then

$$\begin{aligned} &(30.3.9)' \quad & dz^\alpha/dt&=-z^\alpha/t+(\partial^2 U/\partial\xi\,\partial\xi)\xi^\alpha+(\partial^2 U/\partial\xi\,\partial z)z^\alpha+Z^\alpha,\\ && d\xi^\alpha/dt&=-(\partial^2 U/\partial z\,\partial\xi)\xi^\alpha-(\partial^2 U/\partial z\,\partial z)z^\alpha+\Xi^\alpha, \quad |\alpha|=k, \end{aligned}$$

where Z^α and Ξ^α are sums of terms in $D^\alpha_{w,\eta}U'(t,z(t,w,\eta),\xi(t,w,\eta))$ where U is differentiated altogether at least three times. By hypothesis

$$|D^\alpha_{\eta,w}(z(t,w,\eta),\xi(t,w,\eta))|\leqq C_\alpha t^{a(|\alpha|)}, \quad |\alpha|<k,$$

if $a(j)=\max(\mu(j),0)$, and by (30.3.4) we have

$$|D^\alpha_{z,\xi}U'_{z,\xi}(t,z,\xi)|\leqq C_\alpha t^{b(|\alpha|)}$$

if $b(k)=\mu(k)-1$. Since $\mu(1)<0$ and μ is increasing the condition (30.1.11) is valid by the remarks at the end of Section 30.1. Hence

$$|Z^\alpha|+|\Xi^\alpha|\leq Ct^{\mu(k)-1}, \quad |\alpha|=k.$$

If $N(t)$ is defined as before but with summation for $|\alpha|=k$, then scalar multiplication of (30.3.9)′ by z^α and ζ^α gives after summation

$$N'(t)\leq CN(t)t^{1-m(2)}+C_k t^{k-m(k+1)}.$$

We multiply by the integrating factor $E(t)=\exp(-Ct^{2-m(2)}/(2-m(2)))$ and obtain

$$(E(t)N(t))'\leq C_k E(t)t^{k-m(k+1)}\leq C_k t^{k-m(k+1)}.$$

Since $N(T)=0$ and $E(t)^{-1}$ is bounded for $t>1$, it follows that

$$N(t)\leq C' \quad \text{if } k<\kappa; \qquad N(t)\leq C_k' t^{\mu(k)} \quad \text{if } k\geq\kappa,$$

for $\mu(k)<0$ when $k<\kappa$ and $\mu(k)>0$ when $k\geq\kappa$. If $k<\kappa$ we return to the equations (30.3.9)′ and obtain as in the case $k=1$

$$|d(tz^\alpha)/dt|\leq Ct^{-\delta}, \qquad |d\zeta^\alpha/dt|\leq Ct^{-1-\delta}.$$

Since $z^\alpha=0$ and $\zeta^\alpha=0$ when $t=T$ it follows that

$$|tz^\alpha|\leq Ct^{1-\delta}/(1-\delta), \quad |\zeta^\alpha|\leq CT^{-\delta}/\delta \qquad \text{when } t>T,$$

which proves (30.3.7) and (30.3.8).

When studying (30.2) we shall also need estimates for the t derivatives of $(z(t,w,\eta), \zeta(t,w,\eta))$. It will suffice to have them when $w=0$.

Lemma 30.3.2. *When* $\tau>0$ *we have*

$$|D_\eta^\alpha D_t^\tau(z(t,0,\eta), \zeta(t,0,\eta))|\leq C_{\alpha\tau}t^{|\alpha|-m(|\alpha|+\tau)}, \quad t>T, \tag{30.3.10}$$

where (z,ζ) *is defined as in Lemma* 30.3.1.

Proof. Introducing $s=\log t$ as a new variable we find that

$$t^\tau D_t^\tau=(tD_t+(\tau-1)i)\dots(tD_t+i)(tD_t),$$

Since $|\alpha|+\tau-m(|\alpha|+\tau)$ is an increasing function of τ, the estimate (30.3.10) is therefore equivalent to

$$|D_\eta^\alpha(tD_t)^\tau(z(t,0,\eta), \zeta(t,0,\eta))|\leq C_{\alpha\tau}t^{|\alpha|+\tau-m(|\alpha|+\tau)}, \quad t>T. \tag{30.3.10′}$$

This is symmetric in the derivatives with respect to η and $\log t$ and therefore suitable for application of the calculus lemmas in Section 30.1. Now

$$t\partial z/\partial t=-z+t\partial U/\partial\zeta, \quad t\partial\zeta/\partial t=-t\partial U/\partial z, \tag{30.3.1″}$$

and we have

$$|D_{z,\zeta}^\alpha(tD_t)^\tau(tU)|\leq C_{\alpha\tau}t^{|\alpha|+\tau-m(|\alpha|+\tau)}. \tag{30.3.11}$$

The proof is like that of (30.3.4). The derivative to estimate is a linear combination of terms of the form

$$t^{|\alpha'|+|\beta'|} D_x^{\alpha'+\beta'} D_\xi^{\alpha''} V^L(x,\xi); \quad x=t(P_0'(\xi)+z),$$

where $|\alpha'|+|\alpha''|\leqq|\alpha|$ and $|\beta'|\leqq\tau$, for the $\tau-|\beta'|$ derivatives with respect to t which do not fall on V^L reduce the exponent of t. By (30.3.2) such a term can be estimated by t raised to the power

$$|\alpha'|+|\beta'|-m(|\alpha'|+|\beta'|),$$

which gives (30.3.11) since $k-m(k)$ is increasing. By Lemma 30.3.1 with $w=0$ we have

$$|D_\eta^\alpha z(t,0,\eta)|\leqq C_\alpha t^{\mu(|\alpha|)}, \quad |D_\eta^\alpha \xi(t,0,\eta)|\leqq C_\alpha t^{\max(\mu(|\alpha|),0)}.$$

Since $\mu(|\alpha|)=|\alpha|+1-m(|\alpha|+1)$ this gives the desired estimate for the term z in (30.3.1)″ when $\tau=1$. The estimate (30.3.10)′ follows for $\tau=1$ if we apply Lemma 30.1.5 with $a(k)=\max(\mu(k),0)$ and $b(k)=k+1-m(k+1)=\mu(k)$, which is legitimate since (30.1.11) is fulfilled. (See the end of Section 30.1.) Now assume that $\tau\geqq 2$ and that (30.3.10) has already been proved for smaller values of τ. We apply $D_\eta^\alpha(tD_t)^{\tau-1}$ to (30.3.1)″ and observe that

$$|D_\eta^\alpha(tD_t)^{\tau-1} z|\leqq C_{\alpha\tau} t^{|\alpha|+\tau-1-m(|\alpha|+\tau-1)}\leqq C_\alpha t^{|\alpha|+\tau-m(|\alpha|+\tau)},$$

by the inductive hypothesis. Now we apply Lemma 30.1.5 to the variables η and $\log t$ noting that

$$|D_\eta^\alpha(tD_t)^{\tau'}(z(t,0,\eta),\xi(t,0,\eta))|\leqq C_{\alpha\tau} t^{a(|\alpha|+\tau')}$$

if $\tau'<\tau$, with a as above, and that

$$|D_{z,\xi}^\alpha(tD_t)^{\tau'} t U'_{z,\xi}|\leqq C_{\alpha\tau'} t^{b(|\alpha|+\tau')}$$

with $b=\mu$ again. Hence

$$|D_\eta^\alpha(tD_t)^{\tau-1}(tD_t)(z(t,0,\eta),\xi(t,0,\eta))|\leqq C_{\alpha\tau} t^{b(|\alpha|+\tau-1)}$$

which completes the proof of (30.3.10).

Using Lemma 30.3.1 only we shall now solve the time dependent Hamilton-Jacobi equation (30.3).

Theorem 30.3.3. *Assume that $V^L(x,\xi)=\sum V_\alpha^L(x)\xi^\alpha$ is a polynomial in ξ of arbitrary order with real coefficients satisfying (30.1.7)′ with $\kappa=2$ and some δ with $0<\delta<\frac{1}{3}$. Let $P_0(\xi)$ be a real valued non-constant polynomial. Then there is a real valued C^∞ function $W(\xi,t)$ such that for ξ in any compact subset K of $\Omega=\{\xi;\, P_0'(\xi)\neq 0\}$ we have for some t_K*

(30.3.12) $\quad \partial W/\partial t=P_0(\xi)+V^L(\partial W/\partial\xi,\xi) \quad$ *if $\xi\in K$ and $|t|>t_K$,*

(30.3.13) $\quad |D_\xi^\alpha(W(\xi,t)-tP_0(\xi))|\leqq C_{\alpha,K}|t|^{1+|\alpha|-m(|\alpha|)} \quad$ *for all α, if $\xi\in K$.*

Proof. Let $\Omega_0 \Subset \Omega_1 \Subset \Omega_2 \Subset \ldots \Subset \Omega$ be open sets with union Ω. First we solve the Cauchy problem for (30.3.12) with data

$$W(\xi, t) = t P_0(\xi), \quad \xi \in \Omega_2, \ t = t_1, \tag{30.3.14}$$

where t_1 is large. This means that we solve the Hamilton equations (30.3.1) with initial data $\xi = \eta$, $x = t_1 P_0'(\eta)$, $\eta \in \Omega_2$, that is, we solve (30.3.1)′ with initial data

$$\xi = \eta, \quad z = 0 \quad \text{when } t = t_1. \tag{30.3.15}$$

If t_1 is large enough the solution exists for $t > t_1$ by Lemma 30.3.1. From (30.3.7) it follows that the map

$$\Omega_2 \ni \eta \mapsto \xi(t, 0, \eta)$$

has an inverse defined in Ω_1 if $t > t_1$ and t_1 is large enough. Since

$$|D_\eta^\alpha \xi(t, 0, \eta)| \leq C_\alpha t^{a(|\alpha|)}, \quad a(j) = \max(\mu(j), 0),$$

we have an estimate of the same form for the inverse. Now

$$|D_\eta^\alpha z(t, 0, \eta)| \leq C_\alpha t^{\mu(|\alpha|)}$$

so another application of Lemma 30.1.5, with $b = \mu$ this time, gives that $z(t, 0, \eta) = Z(\xi(t, 0, \eta), t)$ where

$$|D_\xi^\alpha Z(\xi, t)| \leq C'_\alpha t^{\mu(|\alpha|)}.$$

Thus the solution of (30.3.12), (30.3.14) exists at least in $\Omega_1 \times [t_1, \infty)$, and since $\partial W(\xi, t)/\partial \xi = t(P_0'(\xi) + Z(\xi, t))$ we have

$$\begin{aligned} |D_\xi^\alpha (W(\xi, t) - t P_0(\xi))| &\leq C''_\alpha t^{1 + \mu(|\alpha| - 1)} \\ &= C''_\alpha t^{1 + |\alpha| - m(|\alpha|)} \end{aligned} \tag{30.3.16}$$

when $\alpha \neq 0$. The equation (30.3.12) gives

$$\left| \frac{\partial}{\partial t} (W(\xi, t) - t P_0(\xi)) \right| = |V^L(\partial W/\partial \xi, \xi)| \leq C t^{-m(0)}$$

and using (30.3.14) we therefore obtain (30.3.13) also when $\alpha = 0$.

Suppose now that we already have a solution W_j of (30.3.12) defined in $\Omega_j \times [t_j, \infty)$ which satisfies (30.3.13) there. With $\chi \in C_0^\infty(\Omega_j)$ equal to 1 in a neighborhood of $\bar{\Omega}_{j-1}$ we shall solve the Cauchy problem for the Hamilton-Jacobi equation (30.3.12) with initial data

$$W(\xi, t) = \chi(\xi) W_j(\xi, t) + (1 - \chi(\xi)) t P_0(\xi), \quad \xi \in \Omega_{j+2}, \ t = t_{j+1}, \tag{30.3.14$'$}$$

where t_{j+1} will be chosen large. This means that the equations (30.3.1)′ must be solved with initial data

$$\begin{gathered} \xi = \eta \in \Omega_{j+2}, \\ z = \psi(\eta) = t^{-1} \partial(\chi(\eta)(W_j(\eta, t) - t P_0(\eta)))/\partial \eta; \quad t = t_{j+1}. \end{gathered} \tag{30.3.15$'$}$$

With C_α denoting different constants in different estimates we have by (30.3.13)

(30.3.16) $$|D^\alpha \psi(\eta)| \leqq C_\alpha t_{j+1}^{\mu(|\alpha|)},$$

for μ is increasing. With the notation in Lemma 30.3.1 the desired solution of (30.3.1)′ is $(\xi(t, \psi(\eta), \eta), z(t, \psi(\eta), \eta))$ and it exists for $t \geqq t_{j+1}$ if t_{j+1} is large enough. We claim that for $t \geqq t_{j+1}$

(30.3.17) $$|D_\eta^\alpha \xi(t, \psi(\eta), \eta)| \leqq C_\alpha t^{a(|\alpha|)}, \quad a(k) = \max(\mu(k), 0),$$

(30.3.18) $$|D_\eta^\alpha z(t, \psi(\eta), \eta)| \leqq C_\alpha t^{\mu(|\alpha|)}.$$

Since

$$|D_{w,\eta}^\alpha (z(t, w, \eta), \xi(t, w, \eta))| \leqq C_\alpha t^{a(|\alpha|)}$$

this follows from Lemma 30.1.5 and (30.3.16) except when $|\alpha| \leqq 1$ in (30.3.18). By (30.3.8)

$$|z(t, \psi(\eta), \eta)| \leqq |\psi(\eta)| t_{j+1}/t + C t^{\mu(0)}, \quad t \geqq t_{j+1},$$

and $t_{j+1}^{\mu(0)+1} \leqq t^{\mu(0)+1}$ since $\mu(0)+1 = 2 - m(1) > 0$ so (30.3.18) follows for $\alpha = 0$. Similarly we can estimate

$$D_{\eta_j} z(t, \psi(\eta), \eta) = \partial z/\partial \eta_j + \sum \partial z/\partial w_k \, \partial \psi_k/\partial \eta_j$$

using (30.3.8) and (30.3.16), which gives the bound

$$C t^{\mu(1)} + C t_{j+1}^{\mu(1)} t_{j+1}/t \leqq C' t^{\mu(1)}$$

since $\mu(1)+1 = 3 - m(2) > 0$. This completes the proof of (30.3.18). Also note that

(30.3.19) $$|W(\xi, t) - t P_0(\xi)| = |\chi(\xi)| \, |W_j(\xi, t) - t P_0(\xi)| \leqq C t^{1-m(0)}, \quad t = t_{j+1}.$$

The argument used in the first part of the construction now shows with no further change that if t_{j+1} is large enough there is a solution W_{j+1} of (30.3.12) in $\Omega_{j+1} \times [t_{j+1}, \infty)$ which is equal to W_j in $\Omega_{j-1} \times [t_{j+1}, \infty)$ and satisfies (30.3.13) in $\Omega_{j+1} \times [t_{j+1}, \infty)$ for some new constants. The functions W_j in $\Omega_{j-1} \times [t_j, \infty)$ define together a solution W of (30.3.12) in the union $\bigcup (\Omega_{j-1} \times [t_j, \infty))$. We can extend W to a C^∞ function vanishing outside $\Omega \times \mathbb{R}$ so that $W(\xi, t) = 0$ when $t < 1$ or $t < t_j/2$, $\xi \notin \Omega_{j-1}$ for some j. The construction can then be joined to an analogous one for large negative t. This completes the proof of Theorem 30.3.3.

Next we shall construct a Lagrangian Λ contained in the set where

(30.3.20) $$P_0(\xi) + V^L(x, \xi) = \lambda.$$

In doing so we assume that λ is not a critical value of P_0 and that $|P_0| \to \infty$ at ∞. Thus

$$M_\lambda = \{\xi;\ P_0(\xi) = \lambda\}$$

is a compact C^∞ manifold. If $V^L = 0$ then the normal bundles

$$N_\pm(M_\lambda) = \{(t P_0'(\xi), \xi);\ t \gtrless 0, \ \xi \in M_\lambda\}$$

are Lagrangians with the required properties and homogeneous with respect to x. We want to choose Λ as close to $N_\pm(M_\lambda)$ as possible at infinity. In doing so it suffices to consider $N_+(M_\lambda)$, for replacing P_0 and λ by $-P_0$ and $-\lambda$ interchanges N_+ and N_-. We assume that V^L is κ-admissible for some $\kappa \geqq 2$.

Let T be a large positive number and set

$$\Sigma_{\lambda,T} = \{(x, \xi);\ x = TP_0'(\xi),\ P_0(\xi) + V^L(x, \xi) = \lambda\}.$$

This is the part of the Lagrangian graph of $\xi \mapsto TP_0'(\xi)$ where $\xi \in M_{\lambda,T}$,

$$M_{\lambda,T} = \{\xi;\ P_0(\xi) + V^L(TP_0'(\xi), \xi) = \lambda\}.$$

Lemma 30.3.4. *$M_{\lambda,T}$ is a C^∞ hypersurface if T is large enough. If $\xi^0 \in M_\lambda$ and $\partial P_0(\xi^0)/\partial \xi_1 \neq 0$ then there is a neighborhood of ξ^0 where M_λ and $M_{\lambda,T}$ are defined by*

$$\xi_1 = \Xi(\xi') \quad \text{resp.} \quad \xi_1 = \Xi_T(\xi').$$

Here $\xi' = (\xi_2, \dots, \xi_n)$, Ξ and Ξ_T are in C^∞, and

$$|D^\alpha_{\xi'}(\Xi(\xi') - \Xi_T(\xi'))| \leqq C_\alpha T^{|\alpha| - m(|\alpha|)}. \tag{30.3.21}$$

Proof. It suffices to prove the last statement which is of course also applicable with ξ_1 replaced by another coordinate ξ_j such that $\partial P_0(\xi^0)/\partial \xi_j \neq 0$. The implicit function theorem shows that M_λ can be defined by $\xi_1 = \Xi(\xi')$ near ξ^0. If we multiply the equation above for $M_{\lambda,T}$ by the C^∞ function $(\xi_1 - \Xi(\xi'))/(P_0(\xi) - \lambda)$ it takes the form

$$\xi_1 - \Xi(\xi') = F_T(\xi)$$

where $F_T \in C^\infty$ in a neighborhood of ξ^0 independent of T and satisfies

$$|D^\alpha F_T(\xi)| \leqq C_\alpha T^{|\alpha| - m(|\alpha|)}. \tag{30.3.22}$$

Here we have used again that $k - m(k)$ is increasing so that the largest terms in $D^\alpha F_T(\xi)$ are obtained when all derivatives fall on the x variable in V^L. The right-hand side is small for large T when $|\alpha| \leqq 2$. Hence the implicit function theorem shows that in a fixed neighborhood of ξ^0 the equation $\xi_1 - \Xi(\xi') = F_T(\xi)$ can be solved for ξ_1. Thus $\xi_1 = \Xi_T(\xi')$ where (30.3.21) is valid for $\alpha = 0$. Differentiation of the equation $\Xi_T - \Xi = F_T(\Xi_T, \xi')$ gives

$$D^\alpha_{\xi'}(\Xi_T - \Xi) = D^\alpha_{\xi'} F_T(\Xi_T, \xi').$$

The right-hand side is the sum of $\partial F_T/\partial \xi_1 D^\alpha_{\xi'} \Xi_T$ and terms where Ξ_T is differentiated less than $|\alpha|$ times.. These can be estimated if (30.3.21) is already known for derivatives of lower order, as we may assume. Put $a(j) = \max(0, j - m(j))$ and $b(j) = j - m(j)$. Since (30.3.21) implies

$$|D^\alpha_{\xi'}(\Xi_T(\xi'), \xi')| \leqq C'_\alpha T^{a(|\alpha|)}$$

and (30.1.11) is valid by the remarks at the end of Section 30.1, it follows from Lemma 30.1.5 and the inductive hypothesis that for large T

$$(1 - CT^{1 - m(1)})|D^\alpha_{\xi'}(\Xi_T(\xi') - \Xi(\xi'))| \leqq C''_\alpha T^{b(|\alpha|)}.$$

Here we have used that $|\partial F_T/\partial \xi_1 D^{\alpha}_{\xi'} \Xi| \leqq CT^{b(1)} \leqq CT^{b(|\alpha|)}$. This completes the proof.

From Lemma 30.3.4 it follows at once that $\Sigma_{\lambda,T}$ is an isotropic manifold of dimension $n-1$. The Hamilton field

$$\partial(P_0(\xi)+V^L(x,\xi))/\partial\xi\,\partial/\partial x-\partial V^L(x,\xi)/\partial x\,\partial/\partial\xi$$

is transversal to it when T is large enough, so the union of the orbits starting at $\Sigma_{\lambda,T}$ is locally Lagrangian. In fact, this remains true in the future:

Theorem 30.3.5. *Assume that (30.3.2) and (30.3.3) are valid for some $\kappa \geqq 2$. If T is large enough the union Λ_T^+ of the forward integral curves of the Hamilton field of $P_0(\xi)+V^L(x,\xi)$ starting at $\Sigma_{\lambda,T}$ is a Lagrangian manifold remaining over any given neighborhood of M_λ. If ξ^0 is a point in M_λ where $\partial P_0(\xi)/\partial\xi_1>0$ then Λ_T^+ is defined for ξ in a neighborhood of ξ^0 by*

$$\xi_1=\partial G(x_1,\xi')/\partial x_1, \quad x'=-\partial G(x_1,\xi')/\partial\xi', \tag{30.3.23}$$

where G is a C^∞ function for large x_1 and for ξ' in a neighborhood of $\xi^{0\prime}$,

$$|D^{\alpha'}_{\xi'}D^{\alpha_1}_{x_1}(G(x_1,\xi')-G_0(x_1,\xi'))| \leqq C_\alpha x_1^{1+|\alpha'|-m(|\alpha|)}. \tag{30.3.24}$$

Here $G_0(x_1,\xi')=x_1\Xi(\xi')$ if M_λ is defined by $\xi_1=\Xi(\xi')$ near ξ^0 so that G_0 defines $N_+(M_\lambda)$ locally according to (30.3.23).

Proof. Let us first verify the last statement. Since $P_0(\Xi(\xi'),\xi')-\lambda=0$ we have

$$P_0^{(1)}\partial\Xi/\partial\xi'+\partial P_0/\partial\xi'=0.$$

Hence

$$(x_1,x')=(x_1,-\partial G_0(x_1,\xi')/\partial\xi')=x_1(1,-\partial\Xi/\partial\xi')$$
$$=x_1P_0^{(1)}(\Xi,\xi')^{-1}P_0'(\Xi,\xi')\in N_+(M_\lambda).$$

By definition Λ_T^+ is the union of the orbits

$$(T,\infty)\ni t\mapsto(x(t),\xi(t))$$

where $(x(t),\xi(t))$ is a solution of (30.3.1) starting at a point in $\Sigma_{\lambda,T}$ when $t=T$. We write as usual $x(t)=t(P_0'(\xi)+z(t))$. The initial conditions are then $z(T)=0$ and $\xi(T)\in M_{\lambda,T}$. According to Lemma 30.3.4 we can parametrize the starting values $\xi(T)$ near ξ^0 by

$$\eta'\mapsto(\Xi_T(\eta'),\eta')=\xi_T(\eta').$$

It follows at once from (30.3.7), (30.3.8) that $|\xi(t)-\xi_T(\eta')|\leqq CT^{-\delta}$ and $|z(t)|\leqq Ct^{-\delta}$ on the corresponding orbit. To prove the theorem we must study the map

$$(t,\eta')\mapsto(t(P_0'(\hat\xi(t,\eta'))+\hat z(t,\eta')),\hat\xi(t,\eta'))=(\hat x(t,\eta'),\hat\xi(t,\eta')),$$
$$\hat z(t,\eta')=z(t,0,\xi_T(\eta')),\quad \hat\xi(t,\eta')=\xi(t,0,\xi_T(\eta')),$$

when $t \geqq T$ and η' is close to $\xi^{0\prime}$. We shall estimate the derivatives and show that the map $(t, \eta') \mapsto (\hat{x}_1(t, \eta'), \hat{\xi}'(t, \eta'))$ is invertible. This will show that we have a global Lagrangian with parameters $\hat{x}_1, \hat{\xi}'$, hence that there is a function $G(\hat{x}_1, \hat{\xi}')$ with

$$dG(\hat{x}_1, \hat{\xi}') = \xi_1 d\hat{x}_1 - \langle \hat{x}', d\hat{\xi}' \rangle,$$

for the differential of the right hand side is the symplectic form which vanishes on the flowout of $\Sigma_{\lambda, T}$.

From Lemmas 30.3.1 and 30.3.2 it follows that

$$|D_\eta^\alpha (tD_t)^\tau (z(t,0,\eta), \xi(t,0,\eta))| \leqq C_{\alpha\tau} t^{\mu^+(|\alpha|+\tau)}$$

where $\mu^+(j) = \max(0, j+1-m(j+1))$, and by (30.3.21) we have

$$|D_{\eta'}^\alpha \xi_T(\eta')| \leqq C_\alpha t^{a(j)}, \quad a(j) = \max(0, j - m(j)).$$

If we apply Lemma 30.1.5 with $b = \mu^+$ noting that $b(1) + a(j) = a(j) \leqq b(j)$ we obtain

$$|D_{\eta'}^\alpha (tD_t)^\tau (\hat{x}(t, \eta')/t, \hat{\xi}(t, \eta'))| \leqq C_{\alpha\tau} t^{\mu^+(|\alpha|+\tau)}.$$

Since

$$\hat{x}_1(t, \eta')/t = P_0^{(1)}(\xi(t, \eta')) + \hat{z}_1(t, \eta')$$

is close to $P_0^{(1)}(\xi^0) > 0$ if η' is close to $\xi^{0\prime}$ and T is large, we have a lower bound for $\hat{x}_1/t$ also and conclude using Lemma 30.1.4 that

$$|D_{\eta'}^\alpha (tD_t)^\tau (t/\hat{x}_1(t, \eta'), \hat{x}'(t, \eta')/\hat{x}_1(t, \eta'))| \leqq C_{\alpha\tau} t^{\mu^+(|\alpha|+\tau)}.$$

To discuss the map $(t, \eta') \mapsto (\hat{x}_1(t, \eta'), \hat{\xi}'(t, \eta'))$ it is convenient to use the logarithmic variables

$$s = \log t, \quad X = \log \hat{x}_1(t, \eta') = s + \log(P_0^{(1)}(\xi(t, \eta')) + \hat{z}_1(t, \eta')).$$

Then $\partial/\partial s = t\partial/\partial t$ and $\partial(\xi(t, \eta'), \hat{z}(t, \eta'))/\partial s = O(t^{-\delta})$ by (30.3.1)′, (30.3.4) since $\hat{z} = O(t^{-\delta})$ by (30.3.7). We have $\partial(\xi(t, \eta') - (\Xi_T(\eta'), \eta'))/\partial \eta' = O(T^{-\delta})$ by (30.3.7), and $\partial \hat{z}(t, \eta')/\partial \eta' = O(t^{-\delta})$ by (30.3.8), the difference between the Jacobian matrix $\partial(X, \hat{\xi}')/\partial(s, \eta')$ and $\partial(s + \log P_0^{(1)}(\Xi(\eta'), \eta'), \eta')/\partial(s, \eta')$ is $O(T^{-\delta})$, so the determinant is $1 + O(T^{-\delta})$. When T is large it follows that the map $(s, \eta') \mapsto (X, \hat{\xi}')$ can be inverted. Since

$$|D_{\eta'}^\alpha D_s^\tau (X, \hat{\xi}')| \leqq C_{\alpha\tau} t^{\mu^+(|\alpha|+\tau)}$$

it follows from Lemma 30.1.5 that an estimate of the same form is valid for the inverse. Thus $\hat{x}'$ and ξ_1 are functions of $\hat{x}_1$ and $\hat{\xi}'$ satisfying

$$|D_{\hat{\xi}'}^\alpha (\hat{x}_1 D_{\hat{x}_1})^\tau (\hat{x}'/\hat{x}_1, \xi_1)| \leqq C_{\alpha\tau} t^{\mu^+(|\alpha|+\tau)}.$$

This means that (30.3.23) is valid for a function G_T defined up to a constant term, and that we have

$$|D_{\xi'}^\alpha (x_1 D_{x_1})^\tau (x_1^{-1} \partial G/\partial \xi', \partial G/\partial x_1)| \leqq C_{\alpha\tau} x_1^{\mu^+(|\alpha|+\tau)}.$$

This proves (30.3.24) when $|\alpha| > \kappa$; in that case we have

$$|D_{\xi'}^{\alpha'} D_{x_1}^{\alpha_1} G_0(x_1, \xi')| \leqq C_\alpha x_1^{1-\alpha_1} \leqq C_\alpha x_1^{1+|\alpha'|-m(|\alpha|)}$$

since G_0 is homogeneous in x_1 of degree 1 and $m(|\alpha|)<|\alpha|$ when $|\alpha|>\kappa$. It remains to prove (30.3.24) when $|\alpha|\leqq\kappa$; then there is some cancellation taking place between the two terms.

The estimate (30.3.24) means for $\alpha=0$ that

$$|G(x_1,\xi')-G_0(x_1,\xi')|\leqq Cx_1^{1-\delta},$$

and it follows by integration of the estimate

$$|\partial G(x_1,\xi')/\partial x_1-\partial G_0(x_1,\xi')/\partial x_1|\leqq Cx_1^{-\delta}$$

corresponding to $\alpha'=0$ and $\alpha_1=1$. Now

$$\partial G/\partial x_1-\partial G_0/\partial x_1=\xi_1-\Xi(\xi');$$
$$\partial G/\partial\xi_j-\partial G_0/\partial\xi_j=-x_j-x_1\,\partial\Xi/\partial\xi_j,\quad j\neq 1;$$

where ξ_1 and x' are the functions of x_1 and ξ' defining Λ_T^+. Hence the remaining estimates (30.3.24) are equivalent to

$$(30.3.24)'\qquad |D_{x_1}^{\beta_1}D_{\xi'}^{\alpha'}(\xi_1-\Xi(\xi'))|+|D_{x_1}^{\beta_1}D_{\xi'}^{\alpha'}(x'+x_1\,\partial\Xi/\partial\xi')|/x_1 \leqq Cx_1^{-\delta-\beta_1},\quad \text{if } |\alpha'|+\beta_1<\kappa.$$

Since the map $(t,\eta')\mapsto(\hat{x}_1(t,\eta'),\hat{\xi}'(t,\eta'))$ is invertible and

$$|D_{\hat{x}_1}^{\beta_1}D_{\hat{\xi}'}^{\alpha'}t|\leqq Ct^{1-\beta_1},\quad |D_{\hat{x}_1}^{\beta_1}D_{\hat{\xi}'}^{\alpha'}\eta'|\leqq Ct^{-\beta_1}\quad \text{if } |\alpha'|+\beta_1<\kappa,$$

the estimate (30.3.24)′ can equally well be written in terms of t,η' derivatives,

$$(30.3.24)''\qquad |D_t^{\beta_1}D_{\eta'}^{\alpha'}(\hat{\xi}_1(t,\eta')-\Xi(\hat{\xi}'(t,\eta')))|\leqq Ct^{-\delta-\beta_1},$$
$$|D_t^{\beta_1}D_{\eta'}^{\alpha'}(\hat{x}'(t,\eta')+\hat{x}_1\,\partial\Xi(\hat{\xi}'(t,\eta'))/\partial\hat{\xi}')|\leqq Ct^{1-\delta-\beta_1},\quad |\alpha'|+\beta_1<\kappa.$$

By Lemma 30.3.2 and Lemma 30.3.4 we have

$$|D_tD_{\eta'}^{\alpha'}(tD_t)^{\beta_1}\hat{\xi}(t,\eta')|\leqq Ct^{-1-\delta},\quad \text{if } |\alpha'|+\beta_1<\kappa.$$

Hence $\hat{\xi}(t,\eta')$ has a limit $\hat{\xi}(\infty,\eta')$ in $C^{\kappa-1}$ as $t\to\infty$, and if we write

$$\hat{\xi}(t,\eta')=\hat{\xi}(\infty,\eta')+\rho(t,\eta')$$

then

$$(30.3.25)\qquad |D_{\eta'}^{\alpha}D_t^{\beta_1}\rho(t,\eta')|\leqq Ct^{-\delta-\beta_1},\quad |\alpha'|+\beta_1<\kappa.$$

Similarly it follows from Lemmas 30.3.2 and 30.3.4 that

$$(30.3.26)\qquad |D_{\eta'}^{\alpha'}D_t^{\beta_1}\hat{z}(t,\eta')|\leqq Ct^{-\delta-\beta_1},\quad |\alpha'|+\beta_1<\kappa.$$

We have

$$\hat{x}_j(t,\eta')+\hat{x}_1\,\Xi^{(j)}(\hat{\xi}'(t,\eta'))=t(P_0^{(j)}(\hat{\xi}(t,\eta'))+P_0^{(1)}(\hat{\xi}(t,\eta'))\Xi^{(j)}(\hat{\xi}'(t,\eta'))$$
$$+\hat{z}_j(t,\eta')+\hat{z}_1(t,\eta')\Xi^{(j)}(\hat{\xi}'(t,\eta'))).$$

For the terms involving $\hat{z}$ it is clear that we have estimates of the form (30.3.24)″. Now $P_0^{(j)}(\xi)+P_0^{(1)}(\xi)\Xi^{(j)}(\xi')=0$ on M_λ since $P_0(\Xi,\xi')=0$. To complete the proof of (30.3.24)″ it therefore remains only to show that if $F\in C^\infty$ and $F=0$ on M_λ then

$$|D_{\eta'}^{\alpha'} D_t^{\beta_1} F(\xi(t,\eta'))| \leq C t^{-\delta-\beta_1}, \quad |\alpha'|+\beta_1<\kappa. \tag{30.3.27}$$

We can write $F(\xi+\rho)-F(\xi)=\rho H(\xi,\rho)$ where $H\in C^\infty$. Since $\xi(\infty,\eta')\in M_\lambda$ it follows that

$$F(\xi(t,\eta'))=\rho(t,\eta')H(\xi(\infty,\eta'),\rho(t,\eta')).$$

The derivatives of $\xi(t,\eta')$, $\rho(t,\eta')$ of order $<\kappa$ with respect to η' and $\log t$ are uniformly bounded and so are those of $t^\delta\rho$. This proves (30.3.27) and completes the proof of the theorem.

The following theorem shows how G is transformed if one passes to a representation involving another privileged variable than x_1.

Theorem 30.3.6. *There is a unique function $\psi\in C^\infty(\Lambda_T^\ddagger)$ such that*

$$d\psi=\sum x_j d\xi_j \quad on\ \Lambda_T^\ddagger, \qquad \psi=-TV^L \quad on\ \Sigma_{\lambda,T}. \tag{30.3.28}$$

With the notation in Theorem 30.3.5 we have if G is suitably normalized

$$G=x_1\xi_1-\psi \quad on\ \Lambda_T^\ddagger \tag{30.3.29}$$

when x_1 is large and ξ is close to ξ^0.

Proof. The form $\sum x_j d\xi_j$ on $\Lambda_T^\ddagger$ is closed since $\Lambda_T^\ddagger$ is Lagrangian. The restriction to $\Sigma_{\lambda,T}$ is

$$T\sum P_0^{(j)}(\xi)\,d\xi_j=Td(P_0(\xi)-\lambda)=-TdV^L(x,\xi).$$

Thus there is a unique function ψ satisfying (30.3.28), and

$$d\psi/dt=\langle x, d\xi/dt\rangle=\langle x, -\partial V^L(x,\xi)/\partial x\rangle$$

along the Hamilton orbits in $\Lambda_T^\ddagger$. Now

$$dG=\xi_1\,dx_1-\langle x', d\xi'\rangle=d(x_1\xi_1-\psi) \quad \text{on } \Lambda_T^\ddagger$$

which proves (30.3.29) if the undetermined constant in G is chosen correctly. The proof is complete.

If V^L vanishes for large $|x|$ then ψ is constant on the Hamilton orbits of $P_0+V^L-\lambda=P_0-\lambda$ in $\Lambda_T^\ddagger$ far away. These stay fixed over M_λ, so there is a C^∞ function ψ_∞ on M_λ such that $\psi(x,\xi)=\psi_\infty(\xi)$ for $(x,\xi)\in\Lambda_T^\ddagger$ and $|x|$ large. Moreover, $\xi_1=\Xi(\xi')$ on $\Lambda_T^\ddagger$ for large $|x|$ if ξ is close to ξ^0, so (30.3.29) gives

$$G(x_1,\xi')=x_1\,\Xi(\xi')-\psi_\infty(\Xi(\xi'),\xi').$$

Hence $\Lambda_T^\ddagger$ is for large x_1 and ξ close to ξ^0 equal to

$$\{(x_1, -x_1\,\partial\Xi/\partial\xi'+\partial\psi_\infty(\Xi(\xi'),\xi')/\partial\xi', \Xi(\xi'),\xi')\}.$$

Now $(1, -\partial\Xi/\partial\xi')$ is the normal direction of M_λ at $(\Xi(\xi'), \xi')$. The map

$$T^*(\mathbb{R}^n)_{M_\lambda} \to T^*(M_\lambda) \tag{30.3.30}$$

obtained by restricting covectors to $T(M_\lambda)$ annihilates the normal direction and sends $\Lambda_\mp^\ddagger$ to the graph of the differential of ψ_∞. Thus we have

Theorem 30.3.7. *If V^L vanishes for large $|x|$ then the function in Theorem 30.3.6 is defined by a function ψ_∞ on M_λ when $|x|$ is large. $\Lambda_\mp^\ddagger$ is then equal to the part in the direction $N_+(M_\lambda)$ of the inverse image of the Lagrangian*

$$\{(d\psi_\infty(\xi), \xi), \xi \in M_\lambda\} \subset T^*(M_\lambda)$$

under the natural map (30.3.30). *With the notation in Theorem* 30.3.5

$$G(x_1, \xi') = x_1 \Xi(\xi') - \psi_\infty(\Xi(\xi'), \xi').$$

The end effect of a perturbation V^L of compact support is thus a shift of the Lagrangian $N_+(M_\lambda)$ by the graph of the differential of ψ_∞.

30.4. Modified Wave Operators

As observed at the end of Section 14.4 one cannot expect the wave operators

$$W_\pm = \lim_{t \to \pm\infty} e^{itH} e^{-itH_0}$$

of Theorem 14.4.6 to exist for long range perturbations. However, in this section we shall prove the existence of the limit if e^{-itH_0} is replaced by the unitary convolution operator $e^{-iW(D,t)}$ where W is obtained from Theorem 30.3.3. In this context ellipticity of the unperturbed and the perturbed operator is completely irrelevant, so we list now the weakened admissibility conditions which we shall need.

(i) $P_0(\xi)$ is a real valued non-constant polynomial of order m.

(ii) $V(x, D) = \sum_{|\alpha| \leq \mu} V_\alpha(x) D^\alpha$ has measurable coefficients,

$$\sum_\alpha \int |V_\alpha(x)|^2 (1+|x|^2)^{-M} dx < \infty$$

for some M, which makes $V(x, D)$ continuous from $\mathscr{S}$ to L^2.

(iii) $V(x, D)$ is symmetric, that is,

$$(V(x,D)u, v) = (u, V(x,D)v); \qquad u, v \in \mathscr{S}.$$

(iv) $V(x, D) = V^L(x, D) + V^S(x, D)$ where V^L has real coefficients satisfying (30.1.7)′ with $\varkappa = 2$, and for almost every ξ with $P_0'(\xi) \neq 0$ there is a conic neighborhood Γ of $P_0'(\xi)$ such that

$$\sum_\alpha \int_1^\infty \Big(\sup_{t<|x|<2t} \int_{|x-y|<1,\, y\in\Gamma} |V_\alpha^S(y)|^2\, dy\Big)^{\frac12} dt < \infty.$$

(v) The symmetric operator $P_0(D)+V(x,D)$ with domain $\mathscr{S}$ has a self-adjoint extension H in $L^2(\mathbf{R}^n)$.

By Theorem 30.2.1 these conditions are fulfilled for admissible perturbations of elliptic operators. However, allowing V_α^S to be quite large in some directions may make the existence of self-adjoint extensions problematic in general.

Theorem 30.4.1. *Assume that the preceding conditions* (i)–(v) *are fulfilled, and let $W(\xi,t)$ have the properties in Theorem 30.3.3. Then the limits*

$$\tilde W_\pm u = \lim_{t\to\pm\infty} e^{itH} e^{-iW(D,t)} u, \quad u\in L^2(\mathbf{R}^n), \tag{30.4.1}$$

exist in L^2 norm. The modified wave operators $\tilde W_\pm$ are isometric and intertwine the self-adjoint operator H_0 defined by $P_0(D)$ and H,

$$e^{isH}\tilde W_\pm = \tilde W_\pm e^{isH_0}, \quad s\in\mathbf{R}. \tag{30.4.2}$$

Proof. If we admit the existence of the limits it is obvious that $\tilde W_\pm$ are isometric, and we have

$$\begin{aligned} e^{isH}\tilde W_\pm e^{-isH_0}u &= \lim_{t\to\pm\infty} e^{i(t+s)H} e^{-iW(D,t)} e^{-isH_0} u\\ &= \lim_{t\to\pm\infty} e^{itH} e^{-iW(D,t)} e^{i(W(D,t)-W(D,t-s))} e^{-isH_0} u. \end{aligned}$$

The Fourier transform of $e^{i(W(D,t)-W(D,t-s))}e^{-isH_0}u$ is

$$\exp i(W(\xi,t)-W(\xi,t-s)-sP_0(\xi))\,\hat u(\xi). \tag{30.4.3}$$

If ξ is in a compact subset of

$$\Omega = \{\xi; P_0'(\xi) \neq 0\}$$

and t is large then

$$\partial W(\xi,t)/\partial t - P_0(\xi) = V^L(\partial W(\xi,t)/\partial\xi, \xi) = O(t^{-m(0)})$$

by (30.3.12), for

$$|\partial W(\xi,t)/\partial\xi - tP_0'(\xi)| \le Ct^{2-m(1)} = Ct^{1-\delta}$$

by (30.3.13). The complement of Ω is of measure 0, so (30.4.3) converges to $\hat u(\xi)$ almost everywhere and is bounded by $|\hat u|$. Hence

$$e^{i(W(D,t)-W(D,t-s))}e^{-isH_0}u \to u$$

in L^2, which proves (30.4.2).

To prove the existence of the limits (30.4.1) we need a lemma.

Lemma 30.4.2. *Let $\hat{u} \in C_0^\infty(\Omega)$ and let ω be a neighborhood of the set $\{P_0'(\xi);\ \xi \in \operatorname{supp} \hat{u}\}$. Then $u_t = e^{-iW(D,t)} u$ is concentrated in $t\omega$ in the sense that*

$$|u_t(x)| \leqq C_N(|x|+|t|)^{-N}, \quad N=1,2,\ldots, \text{ if } x/t \notin \omega. \tag{30.4.4}$$

Proof. By definition

$$u_t(x) = (2\pi)^{-n} \int e^{i(\langle x,\xi\rangle - W(\xi,t))} \hat{u}(\xi)\, d\xi.$$

Writing $R(\xi,t) = W(\xi,t) - t P_0(\xi)$ we have by (30.3.13)

$$|D_\xi^\alpha R(\xi,t)| \leqq C_\alpha t^{1+|\alpha|-m(|\alpha|)}, \quad \xi \in \operatorname{supp} \hat{u}. \tag{30.4.5}$$

Recall that

$$1+|\alpha|-m(|\alpha|) = 1-\delta, \quad |\alpha| \leqq 2;$$

$$1+|\alpha|-m(|\alpha|) = 1+|\alpha|-1-(1+\delta)|\alpha|/2 = c|\alpha|, \quad |\alpha| \geqq 2;$$

where $c=(1-\delta)/2<\frac{1}{2}$. We shall introduce $t^c \xi$ as a new integration variable to make the derivatives of R of order $\geqq 2$ bounded, but since $\hat{u}(\xi/t^c)$ has increasing support we must then decompose by means of a partition of unity

$$\sum \chi_g(\xi) = 1,$$

chosen according to Theorem 1.4.6. Thus g runs over all lattice points, $\chi_0 \in C_0^\infty$ and $\chi_g(\xi) = \chi_0(\xi - g)$. Set

$$\hat{u}_g(\xi) = \hat{u}(\xi/t^c)\, \chi_g(\xi)$$

which is then uniformly bounded in C_0^∞ after translation by g. We have

$$u_t(x) = (2\pi)^{-n} \sum_g \int e^{i\phi(x,\xi,t)}\, \hat{u}_g(\xi)\, d\xi / t^{nc} \tag{30.4.6}$$

where

$$\phi(x,\xi,t) = \langle x, \xi/t^c\rangle - W(\xi/t^c, t).$$

The number of terms with $\hat{u}_g \neq 0$ is $\leqq C t^{nc}$. Now

$$t^c \phi_\xi'(x,\xi,t) = x - W_\xi'(\xi/t^c, t) = x - t P_0'(\xi/t^c) - R_\xi'(\xi/t^c, t)$$

and $|R_\xi'(\xi,t)| \leqq C t^{1-\delta}$ if $\xi \in \operatorname{supp} \hat{u}$. If $x/t \notin \omega$ we have

$$|x|+|t| \leqq C|x - t P_0'(\xi)|, \qquad \xi \in \operatorname{supp} \hat{u},$$

hence

$$|x|+|t| \leqq 2C t^c |\phi_\xi'(x,\xi,t)|, \quad \xi \in \operatorname{supp} \hat{u}_g,$$

if t is sufficiently large. We also have uniform upper bounds for all ξ derivatives of $t^c \phi(x,\xi,t)/(|x|+|t|)$. From Theorem 7.7.1 we therefore obtain for every N

$$|u_t(x)| \leqq C_N((|x|+|t|)/|t|^c)^{-N}$$

and this proves (30.4.4) since

$$|x|^{1-c} \leqq (|x|+|t|)/|t|^c$$

when $t=1$, hence for every t by homogeneity.

Continued Proof of Theorem 30.4.1. Let Ω_1 be the set of all $\xi\in\Omega$ such that V^S satisfies the condition (iv) in a conic neighborhood Γ of $P_0'(\xi)$. This is of course an open set, and by hypothesis the complement is of measure 0. It suffices to prove the existence of the limit (30.4.1) when $\hat{u}\in C_0^\infty(\Omega_1)$, thus $u\in\mathscr{S}$, for this is a dense subset of L^2. Then

$$e^{itH}e^{-iW(D,t)}u=e^{itH}u_t$$

where u_t is defined as in Lemma 30.4.2. Thus u_t is a C^∞ function of t with values in $\mathscr{S}$, which proves the existence of the derivative

$$\frac{d}{dt}(e^{itH}e^{-iW(D,t)}u)=e^{itH}(iHu_t-\partial u_t/\partial t)=ie^{itH}(P_0(D)+V(x,D)-W_t'(D,t))u_t.$$

The existence of the limits (30.4.1) will follow if we show that

$$\int_{-\infty}^{\infty}\|V^S(x,D)u_t\|\,dt<\infty, \tag{30.4.7}$$

$$\int_{-\infty}^{\infty}\|(V^L(x,D)-V^L(W_\xi'(D,t),D))u_t\|\,dt<\infty. \tag{30.4.8}$$

Here we have used that (30.3.12) is valid in $\operatorname{supp}\hat{u}$ if $|t|$ is large.

The proof of (30.4.7) does not differ much from that of Theorem 14.4.6. Choose a neighborhood $\omega\Subset\mathbf{R}^n\setminus 0$ of $\{P_0'(\xi);\ \xi\in\operatorname{supp}\hat{u}\}$ such that (iv) is valid in the cone Γ generated by ω. In view of (ii) and Lemma 30.4.2, applied to $D^\alpha u$, we have for $|t|\geqq 1$

$$\int_{x/t\notin\omega}|V^S(x,D)u_t|^2\,dx\leqq C\sup(1+|x|)^{2M}(|x|+|t|)^{-2N}\leqq C/t^4,$$

if $N\geqq M+2$. Let B_x be the unit ball with center at x. Then it follows from (iv) that

$$\int_{t\omega\cap B_x}|V^S u_t|^2\,dy\leqq CI(t)^2\sum_{|\alpha|\leqq\mu}\sup_{B_x}|D^\alpha u_t|^2$$

where $I(t)$ is an integrable function of t. If $2B_x$ is the ball with radius 2 and center at x, we have by Lemma 7.6.3

$$\sum_{|\alpha|\leqq\mu}\sup_{B_x}|D^\alpha u_t|^2\leqq C\sum_{|\alpha|\leqq\mu+s}\int_{2B_x}|D^\alpha u_t|^2\,dy$$

if s is an integer $>n/2$. The integral of the right-hand side with respect to x is a finite constant since $\|D^\alpha u_t\|=\|D^\alpha u\|$ for every t. Hence

$$\|V^S(x,D)u_t\|\leqq C(I(t)+t^{-2}),\quad |t|>1,$$

and since $V^S(x,D)u_t$ is a continuous function of t with values in L^2, we have proved (30.4.7).

We have now arrived at the main point, the proof of (30.4.8). Writing $u^\alpha=D^\alpha u$ we have

$$\begin{aligned}(V^L(x,D)-V^L(W_\xi'(D,t),D))e^{-iW(D,t)}u\\=\sum_\alpha(V_\alpha^L(x)-V_\alpha^L(W_\xi'(D,t)))e^{-iW(D,t)}u^\alpha.\end{aligned} \tag{30.4.9}$$

To estimate the right-hand side we shall use the fact that

$$(x_j - W'_{\xi_j}(D,t))e^{-iW(D,t)}u^a = e^{-iW(D,t)}(x_j u^a) \tag{30.4.10}$$

which just means that the symbol of the commutator of multiplication by x_j and $e^{-iW(D,t)}$ is equal to $W'_{\xi_j}(D,t)e^{-iW(D,t)}$. The Fourier transform of $x_j u^a$ is $-D_j\widehat{u^a}$ so the support is contained in that of $\hat{u}$. We shall assume that $\operatorname{supp}\hat{u}$ is so close to a point ξ_0 with $x_0 = P'_0(\xi_0) \neq 0$ that

$$|P'_0(\xi) - x_0| < |x_0|/3, \qquad \xi \in \operatorname{supp}\chi_0,$$

for some $\chi_0 \in C_0^\infty$ equal to 1 in $\operatorname{supp}\hat{u}$. Choose $\chi \in C_0^\infty(\mathbb{R}^n)$ equal to 1 when $|x - x_0| < |x_0|/3$ so that $|x - x_0| < |x_0|/2$ if $x \in \operatorname{supp}\chi$. By Lemma 30.4.2 it suffices to estimate the product of (30.4.9) by $\chi(x/t)$, for the product by $1 - \chi(x/t)$ can be estimated by $(|t| + |x|)^{-N}$ for any N.

Let $f \in C^\infty(\mathbb{R}^n)$ be any function such as V_a^L with

$$|D^\alpha f(x)| \leq C_\alpha (1 + |x|)^{-m(|\alpha|)} \tag{30.4.11}$$

for all α. We can write

$$f(x) - f(y) = \sum (x_j - y_j) f_j(x,y), \qquad f_j(x,y) = \int_0^1 (\partial_j f)(y + s(x-y))\, ds.$$

Differentiation can be made under the integral sign, so we have for $t > 1$

$$|D^\alpha_{x,y} f_j(x,y)| \leq C_\alpha t^{-m(|\alpha|+1)} \tag{30.4.12}$$
$$\text{if } |x/t - x_0| < |x_0|/2, \quad |y/t - x_0| < |x_0|/2.$$

We recall that

$$m(|\alpha|+1) = 1 + (1+\delta)(|\alpha|+1)/2 \quad \text{if } \alpha \neq 0, \quad m(1) = 1 + \delta.$$

Now we introduce the symbol

$$G_{jt}(x,\xi) = \chi(x/t)\chi_0(\xi) f_j(x, \partial W(\xi,t)/\partial\xi).$$

For large t we have $|\partial W(\xi,t)/\partial\xi - t x_0| < t|x_0|/2$ if $\xi \in \operatorname{supp}\chi_0$, and

$$\sum_j G_{jt}(x,\xi)(x_j - \partial W(\xi,t)/\partial\xi_j) = \chi(x/t)(f(x) - f(\partial W(\xi,t)/\partial\xi))\chi_0(\xi). \tag{30.4.13}$$

As soon as we have information on the symbol properties of G_{jt} we can use this together with (30.4.10) to gain estimates of the sum in (30.4.9).

Lemma 30.4.3. *When t is large we have for all α and β*

$$|D_x^\beta D_\xi^\alpha G_{jt}(x,\xi)| \leq C_{\alpha\beta} t^{|\alpha| - m(|\alpha|+|\beta|+1)}. \tag{30.4.14}$$

Proof. Since $|\alpha| + |\beta| - m(|\alpha|+|\beta|+1)$ increases with $|\alpha|$ and $|\beta|$ it is sufficient to prove such estimates for the derivatives of $f_j(x, \partial W(\xi,t)/\partial\xi)$ when $x/t \in \operatorname{supp}\chi$ and $\xi \in \operatorname{supp}\chi_0$. When $\alpha = 0$ they follow at once from (30.4.12). By

(30.3.13) we have in supp χ_0

$$|D_\xi^\alpha \partial W(\xi,t)/\partial\xi| \leqq C_\alpha t^{a(|\alpha|)}, \qquad a(k)=1+\max(0,k+1-m(k+1)),$$

and for f_j we have the estimate (30.4.12). To estimate $D_x^\beta D_\xi^\alpha f_j(x,\partial W(\xi,t)/\partial\xi)$ we can apply Lemma 30.1.5 with $b(k)=-m(k+1+|\beta|)$ and

$$c(k)=\max(b(1)+a(k), b(k)+k\,a(1))=\max(b(1)+a(k), b(k)+k).$$

We have $c(1)=b(1)+1$, and when $k>1$ then

$$\begin{aligned}a(k)+b(1)-b(k)-k&=k+2-m(k+1)-m(2+|\beta|)+m(k+1+|\beta|)-k\\&=1-(1+\delta)(k+1+2+|\beta|-k-1-|\beta|)/2=-\delta\end{aligned}$$

since all arguments are in the linear part of m. Hence

$$c(k)=b(k)+k=k-m(k+1+|\beta|) \quad \text{if } k\geqq 1,$$

which proves (30.4.14).

End of Proof of Theorem 30.4.1. The double sequence $k-m(j+k+1)$ is convex and has slope $\leqq(1-\delta)/2$ resp. $\leqq-(1+\delta)/2$ in k and in j. Hence $k-m(j+k+1)\leqq -m(1)+k(1-\delta)/2-j(1+\delta)/2$, which proves that

$$|D_x^\beta D_\xi^\alpha G_{jt}(x,\xi)|\leqq C_{\alpha\beta}\, t^{-1-\delta}\, t^{(1-\delta)|\alpha|/2-(1+\delta)|\beta|/2}.$$

Since $|x|/t$ is bounded above and below in the support we can replace t by $(1+|x|)$ in the last factor and conclude that $G_{jt}t^{1+\delta}$ is uniformly bounded in the symbol space $S^0_{(1+\delta)/2,(1-\delta)/2}$ with the x and ξ variables interchanged. This just means that we have the conjugate by the Fourier transformation of the adjoint of a pseudo-differential operator in one of the standard classes discussed before Theorem 18.1.35, so it follows that

(30.4.15) $$\|G_{jt}(x,D)\|\leqq C t^{-1-\delta}.$$

(Alternatively we could use the results on the advanced calculus in Section 18.6.) By (30.4.13) we have

$$\sum_j G_{jt}(x,D)(x_j-W'_{\xi_j}(D,t))=\chi(x/t)(f(x)-f(W'_\xi(D,t)))\,\chi_0(D)-i\sum_j G'_{jt\xi_j}(x,D).$$

Since $1-m(2)=-1-\delta$ it follows from (30.4.14) that

$$|D_x^\beta D_\xi^\alpha D_{\xi_j} G_{jt}(x,\xi)|\leqq C_\alpha t^{-1-\delta}\, t^{(1-\delta)|\alpha|/2-(1+\delta)|\beta|/2}.$$

The norm of the corresponding operator is therefore $\leqq C t^{-1-\delta}$. If $\chi_0=1$ in supp $\hat{v}$ it follows that

$$\|\chi(x/t)(f(x)-f(W'_\xi(D,t)))\,v\|\leqq C t^{-1-\delta}(\|(x-W'_\xi(D,t))\,v\|+\|v\|).$$

If we apply this result with $f(x)=V_\alpha^L(x)$ and $v=D^\alpha u_t$, it follows in view of (30.4.10) and Lemma 30.4.2 that (30.4.8) is valid. This completes the proof of Theorem 30.4.1.

30.5. Distorted Fourier Transforms and Asymptotic Completeness

In this section we shall modify the definition of distorted Fourier transforms in Section 14.6 and establish a connection with the modified wave operators analogous to Theorem 14.6.5. To motivate the arguments it is useful to recall the definitions used in Section 14.6 in the case of short range perturbations and recast them in a form which is now more appropriate.

Thus let V be a perturbation which even vanishes outside a compact set. When λ is not in the set $Z(P_0)$ of critical values of P_0 or the set Λ of eigenvalues of $P_0(D)+V(x,D)$ we defined for $f\in B$ another $f_0\in B$ by the equation

$$u=R(\lambda+i0)f=R_0(\lambda+i0)f_0$$

and set on $M_\lambda=\{\xi; P_0(\xi)=\lambda\}$

$$F_+f(\xi)=\hat{f}_0(\xi).$$

Here R_0 is the resolvent of the unperturbed operator. Thus

$$(P_0(D)+V(x,D)-\lambda)u=f; \quad (P_0(D)-\lambda)u=f-V(x,D)u=f_0.$$

If f has compact support it follows that f_0 has compact support and that

$$\hat{u}(\xi)=(P_0(\xi)-\lambda-i0)^{-1}\hat{f}_0(\xi)$$

is a conormal distribution with respect to M_λ. The restriction of $\hat{f}_0$ to M_λ is determined by the asymptotic behavior of u at infinity. In fact, let $\chi\in C_0^\infty$ and assume that M_λ is defined by $\xi_1=\Xi(\xi')$ in $\operatorname{supp}\chi$ and that $\partial P_0(\xi)/\partial\xi_1>0$ there. For the Fourier transform of $\chi(D)u$ with respect to $x'=(x_2,\dots,x_n)$ we then obtain by Lemma 14.2.1 that in $L^2(\mathbb{R}^{n-1})$ norm

$$e^{-ix_1\Xi(\xi')}\widehat{\chi(D)u}(x_1,\xi')\to i(\chi(\xi)/P_0^{(1)}(\xi))F_+f(\xi), \quad \xi_1=\Xi(\xi'), \tag{30.5.1}$$

when $x_1\to\infty$. We shall prove that the limit of the left-hand side of (30.5.1) still exists in $L^2(\mathbb{R}^{n-1})$ for perturbations of long range provided that $x_1\Xi(\xi')$ is replaced by the function G defining the distorted normal bundle $\Lambda_\mp^\ddagger$ of M_λ in Theorem 30.3.5. When V has compact support, this follows from Theorem 30.3.7, for $G=x_1\Xi(\xi')-\psi_\infty(\Xi(\xi'),\xi')$ for large x_1, hence

$$e^{-iG(x_1,\xi')}\widehat{\chi(D)u}(x_1,\xi')\to i(\chi(\xi)/P_0^{(1)}(\xi))\tilde{F}_+f(\xi), \quad \xi_1=\Xi(\xi'), \tag{30.5.1$'$}$$

when $x_1\to\infty$, if

$$\tilde{F}_+f=e^{i\psi_\infty}F_+f \quad \text{on } M_\lambda. \tag{30.5.2}$$

Assuming from now on just that V is 2-admissible we shall use (30.5.1)$'$ to define a modified distorted Fourier transform $\tilde{F}_+$ although (30.5.2) is no longer meaningful then. Parseval's formula will follow from (30.2.31), and asymptotic completeness is an easy consequence.

To prove the existence of the limit (30.5.1)′ when $f\in B$ and $u=R(\lambda+i0)f$ we first recall that it follows from Theorem 30.2.6 that for every T

$$\int_{|x|<R,\,x_1<T} |\chi(D)u|^2\,dx/R\to 0,\qquad R\to\infty. \tag{30.5.3}$$

We shall use this property to identify $\chi(D)u$ as solution of an equation obtained by factoring the equation

$$(P_0(D)+V^L(x,D)-\lambda)u=f-V^S(x,D)u\in P$$

to a first order differential equation with respect to x_1. When making this factorization we may assume that V^L vanishes when $|x|<R$ for some large R, hence that V^L is small everywhere, for we can add to V^S the product of V^L by any cutoff function. We may still assume that V^L has fixed bounds (30.1.7)′.

Lemma 30.5.1. *Let K be a compact neighborhood of a point $\xi^0\in M_\lambda$ such that $P_0^{(1)}(\xi)>0$ in K and $M_\lambda\cap K\ni\xi\mapsto\xi'\in\mathbb{R}^{n-1}$ is bijective with range K', hence $\xi_1=\Xi(\xi')$ on M_λ in a neighborhood of $M_\lambda\cap K$. If V^L satisfies (30.1.7)′ and vanishes when $|x|<R$ for some sufficiently large R then the equation $P_0(\xi)+V^L(x,\xi)-\lambda=0$ has for ξ' in a neighborhood of K' a unique solution $\xi_1=a(x,\xi')$ close to $\Xi(\xi')$, and we have*

$$|D_x^\beta D_{\xi'}^\alpha(a(x,\xi')-\Xi(\xi'))|\le C_{\alpha\beta}(1+|x|)^{-m(|\beta|)},\qquad \xi'\in K'. \tag{30.5.4}$$

Proof. When ξ' is close to K' we want to find a small solution $\xi_1(x,\xi')$ of the equation

$$P_0(\xi_1+\Xi(\xi'),\xi')+V^L(x,\xi_1+\Xi(\xi'),\xi')=\lambda.$$

Now $\xi_1/(P_0(\xi_1+\Xi(\xi'),\xi')-\lambda)$ is a C^∞ function near $K'\times\{0\}$, and multiplication by it puts our equation in the form

$$\xi_1=F(x,\xi) \tag{30.5.5}$$

where $F(x,\xi)=-V^L(x,\xi_1+\Xi(\xi'),\xi')\,\xi_1/(P_0(\xi_1+\Xi(\xi'),\xi')-\lambda)$. From (30.1.7)′ we obtain if ξ is close to $K'\times\{0\}$

$$|D_x^\beta D_\xi^\alpha F(x,\xi)|\le C_{\alpha\beta}(1+|x|)^{-m(|\beta|)}. \tag{30.5.6}$$

In particular it follows if V^L is sufficiently small that there is a unique small solution of the equation (30.5.5) for any ξ' close to K', and

$$|\xi_1|\le C_{00}(1+|x|)^{-m(0)}.$$

Assume that we have already proved for a certain positive integer k

$$|D_x^\beta D_{\xi'}^\alpha\,\xi_1(x,\xi')|\le C_{\alpha\beta}(1+|x|)^{-m(|\beta|)},\qquad |\alpha+\beta|<k. \tag{30.5.7}$$

If $|\alpha+\beta|=k$ it follows by differentiation of (30.5.5) that

$$(1-\partial F/\partial\xi_1)\,D_x^\beta D_{\xi'}^\alpha\,\xi_1=D_x^\beta D_{\xi'}^\alpha\,F(x,\xi)-\partial F/\partial\xi_1\,D_x^\beta D_{\xi'}^\alpha\,\xi_1.$$

Here $(1-\partial F/\partial\xi_1)>\frac{1}{2}$, say. In the right-hand side we have one term where no derivative falls on ξ_1, but in all other terms a derivative of F, of order β_0 with respect to x, is multiplied by $j\geqq 2$ derivatives of ξ_1, of order $\beta_1,\dots,\beta_j$ with respect to x. Since the order of these inner derivatives is $<|\alpha+\beta|$ and $\beta=\beta_0+\dots+\beta_j$, we obtain by (30.5.6) and the inductive hypothesis a bound of the form

$$C(1+|x|)^{-m(|\beta_0|)-m(|\beta_1|)-\dots-m(|\beta_j|)}$$

for such a term. Since the only extreme points in the simplex

$$\{t\in\mathbf{R}^{j+1}; 0\leqq t_i, t_0+\dots+t_j=|\beta|\}$$

are points with one coordinate equal to $|\beta|$ and the others 0, we have the bound $C(1+|x|)^{-m(|\beta|)-jm(0)}$ in view of the convexity. This completes the inductive proof of (30.5.7) which is equivalent to (30.5.4).

The quotient

$$Q_0(\xi)=(P_0(\xi)-\lambda)/(\xi_1-\Xi(\xi'))$$

is a polynomial of degree $m-1$ in ξ_1 with coefficients in C^∞ in a neighborhood of K'. It is clear that $Q_0\neq 0$ in K since the zeros of $P_0-\lambda$ in K have been divided out.

Lemma 30.5.2. *The quotient*

$$Q(x,\xi)=(P_0(\xi)+V^L(x,\xi)-\lambda)/(\xi_1-a(x,\xi'))$$

is a polynomial in ξ_1 of degree $m-1$, and

$$|D_\xi^\alpha D_x^\beta(Q(x,\xi)-Q_0(\xi))|\leqq C_{\alpha\beta}(1+|x|)^{-m(|\beta|)},\quad \xi\in K. \tag{30.5.8}$$

Proof. By the definition of Q_0 we have

$$\begin{aligned}&P_0(\xi)+V^L(x,\xi)-\lambda-Q_0(\xi)(\xi_1-a(x,\xi'))\\&\quad=V^L(x,\xi)+Q_0(\xi)(a(x,\xi')-\Xi(\xi'))=\sum_0^m \xi_1^j\, r_j(x,\xi'),\end{aligned}$$

and it follows from (30.1.7)′ and (30.5.4) that

$$|D_\xi^\alpha D_x^\beta r_j(x,\xi')|\leqq C_{\alpha\beta}(1+|x|)^{-m(|\beta|)},\quad \xi'\in K'.$$

If we write

$$Q(x,\xi)-Q_0(\xi)=\sum_0^{m-1} q_j(x,\xi')\,\xi_1^j$$

then $q_{m-1}=r_m$ and $q_{j-1}-a q_j=r_j$ if $0<j<m$. If q_j satisfies an estimate of the form (30.5.8), we also have one for Ξq_j, and $(a-\Xi)q_j$ has an even better bound in view of Lemma 30.1.4 and (30.5.4). By induction for decreasing j we obtain (30.5.8).

Lemma 30.5.3. *If V^L vanishes when $|x|<R$ and R is large enough then*

$$|D_\xi^\alpha D_x^\beta(1/Q-1/Q_0)|\leqq C_{\alpha\beta}(1+|x|)^{-m(|\beta|)},\quad \xi\in K. \tag{30.5.9}$$

Proof. Since $|Q-Q_0|<C(1+|x|)^{-m(0)}$ we have $|Q-Q_0|<Q_0/2$ if V^L vanishes in a large enough set, and then it follows that

$$|1/Q-1/Q_0|<|Q-Q_0|/|QQ_0|<2C(1+|x|)^{-m(0)}/|Q_0|^2.$$

Thus $f=1/Q-1/Q_0$ has the required bound. Since

$$Qf=1-Q/Q_0=(Q_0-Q)(1/Q_0)$$

we have by (30.5.8)

$$|D_\xi^\alpha D_x^\beta(Qf)|\leq C_{\alpha\beta}(1+|x|)^{-m(|\beta|)}.$$

If (30.5.9) is proved for derivatives of order $<|\alpha|+|\beta|$ it follows in view of Lemma 30.1.4 that

$$|QD_\xi^\alpha D_x^\beta f|\leq C'_{\alpha\beta}(1+|x|)^{-m(|\beta|)}$$

which completes the proof.

Now recall that we plan to study $u=R(\lambda+i0)f$ when $f\in B$ and $\lambda\in\mathbf{R}\setminus(Z(P_0)\cup\Lambda)$. Then we have

$$(P_0(D)+V^L(x,D)-\lambda)u=f-V^S(x,D)u,$$

and multiplication by $(Q^{-1}\chi)(x,D)$ gives if $\chi\in C_0^\infty(K)$

$$(D_1-a(x,D'))\chi(D)u=f_\chi \tag{30.5.10}$$

where

$$f_\chi=(Q^{-1}\chi)(x,D)(f-V^S(x,D)u)+R(x,D)u, \tag{30.5.11}$$

$$R(x,D)=(D_1-a(x,D'))\chi(D)-(Q^{-1}\chi)(x,D)(P_0(D)+V^L(x,D)-\lambda). \tag{30.5.12}$$

With G_δ defined by (30.2.6) we have

$$R\in S((1+|x|)^{-1-\delta},G_\delta) \tag{30.5.13}$$

for the leading terms cancel in the asymptotic series for the symbol, and in the expansion of the symbol of

$$-(Q^{-1}\chi)(x,D)V^L(x,D)$$

all other terms are in this symbol class by (30.5.8) and (30.1.7)′. Hence $f_\chi\in B$, and since $-(Q^{-1}\chi)(x,D)V^S+R(x,D)$ is a compact operator from

$$\{u;D^\alpha u\in B^*;|\alpha|\leq m\}$$

to B, it follows from the uniformity in λ and Theorem 30.2.10 that f_χ is a continuous function of λ with values in B, if $\operatorname{supp}\chi$ is close to ξ^0, $\lambda_0=P_0(\xi^0)\notin Z(P_0)\cup\Lambda$, and λ is close to λ_0. Then we have

$$\|f_\chi\|_B\leq C\|f\|_B.$$

Also recall that $\chi(D)u$ satisfies (30.5.3). We shall now switch to a study of (30.5.10). Choose $\psi\in C_0^\infty(\mathbf{R}^{n-1})$ so that $\psi(\xi')=1$ in a neighborhood of K' and

a, Ξ are defined near supp ψ. Then $\tilde a=\psi a$, $\tilde\Xi=\psi\Xi$ are defined in $\mathbb{R}^{2n-1}$, (30.5.4) is valid for $\tilde a-\tilde\Xi$, and we can replace a by $\tilde a$ in (30.5.10) provided that supp χ is so small that $\psi=1$ there.

Lemma 30.5.4. *If $v\in B^*$, $g\in B$, and*

$$D_1 v-\tilde a(x,D')v=g, \tag{30.5.14}$$

$$\int_{|x|<R,\, x_1<T} |v|^2\,dx=o(R) \quad \textit{for every } T, \tag{30.5.15}$$

then

$$\|v(x_1,.)\|_{L^2}\le C\int_{-\infty}^{x_1}\|g(t,.)\|_{L^2}\,dt. \tag{30.5.16}$$

For every $g\in L^1(\mathbb{R},L^2(\mathbb{R}^{n-1}))$ there is a unique solution (30.5.14) *satisfying* (30.5.16).

Proof. Recall that by Theorem 14.1.2 we have

$$B\subset L^1(\mathbb{R},L^2(\mathbb{R}^{n-1})),\qquad \text{hence } B^*\supset L^\infty(\mathbb{R},L^2(\mathbb{R}^{n-1})).$$

Hence (30.5.15) follows from (30.5.16) so the uniqueness in the last statement follows if we show that (30.5.14), (30.5.15) imply (30.5.16) and take $g=0$. In doing so we first assume that v has compact support and that v is a C^1 function of x_1 with values in $L^2(\mathbb{R}^{n-1})$. Then

$$\begin{aligned}\frac{\partial}{\partial x_1}\|v(x_1)\|^2&=2\operatorname{Re}(iD_1 v,v)=2\operatorname{Re}(ig,v)+2\operatorname{Re}(i\tilde a(x,D')v,v)\\&\le 2\|g(x_1)\|\,\|v(x_1)\|+2C(1+|x_1|)^{-m(1)}\|v(x_1)\|^2,\end{aligned}$$

where we have written for example $v(x_1)$ for $v(x_1,.)\in L^2(\mathbb{R}^{n-1})$. In fact,

$$\tilde a(x,D')-\tilde a(x,D')^*\in \operatorname{Op} S((1+|x|)^{-m(1)},G_\delta)\subset \operatorname{Op} S((1+|x_1|)^{-m(1)},G_\delta)$$

so the norm as operator in $L^2(\mathbb{R}^{n-1})$ for fixed x_1 is $\le 2C(1+|x_1|)^{-m(1)}$. Hence

$$\partial\|v(x_1)\|/\partial x_1\le\|g(x_1)\|+C(1+|x|)^{-m(1)}\|v(x_1)\|.$$

Multiplication by $e^{-CE(x_1)}$ where

$$E(x_1)=\int_{-\infty}^{x_1}(1+|t|)^{-m(1)}\,dt$$

gives after integration

$$\|v(x_1)\|\le\int_{-\infty}^{x_1}e^{C(E(x_1)-E(t))}\|g(t)\|\,dt$$

which proves (30.5.16) since E is bounded.

If v is just assumed to have compact support we can still conclude from (30.5.14) that $D_1 v\in L^2$ so v is continuous with values in L^2. Set

$$v_\varepsilon(x)=\int_0^\varepsilon v(x_1-t,x')\,dt/\varepsilon.$$

Applying the part of the lemma which is already proved we obtain

$$\|v_\varepsilon(x_1,.)\|_{L^2} \leq C\left(\int_{-\infty}^{x_1} \|g(t,.)\|\, dt + \int_{-\infty}^{x_1} \|\tilde{a}(x,D')(v_\varepsilon - v)\|\, dt\right).$$

The last integral converges to 0 when $\varepsilon \to 0$ since $v_\varepsilon \to v$ in L^2 and $\tilde{a}(x,D')$ is L^2 continuous. Hence (30.5.16) remains valid in this case.

Now we drop the assumption that v has compact support and just assume (30.5.15). Let $\Psi \in C_0^\infty$ be equal to 1 in the unit ball and set

$$v^R(x) = \Psi(x/R)\, v(x).$$

Then

(30.5.17) $$(D_1 - \tilde{a}(x,D'))\, v^R(x) - \Psi(x/R)\, g = R^{-1} v(x)(\partial_1 \Psi)(x/R) - [\tilde{a}(x,D'), \Psi(x/R)]\, v.$$

The symbol of

$$R^{\frac{1}{2}}([\tilde{a}(x,D'), \Psi(x/R) - 1] + i R^{-1} \sum_2^n \tilde{a}^{(j)}(x,D')(\partial_j \Psi)(x/R))$$

is bounded in $S((1+|x|)^{-\frac{1}{2}}, G_s)$ when $R \to \infty$ since $R^{\frac{1}{2}}(\Psi(x/R)-1)$ and $R^{-\frac{1}{2}}(\partial_j \Psi)(x/R)$ are bounded in $S((1+|x|)^{\pm\frac{1}{2}}, G_1)$ respectively. Application to $v \in B^*$ gives functions bounded in B, hence in $L^1(L^2)$. Now $R^{-1}(\partial_j \Psi)(x/R)\, v \to 0$ in $L^1((-\infty, T), L^2(\mathbb{R}^{n-1}))$ by Cauchy-Schwarz' inequality and (30.5.15), so the right-hand side of (30.5.17) tends to 0 in this space. Hence (30.5.16) applied to v^R gives (30.5.16) when $R \to \infty$. The proof of (30.5.16) is now complete.

For the adjoint operator L^* of $L = D_1 - \tilde{a}(x,D')$ we have by the first part of the proof

(30.5.16)′ $$\|\phi(x_1)\| \leq C \int_{x_1}^{\infty} \|(L^*\phi)(t)\|\, dt, \qquad \phi \in C_0^\infty(\mathbb{R}^n).$$

To solve the equation $Lv = g \in L^1(\mathbb{R}, L^2(\mathbb{R}^{n-1}))$ we must find v so that

$$(v, L^*\phi) = (g, \phi), \qquad \phi \in C_0^\infty(\mathbb{R}^n).$$

By the estimate (30.5.16)′ we have

$$|(g,\phi)| \leq C \iint_{t > x_1} \|(L^*\phi)(t)\|\, \|g(x_1)\|\, dt\, dx_1.$$

Hence the antilinear map $L^*\phi \mapsto (g,\phi)$ can be extended from $L^* C_0^\infty$ to an antilinear form on L^2_{comp} with the same bound, that is, we can find $v \in L^2_{\text{loc}}$ with $Lv = g$ and

$$|(v,\phi)| \leq C \iint_{t > x_1} \|\phi(t)\|\, \|g(x_1)\|\, dt\, dx_1, \qquad \phi \in L^2_{\text{comp}}.$$

But this means that

$$\|v(t)\| \leq C \int_{-\infty}^{t} \|g(x_1)\|\, dx_1$$

for almost all t and completes the proof of the lemma.

Remark. The results of Section 30.2 can equally well be developed with Lemma 30.5.4 as starting point instead of Proposition 30.2.4. However, the proofs chosen in Section 30.2 seem preferable since they work directly in a full neighborhood of M_λ and give better geometrical insight. A full development of the approach in Lemma 30.5.4 also requires factorization of the symbol P_0+V^L-z for $\operatorname{Im} z \neq 0$.

If the number T in Theorem 30.3.5 is chosen large enough then the function G there is defined for $\xi' \in \operatorname{supp}\psi$ and $x_1 \geqq x_1^0$, say. Since $P_0(\xi) + V^L(x,\xi)-\lambda=0$ in $\Lambda_T^\ddagger$ we have in this set

$$\partial G(x_1,\xi')/\partial x_1 - a(x_1, -\partial G/\partial\xi', \xi')=0. \tag{30.5.18}$$

Set $\tilde G(x_1,\xi')=\psi(\xi')\,G(x_1,\xi')$, defined as 0 outside $\operatorname{supp}\psi$. Then $\tilde G$ is defined when $x_1 \geqq x_1^0$, and $\tilde G = G$ when $\psi=1$. By (30.3.24) we have

$$|D_{\xi'}^{\alpha'} D_{x_1}^{\alpha_1}(\tilde G(x_1,\xi')-x_1\tilde\Xi(\xi'))| \leqq C_\alpha x_1^{1+|\alpha'|-m(|\alpha|)}. \tag{30.5.19}$$

Now we introduce the functions

$$\begin{aligned} A_1(x,\xi)&=\xi_1-\partial\tilde G(x_1,\xi')/\partial x_1,\\ A_j(x,\xi)&=x_j+\partial\tilde G(x_1,\xi')/\partial\xi_j, \quad j\neq 1,\end{aligned} \tag{30.5.20}$$

which vanish on the Lagrangian $\Lambda_T^\ddagger$ over K when $x_1>x_1^0$. Their Poisson brackets are zero, and since operators with symbols depending only on x_1 and ξ' commute it follows that

$$[A_j(x,D), A_k(x,D)]=0; \quad j,\ k=1,\dots,n. \tag{30.5.21}$$

We shall estimate

$$v_j=A_j(x,D)\,v \tag{30.5.22}$$

first for $j\neq 1$, then for $j=1$, when v is a solution of an equation closely related to (30.5.14). The symbol of the operator there is

$$\xi_1-\tilde a(x,\xi')=A_1(x,\xi)+G_{(1)}(x_1,\xi')-a(x,\xi')$$

when $\psi(\xi')=1$. It follows from (30.5.18) that $G_{(1)}(x_1,\xi')-a(x,\xi')$ vanishes when $A_j(x,\xi')=0$ for $j\neq 1$. We shall use this to factor the symbol. In stating the result it is convenient to use the metric

$$g_\delta=|dx'|^2(1+|x|)^{-1-\delta}+|d\xi'|^2(1+|x|)^{1-\delta}$$

where x_1 occurs as a parameter. It is of course the same metric as in the proof of Theorem 30.4.1 although ξ_1 does not occur now and x_1 is relegated to the role of a parameter. However, this makes it obvious that we have a uniformly σ temperate metric.

Lemma 30.5.5. *$D_{\xi'}^{\alpha} D_{x'}^{\beta} A_j(x,\xi)$ is uniformly bounded in $S((1+|x|)^{1-|\beta|}, g_\delta)$ when $j>1$, $|\alpha+\beta|\leqq 1$ and $x_1\geqq x_1^0$. There exist functions $B_j(x,\xi')$ such that $D_{\xi'}^{\alpha} D_{x'}^{\beta} B_j$*

is bounded in $S(x_1^{-\delta}(1+|x|)^{-1-|\beta|}, g_\delta)$ *when* $|\alpha+\beta|\leqq 1$, $x_1\geqq x_1^0$, *and*

$$G_{(1)}(x_1,\xi')-a(x,\xi')=\sum_2^n B_j(x,\xi')\,A_j(x,\xi') \tag{30.5.23}$$

when $\psi(\xi')=1$.

Proof. The statement on A_j is obvious when $|\beta|=1$. When $\beta=0$ it follows from the fact that by (30.5.19)

$$|D_{\xi'}^\alpha A_k(x,\xi')|\leqq C_\alpha x_1^{\max(2+|\alpha|-m(1+|\alpha|),\,1)}, \tag{30.5.24}$$

for the exponent is $\leqq 1+(1-\delta)(|\alpha|-1)/2$ when $|\alpha|\geqq 1$. It suffices to prove that there exist functions B_j satisfying (30.5.23) and the required estimates when $\xi'\in\operatorname{supp}\psi$, for if they are multiplied by ψ all the required properties will follow.

By (30.5.18) we may replace $G_{(1)}(x_1,\xi')$ by $a(x_1,-\partial\tilde G/\partial\xi',\xi')$ in (30.5.23) Taylor's formula gives

$$a(x_1,-\partial\tilde G/\partial\xi',\xi')-a(x,\xi')=\sum_2^n c_j(x,\xi')\,A_j(x,\xi')$$

where

$$c_j(x,\xi')=-\int_0^1 a_{(j)}(x_1,x_2-s\,A_2(x,\xi'),\dots,x_n-s\,A_n(x,\xi'),\xi')\,ds. \tag{30.5.25}$$

The solution $B_j=c_j$ of (30.5.23) will be used when $\sum A_k^2<4x_1^2$. When $\sum A_k^2>x_1^2$ we shall also use the solution

$$d_j(x,\xi')=(G_{(1)}(x_1,\xi')-a(x,\xi'))\,A_j(x,\xi')\Big/\sum_2^n A_k(x,\xi')^2.$$

To piece the two solutions together we choose $\phi\in C_0^\infty(\{x';|x'|<2\})$ equal to 1 in the unit ball and set

$$\Phi(x,\xi')=\phi(A_2(x,\xi')/x_1,\dots,A_n(x,\xi')/x_1).$$

First we shall estimate c_j in $\operatorname{supp}\Phi$. Differentiation of (30.5.25) gives

$$D_{x'}^\beta c_j(x,\xi')=-\int_0^1 (1-s)^{|\beta|}\,a_{(j,\beta)}(x_1,x_2-s\,A_2,\dots,x_n-s\,A_n,\xi')\,ds.$$

In view of (30.5.4) we have

$$|D_{x'}^{\beta'}D_{\xi'}^{\alpha'}a_{(j,\beta)}|\leqq C_{\alpha'\beta'}\,x_1^{-m(|\beta'|+|\beta|+1)},$$

so we can apply Lemma 30.1.5 with $x_1=t$ and $a(i)=\max(1,2+i-m(i+1))$, $b(i)=-m(i+|\beta|+1)$, noting that for $i>1$

$$\begin{aligned} b(i)+i\,a(1)-b(1)-a(i)&=-m(i+|\beta|+1)+i+m(|\beta|+2)-2-i+m(1+i)\\ &=m(1+i)-m(i+|\beta|+1)+m(|\beta|+2)-2\\ &\geqq m(2)-2>0.\end{aligned}$$

Thus $c(i)=b(i)+i$ with the notation of Lemma 30.1.5, and it follows that

$$|D_{\xi'}^\alpha D_{x'}^\beta c_j(x,\xi')|\leqq C_{\alpha\beta}\,x_1^{|\alpha|-1-(|\alpha|+|\beta|+1)(1+\delta)/2},\qquad |\alpha+\beta|>0,$$

in supp Φ provided that $a_{(j)}$ is independent of ξ'; if not we obtain terms with an even better estimate when some derivatives fall on the ξ' arguments. It is also obvious that in supp Φ we have

$$|c_j(x,\xi')| \leq C x_1^{-1-\delta}.$$

Since A_j/x_1 is bounded in $S(1,g_\delta)$ in supp Φ it is clear that Φ is bounded in $S(1,g_\delta)$. If we differentiate once and use the statement on A_j already proved in the lemma it follows that $D_{\xi'}^\alpha D_{x'}^\beta \Phi$ is bounded in $S((1+|x|)^{-|\beta|},g_\delta)$ when $|\alpha+\beta|=1$, which proves that Φc_j has the estimates stated for B_j in the lemma.

We shall now estimate d_j in $\operatorname{supp}(1-\Phi)$. Writing

$$G_{(1)}-a=G_{(1)}-\Xi+\Xi-a$$

we obtain using (30.3.24) and (30.5.4)

$$|D_{\xi'}^\alpha(G_{(1)}-a)| \leq C_\alpha x_1^{1+|\alpha|-m(|\alpha|+1)},$$

for $|x|^{-m(0)} \leq x_1^{-m(0)} \leq x_1^{1+|\alpha|-m(|\alpha|+1)}$. We also have by (30.5.4)

$$|D_{\xi'}^\alpha D_{x'}^\beta(G_{(1)}-a)| \leq C_{\alpha\beta}(1+|x|)^{-m(|\beta|)}, \quad \beta \neq 0.$$

Thus $D_{\xi'}^\alpha D_{x'}^\beta(G_{(1)}-a)$ is bounded in $S(x_1^{-\delta}(1+|x|)^{-|\beta|},g_\delta)$ when $|\alpha+\beta| \leq 1$. To prove that $(1-\Phi)d_j$ has the required bounds it suffices now to show that

$$D_{\xi'}^\alpha D_{x'}^\beta \left((1-\Phi)\left(\sum_2^n A_k^2\right)^{-1}\right) \text{ is bounded in } S((1+|x|)^{-2-|\beta|},g_\delta),$$

when $|\alpha+\beta| \leq 1$. This follows at once from the properties of A_k listed in the theorem, which completes the proof.

If $\psi=1$ in supp $\hat{v}$ it follows from (30.5.14) and (30.5.23) that

$$A_1(x,D)v+\sum_2^n B_j(x,D')A_j(x,D')v=g+T(x,D')v, \tag{30.5.26}$$

$$T(x,D')=\sum_2^n (B_j(x,D')A_j(x,D')-(B_jA_j)(x,D')). \tag{30.5.27}$$

The following lemma gives an estimate for T and for the operators

$$\begin{aligned} R_{kj}(x,D') &= [A_k(x,D'),B_j(x,D')], \\ R_j(x,D') &= B_j(x,D')A_j(x,D')-A_j(x,D')B_j^*(x,D'). \end{aligned} \tag{30.5.28}$$

Lemma 30.5.6. *For $x_1 \geq x_1^0$ we have with operator norm in $L^2(\mathbb{R}^{n-1})$*

$$\begin{aligned} &\|T(x,D')\| + \sum_{j,k=2}^n \|R_{kj}(x,D')\| \\ &\quad + \sum_2^n (\|B_j(x,D')\| + \|R_j(x,D')\|) \leq C x_1^{-1-\delta}. \end{aligned} \tag{30.5.29}$$

Proof. It follows from Lemma 30.5.5 that the symbols of T, R_{kj}, B_j and $R_j = -R_{jj} + A_j(B_j - B_j^*)$ are bounded in $S(x_1^{-\delta}(1+|x|)^{-1}, g_\delta)$, hence in $S(x_1^{-1-\delta}, g_\delta)$. By Theorem 18.6.3 this implies (30.5.29).

For our solution $v = \chi(D)u$ of (30.5.10) we know by Lemma 30.5.4 that $v \in L^\infty(\mathbb{R}, L^2(\mathbb{R}^{n-1}))$, and it follows from (30.5.29) that v satisfies (30.5.26) with right-hand side $h = f_\chi + T(x, D')\, v \in L^1((x_1^0, \infty), L^2(\mathbb{R}^{n-1}))$. Below we shall study (30.5.26) when the right-hand side h and $v(x_1^0)$ are small at infinity. The following lemma will allow us to use density arguments to pass to arbitrary $h \in L^1(\mathbb{R}, L^2(\mathbb{R}^{n-1}))$ and $v_0 \in L^2(\mathbb{R}^{n-1})$.

Lemma 30.5.7. *For every $h \in L^1_{\text{loc}}([x_1^0, \infty), L^2(\mathbb{R}^{n-1}))$ and $v_0 \in L^2(\mathbb{R}^{n-1})$ there is a unique solution in $C([x_1^0, \infty), L^2(\mathbb{R}^{n-1}))$ of the Cauchy problem*

$$Lv = A_1(x, D)\, v + \sum_2^n B_j(x, D')\, A_j(x, D')\, v = h \quad \text{for } x_1 > x_1^0, \qquad v = v_0 \quad \text{when } x_1 = x_1^0, \tag{30.5.30}$$

and we have the estimate

$$\|v(x_1)\| \le C\left(\int_{x_1^0}^{x_1} \|h(t)\|\, dt + \|v_0\|\right), \quad x_1 > x_1^0. \tag{30.5.31}$$

Proof. The existence proof in Lemma 30.5.4 only used the fact that

$$\int \|\tilde{a}(x, D') - \tilde{a}(x, D')^*\|\, dx_1 < \infty,$$

where the norm is operator norm in $L^2(\mathbb{R}^{n-1})$. Since $\tilde{G}_{(1)}(x, D')$ is self adjoint and

$$\|B_j(x, D')\, A_j(x, D') - (B_j(x, D')\, A_j(x, D'))^*\| = \|R_j(x, D')\| < C x_1^{-1-\delta}$$

by (30.5.29), it follows that for arbitrary $\psi \in L^1_{\text{comp}}([x_1^0, \infty), L^2(\mathbb{R}^{n-1}))$ the equation $L^* \phi = \psi$ has a unique solution when $x_1 > x_1^0$ such that

$$\|\phi(x_1)\| \le C \int_{x_1}^\infty \|\psi(t)\|\, dt.$$

Thus $\phi(x_1) = 0$ for large x_1. If v satisfies (30.5.30) then

$$\int_{x_1^0}^\infty (v(t), \psi(t))\, dt = \int_{x_1^0}^\infty (h(t), \phi(t))\, dt - i(v_0, \phi(x_1^0)). \tag{30.5.31$'$}$$

The right-hand side depends linearly on ψ and can be estimated by

$$\left(\int_{x_1^0}^{x_1} \|h(t)\|\, dt + \|v_0\|\right) C \int_{x_1^0}^\infty \|\psi(t)\|\, dt \quad \text{if } \psi = 0 \text{ in } (x_1, \infty),$$

so (30.5.31)′ defines an element $v \in L^\infty_{\text{loc}}([x_1^0, \infty), L^2(\mathbb{R}^{n-1}))$ satisfying (30.5.31). When $\psi = L^* \phi$ with $\phi \in C_0^\infty((x_1^0, \infty) \times \mathbb{R}^{n-1})$ it follows that $Lv = h$. Hence v is continuous with values in $L^2(\mathbb{R}^{n-1})$. Taking $\psi \in C_0^\infty$ we obtain $v(x_1^0) = v_0$.

It follows from Lemma 30.5.7 that any solution of (30.5.30) with $h\in L^1(L^2)$ can be uniformly approximated by a solution of (30.5.30) with h and v_0 of compact support. This limits the growth when $x'\to\infty$:

Lemma 30.5.8. *If in addition to the hypothesis in Lemma 30.5.7 we have $|x'|h\in L^1(L^2)$ and $|x'|v_0\in L^2$ then*

$$(\int |x'|^2 |v(x_1,x')|^2\,dx')^{\frac12} \leq C\int_{x_1^0}^{x_1} dt(\int (|x'|^2+(x_1-t)^2)|h(t,x')|^2\,dx')^{\frac12}$$
$$+C(x_1-x_1^0)\|v_0\|+C\||x'|v_0\|.$$

Proof. Let $1<j\leq n$ and set $\psi_\varepsilon(x)=x_j(1+|\varepsilon x'|^2)^{-\frac12}$. For fixed ε this is a bounded function, and $\partial\psi_\varepsilon(x)/\partial x_k$ is uniformly bounded in $S(1,g_\delta)$. We have

$$L(\psi_\varepsilon v)=\psi_\varepsilon h+[A_1,\psi_\varepsilon]v+\sum(B_j[A_j,\psi_\varepsilon]+[B_j,\psi_\varepsilon]A_j)v.$$

For $j\neq 1$ the symbol of $[A_j,\psi_\varepsilon]$ is bounded in $S((1+|x|),g_\delta)$ and that of $[B_j,\psi_\varepsilon]$ is bounded in $S(x_1^{-\delta}(1+|x|)^{-1},g_\delta)$, so the symbol of the sum is bounded in $S(x_1^{-\delta},g_\delta)$, all by Lemma 30.5.5. Moreover, $[A_1,\psi_\varepsilon]$ is bounded in $S(1,g_\delta)$. Hence it follows from (30.5.31) that

$$\|\psi_\varepsilon v(x_1)\|\leq C\left(\int_{x_1^0}^{x_1}(\|\psi_\varepsilon h(t)\|+\|v(t)\|)\,dt+\|\psi_\varepsilon v_0\|\right).$$

When $\varepsilon\to 0$ we obtain by another application of (30.5.31)

$$\|x_j v(x_1)\|\leq C\left(\int_{x_1^0}^{x_1}\|x_j h(t)\|\,dt+C\int_{x_1^0}^{x_1}(x_1-t)\,\|h(t)\|\,dt+C(x_1-x_1^0)\,\|v_0\|+\|x_j v_0\|\right).$$

This completes the proof of the lemma.

We are now prepared to determine the asymptotic behavior of the solutions of (30.5.30).

Lemma 30.5.9. *Let v be a solution of (30.5.30) and assume that $|x|h\in L^1((x_1^0,\infty),L^2(\mathbb{R}^{n-1}))$ and that $(1+|x'|)v_0\in L^2(\mathbb{R}^{n-1})$. Then*

$$\sum_2^n \|A_j(x,D')v\|_{L^2(\mathbb{R}^{n-1})}\leq CM,\quad x_1>x_1^0, \tag{30.5.32}$$

where

$$M=\int_{x_1^0}^{\infty}\sum_2^n\|A_j(x,D')h\|_{L^2(\mathbb{R}^{n-1})}\,dx_1+\sum_2^n\|A_j(x_1^0,x',D')v_0\|_{L^2(\mathbb{R}^{n-1})}, \tag{30.5.33}$$

and we have

$$\|A_1(x,D)\|_{L^2(\mathbb{R}^{n-1})}\leq C(\|h(x_1)\|+x_1^{-1-\delta}M). \tag{30.5.34}$$

The limit

$$v_\infty = \lim_{x_1\to\infty} e^{-i\tilde{G}(x_1,D')} v(x_1,.) \tag{30.5.35}$$

exists in $L^2(\mathbb{R}^{n-1})$ *and*

$$\left(\int (1+|x'|)^2 |v_\infty(x')|^2\,dx'\right)^{\frac12} \leqq C(\|(1+|x|)h\|_{L^1(L^2)} + \|(1+|x'|)v_0\|). \tag{30.5.36}$$

Proof. With $v_k = A_k(x,D)v$ as in (30.5.22) we have for $k \neq 1$

$$A_1(x,D)v_k + \sum_2^n B_j(x,D')A_j(x,D')v_k + \sum_2^n R_{kj}(x,D')v_j = A_k(x,D')h = h_k$$

where R_{kj} is defined by (30.5.28), for A_j and A_k commute. As in the proof of Lemma 30.5.7 we have a quite harmless perturbation of the equation (30.5.14) but to emphasize this point we repeat the first part of the proof of Lemma 30.5.4. It follows from Lemma 30.5.8 that v_k is locally bounded with values in L^2. Writing $V(x_1)^2 = \sum_2^n \|v_k(x_1)\|^2$, we have

$$\begin{aligned}\partial V(x_1)^2/\partial x_1 &= 2\,\mathrm{Re}\sum (iD_1 v_k, v_k)\\ &= 2\,\mathrm{Re}\Big(\sum(ih_k,v_k) - \sum(iR_{kj}(x,D')v_j,v_k)\Big) - i\sum(R_j(x,D')v_k,v_k).\end{aligned}$$

With $H(x_1)^2 = \sum_2^n \|h_k(x_1)\|^2$ we conclude using (30.5.29) that

$$\partial V(x_1)/\partial x_1 \leqq H(x_1) + C(1+|x_1|)^{-1-\delta} V(x_1)$$

which gives after integration

$$V(x_1) \leqq C'\left(\int_{x_1^0}^{x_1} H(t)\,dt + V(x_1^0)\right).$$

This implies (30.5.32). If we use the equation

$$A_1(x,D)v = h - \sum_2^n B_j(x,D')A_j(x,D')v$$

and the estimate (30.5.29) of $\|B_j\|$, we obtain (30.5.34).

Now we set

$$w = e^{-i\tilde{G}(x_1,D')}v.$$

Then (30.5.32) and (30.5.34) can be rewritten in the form

$$\begin{aligned}&\sum_2^n \|x_j w\|_{L^2(\mathbb{R}^{n-1})} \leqq CM,\\ &\|\partial w/\partial x_1\|_{L^2(\mathbb{R}^{n-1})} \leqq C(\|h(x_1)\| + x_1^{-1-\delta}M).\end{aligned}$$

It follows at once that w has a limit v_∞ as $x_1 \to \infty$ which satisfies (30.5.36). The proof is complete.

In view of the estimate (30.5.31) it follows at once that the limit (30.5.35) exists in L^2 for every $v_0 \in L^2(\mathbb{R}^{n-1})$ and every $h \in L^1((x_1^0, \infty), L^2(\mathbb{R}^{n-1}))$. Returning to the equation (30.5.10) we have therefore proved that if $u = R(\lambda + i0)f$ where $f \in B$ and $\lambda \in \mathbb{R} \setminus (Z(P_0) \cup \Lambda)$, then $e^{-iG(x_1, D')}\chi(D)u$ has a limit in L^2 as $x_1 \to \infty$. Taking Fourier transforms in x' we have proved the existence of the limit in (30.5.1)'. (The reason why we had to replace equation (30.5.14) by equation (30.5.26) in the arguments above is that we see no simple way of describing the right-hand sides of (30.5.14) for which limits exist in the strong sense of Lemma 30.5.9; commutation with A_k leads to some error terms which cannot be easily controlled.)

To be able to use (30.5.1)' to define $\tilde{F}_+ f$ we must show that the limits obtained from different local representations of $\Lambda_T^\pm$ and different cutoff functions χ are compatible with (30.5.1)'. We must also prove a relation between $\tilde{F}_+$, a modified wave operator $\tilde{W}_+$ and the Fourier transformation analogous to Theorem 14.6.5. All this will be done by passage to the limit from the case of perturbations of compact support.

Let $\rho \in C_0^\infty(\mathbb{R}^n)$ be real valued and equal to 1 in the unit ball, and set for large j

$$V_j(x, D)u = \rho(x/j)\, V(x, D)\, \rho(x/j)\, u.$$

V_j is then a symmetric operator, and $V_j = V_j^S + V_j^L$ where

$$V_j^L(x, \xi) = \rho(x/j)^2\, V^L(x, \xi)$$

satisfies (30.1.7)' uniformly with respect to j, and V_j^S satisfies (30.1.6) uniformly too. It follows that Theorem 30.2.5 is valid uniformly for $H_j = P_0(D) + V_j(x, D)$ when j is large. This is a self-adjoint operator with domain $H_{(m)}$. Fix a compact interval $I \subset \mathbb{R} \setminus (Z(P_0) \cup \Lambda)$. The compactness argument used in the proof of Theorem 30.2.10 shows that for large j the boundary values $R_j(\lambda \pm i0)$ of the resolvent of H_j exist when $\lambda \in I$, and that

$$\sum_{|\alpha| \le m} \|D^\alpha R_j(\lambda \pm i0) f\|_{B^*} \le C \|f\|_B \quad \text{if } f \in B,\ \lambda \in I.$$

When $j \to \infty$ we have $D^\alpha R_j(\lambda \pm i0)f \to D^\alpha R(\lambda \pm i0)f$ weakly in B^*, so with the notation in (30.5.11)

$$(Q^{-1}\chi)(x, D)(f - V^S(x, D)\, R_j(\lambda + i0)f) + R(x, D)\, R_j(\lambda + i0)f \to f_\chi$$

in B when $j \to \infty$, uniformly with respect to λ. We can trace this convergence through the proofs of Lemmas 30.5.4 to 30.5.9 and conclude that if G_j is the function replacing G when V^L is replaced by V_j^L then

$$\lim_{j \to \infty} \lim_{x_1 \to \infty} e^{-iG_j(x_1, \xi')}\, \widehat{\chi(D)\, R_j(\lambda + i0) f}(x_1, \xi') = \lim_{x_1 \to \infty} e^{-iG(x_1, \xi')}\, \widehat{\chi(D)\, R(\lambda + i0) f}(x_1, \xi'). \tag{30.5.37}$$

Here the number T in Theorem 30.3.5 is chosen large enough to apply to V_j^L for all j and all $\lambda \in I$. We leave the details of the proof to the reader.

At the beginning of the section we observed that $\tilde{F}_+^j f$ can be defined according to (30.5.2) so that with convergence in $L^2(\mathbb{R}^{n-1})$, unform in λ,

$$\lim_{x_1\to\infty} e^{-iG_j(x_1,\xi')}\overline{\chi(D)R_j(\lambda+i0)f}(x_1,\xi') = i(\chi(\xi)/P_0^{(1)}(\xi))\tilde{F}_+^j f(\xi) \quad \text{if } P_0(\xi)=\lambda\in I.$$

Hence it follows from (30.5.37) that $\chi\tilde{F}_+^j$ converges in $L^2(M_\lambda)$ for every $\lambda\in I$ when $j\to\infty$. For large enough T the same result is of course valid for every term in a fixed sufficiently fine partition of unity $\sum\chi_\nu=1$ on the compact set $P_0^{-1}(I)$. Thus

$$\lim \tilde{F}_+^j f=\tilde{F}_+ f \tag{30.5.38}$$

exists in $L^2(M_\lambda)$ for every $\lambda\in I$ if $f\in B$, and $\tilde{F}_+ f$ is characterized by the validity of (30.5.1)′ for each term in the partition of unity on $P_0^{-1}(I)$, with G replaced by a local defining phase function of $\Lambda_T^\pm$ depending on $x_\nu\to\pm\infty$ and ξ_k for $k\neq\nu$, for some $\nu=1,\dots,n$. Since $\tilde{F}_+^j f\in L^2(P_0^{-1}(I))$ we have $\tilde{F}_+ f\in L^2(P_0^{-1}(I))$.

We shall now prove that

$$2\operatorname{Im}(R(\lambda+i0)f,f)=(2\pi)^{1-n}\int_{M_\lambda}|\tilde{F}_+ f|^2\,dS/|P_0'(\xi)|;\quad f\in B,\quad \lambda\in I. \tag{30.5.39}$$

To do so it suffices in view of (30.5.38) to show that

$$2\operatorname{Im}(R_j(\lambda+i0)f,f)=(2\pi)^{1-n}\int_{M_\lambda}|\tilde{F}_+^j f|^2\,dS/|P_0'(\xi)|;\quad f\in B,\ \lambda\in I. \tag{30.5.39$'$}$$

In the right-hand side we may replace $\tilde{F}_+^j$ by F_+^j in view of (30.5.2). Set

$$u=R_j(\lambda+i0)f.$$

Then

$$(P_0(D)-\lambda)u=f-V_j(x,D)u=f_0,\quad F_+^j f(\xi)=\hat{f}_0(\xi),\quad \xi\in M_\lambda.$$

We have $u=R_0(\lambda+i0)f_0$ and Theorem 14.3.8 with $v_-=0$, $v_+=2\pi i\hat{f}_0/|P_0'|$ proves (30.5.39)′. (See also the proof of (14.6.1)′.)

If we extend the definition of $\tilde{F}_+ f$ so that $\tilde{F}_+ f(\xi)=0$ when $P_0(\xi)\notin I$, it follows from (30.2.31) and (30.5.39) that

$$\|E(I)f\|^2=(2\pi)^{-n}\int|\tilde{F}_+ f|^2\,d\xi,\quad f\in B. \tag{30.5.40}$$

Here $E(I)$ is the spectral projection of H corresponding to the interval I. Hence the map $f\mapsto\tilde{F}_+ f$ can be extended to L^2 so that (30.5.40) remains valid, which means that $\tilde{F}_+$ vanishes on $E(\mathbb{R}\setminus I)L^2$ and is isometric on $E(I)L^2$.

We shall now examine the modified wave operator $\tilde{W}^j$ corresponding to H_j. The proof of Theorem 30.3.3 is applicable with fixed compact sets, cutoff functions and so on and gives a function $W^j(\xi,t)$ satisfying (30.3.12) with V^L replaced by V_j^L and satisfying (30.3.13) uniformly. Moreover, $W^j(\xi,t)=W(\xi,t)$ for large j on any compact subset of $\Omega_0\times\mathbb{R}$ where

$$\Omega_0=\{\xi;\,P_0'(\xi)\neq0\}.$$

If ξ belongs to a compact subset of Ω_0 then $\partial W^j(\xi,t)/\partial t = P_0(\xi)$ when t is larger than some number depending on j, hence

$$W^j(\xi,t) = t P_0(\xi) + \phi_j(\xi)$$

where $\phi_j \in C^\infty(\Omega_0)$. For the modified wave operator $\tilde{W}_+^j$ corresponding to H_j, defined by (30.4.1), it follows that

$$\tilde{W}_+^j = W_+^j e^{-i\phi_j(D)} \tag{30.5.41}$$

where W_+^j is the wave operator of Chapter XIV. Since $W_+^j = F_+^{j*} F$ by Theorem 14.6.5 we conclude that when $\hat{u} \in C_0^\infty(P_0^{-1}(I))$ and $v \in L^2$ we have

$$(\tilde{W}_+^j u, v) = (F_+^{j*} F e^{-i\phi_j(D)} u, v) = (e^{-i\phi_j(\xi)} F u, F_+^j v)$$
$$= (e^{-i\phi_j(\xi)} F u, e^{-i\psi_j(\xi)} \tilde{F}_+^j v)$$

where ψ_j is the function ψ_∞ of (30.5.2) corresponding to V_j^L. With $M_j(\xi) = \exp i(\psi_j(\xi) - \phi_j(\xi))$ it follows that

$$(\tilde{W}_+^j u, v) = (M_j F u, \tilde{F}_+^j v) \quad \text{if } \hat{u} \in C_0^\infty(P_0^{-1}(I)) \text{ and } v \in L^2. \tag{30.5.42}$$

Before letting $j \to \infty$ in (30.5.42) we must show that

$$\tilde{W}_+^j u \to \tilde{W}_+ u \quad \text{in } L^2 \text{ if } \hat{u} \in C_0^\infty(\Omega_0). \tag{30.5.43}$$

By Theorem 30.4.1 we have

$$\tilde{W}_+ u = \lim_{t \to +\infty} e^{itH} e^{-iW(D,t)}$$

and the proof shows that

$$\tilde{W}_+^j u = \lim_{t \to +\infty} e^{itH_j} e^{-iW^j(D,t)} u$$

uniformly with respect to j. To prove (30.5.43) it suffices to show that

$$e^{itH_j} e^{-iW^j(D,t)} u \to e^{itH} e^{-iW(D,t)} u$$

for every fixed t when $j \to \infty$. For large j we have $e^{-iW^j(D,t)} u = e^{-iW(D,t)} u$, so this follows if we show that

$$e^{itH_j} v \to e^{itH} v, \quad \hat{v} \in C_0^\infty(\Omega_0). \tag{30.5.44}$$

The domains of H_j and of H are equal to $H_{(m)}$, and

$$(H_j - H) f \to 0 \quad \text{in } L^2 \text{ for every } f \in H_{(m)}.$$

Hence the convergence is uniform with respect to f if f belongs to a compact subset of $H_{(m)}$. It follows that

$$(H_j - H) e^{-itH} f \to 0 \quad \text{in } L^2$$

uniformly for t in a bounded interval, for $t \mapsto e^{-itH} f$ is a continuous function with values in the domain $H_{(m)}$ of H. Thus the derivative of

$e^{itH_j}e^{-itH}f$ tends uniformly to 0 on any compact set, hence $e^{itH_j}e^{-itH}f \to f$ for every t. If we take $f=e^{itH}v$ we have proved (30.5.44).

Let us now take $v\in B$ in (30.5.42) and let $j\to\infty$ using (30.5.43) and (30.5.38). Since $|M_j|=1$ we can find a weak limit M in L^∞, $|M|\leqq 1$, and obtain

$$(30.5.42)' \qquad (\tilde W_+ u, v)=(MFu, \tilde F_+ v) \quad \text{if } \hat u\in C_0^\infty(P_0^{-1}(I)) \text{ and } v\in B.$$

(In the right-hand side we have scalar product in $L^2(d\xi/(2\pi)^n)$ and in the left-hand side in $L^2(dx)$, as in the diagram (14.6.8).) Taking the supremum of the absolute value when $\|v\|\leqq 1$ we obtain

$$\|\tilde W_+ u\|\leqq \|MFu\| \quad \text{if } \hat u\in C_0^\infty(P_0^{-1}(I)).$$

But $\tilde W_+$ is isometric, so this means that

$$\|M\hat u\|=\|\hat u\| \quad \text{if } \hat u\in C_0^\infty(P_0^{-1}(I)),$$

hence that $|M|=1$ almost everywhere in $P_0^{-1}(I)$. From (30.5.42)' we obtain

$$(30.5.42)'' \qquad \tilde W_+ u=\tilde F_+^* MFu \quad \text{if } u\in L^2 \text{ and } \operatorname{supp}\hat u\subset P_0^{-1}(I).$$

If $v\in E(I)L^2$ we have by (30.5.40)

$$v=\tilde F_+^*\tilde F_+ v=\tilde F_+^* MFu \quad \text{if } u=F^*\overline{M}\tilde F_+ v$$

for $|M|^2=1$ in $\operatorname{supp}\tilde F_+ v$. Hence

$$v=\tilde W_+ u$$

by (30.5.42)'', so v is in the range of $\tilde W_+$. Since I is an arbitrary interval $\subset \mathbb{R}\setminus(Z(P_0)\cup\Lambda)$, we have now reached the goal of this section:

Theorem 30.5.10. *Let $V(x,D)$ be a 2-admissible perturbation of the elliptic self-adjoint operator $P_0(D)$ of order m. Then the range of the modified wave operator $\tilde W_\pm$ of Theorem 30.4.1 is equal to the orthogonal complement of the space spanned by the eigenvectors, dicussed in Theorem 30.2.10. Any other modified wave operator is of the form $\tilde W_\pm M_\pm(D)$ where $M_\pm$ is a measurable function of absolute value 1 almost everywhere. The scattering operator $S=\tilde W_+^*\tilde W_-$ is unitary and commutes with the operator H_0 defined by $P_0(D)$.*

Proof. We have proved all statements referring to the plus sign. The others follow if we apply the results to the complex conjugate of $P_0(D)+V(x,D)$. The proof is complete.

At this point we may adopt as a new definition of $\tilde F_\pm$

$$(30.5.45) \qquad \tilde W_\pm=\tilde F_\pm^* F \quad \text{or equivalently } \tilde F_\pm^*=\tilde W_\pm F^*$$

which makes the diagram (14.6.8) commutative with the modified definitions. From (30.5.42)'' it follows that in $P_0^{-1}(I)$ the new definition of $\tilde F_\pm u$ is equal to the old definition multiplied by a function of absolute value 1.

Notes

It was proved by Dollard [1,2] that the wave operators $W_\pm$ of Theorem 14.4.6 do not exist for the Schrödinger operator with Coulomb potential but that there is a modification which largely serves the same purpose. His proofs relied on the explicit eigenfunction expansion for this operator. The result was generalized to the Schrödinger operator with successively more general potentials by Amrein-Martin-Misra [1], Buslaev-Matveev [1], Alsholm-Kato [1] and Alsholm [1]. In Hörmander [34] the existence of modified wave operators was established for perturbations of any operator $P_0(D)$ with $\det P_0'' \not\equiv 0$ under a hypothesis somewhat stronger than the admissibility in Definition 30.1.3. (A serious flaw was that the class of perturbations allowed was not linear.) Thus the existence of modified wave operators is proved in Section 30.4 under the weakest hypotheses yet known to be sufficient.

For the Schrödinger equation with a long range potential the "limiting amplitude principle" was established by Ikebe and Saito [1]. Agmon [7] has extended this result to elliptic operators of arbitrary order, and Section 30.2 is essentially based on his work. A technical difference is that we apply the energy integral method directly to the operator considered without prior factorization. However, the factorization used by Agmon reappears in Section 30.5 when we study the asymptotic behavior of outgoing (incoming) solutions.

The discussion of the Hamilton flow in Section 30.3 is an improvement of that in Hörmander [34] used to construct a solution of

$$\partial W(\xi,t)/\partial t = P_0(\xi) + V^L(\partial W(\xi,t)/\partial \xi, \xi)$$

where V^L is the long range part of the perturbation. W is used in Section 30.4 to define the modified wave operators. (Most work on such operators is limited by the use of approximate solutions of this equation obtained by an iteration method.) If u is an outgoing (incoming) solution of the Schrödinger equation

$$(-\Delta + V(x) - \lambda)u = 0$$

it was proved by Saito [1, 2] and Kitada [1], under more restrictive conditions on V than ours, that

$$r^{(n-1)/2}\, u(r\omega)\exp(-i\psi_\pm(r,\omega))$$

has a limit in $L^2(S^{n-1})$ when $r \to \infty$ for an appropriate choice of the phase $\psi_\pm$. Kitada obtained $\psi_\pm$ as a Legendre transform of the function W constructed in Hörmander [34]. An analysis of his construction shows that it actually gives a Lagrangian manifold contained in the constant energy surface

$$P_0(\xi) + V^L(x,\xi) = \lambda,$$

which approximates the normal bundle of the level surface M_λ: $P_0(\xi)=\lambda$ as closely as possible far away. The Hamilton flow involved in the construction is the same as in the construction of the function W above. (One cannot work with the x coordinates as parameters on the Lagrangian unless M_λ is strictly convex, so we parametrize locally by one x_j coordinate and the ξ_k coordinates with $k \neq j$.) In Section 30.5 this leads to the asymptotic behavior of incoming and outgoing solutions, hence to the definition of modified distorted Fourier transforms and the proof of asymptotic completeness. The earlier results from Amrein-Martin-Misra [1] to Kitada [1] use stronger conditions on the potential, and this is also true of the work of Enss [2] which depends only on time dependent scattering theory. (We refer to Kitada [1] for further references.) Agmon's unpublished work has been the main inspiration also for Section 30.5. He uses only slightly stronger conditions on the perturbation but obtains on the other hand stronger results on the scattering operator.

Bibliography

Agmon, S.: [1] The coerciveness problem for integro-differential forms. J. Analyse Math. 6, 183–223 (1958).

– [2] Spectral properties of Schrödinger operators. Actes Congr. Int. Math. Nice 2, 679–683 (1970).

– [3] Spectral properties of Schrödinger operators and scattering theory. Ann. Scuola Norm. Sup. Pisa (4) 2, 151–218 (1975).

– [4] Unicité et convexité dans les problèmes différentiels. Sém. Math. Sup. No 13, Les Presses de l'Univ. de Montreal, 1966.

– [5] Lectures on elliptic boundary value problems. van Nostrand Mathematical Studies 2, Princeton, N.J. 1965.

– [6] Problèmes mixtes pour les équations hyperboliques d'ordre supérieur. Coll. Int. CNRS 117, 13–18, Paris 1962.

– [7] Some new results in spectral and scattering theory of differential operators on $\mathbb{R}^n$. Sém. Goulaouic-Schwartz 1978–1979, Exp. II, 1–11.

Agmon, S., A. Douglis and L. Nirenberg: [1] Estimates near the boundary for solutions of elliptic partial differential equations satisfying general boundary conditions. I. Comm. Pure Appl. Math. 12, 623–727 (1959); II. Comm. Pure Appl. Math. 17, 35–92 (1964).

Agmon, S. and L. Hörmander: [1] Asymptotic properties of solutions of differential equations with simple characteristics. J. Analyse Math. 30, 1–38 (1976).

Agranovich, M.S.: [1] Partial differential equations with constant coefficients. Uspehi Mat. Nauk 16:2, 27–94 (1961). (Russian; English translation in Russian Math. Surveys 16:2, 23–90 (1961).)

Ahlfors, L. and M. Heins: [1] Questions of regularity connected with the Phragmén-Lindelöf principle. Ann. of Math. 50, 341–346 (1949).

Airy, G.B.: [1] On the intensity of light in a neighborhood of a caustic. Trans. Cambr. Phil. Soc. 6, 379–402 (1838).

Alinhac, S.: [1] Non-unicité du problème de Cauchy. Ann. of Math. 117, 77–108 (1983).

– [2] Non-unicité pour des opérateurs différentiels à caractéristiques complexes simples. Ann. Sci. École Norm. Sup. 13, 385–393 (1980).

– [3] Uniqueness and non-uniqueness in the Cauchy problem. Contemporary Math. 27, 1–22 (1984).

Alinhac, S. and M.S. Baouendi: [1] Uniqueness for the characteristic Cauchy problem and strong unique continuation for higher order partial differential inequalities. Amer. J. Math. 102, 179–217 (1980).

Alinhac, S. and C. Zuily: [1] Unicité et non-unicité du problème de Cauchy pour des opérateurs hyperboliques à caractéristiques doubles. Comm. Partial Differential Equations 6, 799–828 (1981).

Alsholm, P.K.: [1] Wave operators for long range scattering. Mimeographed report, Danmarks Tekniske Højskole 1975.

Alsholm, P.K. and T. Kato: [1] Scattering with long range potentials. In Partial Diff. Eq., Proc. of Symp. in Pure Math. 23, 393–399. Amer. Math. Soc. Providence, R.I. 1973.

Amrein, W.O., Ph.A. Martin and P. Misra: [1] On the asymptotic condition of scattering theory. Helv. Phys. Acta 43, 313–344 (1970).

Andersson, K.G.: [1] Propagation of analyticity of solutions of partial differential equations with constant coefficients. Ark. Mat. 8, 277–302 (1971).

Andersson, K.G. and R.B. Melrose: [1] The propagation of singularities along glidings rays. Invent. Math. 41, 197–232 (1977).

Arnold, V.I.: [1] On a characteristic class entering into conditions of quantization. Funkcional. Anal. i Priložen. 1, 1–14 (1967) (Russian); also in Functional Anal. Appl. 1, 1–13 (1967).

Aronszajn, N.: [1] Boundary values of functions with a finite Dirichlet integral. Conference on Partial Differential Equations 1954, University of Kansas, 77–94.

– [2] A unique continuation theorem for solutions of elliptic partial differential equations or inequalities of second order. J. Math. Pures Appl. 36, 235–249 (1957).

Aronszajn, N., A. Krzywcki and J. Szarski: [1] A unique continuation theorem for exterior differential forms on Riemannian manifolds. Ark. Mat. 4, 417–453 (1962).

Asgeirsson, L.: [1] Über eine Mittelwerteigenschaft von Lösungen homogener linearer partieller Differentialgleichungen 2. Ordnung mit konstanten Koeffizienten. Math. Ann. 113, 321–346 (1937).

Atiyah, M.F.: [1] Resolution of singularities and division of distributions. Comm. Pure Appl. Math. 23, 145–150 (1970).

Atiyah, M.F. and R. Bott: [1] The index theorem for manifolds with boundary. Proc. Symp. on Differential Analysis, 175–186. Oxford 1964.

– [2] A Lefschetz fixed point formula for elliptic complexes. I. Ann. of Math. 86, 374–407 (1967).

Atiyah, M.F., R. Bott and L. Gårding: [1] Lacunas for hyperbolic differential operators with constant coefficients. I. Acta Math. 124, 109–189 (1970).

– [2] Lacunas for hyperbolic differential operators with constant coefficients. II. Acta Math. 131, 145–206 (1973).

Atiyah, M.F., R. Bott and V.K. Patodi: [1] On the heat equation and the index theorem. Invent. Math. 19, 279–330 (1973).

Atiyah, M.F. and I.M. Singer: [1] The index of elliptic operators on compact manifolds Bull. Amer. Math. Soc. 69, 422–433 (1963).

– [2] The index of elliptic operators. I, III. Ann. of Math. 87, 484–530 and 546–604 (1968).

Atkinson, F.V.: [1] The normal solubility of linear equations in normed spaces. Mat. Sb. 28 (70), 3–14 (1951) (Russian).

Avakumovič, V.G.: [1] Über die Eigenfunktionen auf geschlossenen Riemannschen Mannigfaltigkeiten. Math. Z. 65, 327–344 (1956).

Bang, T.: [1] Om quasi-analytiske funktioner. Thesis, Copenhagen 1946, 101 pp.

Baouendi, M.S. and Ch. Goulaouic: [1] Nonanalytic-hypoellipticity for some degenerate elliptic operators. Bull. Amer. Math. Soc. 78, 483–486 (1972).

Beals, R.: [1] A general calculus of pseudo-differential operators. Duke Math. J. 42, 1–42 (1975).

Beals, R. and C. Fefferman: [1] On local solvability of linear partial differential equations. Ann. of Math. 97, 482–498 (1973).

– [2] Spatially inhomogeneous pseudo-differential operators I. Comm. Pure Appl. Math. 27, 1–24 (1974).

Beckner, W.: [1] Inequalities in Fourier analysis. Ann. of Math. 102, 159–182 (1975).

Berenstein, C.A. and M.A. Dostal: [1] On convolution equations I. In L'anal. harm. dans le domaine complexe. Springer Lecture Notes in Math. 336, 79–94 (1973).

Bernstein, I.N.: [1] Modules over a ring of differential operators. An investigation of the fundamental solutions of equations with constant coefficients. Funkcional. Anal. i Priložen. 5:2, 1–16 (1971) (Russian); also in Functional Anal. Appl. 5, 89–101 (1971).

Bernstein, I.N. and S.I. Gelfand: [1] Meromorphy of the function P^λ. Funkcional. Anal. i Priložen. 3:1, 84–85 (1969) (Russian); also in Functional Anal. Appl. 3, 68–69 (1969).

Bernstein, S.: [1] Sur la nature analytique des solutions des équations aux dérivées partielles du second ordre. Math. Ann. 59, 20–76 (1904).

Beurling, A.: [1] Quasi-analyticity and general distributions. Lectures 4 and 5, Amer. Math. Soc. Summer Inst. Stanford 1961 (Mimeographed).

– [2] Sur les spectres des fonctions. Anal. Harm. Nancy 1947, Coll. Int. XV, 9–29.

– [3] Analytic continuation across a linear boundary. Acta Math. 128, 153–182 (1972).

Björck, G.: [1] Linear partial differential operators and generalized distributions. Ark. Mat. 6, 351–407 (1966).

Björk, J.E.: [1] Rings of differential operators. North-Holland Publ. Co. Math. Library series 21 (1979).

Bochner, S.: [1] Vorlesungen über Fouriersche Integrale. Leipzig 1932.

Boman, J.: [1] On the intersection of classes of infinitely differentiable functions. Ark. Mat. 5, 301–309 (1963).

Bonnesen, T. and W. Fenchel: [1] Theorie der konvexen Körper. Erg. d. Math. u. ihrer Grenzgeb. 3, Springer Verlag 1934.

Bony, J.M.: [1] Une extension du théorème de Holmgren sur l'unicité du problème de Cauchy. C.R. Acad. Sci. Paris 268, 1103–1106 (1969).

– [2] Extensions du théorème de Holmgren. Sém. Goulaouic-Schwartz 1975–1976, Exposé no. XVII.

– [3] Equivalence des diverses notions de spectre singulier analytique. Sém. Goulaouic-Schwartz 1976–1977, Exposé no. III.

Bony, J.M. and P. Schapira: [1] Existence et prolongement des solutions holomorphes des équations aux dérivées partielles. Invent. Math. 17, 95–105 (1972).

Borel, E.: [1] Sur quelques points de la théorie des fonctions. Ann. Sci. École Norm. Sup. 12 (3), 9–55 (1895).

Boutet de Monvel, L.: [1] Comportement d'un opérateur pseudo-différentiel sur une variété à bord. J. Analyse Math. 17, 241–304 (1966).

– [2] Boundary problems for pseudo-differential operators. Acta Math. 126, 11–51 (1971).

– [3] On the index of Toeplitz operators of several complex variables. Invent. Math. 50, 249–272 (1979).

– [4] Hypoelliptic operators with double characteristics and related pseudo-differential operators. Comm. Pure Appl. Math. 27, 585–639 (1974).

Boutet de Monvel, L., A. Grigis and B. Helffer: [1] Paramétrixes d'opérateurs pseudo-différentiels à caractéristiques multiples. Astérisque 34–35, 93–121 (1976).

Boutet de Monvel, L. and V. Guillemin: [1] The spectral theory of Toeplitz operators. Ann. of Math. Studies 99 (1981).

Brézis, H.: [1] On a characterization of flow-invariant sets. Comm. Pure Appl. Math. 23, 261–263 (1970).

Brodda, B.: [1] On uniqueness theorems for differential equations with constant coefficients. Math. Scand. 9, 55–68 (1961).

Browder, F.: [1] Estimates and existence theorems for elliptic boundary value problems. Proc. Nat. Acad. Sci. 45, 365–372 (1959).

Buslaev, V.S. and V.B. Matveev: [1] Wave operators for the Schrödinger equation with a slowly decreasing potential. Theor. and Math. Phys. 2, 266–274 (1970). (English translation.)

Calderón, A.P.: [1] Uniqueness in the Cauchy problem for partial differential equations. Amer. J. Math. 80, 16–36 (1958).

– [2] Existence and uniqueness theorems for systems of partial differential equations. Fluid Dynamics and Applied Mathematics (Proc. Symp. Univ. of Maryland 1961), 147–195. New York 1962.

– [3] Boundary value problems for elliptic equations. Outlines of the joint Soviet-American symposium on partial differential equations, 303–304, Novosibirsk 1963.

Calderón, A.P. and R. Vaillancourt: [1] On the boundedness of pseudo-differential operators. J. Math. Soc. Japan 23, 374–378 (1972).

– [2] A class of bounded pseudo-differential operators. Proc. Nat. Acad. Sci. U.S.A. 69, 1185–1187 (1972).

Calderón, A.P. and A. Zygmund: [1] On the existence of certain singular integrals. Acta Math. 88, 85–139 (1952).

Carathéodory, C.: [1] Variationsrechnung und partielle Differentialgleichungen Erster Ordnung. Teubner, Berlin, 1935.

Carleman, T.: [1] Sur un problème d'unicité pour les systèmes d'équations aux dérivées partielles à deux variables indépendentes. Ark. Mat. Astr. Fys. 26B No 17, 1–9 (1939).

– [2] L'intégrale de Fourier et les questions qui s'y rattachent. Publ. Sci. Inst. Mittag-Leffler, Uppsala 1944.

– [3] Propriétés asymptotiques des fonctions fondamentales des membranes vibrantes. C.R. Congr. des Math. Scand. Stockholm 1934, 34–44 (Lund 1935).

Catlin, D.: [1] Necessary conditions for subellipticity and hypoellipticity for the $\bar\partial$Neumann problem on pseudoconvex domains. In Recent developments in several complex variables. Ann. of Math. Studies 100, 93–100 (1981).

Cauchy, A.: [1] Mémoire sur l'intégration des équations linéaires. C.R. Acad. Sci. Paris 8 (1839). In Œuvres IV, 369–426, Gauthier-Villars, Paris 1884.

Cerezo, A., J. Chazarain and A. Piriou: [1] Introduction aux hyperfonctions. Springer Lecture Notes in Math. 449, 1–53 (1975).

Chaillou, J.: [1] Hyperbolic differential polynomials and their singular perturbations. D. Reidel Publ. Co. Dordrecht, Boston, London 1979.

Chazarain, J.: [1] Construction de la paramétrix du problème mixte hyperbolique pour l'équation des ondes. C.R. Acad. Sci. Paris 276, 1213–1215 (1973)

– [2] Formules de Poisson pour les variétés riemanniennes. Invent. Math. 24, 65–82 (1974).

Chazarain, J. and A. Piriou: [1] Introduction à la théorie des équations aux dérivées partielles linéaires. Gauthier-Villars 1981.

Chester, C., B. Friedman and F. Ursell: [1] An extension of the method of steepest descent. Proc. Cambr. Phil. Soc. 53, 599–611 (1957).

Cohen, P.: [1] The non-uniqueness of the Cauchy problem. O.N.R. Techn. Report 93, Stanford 1960.

– [2] A simple proof of the Denjoy-Carleman theorem. Amer. Math. Monthly 75, 26–31 (1968).

– [3] A simple proof of Tarski's theorem on elementary algebra. Mimeographed manuscript, Stanford University 1967, 6 pp.

Colin de Verdière. Y.: [1] Sur le spectre des opérateurs elliptiques à bicharactéristiques toutes périodiques. Comment. Math. Helv. 54. 508–522 (1979).

Cook, J.: [1] Convergence to the Møller wave matrix. J. Mathematical Physics 36, 82–87 (1957).

Cordes, H.O.: [1] Über die eindeutige Bestimmtheit der Lösungen elliptischer Differentialgleichungen durch Anfangsvorgaben. Nachr. Akad. Wiss. Göttingen Math.-Phys. Kl. IIa, No. 11, 239–258 (1956).

Cotlar, M.: [1] A combinatorial inequality and its application to L^2 spaces. Rev. Math. Cuyana 1, 41–55 (1955).

Courant, R. and D. Hilbert: [1] Methoden der Mathematischen Physik II. Berlin 1937.

Courant, R. and P.D. Lax: [1] The propagation of discontinuities in wave motion. Proc. Nat. Acad. Sci. 42, 872–876 (1956).

De Giorgi, E.: [1] Un esempio di non-unicitá della soluzione del problema di Cauchy relativo ad una equazione differenziale lineare a derivate parziali ti tipo parabolico. Rend. Mat. 14, 382–387 (1955).

– [2] Solutions analytiques des équations aux dérivées partielles à coefficients constants. Sém. Goulaouic-Schwartz 1971–1972, Exposé 29.

Deič, V.G., E.L. Korotjaev and D.R. Jafaev: [1] The theory of potential scattering with account taken of spatial anisotropy. Zap. Naučn. Sem. Leningrad Otdel. Mat. Inst. Steklov 73, 35–51 (1977).

Dencker, N.: [1] On the propagation of singularities for pseudo-differential operators of principal type. Ark. Mat. 20, 23–60 (1982).

– [2] The Weyl calculus with locally temperate metrics and weights. Ark. Mat. 24, 59–79 (1986).

Dieudonné, J.: [1] Sur les fonctions continus numériques définies dans un produit de deux espaces compacts. C.R. Acad. Sci. Paris 205, 593–595 (1937).

Dieudonné, J. and L. Schwartz: [1] La dualité dans les espaces $(\mathscr{F})$ et $(\mathscr{L}\mathscr{F})$. Ann. Inst. Fourier (Grenoble) 1, 61–101 (1949).

Dollard, J.D.: [1] Asymptotic convergence and the Coulomb interaction. J. Math. Phys. 5, 729–738 (1964).

– [2] Quantum mechanical scattering theory for short-range and Coulomb interactions. Rocky Mountain J. Math. 1, 5–88 (1971).

Douglis, A. and L. Nirenberg: [1] Interior estimates for elliptic systems of partial differential equations. Comm. Pure Appl. Math. 8, 503–538 (1955).

Duistermaat, J.J.: [1] Oscillatory integrals, Lagrange immersions and unfolding of singularities. Comm. Pure Appl. Math. 27, 207–281 (1974).

Duistermaat, J.J. and V.W. Guillemin: [1] The spectrum of positive elliptic operators and periodic bicharacteristics. Invent. Math. 29, 39–79 (1975).

Duistermaat, J.J. and L. Hörmander: [1] Fourier integral operators II. Acta Math. 128, 183–269 (1972).

Duistermaat, J.J. and J. Sjöstrand: [1] A global construction for pseudo-differential operators with non-involutive characteristics. Invent. Math. 20, 209–225 (1973).

DuPlessis, N.: [1] Some theorems about the Riesz fractional integral. Trans. Amer. Math. Soc. 80, 124–134 (1955).

Egorov, Ju.V.: [1] The canonical transformations of pseudo-differential operators. Uspehi Mat. Nauk 24:5, 235–236 (1969).

– [2] Subelliptic pseudo-differential operators. Dokl. Akad. Nauk SSSR 188, 20–22 (1969); also in Soviet Math. Doklady 10, 1056–1059 (1969).

– [3] Subelliptic operators. Uspehi Mat. Nauk 30:2, 57–114 and 30:3, 57–104 (1975); also in Russian Math. Surveys 30:2, 59–118 and 30:3, 55–105 (1975).

Ehrenpreis, L.: [1] Solutions of some problems of division I. Amer. J. Math. 76, 883–903 (1954).

– [2] Solutions of some problems of division III. Amer. J. Math. 78, 685–715 (1956).

– [3] Solutions of some problems of division IV. Amer. J. Math. 82, 522–588 (1960).

– [4] On the theory of kernels of Schwartz. Proc. Amer. Math. Soc. 7, 713–718 (1956).

– [5] A fundamental principle for systems of linear differential equations with constant coefficients, and some of its applications. Proc. Intern. Symp. on Linear Spaces, Jerusalem 1961, 161–174.

– [6] Fourier analysis in several complex variables. Wiley-Interscience Publ., New York, London, Sydney, Toronto 1970.

– [7] Analytically uniform spaces and some applications. Trans. Amer. Math. Soc. 101, 52–74 (1961).

– [8] Solutions of some problems of division V. Hyperbolic operators. Amer. J. Math. 84, 324–348 (1962).

Enqvist, A.: [1] On fundamental solutions supported by a convex cone. Ark. Mat. 12, 1–40 (1974).

Enss, V.: [1] Asymptotic completeness for quantum-mechanical potential scattering. I. Short range potentials. Comm. Math. Phys. 61, 285–291 (1978).

– [2] Geometric methods in spectral and scattering theory of Schrödinger operators.

In Rigorous Atomic and Molecular Physics, G. Velo and A. Wightman ed., Plenum, New York, 1980–1981 (Proc. Erice School of Mathematical Physics 1980).

Eškin, G.I.: [1] Boundary value problems for elliptic pseudo-differential equations. Moscow 1973; Amer. Math. Soc. Transl. of Math. Monographs 52, Providence, R.I. 1981.

– [2] Parametrix and propagation of singularities for the interior mixed hyperbolic problem. J. Analyse Math. 32, 17–62 (1977).

– [3] General initial-boundary problems for second order hyperbolic equations. In Sing. in Boundary Value Problems. D. Reidel Publ. Co., Dordrecht, Boston, London 1981, 19–54.

– [4] Initial boundary value problem for second order hyperbolic equations with general boundary conditions I. J. Analyse Math. 40, 43–89 (1981).

Fedosov, B.V.: [1] A direct proof of the formula for the index of an elliptic system in Euclidean space. Funkcional. Anal. i Priložen. 4:4, 83–84 (1970) (Russian); also in Functional Anal. Appl. 4, 339–341 (1970).

Fefferman, C.L.: [1] The uncertainty principle. Bull. Amer. Math. Soc. 9, 129–206 (1983).

Fefferman, C. and D.H. Phong: [1] On positivity of pseudo-differential operators. Proc. Nat. Acad. Sci. 75, 4673–4674 (1978).

– [2] The uncertainty principle and sharp Gårding inequalities. Comm. Pure Appl. Math. 34, 285–331 (1981).

Fredholm, I.: [1] Sur l'intégrale fondamentale d'une équation différentielle elliptique à coefficients constants. Rend. Circ. Mat. Palermo 25, 346–351 (1908).

Friedlander, F.G.: [1] The wave front set of the solution of a simple initial-boundary value problem with glancing rays. Math. Proc. Cambridge Philos. Soc. 79, 145–159 (1976).

Friedlander, F.G. and R.B. Melrose: [1] The wave front set of the solution of a simple initial-boundary value problem with glancing rays. II. Math. Proc. Cambridge Philos. Soc. 81, 97–120 (1977).

Friedrichs, K.: [1] On differential operators in Hilbert spaces. Amer. J. Math. 61, 523–544 (1939).

– [2] The identity of weak and strong extensions of differential operators. Trans. Amer. Math. Soc. 55, 132–151 (1944).

– [3] On the differentiability of the solutions of linear elliptic differential equations. Comm. Pure Appl. Math. 6, 299–326 (1953).

– [4] On the perturbation of continuous spectra. Comm. Pure Appl. Math. 1, 361–406 (1948).

Friedrichs, K. and H. Lewy: [1] Über die Eindeutigkeit und das Abhängigkeitsgebiet der Lösungen beim Anfangswertproblem linearer hyperbolischer Differentialgleichungen. Math. Ann. 98, 192–204 (1928).

Fröman, N. and P.O. Fröman: [1] JWKB approximation. Contributions to the theory. North-Holland Publ. Co. Amsterdam 1965.

Fuglede, B.: [1] A priori inequalities connected with systems of partial differential equations. Acta Math. 105, 177–195 (1961).

Gabrielov, A.M.: [1] A certain theorem of Hörmander. Funkcional. Anal. i Priložen. 4:2, 18–22 (1970) (Russian); also in Functional Anal. Appl. 4, 106–109 (1970).

Gårding, L.: [1] Linear hyperbolic partial differential equations with constant coefficients. Acta Math. 85, 1–62 (1951).

– [2] Dirichlet's problem for linear elliptic partial differential equations. Math. Scand. 1, 55–72 (1953).

– [3] Solution directe du problème de Cauchy pour les équations hyperboliques. Coll. Int. CNRS, Nancy 1956, 71–90.

– [4] Transformation de Fourier des distributions homogènes. Bull. Soc. Math. France 89, 381–428 (1961).

– [5] Local hyperbolicity. Israel J. Math. 13, 65–81 (1972).

– [6] Le problème de la dérivée oblique pour l'équation des ondes. C.R. Acad. Sci. Paris 285, 773–775 (1977). Rectification C.R. Acad. Sci. Paris 285, 1199 (1978).
– [7] On the asymptotic distribution of the eigenvalues and eigenfunctions of elliptic differential operators. Math. Scand. 1, 237–255 (1953).
Gårding, L. and J.L. Lions: [1] Functional analysis. Nuovo Cimento N. 1 del Suppl. al Vol. (10)14, 9–66 (1959).
Gårding, L. and B. Malgrange: [1] Opérateurs différentiels partiellement hypoelliptiques et partiellement elliptiques. Math. Scand. 9, 5–21 (1961).
Gask, H.: [1] A proof of Schwartz' kernel theorem. Math. Scand. 8, 327–332 (1960).
Gelfand, I.M. and G.E. Šilov: [1] Fourier transforms of rapidly increasing functions and questions of uniqueness of the solution of Cauchy's problem. Uspehi Mat. Nauk 8:6, 3–54 (1953) (Russian); also in Amer. Math. Soc. Transl. (2) 5, 221–274 (1957).
– [2] Generalized functions. Volume 1: Properties and operations. Volume 2: Spaces of fundamental and generalized functions. Academic Press, New York and London 1964, 1968.
Gevrey, M.: [1] Démonstration du théorème de Picard-Bernstein par la méthode des contours successifs; prolongement analytique. Bull. Sci. Math. 50, 113–128 (1926).
Glaeser, G.: [1] Etude de quelques algèbres Tayloriennes. J. Analyse Math. 6, 1–124 (1958).
Godin, P.: [1] Propagation des singularités pour les opérateurs pseudo-différentiels de type principal à partie principal analytique vérifiant la condition (*P*), en dimension 2. C.R. Acad. Sci. Paris 284, 1137–1138 (1977).
Gorin, E.A.: [1] Asymptotic properties of polynomials and algebraic functions of several variables. Uspehi Mat. Nauk 16:1, 91–118 (1961) (Russian); also in Russian Math. Surveys 16:1, 93–119 (1961).
Grubb, G.: [1] Boundary problems for systems of partial differential operators of mixed order. J. Functional Analysis 26, 131–165 (1977).
– [2] Problèmes aux limites pseudo-différentiels dépendant d'un paramètre. C.R. Acad. Sci. Paris 292, 581–583 (1981).
Grušin, V.V.: [1] The extension of smoothness of solutions of differential equations of principal type. Dokl. Akad. Nauk SSSR 148, 1241–1244 (1963) (Russian); also in Soviet Math. Doklady 4, 248–252 (1963).
– [2] A certain class of hypoelliptic operators. Mat. Sb. 83, 456–473 (1970) (Russian); also in Math. USSR Sb. 12, 458–476 (1970).
Gudmundsdottir, G.: [1] Global properties of differential operators of constant strength. Ark. Mat. 15, 169–198 (1977).
Guillemin, V.: [1] The Radon transform on Zoll surfaces. Advances in Math. 22, 85–119 (1976).
– [2] Some classical theorems in spectral theory revisited. Seminar on sing. of sol. of diff. eq., Princeton University Press, Princeton, N.J., 219–259 (1979).
– [3] Some spectral results for the Laplace operator with potential on the *n*-sphere. Advances in Math. 27, 273–286 (1978).
Guillemin, V. and D. Schaeffer: [1] Remarks on a paper of D. Ludwig. Bull. Amer. Math. Soc. 79, 382–385 (1973).
Guillemin, V. and S. Sternberg: [1] Geometrical asymptotics. Amer. Math. Soc. Surveys 14, Providence, R.I. 1977.
Gurevič, D.I.: [1] Counterexamples to a problem of L. Schwartz. Funkcional. Anal. i Priložen. 9:2, 29–35 (1975) (Russian); also in Functional Anal. Appl. 9, 116–120 (1975).
Hack, M.N.: [1] On convergence to the Møller wave operators. Nuovo Cimento (10) 13, 231–236 (1959).
Hadamard, J.: [1] Le problème de Cauchy et les équations aux dérivées partielles linéaires hyperboliques. Paris 1932.
Haefliger, A.: [1] Variétés feuilletées. Ann. Scuola Norm. Sup. Pisa 16, 367–397 (1962).

Hanges, N.: [1] Propagation of singularities for a class of operators with double characteristics. Seminar on singularities of sol. of linear partial diff. eq., Princeton University Press, Princeton, N.J. 1979, 113-126.

Hardy, G.H. and J.E. Littlewood: [1] Some properties of fractional integrals. (I) Math. Z. 27, 565-606 (1928); (II) Math. Z. 34, 403-439 (1931-32).

Hausdorff, F.: [1] Eine Ausdehnung des Parsevalschen Satzes über Fourierreihen. Math. Z. 16, 163-169 (1923).

Hayman, W.K. and P.B. Kennedy: [1] Subharmonic functions I. Academic Press, London, New York, San Francisco 1976.

Hedberg, L.I.: [1] On certain convolution inequalities. Proc. Amer. Math. Soc. 36, 505-510 (1972).

Heinz, E.: [1] Über die Eindeutigkeit beim Cauchyschen Anfangswertproblem einer elliptischen Differentialgleichung zweiter Ordnung. Nachr. Akad. Wiss. Göttingen Math.-Phys. Kl. IIa No. 1, 1-12 (1955).

Helffer, B.: [1] Addition de variables et applications à la régularité. Ann. Inst. Fourier (Grenoble) 28:2, 221-231 (1978).

Helffer, B. and J. Nourrigat: [1] Caractérisation des opérateurs hypoelliptiques homogènes invariants à gauche sur un groupe de Lie nilpotent gradué. Comm. Partial Differential Equations 4:8, 899-958 (1979).

Herglotz, G.: [1] Über die Integration linearer partieller Differentialgleichungen mit konstanten Koeffizienten I-III. Berichte Sächs. Akad. d. Wiss. 78, 93-126, 287-318 (1926); 80, 69-114 (1928).

Hersh, R.: [1] Boundary conditions for equations of evolution. Arch. Rational Mech. Anal. 16, 243-264 (1964).

– [2] On surface waves with finite and infinite speed of propagation. Arch. Rational Mech. Anal. 19, 308-316 (1965).

Hirzebruch, F.: [1] Neue topologische Methoden in der algebraischen Geometrie. Springer Verlag, Berlin-Göttingen-Heidelberg 1956.

Hlawka, E.: [1] Über Integrale auf konvexen Körpern. I. Monatsh. Math. 54, 1-36 (1950).

Holmgren, E.: [1] Über Systeme von linearen partiellen Differentialgleichungen. Öfversigt af Kongl. Vetenskaps-Akad. Förh. 58, 91-103 (1901).

– [2] Sur l'extension de la méthode d'intégration de Riemann. Ark. Mat. Astr. Fys. 1, No 22, 317-326 (1904).

Hörmander, L.: [1] On the theory of general partial differential operators. Acta Math. 94, 161-248 (1955).

– [2] Local and global properties of fundamental solutions. Math. Scand. 5, 27-39 (1957).

– [3] On the regularity of the solutions of boundary problems. Acta Math. 99, 225-264 (1958).

– [4] On interior regularity of the solutions of partial differential equations. Comm. Pure Appl. Math. 11, 197-218 (1958).

– [5] On the division of distributions by polynomials. Ark. Mat. 3, 555-568 (1958).

– [6] Differentiability properties of solutions of systems of differential equations. Ark. Mat. 3, 527-535 (1958).

– [7] Definitions of maximal differential operators. Ark. Mat. 3, 501-504 (1958).

– [8] On the uniqueness of the Cauchy problem I, II. Math. Scand. 6, 213-225 (1958); 7, 177-190 (1959).

– [9] Null solutions of partial differential equations. Arch. Rational Mech. Anal. 4, 255-261 (1960).

– [10] Differential operators of principal type. Math. Ann. 140, 124-146 (1960).

– [11] Differential equations without solutions. Math. Ann. 140, 169-173 (1960).

– [12] Hypoelliptic differential operators. Ann. Inst. Fourier (Grenoble) 11, 477-492 (1961).

– [13] Estimates for translation invariant operators in L^p spaces. Acta Math. 104, 93–140 (1960).
– [14] On the range of convolution operators. Ann. of Math. 76, 148–170 (1962).
– [15] Supports and singular supports of convolutions. Acta Math. 110, 279–302 (1963).
– [16] Pseudo-differential operators. Comm. Pure Appl. Math. 18, 501–517 (1965).
– [17] Pseudo-differential operators and non-elliptic boundary problems. Ann. of Math. 83, 129–209 (1966).
– [18] Pseudo-differential operators and hypoelliptic equations. Amer. Math. Soc. Symp. on Singular Integrals, 138–183 (1966).
– [19] An introduction to complex analysis in several variables. D. van Nostrand Publ. Co., Princeton, N.J. 1966.
– [20] Hypoelliptic second order differential equations. Acta Math. 119, 147–171 (1967).
– [21] On the characteristic Cauchy problem. Ann. of Math. 88, 341–370 (1968).
– [22] The spectral function of an elliptic operator. Acta Math. 121, 193–218 (1968).
– [23] Convolution equations in convex domains. Invent. Math. 4, 306–317 (1968).
– [24] On the singularities of solutions of partial differential equations. Comm. Pure Appl. Math. 23, 329–358 (1970).
– [25] Linear differential operators. Actes Congr. Int. Math. Nice 1970, 1, 121–133.
– [26a] The calculus of Fourier integral operators. Prospects in math. Ann. of Math. Studies 70, 33–57 (1971).
– [26] Fourier integral operators I. Acta Math. 127, 79–183 (1971).
– [27] Uniqueness theorems and wave front sets for solutions of linear differential equations with analytic coefficients. Comm. Pure Appl. Math. 24, 671–704 (1971).
– [28] A remark on Holmgren's uniqueness theorem. J. Diff. Geom. 6, 129–134 (1971).
– [29] On the existence and the regularity of solutions of linear pseudo-differential equations. Ens. Math. 17, 99–163 (1971).
– [30] On the singularities of solutions of partial differential equations with constant coefficients. Israel J. Math. 13, 82–105 (1972).
– [31] On the existence of real analytic solutions of partial differential equations with constant coefficients. Invent. Math. 21, 151–182 (1973).
– [32] Lower bounds at infinity for solutions of differential equations with constant coefficients. Israel J. Math. 16, 103–116 (1973).
– [33] Non-uniqueness for the Cauchy problem. Springer Lecture Notes in Math. 459, 36–72 (1975).
– [34] The existence of wave operators in scattering theory. Math. Z. 146, 69–91 (1976).
– [35] A class of hypoelliptic pseudo-differential operators with double characteristics. Math. Ann. 217, 165–188 (1975).
– [36] The Cauchy problem for differential equations with double characteristics. J. Analyse Math. 32, 118–196 (1977).
– [37] Propagation of singularities and semiglobal existence theorems for (pseudo-) differential operators of principal type. Ann. of Math. 108, 569–609 (1978).
– [38] Subelliptic operators. Seminar on sing. of sol. of diff. eq. Princeton University Press, Princeton, N.J., 127–208 (1979).
– [39] The Weyl calculus of pseudo-differential operators. Comm. Pure Appl. Math. 32, 359–443 (1979).
– [40] Pseudo-differential operators of principal type. Nato Adv. Study Inst. on Sing. in Bound. Value Problems. Reidel Publ. Co., Dordrecht, 69–96 (1981).
– [41] Uniqueness theorems for second order elliptic differential equations. Comm. Partial Differential Equations 8, 21–64 (1983).
– [42] On the index of pseudo-differential operators. In Elliptische Differentialgleichungen Band II, Akademie-Verlag, Berlin 1971, 127–146.
– [43] L^2 estimates for Fourier integral operators with complex phase. Ark. Mat. 21, 297–313 (1983).
– [44] On the subelliptic test estimates. Comm. Pure Appl. Math. 33, 339–363 (1980).

Hurwitz, A.: [1] Über die Nullstellen der Bessel'schen Funktion. Math. Ann. 33, 246–266 (1889).

Iagolnitzer, D.: [1] Microlocal essential support of a distribution and decomposition theorems – an introduction. In Hyperfunctions and theoretical physics. Springer Lecture Notes in Math. 449, 121–132 (1975).

Ikebe, T.: [1] Eigenfunction expansions associated with the Schrödinger operator and their applications to scattering theory. Arch. Rational Mech. Anal. 5, 1–34 (1960).

Ikebe, T. and Y. Saito: [1] Limiting absorption method and absolute continuity for the Schrödinger operator. J. Math. Kyoto Univ. 12, 513–542 (1972).

Ivrii, V.Ja: [1] Sufficient conditions for regular and completely regular hyperbolicity. Trudy Moskov. Mat. Obšč. 33, 3–65 (1975) (Russian); also in Trans. Moscow Math. Soc. 33, 1–65 (1978).

– [2] Wave fronts for solutions of boundary value problems for a class of symmetric hyperbolic systems. Sibirsk. Mat. Ž. 21:4, 62–71 (1980) (Russian); also in Sibirian Math. J. 21, 527–534 (1980).

– [3] On the second term in the spectral asymptotics for the Laplace-Beltrami operator on a manifold with boundary. Funkcional. Anal. i Priložen. 14:2, 25–34 (1980) (Russian); also in Functional Anal. Appl. 14, 98–106 (1980).

Ivrii, V.Ja and V.M. Petkov: [1] Necessary conditions for the correctness of the Cauchy problem for non-strictly hyperbolic equations. Uspehi Mat. Nauk 29:5, 3–70 (1974) (Russian); also in Russian Math. Surveys 29:5, 1–70 (1974).

Iwasaki, N.: [1] The Cauchy problems for effectively hyperbolic equations (general case). J. Math. Kyoto Univ. 25, 727–743 (1985).

Jauch, J.M. and I.I. Zinnes: [1] The asymptotic condition for simple scattering systems. Nuovo Cimento (10) 11, 553–567 (1959).

Jerison D. and C.E. Kenig: [1] Unique continuation and absence of positive eigenvalues for Schrödinger operators. Ann. of Math. 121, 463–488 (1985).

John, F.: [1] On linear differential equations with analytic coefficients. Unique continuation of data. Comm. Pure Appl. Math. 2, 209–253 (1949).

– [2] Plane waves and spherical means applied to partial differential equations. New York 1955.

– [3] Non-admissible data for differential equations with constant coefficients. Comm. Pure Appl. Math. 10, 391–398 (1957).

– [4] Continuous dependence on data for solutions of partial differential equations with a prescribed bound. Comm. Pure Appl. Math. 13, 551–585 (1960).

– [5] Linear partial differential equations with analytic coefficients. Proc. Nat. Acad. Sci. 29, 98–104 (1943).

Jörgens, K. and J. Weidmann: [1] Zur Existenz der Wellenoperatoren. Math. Z. 131, 141–151 (1973).

Kashiwara, M.: [1] Introduction to the theory of hyperfunctions. In Sem. on micro-local analysis, Princeton Univ. Press, Princeton, N.J., 1979, 3–38.

Kashiwara, M. and T. Kawai: [1] Microhyperbolic pseudo-differential operators. I. J. Math. Soc. Japan 27, 359–404 (1975).

Kato, T.: [1] Growth properties of solutions of the reduced wave equation with a variable coefficient. Comm. Pure Appl. Math. 12, 403–425 (1959).

Keller, J.B.: [1] Corrected Bohr-Sommerfeld quantum conditions for nonseparable systems. Ann. Physics 4, 180–188 (1958).

Kitada, H.: [1] Scattering theory for Schrödinger operators with long-range potentials. I: Abstract theory. J. Math. Soc. Japan 29, 665–691 (1977). II: Spectral and scattering theory. J. Math. Soc. Japan 30, 603–632 (1978).

Knapp, A.W. and E.M. Stein: [1] Singular integrals and the principal series. Proc. Nat. Acad. Sci. U.S.A. 63, 281–284 (1969).

Kohn, J.J.: [1] Harmonic integrals on strongly pseudo-convex manifolds I, II. Ann. of Math. 78, 112–148 (1963); 79, 450–472 (1964).

– [2] Pseudo-differential operators and non-elliptic problems. In Pseudo-differential operators, CIME conference, Stresa 1968, 157–165. Edizione Cremonese, Roma 1969.

Kohn, J.J. and L. Nirenberg: [1] On the algebra of pseudo-differential operators. Comm. Pure Appl. Math. 18, 269–305 (1965).

– [2] Non-coercive boundary value problems. Comm. Pure Appl. Math. 18, 443–492 (1965).

Kolmogorov, A.N.: [1] Zufällige Bewegungen. Ann. of Math. 35, 116–117 (1934).

Komatsu, H.: [1] A local version of Bochner's tube theorem. J. Fac. Sci. Tokyo Sect. I-A Math. 19, 201–214 (1972).

– [2] Boundary values for solutions of elliptic equations. Proc. Int. Conf. Funct. Anal. Rel. Topics, 107–121. University of Tokyo Press, Tokyo 1970.

Kreiss, H.O.: [1] Initial boundary value problems for hyperbolic systems. Comm. Pure Appl. Math. 23, 277–298 (1970).

Krzyżański, M. and J. Schauder: [1] Quasilineare Differentialgleichungen zweiter Ordnung vom hyperbolischen Typus. Gemischte Randwertaufgaben. Studia Math. 6, 162–189 (1936).

Kumano-go, H.: [1] Factorizations and fundamental solutions for differential operators of elliptic-hyperbolic type. Proc. Japan Acad. 52, 480–483 (1976).

Kuroda, S.T.: [1] On the existence and the unitary property of the scattering operator. Nuovo Cimento (10) 12, 431–454 (1959).

Lascar, B. and R. Lascar: [1] Propagation des singularités pour des équations hyperboliques à caractéristiques de multiplicité au plus double et singularités Masloviennes II. J. Analyse Math. 41, 1–38 (1982).

Lax, A.: [1] On Cauchy's problem for partial differential equations with multiple characteristics. Comm. Pure Appl. Math. 9, 135–169 (1956).

Lax, P.D.: [2] On Cauchy's problem for hyperbolic equations and the differentiability of solutions of elliptic equations. Comm. Pure Appl. Math. 8, 615–633 (1955).

– [3] Asymptotic solutions of oscillatory initial value problems. Duke Math. J. 24, 627–646 (1957).

Lax, P.D. and L. Nirenberg: [1] On stability for difference schemes: a sharp form of Gårding's inequality. Comm. Pure Appl. Math. 19, 473–492 (1966).

Lebeau, G.: [1] Fonctions harmoniques et spectre singulier. Ann. Sci. École Norm. Sup. (4) 13, 269–291 (1980).

Lelong, P.: [1] Plurisubharmonic functions and positive differential forms. Gordon and Breach, New York, London, Paris 1969.

– [2] Propriétés métriques des variétés définies par une équation. Ann. Sci. École Norm. Sup. 67, 22–40 (1950).

Leray, J.: [1] Hyperbolic differential equations. The Institute for Advanced Study, Princeton, N.J. 1953.

– [2] Uniformisation de la solution du problème linéaire analytique de Cauchy près de la variété qui porte les données de Cauchy. Bull. Soc. Math. France 85, 389–429 (1957).

Lerner, N.: [1] Unicité de Cauchy pour des opérateurs différentiels faiblement principalement normaux. J. Math. Pures Appl. 64, 1–11 (1985).

Lerner, N. and L. Robbiano: [1] Unicité de Cauchy pour des opérateurs de type principal. J. Analyse Math. 44, 32–66 (1984/85).

Levi, E.E.: [1] Caratterische multiple e problema di Cauchy. Ann. Mat. Pura Appl. (3) 16, 161–201 (1909).

Levinson, N.: [1] Transformation of an analytic function of several variables to a canonical form. Duke Math. J. 28, 345–353 (1961).

Levitan, B.M.: [1] On the asymptotic behavior of the spectral function of a self-adjoint differential equation of the second order. Izv. Akad. Nauk SSSR Ser. Mat. 16, 325–352 (1952).

– [2] On the asymptotic behavior of the spectral function and on expansion in eigenfunctions of a self-adjoint differential equation of second order II. Izv. Akad. Nauk SSSR Ser. Mat. 19, 33–58 (1955).

Lewy, H.: [1] An example of a smooth linear partial differential equation without solution. Ann. of Math. 66, 155–158 (1957).
– [2] Extension of Huyghen's principle to the ultrahyperbolic equation. Ann. Mat. Pura Appl. (4) 39, 63–64 (1955).
Lions, J.L.: [1] Supports dans la transformation de Laplace. J. Analyse Math. 2, 369–380 (1952–53).
Lions, J.L. and E. Magenes: [1] Problèmes aux limites non homogènes et applications I–III. Dunod, Paris, 1968–1970.
Łojasiewicz, S.: [1] Sur le problème de division. Studia Math. 18, 87–136 (1959).
Lopatinski, Ya.B.: [1] On a method of reducing boundary problems for a system of differential equations of elliptic type to regular integral equations. Ukrain. Mat. Ž. 5, 123–151 (1953). Amer. Math. Soc. Transl. (2) 89, 149–183 (1970).
Ludwig, D.: [1] Exact and asymptotic solutions of the Cauchy problem. Comm. Pure Appl. Math. 13, 473–508 (1960).
– [2] Uniform asymptotic expansions at a caustic. Comm. Pure Appl. Math. 19, 215–250 (1966).
Luke, G.: [1] Pseudodifferential operators on Hilbert bundles. J. Differential Equations 12, 566–589 (1972).
Malgrange, B.: [1] Existence et approximation des solutions des équations aux dérivées partielles et des équations de convolution. Ann. Inst. Fourier (Grenoble) 6, 271–355 (1955–56).
– [2] Sur une classe d'opérateurs différentiels hypoelliptiques. Bull. Soc. Math. France 85, 283–306 (1957).
– [3] Sur la propagation de la régularité des solutions des équations à coefficients constants. Bull. Math. Soc. Sci. Math. Phys. R.P. Roumanie 3 (53), 433–440 (1959).
– [4] Sur les ouverts convexes par rapport à un opérateur différentiel. C. R. Acad. Sci. Paris 254, 614–615 (1962).
– [5] Sur les systèmes différentiels à coefficients constants. Coll. CNRS 113–122, Paris 1963.
– [6] Ideals of differentiable functions. Tata Institute, Bombay, and Oxford University Press 1966.
Mandelbrojt, S.: [1] Analytic functions and classes of infinitely differentiable functions. Rice Inst. Pamphlet 29, 1–142 (1942).
– [2] Séries adhérentes, régularisations des suites, applications. Coll. Borel, Gauthier-Villars, Paris 1952.
Martineau, A.: [1] Les hyperfonctions de M. Sato. Sém. Bourbaki 1960–1961, Exposé No 214.
– [2] Le "edge of the wedge theorem" en théorie des hyperfonctions de Sato. Proc. Int. Conf. Funct. Anal. Rel. Topics, 95–106. University of Tokyo Press, Tokyo 1970.
Maslov, V.P.: [1] Theory of perturbations and asymptotic methods. Moskov. Gos. Univ., Moscow 1965 (Russian).
Mather, J.: [1] Stability of C^∞ mappings: I. The division theorem. Ann. of Math. 87, 89–104 (1968).
Melin, A.: [1] Lower bounds for pseudo-differential operators. Ark. Mat. 9, 117–140 (1971).
– [2] Parametrix constructions for right invariant differential operators on nilpotent groups. Ann. Global Analysis and Geometry 1, 79–130 (1983).
Melin, A. and J. Sjöstrand: [1] Fourier integral operators with complex-valued phase functions. Springer Lecture Notes in Math. 459, 120–223 (1974).
– [2] Fourier integral operators with complex phase functions and parametrix for an interior boundary value problem. Comm. Partial Differential Equations 1:4, 313–400 (1976).
Melrose, R.B.: [1] Transformation of boundary problems. Acta Math. 147, 149–236 (1981).

– [2] Equivalence of glancing hypersurfaces. Invent. Math. 37, 165–191 (1976).
– [3] Microlocal parametrices for diffractive boundary value problems. Duke Math. J. 42, 605–635 (1975).
– [4] Local Fourier-Airy integral operators. Duke Math. J. 42, 583–604 (1975).
– [5] Airy operators. Comm. Partial Differential Equations 3:1, 1–76 (1978).
– [6] The Cauchy problem for effectively hyperbolic operators. Hokkaido Math. J. To appear.
– [7] The trace of the wave group. Contemporary Math. 27, 127–167 (1984).
Melrose, R.B. and J. Sjöstrand: [1] Singularities of boundary value problems I, II. Comm. Pure Appl. Math. 31, 593–617 (1978); 35, 129–168 (1982).
Mihlin, S.G.: [1] On the multipliers of Fourier integrals. Dokl. Akad. Nauk SSSR 109, 701–703 (1956) (Russian).
Mikusiński, J.: [1] Une simple démonstration du théorème de Titchmarsh sur la convolution. Bull. Acad. Pol. Sci. 7, 715–717 (1959).
– [2] The Bochner integral. Birkhäuser Verlag, Basel and Stuttgart 1978.
Minakshisundaram, S. and Å. Pleijel: [1] Some properties of the eigenfunctions of the Laplace operator on Riemannian manifolds. Canad. J. Math. 1, 242–256 (1949).
Mizohata, S.: [1] Unicité du prolongement des solutions des équations elliptiques du quatrième ordre. Proc. Jap. Acad. 34, 687–692 (1958).
– [2] Systèmes hyperboliques. J. Math. Soc. Japan 11, 205–233 (1959).
– [3] Note sur le traitement par les opérateurs d'intégrale singulière du problème de Cauchy. J. Math. Soc. Japan 11, 234–240 (1959).
– [4] Solutions nulles et solutions non analytiques. J. Math. Kyoto Univ. 1, 271–302 (1962).
– [5] Some remarks on the Cauchy problem. J. Math. Kyoto Univ. 1, 109–127 (1961).
Møller, C.: [1] General properties of the characteristic matrix in the theory of elementary particles. I. Kongl. Dansk. Vidensk. Selsk. Mat.-Fys. Medd. 23, 2–48 (1945).
Morrey, C.B.: [1] The analytic embedding of abstract real-analytic manifolds. Ann. of Math. 68, 159–201 (1958).
Morrey, C.B. and L. Nirenberg: [1] On the analyticity of the solutions of linear elliptic systems of partial differential equations. Comm. Pure Appl. Math. 10, 271–290 (1957).
Moyer, R.D.: [1] Local solvability in two dimensions: Necessary conditions for the principle-type case. Mimeographed manuscript, University of Kansas 1978.
Müller, C.: [1] On the behaviour of the solutions of the differential equation $\Delta U = F(x, U)$ in the neighborhood of a point. Comm. Pure Appl. Math. 7, 505–515 (1954).
Münster, M.: [1] On A. Lax's condition of hyperbolicity. Rocky Mountain J. Math. 8, 443–446 (1978).
– [2] On hyperbolic polynomials with constant coefficients. Rocky Mountain J. Math. 8, 653–673 (1978).
von Neumann, J. and E. Wigner: [1] Über merkwürdige diskrete Eigenwerte. Phys. Z. 30, 465–467 (1929).
Nirenberg, L.: [1] Remarks on strongly elliptic partial differential equations. Comm. Pure Appl. Math. 8, 648–675 (1955).
– [2] Uniqueness in Cauchy problems for differential equations with constant leading coefficients. Comm. Pure Appl. Math. 10, 89–105 (1957).
– [3] A proof of the Malgrange preparation theorem. Liverpool singularities I. Springer Lecture Notes in Math. 192, 97–105 (1971).
– [4] On elliptic partial differential equations. Ann. Scuola Norm. Sup. Pisa (3) 13, 115–162 (1959).
– [5] Lectures on linear partial differential equations. Amer. Math. Soc. Regional Conf. in Math., 17, 1–58 (1972).

Nirenberg, L. and F. Treves: [1] Solvability of a first order linear partial differential equation. Comm. Pure Appl. Math. 16, 331–351 (1963).

– [2] On local solvability of linear partial differential equations. I. Necessary conditions. II. Sufficient conditions. Correction. Comm. Pure Appl. Math. 23, 1–38 and 459–509 (1970); 24, 279–288 (1971).

Nishitani, T.: [1] Local energy integrals for effectively hyperbolic operators I, II. J. Math. Kyoto Univ. 24, 623–658 and 659–666 (1984).

Noether, F.: [1] Über eine Klasse singulärer Integralgleichungen. Math. Ann. 82, 42–63 (1921).

Olejnik, O.A.: [1] On the Cauchy problem for weakly hyperbolic equations. Comm. Pure Appl. Math. 23, 569–586 (1970).

Olejnik, O.A. and E.V. Radkevič: [1] Second order equations with non-negative characteristic form. In Matem. Anal. 1969, ed. R.V. Gamkrelidze, Moscow 1971 (Russian). English translation Plenum Press, New York-London 1973.

Oshima, T.: [1] On analytic equivalence of glancing hypersurfaces. Sci. Papers College Gen. Ed. Univ. Tokyo 28, 51–57 (1978).

Palamodov, V.P.: [1] Linear differential operators with constant coefficients. Moscow 1967 (Russian). English transl. Grundl. d. Math. Wiss. 168, Springer Verlag, New York, Heidelberg, Berlin 1970.

Paley, R.E.A.C. and N. Wiener: [1] Fourier transforms in the complex domain. Amer. Math. Soc. Coll. Publ. XIX, New York 1934.

Pederson, R.: [1] On the unique continuation theorem for certain second and fourth order elliptic equations. Comm. Pure Appl. Math. 11, 67–80 (1958).

– [2] Uniqueness in the Cauchy problem for elliptic equations with double characteristics. Ark. Mat. 6, 535–549 (1966).

Peetre, J.: [1] Théorèmes de régularité pour quelques classes d'opérateurs différentiels. Thesis, Lund 1959.

– [2] Rectification à 1' article "Une caractérisation abstraite des opérateurs différentiels". Math. Scand. 8, 116–120 (1960).

– [3] Another approach to elliptic boundary problems. Comm. Pure Appl. Math. 14, 711–731 (1961).

– [4] New thoughts on Besov spaces. Duke Univ. Math. Series I. Durham, N.C. 1976.

Persson, J.: [1] The wave operator and P-convexity. Boll. Un. Mat. Ital. (5) 18-B, 591–604 (1981).

Petrowsky, I.G.: [1] Über das Cauchysche Problem für Systeme von partiellen Differentialgleichungen. Mat. Sb. 2 (44), 815–870 (1937).

– [2] Über das Cauchysche Problem für ein System linearer partieller Differentialgleichungen im Gebiete der nichtanalytischen Funktionen. Bull. Univ. Moscow Sér. Int. 1, No. 7, 1–74 (1938).

– [3] Sur l'analyticité des solutions des systèmes d'équations différentielles. Mat. Sb. 5 (47), 3–70 (1939).

– [4] On the diffusion of waves and the lacunas for hyperbolic equations. Mat. Sb. 17 (59), 289–370 (1945).

– [5] Some remarks on my papers on the problem of Cauchy. Mat. Sb. 39 (81), 267–272 (1956). (Russian.)

Pham The Lai: [1] Meilleures estimations asymptotiques des restes de la fonction spectrale et des valeurs propres relatifs au laplacien. Math. Scand. 48, 5–31 (1981).

Piccinini, L.C.: [1] Non surjectivity of the Cauchy-Riemann operator on the space of the analytic functions on $\mathbb{R}^n$. Generalization to the parabolic operators. Bull. Un. Mat. Ital. (4) 7, 12–28 (1973).

Pliš, A.: [1] A smooth linear elliptic differential equation without any solution in a sphere. Comm. Pure Appl. Math. 14, 599–617 (1961).

– [2] The problem of uniqueness for the solution of a system of partial differential equations. Bull. Acad. Pol. Sci. 2, 55–57 (1954).

– [3] On non-uniqueness in Cauchy problem for an elliptic second order differential equation. Bull. Acad. Pol. Sci. 11, 95–100 (1963).

Poincaré, H.: [1] Sur les propriétés du potentiel et les fonctions abeliennes. Acta Math. 22, 89–178 (1899).

Povzner, A.Ya.: [1] On the expansion of arbitrary functions in characteristic functions of the operator $-\Delta u + cu$. Mat. Sb. 32 (74), 109–156 (1953).

Radkevič, E.: [1] A priori estimates and hypoelliptic operators with multiple characteristics. Dokl. Akad. Nauk SSSR 187, 274–277 (1969) (Russian); also in Soviet Math. Doklady 10, 849–853 (1969).

Ralston, J.: [1] Solutions of the wave equation with localized energy. Comm. Pure Appl. Math. 22, 807–823 (1969).

– [2] Gaussian beams and the propagation of singularities. MAA Studies in Mathematics 23, 206–248 (1983).

Reed, M. and B. Simon: [1] Methods of modern mathematical physics. III. Scattering theory. Academic Press 1979.

Rempel, S. and B.-W. Schulze: [1] Index theory of elliptic boundary problems. Akademie-Verlag, Berlin (1982).

de Rham, G.: [1] Variétés différentiables. Hermann, Paris, 1955.

Riesz, F.: [1] Sur l'existence de la dérivée des fonctions d'une variable réelle et des fonctions d'intervalle. Verh. Int. Math. Kongr. Zürich 1932, I, 258–269.

Riesz, M.: [1] L'intégrale de Riemann-Liouville et le problème de Cauchy. Acta Math. 81, 1–223 (1949).

– [2] Sur les maxima des formes bilinéaires et sur les fonctionnelles linéaires. Acta Math. 49, 465–497 (1926).

– [3] Sur les fonctions conjuguées. Math. Z. 27, 218–244 (1928).

– [4] Problems related to characteristic surfaces. Proc. Conf. Diff. Eq. Univ. Maryland 1955, 57–71.

Rothschild, L.P.: [1] A criterion for hypoellipticity of operators constructed from vector fields. Comm. Partial Differential Equations 4:6, 645–699 (1979).

Saito, Y.: [1] On the asymptotic behavior of the solutions of the Schrödinger equation $(-\Delta + Q(y) - k^2)V = F$. Osaka J. Math. 14, 11–35 (1977).

– [2] Eigenfunction expansions for the Schrödinger operators with long-range potentials $Q(y) = O(|y|^{-\varepsilon})$, $\varepsilon > 0$. Osaka J. Math. 14, 37–53 (1977).

Sakamoto, R.: [1] E-well posedness for hyperbolic mixed problems with constant coefficients. J. Math. Kyoto Univ. 14, 93–118 (1974).

– [2] Mixed problems for hyperbolic equations I. J. Math. Kyoto Univ. 10, 375–401 (1970), II. J. Math. Kyoto Univ. 10, 403–417 (1970).

Sato, M.: [1] Theory of hyperfunctions I. J. Fac. Sci. Univ. Tokyo I, 8, 139–193 (1959).

– [2] Theory of hyperfunctions II. J. Fac. Sci. Univ. Tokyo I, 8, 387–437 (1960).

– [3] Hyperfunctions and partial differential equations. Proc. Int. Conf. on Funct. Anal. and Rel. Topics, 91–94, Tokyo University Press, Tokyo 1969.

– [4] Regularity of hyperfunction solutions of partial differential equations. Actes Congr. Int. Math. Nice 1970, 2, 785–794.

Sato, M., T. Kawai and M. Kashiwara: [1] Hyperfunctions and pseudodifferential equations. Springer Lecture Notes in Math. 287, 265–529 (1973).

Schaefer, H.H.: [1] Topological vector spaces. Springer Verlag, New York, Heidelberg, Berlin 1970.

Schapira, P.: [1] Hyperfonctions et problèmes aux limites elliptiques. Bull. Soc. Math. France 99, 113–141 (1971).

– [2] Propagation at the boundary of analytic singularities. Nato Adv. Study Inst. on Sing. in Bound. Value Problems. Reidel Publ. Co., Dordrecht, 185–212 (1981).

– [3] Propagation at the boundary and reflection of analytic singularities of solutions of linear partial differential equations. Publ. RIMS, Kyoto Univ., 12 Suppl., 441–453 (1977).

Schechter, M.: [1] Various types of boundary conditions for elliptic equations. Comm. Pure Appl. Math. 13, 407–425 (1960).
– [2] A generalization of the problem of transmission. Ann. Scuola Norm. Sup. Pisa 14, 207–236 (1960).
Schwartz, L.: [1] Théorie des distributions I, II. Hermann, Paris, 1950–51.
– [2] Théorie des noyaux. Proc. Int. Congr. Math. Cambridge 1950, I, 220–230.
– [3] Sur l'impossibilité de la multiplication des distributions. C. R. Acad. Sci. Paris 239, 847–848 (1954).
– [4] Théorie des distributions à valeurs vectorielles I. Ann. Inst. Fourier (Grenoble) 7, 1–141 (1957).
– [5] Transformation de Laplace des distributions. Comm. Sém. Math. Univ. Lund, Tome suppl. dédié à Marcel Riesz, 196–206 (1952).
– [6] Théorie générale des fonctions moyenne-périodiques. Ann. of Math. 48, 857–929 (1947).
Seeley, R.T.: [1] Singular integrals and boundary problems. Amer. J. Math. 88, 781–809 (1966).
– [2] Extensions of C^∞ functions defined in a half space. Proc. Amer. Math. Soc. 15, 625–626 (1964).
– [3] A sharp asymptotic remainder estimate for the eigenvalues of the Laplacian in a domain of $\mathbb{R}^3$. Advances in Math. 29, 244–269 (1978).
– [4] An estimate near the boundary for the spectral function of the Laplace operator. Amer. J. Math. 102, 869–902 (1980).
– [5] Elliptic singular integral equations. Amer Math. Soc. Symp. on Singular Integrals, 308–315 (1966).
Seidenberg, A.: [1] A new decision method for elementary algebra. Ann. of Math. 60, 365–374 (1954).
Shibata, Y.: [1] E-well posedness of mixed initial-boundary value problems with constant coefficients in a quarter space. J. Analyse Math. 37, 32–45 (1980).
Siegel, C.L.: [1] Zu den Beweisen des Vorbereitungssatzes von Weierstrass. In Abhandl. aus Zahlenth. u. Anal., 299–306. Plenum Press, New York 1968.
Sjöstrand, J.: [1] Singularités analytiques microlocales. Prépublications Université de Paris-Sud 82-03.
– [2] Analytic singularities of solutions of boundary value problems. Nato Adv. Study Inst. on Sing. in Bound. Value Prob., Reidel Publ. Co., Dordrecht, 235–269 (1981).
– [3] Parametrices for pseudodifferential operators with multiple characteristics. Ark. Mat. 12, 85–130 (1974).
– [4] Propagation of analytic singularities for second order Dirichlet problems I, II, III. Comm. Partial Differential Equations 5:1, 41–94 (1980), 5:2, 187–207 (1980), and 6:5, 499–567 (1981).
– [5] Operators of principal type with interior boundary conditions. Acta Math. 130, 1–51 (1973).
Sobolev, S.L.: [1] Méthode nouvelle à résoudre le problème de Cauchy pour les équations linéaires hyperboliques normales. Mat. Sb. 1 (43), 39–72 (1936).
– [2] Sur un théorème d'analyse fonctionnelle. Mat. Sb. 4 (46), 471–497 (1938). (Russian; French summary.) Amer. Math. Soc. Transl. (2) 34, 39–68 (1963).
Sommerfeld, A.: [1] Optics. Lectures on theoretical physics IV. Academic Press, New York, 1969.
Stein, E.M.: [1] Singular integrals and differentiability properties of functions. Princeton Univ. Press 1970.
Sternberg, S.: [1] Lectures on differential geometry. Prentice-Hall Inc. Englewood Cliffs, N.J., 1964.
Stokes, G.B.: [1] On the numerical calculation of a class of definite integrals and infinite series. Trans. Cambridge Philos. Soc. 9, 166–187 (1850).

Svensson, L.: [1] Necessary and sufficient conditions for the hyperbolicity of polynomials with hyperbolic principal part. Ark. Mat. 8, 145-162 (1968).

Sweeney, W.J.: [1] The *D*-Neumann problem. Acta Math. 120, 223-277 (1968).

Szegö, G.: [1] Beiträge zur Theorie der Toeplitzschen Formen. Math. Z. 6, 167-202 (1920).

Täcklind, S.: [1] Sur les classes quasianalytiques des solutions des équations aux dérivées partielles du type parabolique. Nova Acta Soc. Sci. Upsaliensis (4) 10, 1-57 (1936).

Tarski, A.: [1] A decision method for elementary algebra and geometry. Manuscript, Berkeley, 63 pp. (1951).

Taylor, M.: [1] Gelfand theory of pseudodifferential operators and hypoelliptic operators. Trans. Amer. Math. Soc. 153, 495-510 (1971).

— [2] Grazing rays and reflection of singularities of solutions to wave equations. Comm. Pure Appl. Math. 29, 1-38 (1976).

— [3] Diffraction effects in the scattering of waves. In Sing. in Bound. Value Problems, 271-316. Reidel Publ. Co., Dordrecht 1981.

— [4] Pseudodifferential operators. Princeton Univ. Press, Princeton, N.J., 1981.

Thorin, O.: [1] An extension of a convexity theorem due to M. Riesz. Kungl Fys. Sällsk. Lund. Förh. 8, No 14 (1939).

Titchmarsh, E.C.: [1] The zeros of certain integral functions. Proc. London Math. Soc. 25, 283-302 (1926).

Treves, F.: [1] Solution élémentaire d'équations aux dérivées partielles dépendant d'un paramètre. C. R. Acad. Sci. Paris 242, 1250-1252 (1956).

— [2] Thèse d'Hörmander II. Sém. Bourbaki 135, 2ᵉ éd. (Mai 1956).

— [3] Relations de domination entre opérateurs différentiels. Acta Math. 101, 1-139 (1959).

— [4] Opérateurs différentiels hypoelliptiques. Ann. Inst. Fourier (Grenoble) 9, 1-73 (1959).

— [5] Local solvability in L^2 of first order linear PDEs. Amer. J. Math. 92, 369-380 (1970).

— [6] Fundamental solutions of linear partial differential equations with constant coefficients depending on parameters. Amer. J. Math. 84, 561-577 (1962).

— [7] Un théorème sur les équations aux dérivées partielles à coefficients constants dépendant de paramètres. Bull. Soc. Math. France 90, 473-486 (1962).

— [8] A new method of proof of the subelliptic estimates. Comm. Pure Appl. Math. 24, 71-115 (1971).

— [9] Introduction to pseudodifferential and Fourier integral operators. Volume 1: Pseudodifferential operators. Volume 2: Fourier integral operators. Plenum Press, New York and London 1980.

Vauthier, J.: [1] Comportement asymptotique des fonctions entières de type exponentiel dans $\mathbb{C}^n$ et bornées dans le domaine réel. J. Functional Analysis 12, 290-306 (1973).

Vekua, I.N.: [1] Systeme von Differentialgleichungen erster Ordnung vom elliptischen Typus und Randwertaufgaben. Berlin 1956.

Veselič, K. and J. Weidmann: [1] Existenz der Wellenoperatoren für eine allgemeine Klasse von Operatoren. Math. Z. 134, 255-274 (1973).

— [2] Asymptotic estimates of wave functions and the existence of wave operators. J. Functional Analysis. 17, 61-77 (1974).

Višik, M.I.: [1] On general boundary problems for elliptic differential equations. Trudy Moskov. Mat. Obšč. 1, 187-246 (1952) (Russian). Also in Amer. Math. Soc. Transl. (2) 24, 107-172 (1963).

Višik, M.I. and G.I. Eškin: [1] Convolution equations in a bounded region. Uspehi Mat. Nauk 20:3 (123), 89-152 (1965). (Russian.) Also in Russian Math. Surveys 20:3, 86-151 (1965).

– [2] Convolution equations in a bounded region in spaces with weighted norms. Mat. Sb. 69 (111), 65–110 (1966) (Russian). Also in Amer. Math. Soc. Transl. (2) 67, 33–82 (1968).

– [3] Elliptic convolution equations in a bounded region and their applications. Uspehi Mat. Nauk 22:1 (133), 15–76 (1967). (Russian.) Also in Russian Math. Surveys 22:1, 13–75 (1967).

– [4] Convolution equations of variable order. Trudy Moskov. Mat. Obšč. 16, 25–50 (1967). (Russian.) Also in Trans. Moskov. Mat. Soc. 16, 27–52 (1967).

– [5] Normally solvable problems for elliptic systems of convolution equations. Mat. Sb. 74 (116), 326–356 (1967). (Russian.) Also in Math. USSR-Sb. 3, 303–332 (1967).

van der Waerden, B.L.: [1] Einführung in die algebraische Geometrie. Berlin 1939.

– [2] Algebra I–II. 4. Aufl. Springer Verlag, Berlin-Göttingen-Heidelberg 1959.

Wang Rou-hwai and Tsui Chih-yung: [1] Generalized Leray formula on positive complex Lagrange-Grassmann manifolds. Res. Report, Inst. of Math., Jilin Univ. 8209, 1982.

Warner, F.W.: [1] Foundations of differentiable manifolds and Lie Groups. Scott, Foresman and Co., Glenview, Ill., London, 1971.

Weinstein, A.: [1] The order and symbol of a distribution. Trans. Amer. Math. Soc. 241, 1–54 (1978).

– [2] Asymptotics of eigenvalue clusters for the Laplacian plus a potential. Duke Math. J. 44, 883–892 (1977).

– [3] On Maslov's quantization condition. In Fourier integral operators and partial differential equations. Springer Lecture Notes in Math. 459, 341–372 (1974).

Weyl, H.: [1] The method of orthogonal projection in potential theory. Duke Math. J. 7, 411–444 (1940).

– [2] Die Idee der Riemannschen Fläche. 3. Aufl., Teubner, Stuttgart, 1955.

– [3] Über gewöhnliche Differentialgleichungen mit Singularitäten und die zugehörigen Entwicklungen willkürlicher Funktionen. Math. Ann. 68, 220–269 (1910).

– [4] Das asymptotische Verteilungsgesetz der Eigenwerte linearer partieller Differentialgleichungen (mit einer Anwendung auf die Theorie der Hohlraumstrahlung). Math. Ann. 71, 441–479 (1912).

Whitney, H.: [1] Analytic extensions of differentiable functions defined in closed sets. Trans. Amer. Math. Soc. 36, 63–89 (1934).

Widom, H.: [1] Eigenvalue distribution in certain homogeneous spaces. J. Functional Analysis 32, 139–147 (1979).

Yamamoto, K.: [1] On the reduction of certain pseudo-differential operators with non-involution characteristics. J. Differential Equations 26, 435–442 (1977).

Zeilon, N.: [1] Das Fundamentalintegral der allgemeinen partiellen linearen Differentialgleichung mit konstanten Koeffizienten. Ark. Mat. Astr. Fys. 6, No 38, 1–32 (1911).

Zerner, M.: [1] Solutions de l'équation des ondes présentant des singularités sur une droite. C. R. Acad. Sci. Paris 250, 2980–2982 (1960).

– [2] Solutions singulières d'équations aux dérivées partielles. Bull. Soc. Math. France 91, 203–226 (1963).

– [3] Domaine d'holomorphie des fonctions vérifiant une équation aux dérivées partielles. C. R. Acad. Sci. Paris 272, 1646–1648 (1971).

Zuily, C.: [1] Uniqueness and non-uniqueness in the Cauchy problem. Progress in Math. 33, Birkhäuser, Boston, Basel, Stuttgart 1983.

Zygmund, A.: [1] On a theorem of Marcinkiewicz concerning interpolation of operators. J. Math. Pures Appl. 35, 223–248 (1956).

Index

to Volumes III and IV (page numbers in *italics* refer to Volume IV)

Index of Notation

to Volumes III and IV (page numbers in *italics* refer to Volume IV)

Spaces of Functions and Distributions, and Their Norms

Spaces of Symbols

Some Special Notation